READINGS IN SOCIOLOGY

READINGS IN SOCIOLOGY

A Biographical Approach

Edited by

BRIGITTE BERGER

Drawings by Robert Binks

BASIC BOOKS, INC., PUBLISHERS

NEW YORK

Library of Congress Catalog Card Number: 73–82233
Paper SBN: 465–06853–7
Cloth SBN: 465–06852–9
Manufactured in the United States of America
DESIGNED BY VINCENT TORRE
74 75 76 77 10 9 8 7 6 5 4 3 2 1

Preface

This reader is primarily intended to be used in conjunction with the introductory textbook *Sociology: A Biographical Approach* by Peter L. and Brigitte Berger. The most distinctive feature of our textbook is indicated by its subtitle. We have organized the material within a biographical frame of reference; different areas of social life are discussed as far as possible in the chronological order in which they are experienced. In addition to this sequence of presentation, we have made an attempt throughout our textbook to relate the individual's immediate experience of society with the remote, impersonal forces and institutions that constitute the major subject matter of sociology. This reader follows the same principle of organization. While our textbook provides the student with a broad theoretical outline applicable to a variety of specific areas, my reader aims to "fill in" this general outline with the details and colors of specific illustrations. Considering this aim, I have attempted throughout to show the relationship between "classical" sociological theory and the sociological analysis of contemporary problems.

In our textbook we have been very careful to try to control our own theoretical bias by giving close attention to approaches in the field with which we do not agree. In other words, we have presented a lot of things that we do not like along with those that we do like. In preparing this reader, I have been somewhat less concerned with such an ideal of objectivity—that is, I have given preference to materials that I like. Thus, I admit an unembarrassed bias toward materials that express a humanistic approach, by which I simply mean an approach to social phenomena in terms of their subjective human meanings. Nevertheless, this is definitely not a sectarian book. The student will find materials expressing a variety of positions within sociology.

Also as in our textbook, there is here a deliberate "ethnocentric" bias; most of my selections deal with the contemporary American scene. I should like to stress, however, that this ethnocentrism is pedagogical and not ideological. I have made no presumption that American society is more important, let alone better, than other societies. But I have felt that the American student would benefit most from illustrations that refer to situations with which he is most probably familiar. Again for pedagogical reasons, in a number of chapters where this seemed indicated the readings alternate between statements of sociological theory and concrete contemporary cases.

All readers of this kind have a built-in weakness: selection is invariably governed by the limitation of space and the desideratum of readability. In other words, as one puts together a reader one gravitates toward materials that can be presented in brief and readable form. It goes without saying that this precludes a large bulk of sociological literature. Further, in presenting abbreviated versions of lengthy presentations, violence is invariably done to some feature of an argument. One can only express the hope that such a reader will stimulate students' interest and eventually send at least the more serious of them to the original sources.

Contents

Chapter 1

THE EXPERIENCE OF SOCIETY

Man in Society BY PETER L. BERGER 4
To Be a Negro BY JAMES WELDON JOHNSON 6
What Is a Social Fact? BY EMILE DURKHEIM 8
The Sociological Imagination BY C. WRIGHT MILLS 10

Chapter 2

THE DISCIPLINE OF SOCIOLOGY

The Stranger BY ALFRED SCHUTZ 20
The Rules of Sociological Method BY EMILE DURKHEIM 25
The Protestant Ethic and the Spirit of Capitalism BY MAX WEBER 30
Two Styles of Research in Current Social Studies BY C. WRIGHT MILLS 36
Invitation to Sociology BY PETER L. BERGER 41

Chapter 3

BECOMING A MEMBER OF SOCIETY—SOCIALIZATION

When I Was a Child BY LILLIAN SMITH 50
The Everyday World of the Child BY MATTHEW SPEIER 55
Mind, Self and Society BY G. H. MEAD 61
Fun Morality: An Analysis of Recent American Child-training Literature BY MARTHA WOLFENSTEIN 69

Chapter 4

WHAT IS AN INSTITUTION? THE CASE OF LANGUAGE

The Lore and Language of Schoolchildren BY IONA AND PETER OPIE 78

Selecting an Identification by Means of Social Categories BY MATTHEW SPEIER 81
Helen Keller Acquires Language BY ANNE MANSFIELD SULLIVAN 84
The Language and Thought of the Child BY JEAN PIAGET 90
Introduction to Nancy Mitford's Noblesse Oblige BY RUSSELL LYNES 93

Chapter 5

THE FAMILY
Family Types and the Urban Villagers BY HERBERT J. GANS 100
Functions of the Crestwood Heights Family BY J. R. SEELEY, R. A. SIM, AND E. W. LOOSLEY 108
In Defense of the Negro Family BY FRANK RIESSMAN 111
Family Structure and Sexual Relations in the Communal Family BY BENNETT M. BERGER, BRUCE M. HACKETT, AND R. MERVYN MILLAR 114
Marriage and the Construction of Reality BY PETER L. BERGER AND HANSFRIED KELLNER 119
After the Party BY JOHN UPDIKE 123

Chapter 6

THE COMMUNITY
The Gold Coast and the Slum BY HARVEY WARREN ZORBAUGH 130
Urbanism and Suburbanism as Ways of Life: A Re-evaluation of Definitions BY HERBERT J. GANS 137
The Metropolis and Mental Life BY GEORG SIMMEL 147
Ethnogenesis and Negro-Americans Today BY LESTER SINGER 156

Chapter 7

THE STRATIFIED COMMUNITY
Social Class in America BY W. LLOYD WARNER, MARCHIA MEECKER, AND KENNETH EELLS 166
The Working Class Subculture: A New View BY S. M. MILLER AND FRANK RIESSMAN 175
White Collar: The American Middle Classes BY C. WRIGHT MILLS 183
Being Refined BY WILLIAM SAROYAN 193

Chapter **8**

THE STRATIFIED SOCIETY

Social Mobility and Personal Identity BY THOMAS LUCKMANN AND PETER L. BERGER 202
The Brutal Bargain BY NORMAN PODHORETZ 208
The Academic Revolution BY CHRISTOPHER JENCKS AND DAVID RIESMAN 211
The Distribution of Personal Income in the United States BY HERMAN P. MILLER 217
Social Mobility and Equal Opportunity BY SEYMOUR MARTIN LIPSET 225

Chapter **9**

WHAT IS SOCIAL CONTROL?
THE CASE OF EDUCATION

The Centrality of Schooling BY CHARLES E. BIDWELL 234
Death at an Early Age BY JONATHAN KOZOL 236
Coming of Age in America BY EDGAR Z. FRIEDENBERG 241
Youth and the Social Order BY F. MUSGROVE 245
The Military Academy as an Assimilating Institution BY SANFORD M. DORNBUSCH 248

Chapter **10**

BUREAUCRACY

Bureaucratization in Industry BY REINHARD BENDIX 260
The Organization Man BY WILLIAM H. WHYTE, JR. 270
Parkinson's Law, or the Rising Pyramid BY C. NORTHCOTE PARKINSON 278
Asylums BY ERVING GOFFMAN 284

Chapter **11**

YOUTH

Centuries of Childhood BY PHILIPPE ARIÈS 292
Population Changes and the Status of the Young BY F. MUSGROVE 296
The Emergence of an Adolescent Subculture in Industrial Society BY JAMES COLEMAN 302
Baltimore BY ELIA KATZ 307

Chapter 12

WORK AND LEISURE

Human Relations in Industry BY WILLIAM FOOTE WHYTE 316
The Ship BY MARIAM G. SHERAR 322
Skill Requirements for Agency Personnel BY IAN LEWIS 330
The Problems and Promise of Leisure BY SEBASTIAN DE GRAZIA 334
Fun BY ORRIN E. KLAPP 343

Chapter 13

POWER

"Power Elite" or "Veto Groups"? BY WILLIAM KORNHAUSER 350
A Country Called Corporate America BY ANDREW HACKER 363
Classes and Parties in American Politics BY SEYMOUR MARTIN LIPSET 372

Chapter 14

DEVIANCE

The Condemnation and Persecution of Hippies BY MICHAEL E. BROWN 386
Problems in the Sociology of Deviance: Social Definitions and Behavior BY RONALD L. AKERS 392
Lower Class Culture as a Generating Milieu of Gang Delinquency BY WALTER B. MILLER 400

Chapter 15

CHANGE

Middletown Faces Both Ways BY ROBERT S. LYND AND HELEN M. LYND 416
Bourgeois and Proletarians BY KARL MARX AND FRIEDRICH ENGELS 421
The Causes for the Progress of the Division of Labor BY EMILE DURKHEIM 430
The "Rationalization" of Education and Training BY MAX WEBER 435
The Blueing of America BY PETER L. BERGER AND BRIGITTE BERGER 439

Chapter 16

OLD AGE, ILLNESS, AND DEATH

Human Obsolescence BY JULES HENRY 447

The Treatment of Tuberculosis as a Bargaining Process BY JULIUS A. ROTH 449

Death and the Social Structure BY ROBERT BLAUNER 458

Chapter 17

VALUES AND ULTIMATE MEANINGS

An American Sacred Ceremony BY W. LLOYD WARNER 468

Belief, Unbelief, and Religion BY THOMAS LUCKMANN 476

There's a New-Time Religion on Campus BY ANDREW M. GREELEY 482

Postscript

WHY SOCIOLOGY?

Sociology and Freedom BY PETER L. BERGER 495

NOTES 505

INDEX 541

READINGS IN
SOCIOLOGY

Chapter 1

THE EXPERIENCE OF SOCIETY

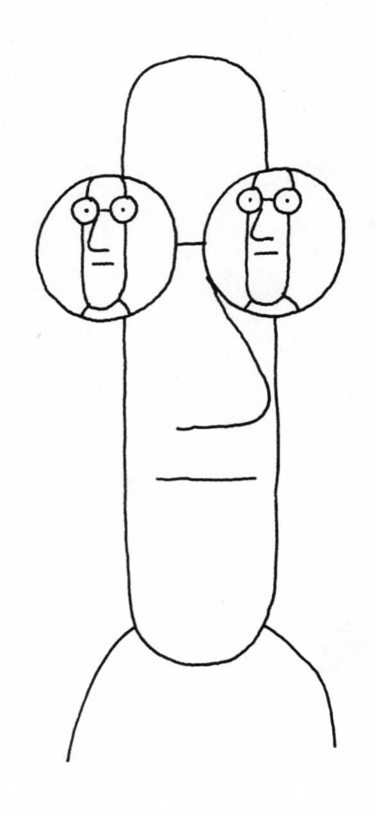

ONE'S social experience occurs in different ways. Thus, the study of this experience involves distinctions. The most important is between the extraordinary and the commonplace. Some encounters with other people in society are surprising, unprecedented events; most are matters of routine. A second important distinction is between the immediate and the remote. Society may be experienced in face-to-face situations, but it may also be experienced as the impact of impersonal forces originating far away from the person affected. These two distinctions are fundamental to sociology.

Only if we understand that most social experience takes place through routines do we understand what it means to say that society consists of structures. Most of our experience of society takes place in ways that are highly ordered and repetitive, in patterns that are predictable. The purpose of sociology is to determine these patterns, the patterns of the routine, the usual, and the ordinary, rather than of the relatively rare, the unusual, and the extraordinary.

Each human being is unique. Thus, in face-to-face situations one encounters distinct personalities. Any human being may also be, however, a representative of some social structure that serves as an impersonal background to the situation. For example, one may meet someone on a train as a person with an interesting face and an endearing smile; one may also meet him as a uniformed representative of the railroad collecting one's fare, or as some other representative of an impersonal collective type (such as a typical advertising executive, long-haired youth, or militant black). The world of immediate, personal encounters—everyday, face-to-face meetings—continuously interacts with the world of remote, impersonal institutions. To understand this interaction in one's own life is to acquire a new kind of self-knowledge. Generally speaking, the worth of sociology for the individual consists in the systematization of this kind of insight.

This theme is stated in the first of the following four selections, from Peter L. Berger's *Invitation to Sociology*. Berger points out that once an individual locates himself on the "map" of society, he may then realize that his own life is crucially affected by realities beyond his control, and initially often beyond his comprehension. Such an experience can be particularly shattering if his location on the "map" is felt to be an undesirable one. The second selection is a moving account of such an experience from *The Autobiography of an Ex-Coloured Man* by James Weldon Johnson, written before World War I. The third selection propounds "classical" theory. It is from *The Rules of Sociological Method,* one of the best-known writings of Emile Durkheim, the most important figure in French sociology. In the passage presented here Durkheim discusses what constitutes a "social fact" and emphasizes the way in which society constrains the individual. The fourth selection is from *The Sociological Imagination* of C. Wright Mills, who may be considered the father of "radical" sociology in its most recent American expression. The book from which the passage is taken was a clarion call of protest against "establishment sociology" as it existed in the 1950s. Mills shows how sociological thinking can enable the individual to see his own life in a broader context.

PETER L. BERGER

MAN IN SOCIETY

AT a certain age children are greatly intrigued by the possibility of locating themselves on a map. It appears strange that one's familiar life should actually have all occurred in an area delineated by a set of quite impersonal (and hitherto unfamiliar) coordinates on the surface of the map. The child's exclamations of "I was there" and "I am here right now" betray the astonishment that the place of last summer's vacation, a place marked in memory by such sharply personal events as the ownership of one's first dog or the secret assembling of a collection of worms, should have specific latitudes and longitudes devised by strangers to one's dog, one's worms, and oneself. This locating of oneself in configurations conceived by strangers is one of the important aspects of what, perhaps euphemistically, is called "growing up." One participates in the real world of grown-ups by having an address. The child who only recently might have mailed a letter addressed "To my Granddaddy" now informs a fellow worm-collector of his exact address—street, town, state and all—and finds his tentative allegiance to the grown-up world view dramatically legitimated by the arrival of the letter.

As the child continues to accept the reality of this world view, he continues to gather addresses—"I'm six years old," "My name is Brown, like my father's, that's because my parents are divorced," "I'm a Presbyterian," "I'm an American," and eventually perhaps "I'm in the special class for the bright kids, that's because my IQ is 130." The horizons of the world, as the grown-ups define it, are determined by the coordinates of remote mapmakers. The child may produce alternate identifications by having himself addressed as father while playing house, as Indian chief or as Davy Crockett, but he will know all the time that he is only playing and that the real facts about himself are those registered by the school authorities. We leave out the quotation marks and thus betray that we too have been trapped into sanity in our childhood—of course, we should write all the key words in quotation marks

Reprinted from *Invitation to Sociology: A Humanistic Perspective* by Peter L. Berger (New York: Doubleday & Co., Anchor Books, 1963), pp. 66–68, by permission of the author and the publisher.

—"know," "real," "facts." The sane child is the one who believes in what it says in the school records. The normal adult is the one who lives within his assigned coordinates.

What is called the commonsense view is actually the grown-up view taken for granted. It is a matter of the school records having become an ontology. One now identifies one's being as a matter of course with the way one is pinpointed on the social map. What this means for one's identity and one's ideas will be pursued further in the next chapter. What interests us at the moment is the way in which such location tells an individual just what he may do and what he can expect of life. To be located in society means to be at the intersection point of specific social forces. Commonly one ignores these forces at one's peril. One moves within society within carefully defined systems of power and prestige. And once one knows how to locate oneself, one also knows that there is not an awful lot that one can do about this.

The way in which lower-class individuals use the pronouns "they" and "them" nicely expresses this consciousness of the heteronomy of one's life. "They" have things fixed in a certain way, "they" call the tune, "they" make the rules. This concept of "them" may not be too easily identified with particular individuals or groups. It is "the system," the map made by strangers, over which one must keep crawling. But it would be a onesided way of looking at "the system" if one assumes that this concept loses its meaning as one moves into the higher levels of society. To be sure, there will be a greater sense of freedom of movement and decision, and realistically so. But the basic coordinates within which one can move and decide have still been drawn by others, most of them strangers, many of them long in their graves. Even the total autocrat exercises his tyranny against constant resistance, not necessarily political resistance, but the resistance of custom, convention and sheer habit. Institutions carry within them a principle of inertia, perhaps founded ultimately on the hard rock of human stupidity. The tyrant finds that even if nobody dares act against him, his orders will still be nullified again and again by simple lack of comprehension. The alien-made fabric of society reasserts itself even against terror. But let us leave the question of tyranny. On the levels occupied by most men, including the writer and (we daresay) almost all the readers of these lines, location in society constitutes a definition of rules that have to be obeyed.

As we have seen, the commonsense view of society understands this. The sociologist does not contradict this understanding. He sharpens it, analyzes its roots, sometimes either modifies or extends it. We shall see later that sociological perspective finally goes beyond the commonsense understanding of "the system" and our captivity in it. But in most specific social situations that the sociologist sets out to analyze he will find little reason to quarrel with the notion that "they" are in charge. On the contrary, "they" will loom larger and in more pervasive fashion over our lives than we thought before the sociological analysis.

JAMES WELDON JOHNSON

TO BE A NEGRO

THERE were some black and brown boys and girls in the school, and several of them were in my class. One of the boys strongly attracted my attention from the first day I saw him. His face was as black as night, but shone as though it were polished; he had sparkling eyes, and when he opened his mouth, he displayed glistening white teeth. It struck me at once as appropriate to call him "Shiny Face," or "Shiny Eyes," or "Shiny Teeth," and I spoke to him often by one of these names to the other boys. These terms were finally merged into "Shiny," and to that name he answered good-naturedly during the balance of his public school days.

"Shiny" was considered without question to be the best speller, the best reader, the best penman—in a word, the best scholar, in the class. He was very quick to catch anything, but, nevertheless, studied hard; thus he possessed two powers very rarely combined in one boy. I saw him year after year, on up into the high school, win the majority of the prizes for punctuality, deportment, essay writing, and declamation. Yet it did not take me long to discover that, in spite of his standing as a scholar, he was in some way looked down upon.

The other black boys and girls were still more looked down upon. Some of the boys often spoke of them as "niggers." Sometimes on the way home from school a crowd would walk behind them repeating:

Nigger, nigger, never die,
Black face and shiny eye.

On one such afternoon one of the black boys turned suddenly on his tormentors and hurled a slate; it struck one of the white boys in the mouth, cutting a slight gash in his lip. At sight of the blood the boy who had thrown the slate ran, and his companions quickly followed. We ran after them pelting them with stones until they separated in several directions. I was very much wrought up over the affair, and went home and told my mother how one of the "niggers" had struck a boy with a slate. I shall never forget how she turned on me. "Don't you ever use that word again," she said, "and don't you ever bother the coloured children at school. You ought to be ashamed of

Reprinted from *The Autobiography of an Ex-Coloured Man* by James Weldon Johnson (New York: Alfred A. Knopf, 1912), pp. 419–21.

yourself." I did hang my head in shame, not because she had convinced me that I had done wrong, but because I was hurt by the first sharp word she had ever given me.

My school-days ran along very pleasantly. I stood well in my studies, not always so well with regard to my behaviour. I was never guilty of any serious misconduct, but my love of fun sometimes got me into trouble. I remember, however, that my sense of humour was so sly that most of the trouble usually fell on the head of the other fellow. My ability to play on the piano at school exercises was looked upon as little short of marvellous in a boy of my age. I was not chummy with many of my mates, but, on the whole, was about as popular as it is good for a boy to be.

One day near the end of my second term at school the principal came into our room and, after talking to the teacher, for some reason said: "I wish all of the white scholars to stand for a moment." I rose with the others. The teacher looked at me and, calling my name, said: "You sit down for the present, and rise with the others." I did not quite understand her, and questioned: "Ma'm?" She repeated, with a softer tone in her voice: "You sit down now, and rise with the others." I sat down dazed. I saw and heard nothing. When the others were asked to rise, I did not know it. When school was dismissed, I went out in a kind of stupor. A few of the white boys jeered me, saying: "Oh, you're a nigger too." I heard some black children say: "We knew he was coloured." "Shiny" said to them: "Come along, don't tease him," and thereby won my undying gratitude.

I hurried on as fast as I could, and had gone some distance before I perceived that "Red Head" was walking by my side. After a while he said to me: "Le' me carry your books." I gave him my strap without being able to answer. When we got to my gate, he said as he handed me my books: "Say, you know my big red agate? I can't shoot with it any more. I'm going to bring it to school for you tomorrow." I took my books and ran into the house. As I passed through the hallway, I saw that my mother was busy with one of her customers; I rushed up into my own little room, shut the door, and went quickly to where my looking-glass hung on the wall. For an instant I was afraid to look, but when I did, I looked long and earnestly. I had often heard people say to my mother: "What a pretty boy you have!" I was accustomed to hear remarks about my beauty; but now, for the first time, I became conscious of it and recognized it. I noticed the ivory whiteness of my skin, the beauty of my mouth, the size and liquid darkness of my eyes, and how the long, black lashes that fringed and shaded them produced an effect that was strangely fascinating even to me. I noticed the softness and glossiness of my dark hair that fell in waves over my temples, making my forehead appear whiter than it really was. How long I stood there gazing at my image I do not know. When I came out and reached the head of the stairs, I heard the lady who had been with my mother going out. I ran downstairs and rushed to where my mother was sitting, with a piece of work in her hands. I buried my head in her lap and blurted out: "Mother, mother, tell me, am I a nigger?" I could not see her face, but I knew the piece of work dropped to the floor and I felt her hands on my head. I looked up into her face and repeated: "Tell me, mother, am I a nigger?" There were tears in her eyes and I could see that she was suffering for me. And then it was that I looked at her critically for the first time. I had thought of her in a childish way only as the most beautiful woman in the world; now I looked at her searching for defects. I could see that her skin was almost brown, that her hair was not so soft as mine, and that she did differ in some way from the other ladies who came to the house; yet, even so, I could see that she was very beautiful, more beautiful than any of them. She must have felt that I was examining her, for she hid her face in my hair and said with difficulty: "No, my darling, you are not a nigger." She went on: "You are as good as anybody; if anyone calls you a nigger, don't notice them." But the more she talked, the less was I reassured, and I stopped her by asking: "Well, mother, am I white? Are you white?" She answered tremblingly: "No, I am not white, but you—your father is one of the greatest men in the country—the best blood of the South is in you—" This suddenly opened up in my heart a fresh chasm of misgiving and fear, and I almost fiercely demanded: "Who is my father? Where is he?" She stroked my hair and said: "I'll tell you about him some day." I sobbed: "I want to know now." She answered, "No, not now."

Perhaps it had to be done, but I have never forgiven the woman who did it so cruelly. It may be that she never knew that she gave me a sword-thrust that day in school which was years in healing.

EMILE DURKHEIM

WHAT IS A SOCIAL FACT?

IN reality there is in every society a certain group of phenomena which may be differentiated from those studied by the other natural sciences. When I fulfil my obligations as brother, husband, or citizen, when I execute my contracts, I perform duties which are defined, externally to myself and my acts, in law and in custom. Even if they conform to my own sentiments and I feel their reality subjectively, such reality is still objective, for I did not create them; I merely inherited them through my education. How many times it

Reprinted from *The Rules of Sociological Method* by Emile Durkheim (Glencoe, Ill.: Free Press, 1950), pp. 1–3, by permission of the publisher.

happens, moreover, that we are ignorant of the details of the obligations incumbent upon us, and that in order to acquaint ourselves with them we must consult the law and its authorized interpreters! Similarly, the church-member finds the beliefs and practices of his religious life ready-made at birth; their existence prior to his own implies their existence outside of himself. The system of signs I use to express my thought, the system of currency I employ to pay my debts, the instruments of credit I utilize in my commercial relations, the practices followed in my profession, etc., function independently of my own use of them. And these statements can be repeated for each member of society. Here, then, are ways of acting, thinking, and feeling that present the noteworthy property of existing outside the individual consciousness.

These types of conduct or thought are not only external to the individual but are, moreover, endowed with coercive power, by virtue of which they impose themselves upon him, independent of his individual will. Of course, when I fully consent and conform to them, this constraint is felt only slightly, if at all, and is therefore unnecessary. But it is, nonetheless, an intrinsic characteristic of these facts, the proof thereof being that it asserts itself as soon as I attempt to resist it. If I attempt to violate the law, it reacts against me so as to prevent my act before its accomplishment, or to nullify my violation by restoring the damage, if it is accomplished and reparable, or to make me expiate it if it cannot be compensated for otherwise.

In the case of purely moral maxims, the public conscience exercises a check on every act which offends it by means of the surveillance it exercises over the conduct of citizens, and the appropriate penalties at its disposal. In many cases the constraint is less violent, but nevertheless it always exists. If I do not submit to the conventions of society, if in my dress I do not conform to the customs observed in my country and in my class, the ridicule I provoke, the social isolation in which I am kept, produce, although in an attenuated form, the same effects as a punishment in the strict sense of the word. The constraint is nonetheless efficacious for being indirect. I am not obliged to speak French with my fellow-countrymen nor to use the legal currency, but I cannot possibly do otherwise. If I tried to escape this necessity, my attempt would fail miserably. As an industrialist, I am free to apply the technical methods of former centuries; but by doing so, I should invite certain ruin. Even when I free myself from these rules and violate them successfully, I am always compelled to struggle with them. When finally overcome, they make their constraining power sufficiently felt by the resistance they offer. The enterprises of all innovators, including successful ones, come up against resistance of this kind.

C. WRIGHT MILLS

THE SOCIOLOGICAL IMAGINATION

NOWADAYS men often feel that their private lives are a series of traps. They sense that within their everyday worlds, they cannot overcome their troubles, and in this feeling, they are often quite correct: What ordinary men are directly aware of and what they try to do are bounded by the private orbits in which they live; their visions and their powers are limited to the close-up scenes of job, family, neighborhood; in other milieux, they move vicariously and remain spectators. And the more aware they become, however vaguely, of ambitions and of threats which transcend their immediate locales, the more trapped they seem to feel.

Underlying this sense of being trapped are seemingly impersonal changes in the very structure of continent-wide societies. The facts of contemporary history are also facts about the success and the failure of individual men and women. When a society is industrialized, a peasant becomes a worker; a feudal lord is liquidated or becomes a businessman. When classes rise or fall, a man is employed or unemployed; when the rate of investment goes up or down, a man takes new heart or goes broke. When wars happen, an insurance salesman becomes a rocket launcher; a store clerk, a radar man; a wife lives alone; a child grows up without a father. Neither the life of an individual nor the history of a society can be understood without understanding both.

Yet men do not usually define the troubles they endure in terms of historical change and institutional contradiction. The well-being they enjoy, they do not usually impute to the big ups and downs of the societies in which they live. Seldom aware of the intricate connection between the patterns of their own lives and the course of world history, ordinary men do not usually know what this connection means for the kinds of men they are becoming and for the kinds of history-making in which they might take part. They do not possess the quality of mind essential to grasp the interplay of man and society, of biography and history, of self and world. They cannot cope with their personal troubles in such ways as to control the structural transformations that usually lie behind them.

Surely it is no wonder. In what period have so many men been so totally exposed at so fast a pace to such earthquakes of change? That Americans have not known such catastrophic changes as have the men and women of other

Reprinted from *The Sociological Imagination* by C. Wright Mills (New York: Oxford University Press, 1959), pp. 3–13.

societies is due to historical facts that are now quickly becoming "merely history." The history that now affects every man is world history. Within this scene and this period, in the course of a single generation, one sixth of mankind is transformed from all that is feudal and backward into all that is modern, advanced, and fearful. Political colonies are freed; new and less visible forms of imperialism installed. Revolutions occur; men feel the intimate grip of new kinds of authority. Totalitarian societies rise, and are smashed to bits—or succeed fabulously. After two centuries of ascendancy, capitalism is shown up as only one way to make society into an industrial apparatus. After two centuries of hope, even formal democracy is restricted to a quite small portion of mankind. Everywhere in the underdeveloped world, ancient ways of life are broken up and vague expectations become urgent demands. Everywhere in the overdeveloped world, the means of authority and of violence become total in scope and bureaucratic in form. Humanity itself now lies before us, the super-nation at either pole concentrating its most co-ordinated and massive efforts upon the preparation of World War Three.

The very shaping of history now outpaces the ability of men to orient themselves in accordance with cherished values. And which values? Even when they do not panic, men often sense that older ways of feeling and thinking have collapsed and that newer beginnings are ambiguous to the point of moral stasis. Is it any wonder that ordinary men feel they cannot cope with the larger worlds with which they are so suddenly confronted? That they cannot understand the meaning of their epoch for their own lives? That—in defense of selfhood—they become morally insensible, trying to remain altogether private men? Is it any wonder that they come to be possessed by a sense of the trap?

It is not only information that they need—in this Age of Fact, information often dominates their attention and overwhelms their capacities to assimilate it. It is not only the skills of reason that they need—although their struggles to acquire these often exhaust their limited moral energy.

What they need, and what they feel they need, is a quality of mind that will help them to use information and to develop reason in order to achieve lucid summations of what is going on in the world and of what may be happening within themselves. It is this quality, I am going to contend, that journalists and scholars, artists and publics, scientists and editors are coming to expect of what may be called the sociological imagination.

1

The sociological imagination enables its possessor to understand the larger historical scene in terms of its meaning for the inner life and the external career of a variety of individuals. It enables him to take into account how individuals, in the welter of their daily experience, often become falsely conscious of their social positions. Within that welter, the framework of modern society is sought, and within that framework the psychologies of a variety of men and women are formulated. By such means the personal uneasiness of individuals is focused upon explicit troubles and the indifference of publics is transformed into involvement with public issues.

The first fruit of this imagination—and the first lesson of the social science that embodies it—is the idea that the individual can understand his own experience and gauge his own fate only by locating himself within his period, that he can know his own chances in life only by becoming aware of those of all individuals in his circumstances. In many ways it is a terrible lesson; in many ways a magnificent one. We do not know the limits of man's capacities for supreme effort or willing degradation, for agony or glee, for pleasurable brutality or the sweetness of reason. But in our time we have come to know that the limits of "human nature" are frighteningly broad. We have come to know that every individual lives, from one generation to the next, in some society; that he lives out a biography, and that he lives it out within some historical sequence. By the fact of his living he contributes, however minutely, to the shaping of this society and to the course of its history, even as he is made by society and by its historical push and shove.

The sociological imagination enables us to grasp history and biography and the relations between the two within society. That is its task and its promise. To recognize this task and this promise is the mark of the classic social analyst. It is characteristic of Herbert Spencer—turgid, polysyllabic, comprehensive; of E. A. Ross—graceful, muckraking, upright; of Auguste Comte and Emile Durkheim; of the intricate and subtle Karl Mannheim. It is the quality of all that is intellectually excellent in Karl Marx; it is the clue to Thorstein Veblen's brilliant and ironic insight, to Joseph Schumpeter's many-sided constructions of reality; it is the basis of the psychological sweep of W. E. H. Lecky no less than of the profundity and clarity of Max Weber. And it is the signal of what is best in contemporary studies of man and society.

No social study that does not come back to the problems of biography, of history and of their intersections within a society has completed its intellectual journey. Whatever the specific problems of the classic social analysts, however limited or however broad the features of social reality they have examined, those who have been imaginatively aware of the promise of their work have consistently asked three sorts of questions:

1. What is the structure of this particular society as a whole? What are its essential components, and how are they related to one another? How does it differ from other varieties of social order? Within it, what is the meaning of any particular feature for its continuance and for its change?

2. Where does this society stand in human history? What are the mechanics by which it is changing? What is its place within and its meaning for the

development of humanity as a whole? How does any particular feature we are examining affect, and how is it affected by, the historical period in which it moves? And this period—what are its essential features? How does it differ from other periods? What are its characteristic ways of history-making?

3. What varieties of men and women now prevail in this society and in this period? And what varieties are coming to prevail? In what ways are they selected and formed, liberated and repressed, made sensitive and blunted? What kinds of "human nature" are revealed in the conduct and character we observe in this society in this period? And what is the meaning for "human nature" of each and every feature of the society we are examining?

Whether the point of interest is a great power state or a minor literary mood, a family, a prison, a creed—these are the kinds of questions the best social analysts have asked. They are the intellectual pivots of classic studies of man in society—and they are the questions inevitably raised by any mind possessing the sociological imagination. For that imagination is the capacity to shift from one perspective to another—from the political to the psychological; from examination of a single family to comparative assessment of the national budgets of the world; from the theological school to the military establishment; from considerations of an oil industry to studies of contemporary poetry. It is the capacity to range from the most impersonal and remote transformations to the most intimate features of the human self—and to see the relations between the two. Back of its use there is always the urge to know the social and historical meaning of the individual in the society and in the period in which he has his quality and his being.

That, in brief, is why it is by means of the sociological imagination that men now hope to grasp what is going on in the world, and to understand what is happening in themselves as minute points of the intersections of biography and history within society. In large part, contemporary man's self-conscious view of himself as at least an outsider, if not a permanent stranger, rests upon an absorbed realization of social relativity and of the transformative power of history. The sociological imagination is the most fruitful form of this self-consciousness. By its use men whose mentalities have swept only a series of limited orbits often come to feel as if suddenly awakened in a house with which they had only supposed themselves to be familiar. Correctly or incorrectly, they often come to feel that they can now provide themselves with adequate summations, cohesive assessments, comprehensive orientations. Older decisions that once appeared sound now seem to them products of a mind unaccountably dense. Their capacity for astonishment is made lively again. They acquire a new way of thinking, they experience a transvaluation of values: in a word, by their reflection and by their sensibility, they realize the cultural meaning of the social sciences.

2

Perhaps the most fruitful distinction with which the sociological imagination works is between "the personal troubles of milieu" and "the public issues of social structure." This distinction is an essential tool of the sociological imagination and a feature of all classic work in social science.

Troubles occur within the character of the individual and within the range of his immediate relations with others; they have to do with his self and with those limited areas of social life of which he is directly and personally aware. Accordingly, the statement and the resolution of troubles properly lie within the individual as a biographical entity and within the scope of his immediate milieu—the social setting that is directly open to his personal experience and to some extent his willful activity. A trouble is a private matter: values cherished by an individual are felt by him to be threatened.

Issues have to do with matters that transcend these local environments of the individual and the range of his inner life. They have to do with the organization of many such milieux into the institutions of an historical society as a whole, with the ways in which various milieux overlap and interpenetrate to form the larger structure of social and historical life. An issue is a public matter: some value cherished by publics is felt to be threatened. Often there is a debate about what that value really is and about what it is that really threatens it. This debate is often without focus if only because it is the very nature of an issue, unlike even widespread trouble, that it cannot very well be defined in terms of the immediate and everyday environments of ordinary men. An issue, in fact, often involves a crisis in institutional arrangements, and often too it involves what Marxists call "contradictions" or "antagonisms."

In these terms, consider unemployment. When, in a city of 100,000, only one man is unemployed, that is his personal trouble, and for its relief we properly look to the character of the man, his skills, and his immediate opportunities. But when in a nation of 50 million employees, 15 million men are unemployed, that is an issue, and we may not hope to find its solution within the range of opportunities open to any one individual. The very structure of opportunities has collapsed. Both the correct statement of the problem and the range of possible solutions require us to consider the economic and political institutions of the society, and not merely the personal situation and character of a scatter of individuals.

Consider war. The personal problem of war, when it occurs, may be how to survive it or how to die in it with honor; how to make money out of it; how to climb into the higher safety of the military apparatus; or how to contribute to the war's termination. In short, according to one's values, to find a set of milieux and within it to survive the war or make one's death in it meaningful. But the structural issues of war have to do with its causes; with what types of men it throws up into command; with its effects upon economic and political, family and religious institutions, with the unorganized irresponsibility of a world of nation-states.

Consider marriage. Inside a marriage a man and a woman may experience personal troubles, but when the divorce rate during the first four years of marriage is 250 out of every 1,000 attempts, this is an indication of a structural issue having to do with the institutions of marriage and the family and other institutions that bear upon them.

Or consider the metropolis—the horrible, beautiful, ugly, magnificent sprawl of the great city. For many upper-class people, the personal solution to "the problem of the city" is to have an apartment with private garage under it in the heart of the city, and forty miles out, a house by Henry Hill, garden by Garrett Eckbo, on a hundred acres of private land. In these two controlled environments—with a small staff at each end and a private helicopter connection—most people could solve many of the problems of personal milieux caused by the facts of the city. But all this, however splendid, does not solve the public issues that the structural fact of the city poses. What should be done with this wonderful monstrosity? Break it all up into scattered units, combining residence and work? Refurbish it as it stands? Or, after evacuation, dynamite it and build new cities according to new plans in new places? What should those plans be? And who is to decide and to accomplish whatever choice is made? These are structural issues; to confront them and to solve them requires us to consider political and economic issues that affect innumerable milieux.

In so far as an economy is so arranged that slumps occur, the problem of unemployment becomes incapable of personal solution. In so far as war is inherent in the nation-state system and in the uneven industrialization of the world, the ordinary individual in his restricted milieu will be powerless—with or without psychiatric aid—to solve the troubles this system or lack of system imposes upon him. In so far as the family as an institution turns women into darling little slaves and men into their chief providers and unweaned dependents, the problem of a satisfactory marriage remains incapable of purely private solution. In so far as the overdeveloped megalopolis and the overdeveloped automobile are built-in features of the overdeveloped society, the issues of urban living will not be solved by personal ingenuity and private wealth.

What we experience in various and specific milieux, I have noted, is often caused by structural changes. Accordingly, to understand the changes of many personal milieux we are required to look beyond them. And the number and variety of such structural changes increase as the institutions within which we live become more embracing and more intricately connected with one another. To be aware of the idea of social structure and to use it with sensibility is to be capable of tracing such linkages among a great variety of milieux. To be able to do that is to possess the sociological imagination.

3

What are the major issues for publics and the key troubles of private individuals in our time? To formulate issues and troubles, we must ask what values are cherished yet threatened, and what values are cherished and supported, by the characterizing trends of our period. In the case both of threat and of support we must ask what salient contradictions of structure may be involved.

When people cherish some set of values and do not feel any threat to them, they experience *well-being*. When they cherish values but *do* feel them to be threatened, they experience a crisis—either as a personal trouble or as a public issue. And if all their values seem involved, they feel the total threat of panic.

But suppose people are neither aware of any cherished values nor experience any threat? That is the experience of *indifference,* which, if it seems to involve all their values, becomes apathy. Suppose, finally, they are unaware of any cherished values, but still are very much aware of a threat? That is the experience of *uneasiness,* of anxiety, which, if it is total enough, becomes a deadly unspecified malaise.

Ours is a time of uneasiness and indifference—not yet formulated in such ways as to permit the work of reason and the play of sensibility. Instead of troubles—defined in terms of values and threats—there is often the misery of vague uneasiness; instead of explicit issues there is often merely the beat feeling that all is somehow not right. Neither the values threatened nor whatever threatens them has been stated; in short, they have not been carried to the point of decision. Much less have they been formulated as problems of social science.

In the 'thirties there was little doubt—except among certain deluded business circles—that there was an economic issue which was also a pack of personal troubles. In these arguments about "the crisis of capitalism," the formulations of Marx and the many unacknowledged re-formulations of his work probably set the leading terms of the issue, and some men came to understand their personal troubles in these terms. The values threatened were plain to see and cherished by all; the structural contradictions that threatened them also seemed plain. Both were widely and deeply experienced. It was a political age.

But the values threatened in the era after World War Two are often neither widely acknowledged as values nor widely felt to be threatened. Much private uneasiness goes unformulated; much public malaise and many decisions of enormous structural relevance never become public issues. For those who accept such inherited values as reason and freedom, it is the uneasiness itself that is the trouble; it is the indifference itself that is the issue. And it is this condition, of uneasiness and indifference, that is the signal feature of our period.

All this is so striking that it is often interpreted by observers as a shift in the very kinds of problems that need now to be formulated. We are frequently told that the problems of our decade, or even the crises of our period, have shifted from the external realm of economics and now have to do with the quality of individual life—in fact with the question of whether there is soon going to be anything that can properly be called individual life. Not child labor but comic books, not poverty but mass leisure, are at the center of concern. Many great public issues as well as many private troubles are described in terms of "the psychiatric"—often, it seems, in a pathetic attempt to avoid the large issues and problems of modern society. Often this statement seems to rest upon a provincial narrowing of interest to the Western societies, or even to the United States—thus ignoring two-thirds of mankind; often, too, it arbitrarily divorces the individual life from the larger institutions within which that life is enacted, and which on occasion bear upon it more grievously than do the intimate environments of childhood.

Problems of leisure, for example, cannot even be stated without considering problems of work. Family troubles over comic books cannot be formulated as problems without considering the plight of the contemporary family in its new relations with the newer institutions of the social structure. Neither leisure nor its debilitating uses can be understood as problems without recognition of the extent to which malaise and indifference now form the social and personal climate of contemporary American society. In this climate, no problems of "the private life" can be stated and solved without recognition of the crisis of ambition that is part of the very career of men at work in the incorporated economy.

It is true, as psychoanalysts continually point out, that people do often have "the increasing sense of being moved by obscure forces within themselves which they are unable to define." But it is *not* true, as Ernest Jones asserted, that "man's chief enemy and danger is his own unruly nature and the dark forces pent up within him." On the contrary: "Man's chief danger" today lies in the unruly forces of contemporary society itself, with its alienating methods of production, its enveloping techniques of political domination, its international anarchy—in a word, its pervasive transformations of the very "nature" of man and the conditions and aims of his life.

It is now the social scientist's foremost political and intellectual task—for here the two coincide—to make clear the elements of contemporary uneasiness and indifference. It is the central demand made upon him by other cultural workmen—by physical scientists and artists, by the intellectual community in general. It is because of this task and these demands, I believe, that the social sciences are becoming the common denominator of our cultural period, and the sociological imagination our most needed quality of mind.

Chapter 2

THE DISCIPLINE OF SOCIOLOGY

THE basic human experiences with which sociology deals are, of course, timeless, but the peculiar perspective of sociology is not; it arose from a very particular historical situation. Human beings always start to think about the routine structures of society when they are subjected to shocks. The great shocks that gave rise to modern sociological thought were the political and social-economic revolutions of the eighteenth and nineteenth centuries in Europe. The word *sociology* as the name of a new intellectual discipline was coined by Auguste Comte (1798–1857), a Frenchman who had the high hope that this new science would enable man not only to understand but to control the current turbulent course of events. In the late nineteenth century, France was still trying to come to terms with the consequences of the revolution of 1789, and it was there and then that sociology reached its first full flowering. In the person of Emile Durkheim (1858–1917), and in the school which he founded, French sociology became a discipline that affected every branch of the science of man.

Germany and the United States were also fertile ground for the growth of sociology. Max Weber (1864–1920), a German, enlisted sociology in a vast scholarly effort to understand the distinctiveness and the underlying dynamics of modern society. Other German sociologists, although differing from Weber, had similar broad historical and even philosophical interests. Sociology in the United States—after an initial period in which it was greatly influenced by Europe—developed in much closer contact with efforts to solve practical social problems. Yet American sociology also produced important new theoretical approaches. Among these is an American tradition in social-psychology that goes back to the work of George Herbert Mead (1863–1931). Another is the development of the structural-functionalist school, which today is mainly associated with the names of Talcott Parsons and Robert Merton.

In general, sociological thinking is characterized by estrangement. In a way, the sociologist becomes a stranger to his own society. He maintains this role of stranger, even if it is only temporary and limited to moments of intellectual reflection, to secure objectivity. The social experience of the stranger is discussed in the first of the following selections, from *The Stranger* by Alfred Schutz, a key figure in the recent development of "phenomenological sociology." Even though Schutz does not explicitly mention the sociologist as a peculiar kind of stranger, the reader should keep this relationship in mind. The second selection returns once again to Emile Durkheim's discussion of the objectivity and constraint of social reality in *The Rules of Sociological Method*. It provides an understanding of the distinctive Durkheimian understanding of society as a reality over and beyond the subjective meanings of individuals. In "classical" sociology Max Weber represents the antithesis to the Durkheimian approach. Weber focused on the way in which subjective meanings relate to social events. The third selection comprises passages from *The Protestant Ethic and the Spirit of Capitalism,* probably Weber's most significant work, in which he tried to show how the peculiar psychological consequences of Protestantism were conducive to the development of modern capitalism. In the fourth selection, C. Wright Mills discusses two key modes of sociological reflection, from the point of view of the individual and from the point of view of the remote social structures that transcend the individual. In the last selection, Peter L. Berger differentiates sociology from other approaches to social reality.

ALFRED SCHUTZ

THE STRANGER

THE present paper intends to study in terms of a general theory of interpretation the typical situation in which a stranger finds himself in his attempt to interpret the cultural pattern of a social group which he approaches and to orient himself within it. For our present purposes the term "stranger" shall mean an adult individual of our times and civilization who tries to be permanently accepted or at least tolerated by the group which he approaches. The outstanding example for the social situation under scrutiny is that of the immigrant, and the following analyses are, as a matter of convenience, worked out with this instance in view. But by no means is their validity restricted to this special case. The applicant for membership in a closed club, the prospective bridegroom who wants to be admitted to the girl's family, the farmer's son who enters college, the city-dweller who settles in a rural environment, the "selectee" who joins the Army, the family of the war worker who moves into a boom town—all are strangers according to the definition just given, although in these cases the typical "crisis" that the immigrant undergoes may assume milder forms or even be entirely absent. Intentionally excluded, however, from the present investigation are certain cases the inclusion of which would require some qualifications in our statements: (*a*) the visitor or guest who intends to establish a merely transitory contact with the group; (*b*) children or primitives; and (*c*) relationships between individuals and groups of different levels of civilization, as in the case of the Huron brought to Europe—a pattern dear to some moralists of the eighteenth century. Furthermore, it is not the purpose of this paper to deal with the processes of social assimilation and social adjustment which are treated in an abundant and, for the most

Reprinted from *Collected Papers* by Alfred Schutz, vol. 2 (The Hague: Martinus Nijhoff, 1964), pp. 91–105, by permission of the publisher.

part, excellent literature [1] but rather with the situation of approaching, which precedes every possible social adjustment and which includes its prerequisites.

As a convenient starting-point we shall investigate how the cultural pattern of group life presents itself to the common sense of a man who lives his everyday life within the group among his fellow men. Following the customary terminology, we use the term "cultural pattern of group life" for designating all the peculiar valuations, institutions, and systems of orientation and guidance (such as the folkways, mores, laws, habits, customs, etiquette, fashions) which, in the common opinion of sociologists of our time, characterize —if not constitute—any social group at a given moment in its history. This cultural pattern, like any phenomenon of the social world, has a different aspect for the sociologist and for the man who acts and thinks within it.[2] The sociologist (as sociologist, not as a man among fellow-men which he remains in his private life) is the disinterested scientific on-looker of the social world. He is disinterested in that he intentionally refrains from participating in the network of plans, means-and-ends relations, motives and chances, hopes and fears, which the actor within the social world uses for interpreting his experiences of it; as a scientist he tries to observe, describe, and classify the social world as clearly as possible in well-ordered terms in accordance with the scientific ideals of coherence, consistency, and analytical consequence. The actor within the social world, however, experiences it primarily as a field of his actual and possible acts and only secondarily as an object of his thinking. In so far as he is interested in knowledge of his social world, he organizes this knowledge not in terms of a scientific system but in terms of relevance to his actions. He groups the world around himself (as the center) as a field of domination and is therefore especially interested in that segment which is within his actual or potential reach. He singles out those of its elements which may serve as means or ends for his "use and enjoyment," [3] for furthering his purposes, and for overcoming obstacles. His interest in these elements is of different degrees, and for this reason he does not aspire to become acquainted with all of them with equal thoroughness. What he wants is *graduated knowledge* of relevant elements, the degree of desired knowledge being correlated with their relevance. Otherwise stated, the world seems to him at any given moment as stratified in different layers of relevance, each of them requiring a different degree of knowledge. To illustrate these strata of relevance we may —borrowing the term from cartography—speak of "isohypses" or "hypsographical contour lines of relevance," trying to suggest by this metaphor that we could show the distribution of the interests of an individual at a given moment with respect both to their intensity and to their scope by connecting elements of equal relevance to his acts, just as the cartographer connects points of equal height by contour lines in order to reproduce adequately the shape of a mountain. The graphical representation of these "contour lines of relevance" would not show them as a single closed field but rather as numerous areas scattered over the map, each of different size and shape. Distinguishing with William James [4] two kinds of knowledge, namely, *"knowledge of acquaintance"* and *"knowledge about,"* we may say that, within the field covered by the contour lines of relevance, there are centers of explicit knowledge *of* what is aimed at; they are surrounded by a halo knowledge *about* what seems to be sufficient; next comes a region in which it will do merely "to put one's trust"; the adjoining foothills are the home of unwarranted hopes and assumptions; between these areas, however, lie zones of complete ignorance.

We do not want to overcharge this image. Its chief purpose has been to illustrate that the knowledge of the man who acts and thinks within the world of his daily life is not homogeneous; it is (1) incoherent, (2) only partially clear, and (3) not at all free from contradictions.

1. It is incoherent because the individual's interests which determine the relevance of the objects selected for further inquiry are themselves not integrated into a coherent system. They are only partially organized under plans of any kind, such as plans of life, plans of work and leisure, plans for every social role assumed. But the hierarchy of these plans changes with the situation and with the growth of the personality; interests are shifted continually and entail an uninterrupted transformation of the shape and density of the relevance lines. Not only the selection of the objects of curiosity but also the degree of knowledge aimed at changes.

2. Man in his daily life is only partially—and we dare say exceptionally—interested in the clarity of his knowledge, i.e., in full insight into the relations between the elements of his world and the general principles ruling those relations. He is satisfied that a well-functioning telephone service is available to him and, normally, does not ask how the apparatus functions in detail and what laws of physics make this functioning possible. He buys merchandise in the store, not knowing how it is produced, and pays with money, although he has only a vague idea of what money really is. He takes it for granted that his fellow-man will understand his thought if expressed in plain language and will answer accordingly, without wondering how this miraculous performance may be explained. Furthermore, he does not search for the truth and does not quest for certainty. All he wants is information on likelihood and insight into the chances or risks which the situation at hand entails for the outcome of his actions. That the subway will run tomorrow as usual is for him almost of the same order of likelihood as that the sun will rise. If by reason of a special interest he needs more explicit knowledge on a topic, a benign modern civilization holds ready for him a chain of information desks and reference libraries.

3. His knowledge, finally, is not consistent. At the same time he may consider statements as equally valid which in fact are incompatible with one another. As a father, a citizen, an employee, and a member of his church he may have the most different and the least congruent opinions on moral, polit-

ical, or economic matters. This inconsistency does not necessarily originate in a logical fallacy. Men's thinking is distributed over subject matters located within different and differently relevant levels, and they are not aware of the modifications they would have to make in passing from one level to another. This and similar problems would have to be explored by a logic of everyday thinking, postulated but not attained by all the great logicians from Leibniz to Husserl and Dewey. Up to now the science of logic has primarily dealt with the logic of science.

The system of knowledge thus acquired—incoherent, inconsistent, and only partially clear, as it is—takes on for the members of the in-group the appearance of a *sufficient* coherence, clarity, and consistency to give anybody a reasonable chance of understanding and of being understood. Any member born or reared within the group accepts the ready-made standardized scheme of the cultural pattern handed down to him by ancestors, teachers, and authorities as an unquestioned and unquestionable guide in all the situations which normally occur within the social world. The knowledge correlated to the cultural pattern carries its evidence in itself—or, rather, it is taken for granted in the absence of evidence to the contrary. It is a knowledge of trustworthy *recipes* for interpreting the social world and for handling things and men in order to obtain the best results in every situation with a minimum of effort by avoiding undesirable consequences. The recipe works, on the one hand, as a precept for actions and thus serves as a scheme of expression: whoever wants to obtain a certain result has to proceed as indicated by the recipe provided for this purpose. On the other hand, the recipe serves as a scheme of interpretation: whoever proceeds as indicated by a specific recipe is supposed to intend the correlated result. Thus it is the function of the cultural pattern to eliminate troublesome inquiries by offering ready-made directions for use, to replace truth hard to attain by comfortable truisms, and to substitute the self-explanatory for the questionable.

This "thinking as usual," as we may call it, corresponds to Max Scheler's idea of the "relatively natural conception of the world" (*relativ natürliche Weltanschauung*); [5] it includes the "of-course" assumptions relevant to a particular social group which Robert S. Lynd describes in such a masterly way —together with their inherent contradictions and ambivalence—as the "Middletown-spirit." [6] Thinking-as-usual may be maintained as long as some basic assumptions hold true, namely: (1) that life and especially social life will continue to be the same as it has been so far, that is to say, that the same problems requiring the same solutions will recur and that, therefore, our former experiences will suffice for mastering future situations; (2) that we may rely on the knowledge handed down to us by parents, teachers, governments, traditions, habits, etc., even if we do not understand its origin and its real meaning; (3) that in the ordinary course of affairs it is sufficient to know something *about* the general type or style of events we may encounter in our life-world in order to manage or control them; and (4) that neither the systems of recipes as schemes of interpretation and expression nor the underlying basic assumptions just mentioned are our private affair, but that they are likewise accepted and applied by our fellow-men.

If only one of these assumptions ceases to stand the test, thinking-as-usual becomes unworkable. Then a "crisis" arises which, according to W. I. Thomas' famous definition, "interrupts the flow of habit and gives rise to changed conditions of consciousness and practice"; or, as we may say, it overthrows precipitously the actual system of relevances. The cultural pattern no longer functions as a system of tested recipes at hand; it reveals that its applicability is restricted to a specific historical situation.

Yet the stranger, by reason of his personal crisis, does not share the above-mentioned basic assumptions. He becomes essentially the man who has to place in question nearly everything that seems to be unquestionable to the members of the approached group.

To him the cultural pattern of the approached group does not have the authority of a tested system of recipes, and this, if for no other reason, because he does not partake in the vivid historical tradition by which it has been formed. To be sure, from the stranger's point of view, too, the culture of the approached group has its peculiar history, and this history is even accessible to him. But it has never become an integral part of his biography, as did the history of his home group. Only the ways in which his fathers and grandfathers lived become for everyone elements of his own way of life. Graves and reminiscences can neither be transferred nor conquered. The stranger, therefore, approaches the other group as a newcomer in the true meaning of the term. At best he may be willing and able to share the present and the future with the approached group in vivid and immediate experience; under all circumstances, however, he remains excluded from such experiences of its past. Seen from the point of view of the approached group, he is a man without a history.

To the stranger the cultural pattern of his home group continues to be the outcome of an unbroken historical development and an element of his personal biography, which for this very reason has been and still is the unquestioned scheme of reference for his "relatively natural conception of the world." As a matter of course, therefore, the stranger starts to interpret his new social environment in terms of his thinking as usual. Within the scheme of reference brought from his home group, however, he finds a ready-made idea of the pattern supposedly valid within the approached group—an idea which necessarily will soon prove inadequate.[7]

EMILE DURKHEIM

THE RULES OF SOCIOLOGICAL METHOD

WE have . . . a set of phenomena, which consists of various ways of behaving, thinking, and feeling, existing independently of the individual and endowed with a persuasive power, or authority, by which they force themselves on him. Consequently, they cannot possibly be confused with either biological phenomena, which consist of organic processes, or with psychological phenomena, which exist only in and through the individual mind. Therefore, this new system constitutes a separate order of things, and the term "social" must be accorded to and reserved for these phenomena.

This term is appropriate, for obviously, since these phenomena are not based on the individual, they must derive from society—either society as a whole or one of the groups which are included in the broader term, such as religious affiliations, political or literary movements, professional bodies, etc. Then again, the qualification applies to these phenomena alone; for the term "social" has a special significance only when it describes things that do not belong in any category of phenomena already specified and listed. It follows that these social phenomena form the proper domain of sociological study.

It is possible that the word "constraint," which we mentioned in defining these phenomena, might frighten off the zealous partisans of free will. They advocate the absolute self-determination of the individual, and find that man's dignity is slighted whenever he is made to feel less than completely his own master. Nowadays, however, it is generally acknowledged that most of our ideas and attitudes do not originate within ourselves, but come to us from the outside. They are able to penetrate our consciousness only by exercising a strong persuasive power. That is the whole meaning of our definition. We know, too, that social constraint is not necessarily incompatible with individuality.

Since the examples we cited previously (legal and moral rules, religious dogma, monetary systems, etc.) all consist of established beliefs and practices, one might be led to think that social phenomena exist only as an aspect of an established social organization. But there are other phenomena that lack these crystallized forms, and yet that have the same objectivity and the same influence over the individual. These may be called "social currents" or mass emotions. For instance, the great waves of enthusiasm, indignation, or pity which sweep through a crowd do not originate in any one particular mind. They

Reprinted from *The Rules of Sociological Method* by Emile Durkheim (Glencoe, Ill.: Free Press, 1950), pp. 18–22, by permission of the publisher.

come to each of us from without and are liable to carry us away in spite of ourselves. In giving way wholeheartedly to these emotions, I may not feel the insistent pressure they exert on me, but the pressure becomes apparent as soon as I try to resist them. When a man tries to oppose a crowd demonstration, the very emotions he wants to repulse turn against him. Now, if this external power of coercion asserts itself so firmly in cases of resistance, then, whether or not we are aware of it, it must also exist when no resistance is offered. In such cases, we are victims of the illusion of having ourselves created something which actually forced itself on us from without.

Though the willingness with which we allow ourselves to be carried away may mask the pressure to which we have been subjected, it does not abolish this pressure (in the same way that air does not become any less heavy because we do not ordinarily feel its weight). Even if we have spontaneously gone along with the mass emotion, the impression we have received is very different from what we should have experienced had we been alone. Once the crowd disperses, once the social currents have ceased to influence us, and we are alone again, the emotions that swept over us may seem strange and even unreal. They might even horrify us by seeming so contrary to our "true nature." We realize, then, that these emotions and feelings were impressed on us to a far greater degree than they were generated by us. Thus, a group of people who are perfectly harmless for the most part may be drawn into lynchings, riots, and other atrocities when gathered into a mob. What we say about these transitory outbursts applies equally to the more permanent currents of opinion on religious, political, literary, or artistic matters which constantly develop around us—in society as a whole, or within more limited groups.

Our definition of the social phenomenon as existing independently of the individual, and exerting a power of constraint over him, can be confirmed by a typical example from everyday life—the way in which children are raised. Examining the facts as they are and have always been, one immediately sees that all education consists of a continual effort to impose certain modes of behavior on the child. We teach him ways of seeing, feeling, and acting which he could never acquire spontaneously. From the very beginning of his life we insist that he eat, drink, and sleep regularly: we require him to be clean, quiet, and obedient. Later on, we compel him to be considerate of others, to respect social customs and conventions, to work hard, etc. If, with time, this pressure ceases to be apparent, it is because it gradually gives rise to habits and tendencies which render such constraint unnecessary, yet which do not replace it, since constraint remains the source of these habits.

It is true that, according to Spencer, a rational education ought to condemn such methods, and allow the child to develop in complete freedom. As this theory of education has never been practiced by any known group, however, it remains simply a personal expression of opinion, and does not contradict the observations I have just made. These observations are particularly interesting when one considers the real object of education, which is the socialization of the individual. This learning process shows in microcosm the historical development of the social being. The pressure to which the child is submitted is the pressure of the social environment, which tends to fashion him in its own image through the medium of parents and teachers.

A social phenomenon cannot be defined by its universality. The fact that the same thought is found in every mind, or that a movement is repeated by every individual, doesn't necessarily make it a social phenomenon. If sociologists in the past have been satisfied to cite this universal quality in a definition of social phenomena, they have mistakenly confused them with what might be called their individual expressions. Actually, it is the collective aspect of the beliefs, tendencies, and practices of a group that characterizes social phenomena. The individual manifestations of collective behavior are something else again. This duality is plainly demonstrated by the fact that these two orders of phenomena are often found dissociated from one another. In fact, some of these ways of acting and thinking acquire, through repetition, a certain rigidity of form which serves to crystallize them and which isolates this behavior from the particular acts that reflect it. Some collective behavior thus assumes a perceptible form, so to speak, and constitutes a reality in its own right, quite distinct from the individual phenomena which make it up. A social phenomenon is not only inherent in the successive acts it engenders but, by a prerogative wholly lacking in the biological field, it can be permanently expressed in a formula or model, which can then be repeated by word of mouth, transmitted through education, or defined in writing. Such is the origin and nature of legal and social rules, popular sayings and proverbs, the professions of faith which set forth religious or political beliefs, the aesthetic standards drawn up by artistic or literary movements, etc. None of these can be entirely reproduced in any of their individual applications since social phenomena can exist in theory alone, even without actually being applied.

Of course, this dissociation we have been discussing may not always be clearly evident. The very fact of its obvious existence in the numerous and important cases just cited, though, is sufficient to prove that a social phenomenon is something quite distinct from its individual manifestations. Even when this dissociation is not readily visible, certain methodological devices may help to disclose it. It is really essential to do so, if we want to separate the social phenomenon from its alloys and examine it in a pure state. There are certain currents of opinion, whose intensity varies with time and place, which impel one group toward a high number of marriages, for example, another toward more suicides, or more or fewer births, etc. These currents are clearly social phenomena. At the first sight they seem inseparable from the forms they take in individual cases, but statistics furnish us with the means of isolating them. They are, in fact, represented with considerable precision by birth, marriage, and suicide rates—that is, by the number obtained by dividing the average annual total of marriages, births, and suicides, by the number of men who are of an age to marry, beget children, or commit suicide. Since each of these figures includes all the individual cases indiscriminately, the particular circumstances which may have had a share in the production of the phenomenon are neutralized and, consequently, do not contribute to its determination. The rate obtained, then, expresses a certain state of the "group mind."

These rates are social phenomena, detached from all foreign components. Their individual manifestations are, of course, to a certain extent social, since they partly reproduce a social model. Each of these manifestations, however, depends a great deal on the mental and physical make-up of the individual and the particular circumstances in which he finds himself. Strictly speaking, then, they are not really social phenomena. The sociologist finds these individual acts of interest, yet they do not form the immediate subject matter of sociology. We could call them socio-psychological, since they belong to two fields of study at once. We find similar compound phenomena in biological organisms, which are in turn studied by the inter-disciplinary natural sciences, such as biochemistry, for example.

One might object that a phenomenon can be social only if it is common to all members of society or, at least, to most of them—that is, only if it is generalized. This may be true, but it is general because it is collective (that is, more or less obligatory) and certainly not collective because it is general. The social phenomenon is a group condition which is found repeated in every individual because it has been imposed on him. It is found in each part because it exists in the whole, rather than in the whole because it exists in the parts. This becomes particularly evident in the beliefs and practices which have been handed down to us, ready-made, by previous generations. We accept and adopt them because, being collective and venerable, they are invested with a special authority that our education has taught us to recognize and respect. It should be noted that the vast majority of social phenomena reach us in this way. Even when the social phenomenon is partly dependent on our direct collaboration, its nature remains unchanged. The mass emotion that surges suddenly through a crowd is not simply the expression of what all the individuals feel in common; it is something quite different, as we have already shown. Such emotion is the result of people being together. It is a product of the actions, reactions and interactions among individuals in a group. If the mass emotion is echoed by each person, it is by virtue of the special force of its collective origin. If all hearts beat in tempo it is not because of a spontaneous and pre-established agreement, but rather because the same force propels them in the same direction. Each one is carried along by all.

We now reach the point where we can present and delineate the domain of sociology with some precision. It comprises only a limited group of phenomena. A social phenomenon is recognized by the externalized power of coercion it exercises, or is capable of exercising, over individuals. The presence of this power is in turn recognized either by the existence of a specific sanction, or by the resistance offered to the individual effort that tends to violate it. However, we can also define a social phenomenon by its diffusion within a group, provided that care is taken to add, as a second essential characteristic, that its existence remains independent of the individual forms it assumes in this diffusion. The last criterion may be easier to apply, in certain cases, than the first. Constraint is easy to ascertain when it is expressed externally through some direct reaction of society, as is the case in law, custom, beliefs, conventions, even in fashions and taste. But when it is only indirect, like the constraint exercised by an economic organization, it is not so readily perceived. Universality, in combination with an external objectivity, may be easier to establish. The second part of our definition of social phenomena will be seen, then, as merely an expansion of the first; for if a mode of behavior whose existence is exterior to the human mind becomes universal, this can be brought about only by an application of coercive pressure by the group.

All the phenomena on which our definition is based have, so far, been "ways of doing," or modes of behavior. There are also, however, "ways of being"; that is, phenomena exhibited in a pattern or structure. These diverse phenomena (e.g., distribution of population, channels of communications, kinds of housing, etc.) present the same characteristic by which we defined the others. The social structure is imposed on the individual in the same way as the behavioral phenomena which we discussed earlier. When we want to know how a society is divided politically, of what these divisions are composed, or what sort of coalition exists among them, physical inspection or geographic observations will be of no help in reaching an answer. These divisions are social even though they have a certain physical basis. We are able to understand this organization only through a study of public law, for it is determined by the law, just as our domestic and civil relations are. This political organization is therefore no less obligatory than the phenomena mentioned earlier. If the population crowds into our cities, instead of scattering throughout the countryside, it is because of a social tendency, a collective drive, which imposes this geographic concentration on individuals. We can no more arbitrarily choose the style of our houses than the type of clothing we do or do not wear; at least, both are equally obligatory. The direction and extent of internal migrations and commerce are determined by the means of communications available. Consequently, it should be necessary to add just one more category to the list of things we have enumerated which present the distinctive sign of a social phenomenon. This addition would be the concept "ways of being." Since the list was not meant to be rigorously exhaustive, this addition would not be absolutely essential.

In fact, it may not even be necessary, for these ways of being, or social structure, are really just a consolidation of social behavior (ways of doing). The political structure of a society is merely the manner in which the component segments have become accustomed to living with one another. If their relationships traditionally have been intimate, the different segments tend to fuse with one another or, in the contrary case, to retain their separate identities. The kind of dwelling is simply the type of house which our peers, and to some extent our ancestors, have been accustomed to building. Means of communications are just the channels which the regular shuttling of commerce and migration have developed. Certainly, if these structural or anatomical phenomena (ways of being) were alone in presenting this permanence they could be considered a separate category. However, a legal regulation is no less permanent than an architectural style, yet the law is a physiological phenomenon (way of doing). A simple moral principle is, assuredly, more flexible than a professional custom, a fashion, or a fad, yet its forms are much more rigid. There is, then, a continuous range of nuances between the extremes marked by society's most structured phenomena and those free-flowing social tendencies which have not yet formed a definite pattern. These varying degrees of consolidation are the only differences that exist between them. Both are just life, more or less crystallized in form. It may be advisable to reserve the term "anatomical" for phenomena that concern the sub-stratum of society. It should be kept in mind, however, that they are similar in nature to the other social phenomena. Our definition will then include the relevant facts if we say: "A social phenomenon is every mode of behavior, fixed or not, that is capable of exerting an externalized power of constraint on the individual," or again, "every mode of behavior that is general throughout a given society, while at the same time existing in its own right, independent of its individual manifestations."

MAX WEBER

THE PROTESTANT ETHIC AND THE SPIRIT OF CAPITALISM

A product of modern European civilization, studying any problem of universal history, is bound to ask himself to what combination of circumstances the fact should be attributed that in Western civilization, and in Western civilization only, cultural phenomena have appeared which (as we like to think) lie in a line of development having *universal* significance and value.

Reprinted from *The Protestant Ethic and the Spirit of Capitalism* by Max Weber, translated by Talcott Parsons (New York: Charles Scribner's Sons, 1958), pp. 13–27, 155–81, by permission of Charles Scribner's Sons and George Allen & Unwin, Ltd.

Only in the West does science exist at a stage of development which we recognize today as valid. Empirical knowledge, reflection on problems of the cosmos and of life, philosophical and theological wisdom of the most profound sort, are not confined to it, though in the case of the last the full development of a systematic theology must be credited to Christianity under the influence of Hellenism, since there were only fragments in Islam and in a few Indian sects. In short, knowledge and observation of great refinement have existed elsewhere, above all in India, China, Babylonia, Egypt. But in Babylonia and elsewhere astronomy lacked—which makes its development all the more astounding—the mathematical foundation which it first received from the Greeks. The Indian geometry had no rational proof; that was another product of the Greek intellect, also the creator of mechanics and physics. The Indian natural sciences, though well developed in observation, lacked the method of experiment, which was, apart from beginnings in antiquity, essentially a product of the Renaissance, as was the modern laboratory. Hence medicine, especially in India, though highly developed in empirical technique, lacked a biological and particularly biochemical foundation. A rational chemistry has been absent from all areas of culture except the West. . . .

Similar statements can be made about other fields. There was printing in China. But a printed literature, designed *only* for print and only possible through it, and, above all, the press and periodicals, have appeared only in the Occident. Institutions of higher education of all possible types, even some superficially similar to our universities, or at least academies, have existed (China, Islam). But a rational, systematic, and specialized pursuit of science, with trained and specialized personnel, has only existed in the West in a sense at all approaching its present dominant place in our culture. Above all is this true of the trained official, the pillar of both the modern State and of the economic life of the West. He forms a type of which there have heretofore only been suggestions, which have never remotely approached its present importance for the social order. Of course the official, even the specialized official, is a very old constituent of the most various societies. But no country and no age has ever experienced, in the same sense as the modern Occident, the absolute and complete dependence of its whole existence, of the political, technical, and economic conditions of its life, on a specially trained *organization* of officials. The most important functions of the everyday life of society have come to be in the hands of technically, commercially, and above all legally trained government officials. . . .

And the same is true of the most fateful force in our modern life, capitalism. The impulse to acquisition, pursuit of gain, of money, of the greatest possible amount of money, has in itself nothing to do with capitalism. This impulse exists and has existed among waiters, physicians, coachmen, artists, prostitutes, dishonest officials, soldiers, nobles, crusaders, gamblers, and beggars. One may say that it has been common to all sorts and conditions of men at all times and in all countries of the earth, wherever the objective possibility of it is or has been given. Capitalism is identical with the pursuit of profit, and forever *renewed* by profit, by means of continuous, rational, capitalistic enterprise. For it must be so: in a wholly capitalistic order of society, an individual capitalistic enterprise which did not take advantage of its opportunities for profit-making would be doomed to extinction. . . .

Now, however, the Occident has developed capitalism in types, forms, and directions which have never existed elsewhere. All over the world there have been merchants, wholesale and retail, local and engaged in foreign trade. Loans of all kinds have been made, and there have been banks with the most various functions, at least comparable to ours of, say, the sixteenth century. Whenever money finances of public bodies have existed, money-lenders have appeared, as in Babylon, Hellas, India, China, Rome. They have financed wars and piracy, contracts and building operations of all sorts. This kind of entrepreneur, the capitalistic adventurer, has existed everywhere. With the exception of trade and credit and banking transactions, their activities were predominantly of an irrational and speculative character, or directed to acquisition by force, above all the acquisition of booty, whether directly in war or in the form of continuous fiscal booty by exploitation of subjects. . . .

But in modern times the Occident has developed, in addition to this, a very different form of capitalism which has appeared nowhere else: the rational capitalistic organization of (formally) free labour. Only suggestions of it are found elsewhere. . . .

Rational industrial organization, attuned to a regular market, and neither to political nor irrationally speculative opportunities for profit, is not, however, the only peculiarity of Western capitalism. The modern rational organization of the capitalistic enterprise would not have been possible without two other important factors in its development: the separation of business from the household, which completely dominates modern economic life, and closely connected with it, rational bookkeeping. . . .

However, all these peculiarities of Western capitalism have derived their significance in the last analysis only from their association with the capitalistic organization of labour. Even what is generally called commercialization, the development of negotiable securities and the rationalization of speculation, the exchanges, etc., is connected with it. For without the rational capitalistic organization of labour, all this, so far as it was possible at all, would have nothing like the same significance, above all for the social structure and all the specific problems of the modern Occident connected with it. Exact calculation—the basis of everything else—is only possible on a basis of free labour.

Hence in a universal history of culture the central problem for us is not, in the last analysis, even from a purely economic viewpoint, the development of capitalistic activity as such, differing in different cultures only in form: the adventurer type, or capitalism in trade, war, politics, or administration as sources of gain. It is rather the origin of this sober bourgeois capitalism with its rational organization of free labour. Or in terms of cultural history, the problem is that of the origin of the Western bourgeois class and of its peculiarities, a problem which is certainly closely connected with that of the origin of the capitalistic organization of labour, but is not quite the same thing. For the bourgeois as a class existed prior to the development of the peculiar modern form of capitalism, though, it is true, only in the Western hemisphere.

Now the peculiar modern Western form of capitalism has been, at first sight, strongly influenced by the development of technical possibilities. Its rationality is today essentially dependent on the calculability of the most important technical factors. But this encouragement was derived from the peculiarities of the social structure of the Occident. We must hence ask, from what parts of that structure was it derived, since not all of them have been of equal importance?

Among those of undoubted importance are the rational structures of law and of administration. For modern rational capitalism has need, not only of the technical means of production, but of a calculable legal system and of administration in terms of formal rules. Without it adventurous and speculative trading capitalism and all sorts of politically determined capitalisms are possible, but no rational enterprise under individual initiative, with fixed capital and certainty of calculations. Such a legal system and such administration have been available for economic activity in a comparative state of legal and formalistic perfection only in the Occident. We must hence inquire where that law came from. Among other circumstances, capitalistic interests have in turn undoubtedly also helped, but by no means alone nor even principally, to prepare the way for the predominance in law and administration of a class of jurists specially trained in rational law. But these interests did not themselves create that law. Quite different forces were at work in this development. And why did not the capitalistic interests do the same in China or India? Why did not the scientific, the artistic, the political, or the economic development there enter upon that path of rationalization which is peculiar to the Occident? For in all the above cases it is a question of the specific and peculiar rationalism of Western culture. Now by this term very different things may be understood. There is, for example, rationalization of mystical contemplation, that is of an attitude which, viewed from other departments of life, is specifically irrational, just as much as there are rationalizations of economic life, of technique, of scientific research, or military training, of law and administration. Furthermore, each one of these fields may be rationalized in terms of very different ultimate values and ends, and what is rational from one point of view may well be irrational from another. Hence rationalizations of the most varied character have existed in various departments of life and in all areas of culture. To characterize their differences from the viewpoint of cultural history it is necessary to know what departments are rationalized, and in what direction. Every such attempt at explanation must, recognizing the fundamental importance of the economic factor, above all take account of the economic conditions. But at the same time the opposite correlation must not be left out of consideration. For though the development of economic rationalism is partly dependent on rational technique and law, it is at the same time determined by the ability and disposition of men to adopt certain types of practical rational conduct. When these types have been obstructed by spiritual obstacles, the development of rational economic conduct has also met serious inner resistance. The magical and religious forces, and the ethical ideas of duty based upon them have in the past always been among the most important formative influences on conduct. . . .

In order to understand the connection between the fundamental religious ideas of ascetic Protestantism and its maxims for everyday economic conduct, it is necessary to examine some of the beliefs of the early Puritans. Waste of time is the first and in principle the deadliest of sins. The span of human life is infinitely short and precious to make sure of one's own election. Loss of time through sociability, idle talk, luxury, even more sleep than is necessary for health, six to at most eight hours, is worthy of absolute moral condemnation. Inactive contemplation is also valueless, or even directly reprehensible if it is at the expense of one's daily work. For it is less pleasing to God than the active performance of His will in a calling.

The usefulness of a calling, and thus its favour in the sight of God, is measured primarily in moral terms, and thus in terms of the importance of the goods produced in it for the community. But a further, and, above all, in practice the most important, criterion is found in private profitableness. For if that God, whose hand the Puritan sees in all the occurrences of life, shows one of His elect a chance of profit, he must do it with a purpose. Hence the faithful Christian must follow the call by taking advantage of the opportunity. "If God show you a way in which you may lawfully get more than in another way (without wrong to your soul or to any other), if you refuse this, and choose the less gainful way, you cross one of the ends of your calling, and you refuse to be God's steward, and to accept His gifts and use them for Him when He requireth it: you may labour to be rich for God, though not for the flesh and sin." Wealth is thus bad ethically only in so far as it is a temptation to idleness and sinful enjoyment of life, and its acquisition is bad only when it is with the purpose of later living merrily and without care. But as a performance of duty in a calling it is not only morally permissible, but actually enjoined. The parable of the servant who was rejected because he did not increase the talent which was entrusted to him seemed to say so directly. To wish to be poor was, it was often argued, the same as wishing to be unhealthy; it is objectionable as a glorification of works and derogatory to the glory of God. Especially begging, on the part of one able to work, is not only the sin of slothfulness, but a violation of the duty of brotherly love according to the Apostle's own word.

The emphasis on the ascetic importance of a fixed calling provided an ethical justification of the modern specialized division of labour. In a similar way the providential interpretation of profit-making justified the activities of the business man.

Let us now clarify the points in which the Puritan idea of the calling and the premium it placed upon ascetic conduct was bound directly to influence the development of a capitalistic way of life. This asceticism turned with all its force against one thing: the spontaneous enjoyment of life and all it had to offer. Its attitude was thus suspicious and often hostile to the aspects of culture without any immediate religious value. The theatre was obnoxious to the Puritans, and with the strict exclusion of the erotic and of nudity from the realm of toleration, a radical view of either literature or art could not exist. The conceptions of idle talk, of superfluities, and of vain ostentation, all designations of an irrational attitude without objective purpose, thus not ascetic, and especially not serving the glory of God, but of man, were always at hand to serve in deciding in favour of sober utility as against any artistic tendencies. This was especially true in the case of decoration of the person, for instance clothing. That powerful tendency toward uniformity of life, which today so immensely aids the capitalistic interest in the standardization of production, had its ideal foundations in the repudiation of all idolatry of the flesh. . . .

On the side of the production of private wealth, asceticism condemned both dishonesty and impulsive avarice. What was condemned as covetousness, Mammonism, etc., was the pursuit of riches for their own sake. For wealth in itself was a temptation. But here asceticism was the power "which ever seeks the good but ever creates evil"; what was evil in its sense was possession and its temptations. For, in conformity with the Old Testament and in analogy to the ethical valuation of good works, asceticism looked upon the pursuit of wealth as an end in itself as highly reprehensible; but the attainment of it as a fruit of labour in a calling was a sign of God's blessing. And even more important: the religious valuation of restless, continuous systematic work in a worldly calling, as the highest means to asceticism, and at the same time the surest and most evident proof of rebirth and genuine faith, must have been the most powerful conceivable lever for the expansion of that attitude toward life which we have here called the spirit of capitalism.

When the limitation of consumption is combined with this release of acquisitive activity, the inevitable practical result is obvious: accumulation of capital through ascetic compulsion to save. The restraints which were imposed upon the consumption of wealth naturally served to increase it by making possible the productive investment of capital. As far as the influence of the Puritan outlook extended, under all circumstances—and this is, of course, much more important than the mere encouragement of capital accumulation—it favoured the development of a rational bourgeois economic life; it was the most important, and above all the only consistent influence in the development of that life. It stood at the cradle of the modern economic man. . . .

A specifically bourgeois economic ethic had grown up. With the consciousness of standing in the fullness of God's grace and being visibly blessed by Him, the bourgeois business man, as long as he remained within the bounds of formal correctness, as long as his moral conduct was spotless and the use to which he put his wealth was not objectionable, could follow his pecuniary interests as he would and feel that he was fulfilling a duty in doing so. The power of religious asceticism provided him in addition with sober, conscientious, and unusually industrious workmen, who clung to their work as to a life purpose willed by God. . . .

One of the fundamental elements of the spirit of modern capitalism, and not only of that but of all modern culture: rational conduct on the basis of the idea of the calling, was born from the spirit of Christian asceticism. The idea that modern labour has an ascetic character is of course not new. Limitation to specialized work, with a renunciation of the Faustian universality of man which it involves, is a condition of any valuable work in the modern world; hence deeds and reununciation inevitably condition each other today. . . .

The Puritan wanted to work in a calling; we are forced to do so. For when asceticism was carried out of monastic cells into everyday life, and began to dominate worldly morality, it did its part in building the tremendous cosmos of the modern economic order. This order is now bound to the technical and economic conditions of machine production which to-day determine the lives of all the individuals who are born into this mechanism, not only those directly concerned with economic acquisition, with irresistible force. Perhaps it will so determine them until the last ton of fossilized coal is burnt. In Baxter's view the care for external goods should only lie on the shoulders of the "saint like a light cloak, which can be thrown aside at any moment." But fate decreed that the cloak should become an iron cage.

C. WRIGHT MILLS

TWO STYLES OF RESEARCH IN CURRENT SOCIAL STUDIES

THE first of these two research-ways might be called the macroscopic. It has a venerable history, reaching notable heights, for example, in the work of

Reprinted from C. Wright Mills, "Two Styles of Research in Current Social Studies," *Philosophy of Science,* vol. 20, pp. 265–75.

Weber and Ostrogorski, Marx and Bryce, Michels, Simmel and Mannheim. These men like to deal with total social structures in a comparative way; their scope is that of the world historian; they attempt to generalize types of historical phenomena, and in a systematic way, to connect the various institutional spheres of a society, and then relate them to prevailing types of men and women. How did the Crusades come about? Are Protestantism and the rise of capitalism related? If so, how? Why is there no socialist movement in the U.S.?

The other way of sociological research might be called the molecular. It is, at first glance, characterized by its usually small-scale problems and by its generally statistical models of verification. Why are 40 per cent more of the women who give marketing advice to their neighbors during a given week on a lower income level than those who gave it during another week? Molecular work has no illustrious antecedents, but, by virtue of historical accident and the unfortunate facts of research finance, has been developed a great deal from studies of marketing and problems connected with media of mass communication. Shying away from social philosophy, it often appears as technique and little else.

Everyone involved in the social studies will recognize these two styles, and by now, a good many will readily agree that "we ought to get the two together." Sometimes this program is put in terms of the statement that the sociologist's ideal task during the next decades is to unite the larger problems and theoretical work of the 19th century, especially that of the Germans, with the research techniques predominant in the 20th century, especially that of the Americans. Within this great dialectic, it is felt, signal and continuous advances in masterful conception and rigorous procedure will be made.

If we inquire more closely into just how the two research-ways differ, we find that there is sometimes a confusion of differences that are non-logical with those that are logical in character. This is revealed, for example, in statements of the difference between the two styles as a political and intellectual dilemma: the more socially or politically significant our problems and work (the more macroscopic), the less rigorous is our solution and the less certain our knowledge (the less molecular).

There is much social truth in such statements; as they have so far been used these two styles of thought do differ in their characteristic value-relevance and political orientation. But this does not mean that any political orientation is inherent in the logic of either style of thought. The evaluative choice of problems characteristic of each of the two methods has not been *necessarily* due to logical capabilities or limitations of either. Molecular work of great political relevance is logically possible; and macroscopic work is not necessarily of broad significance, as a glance at many "political science" monographs proves all too well. No, many of the differences between the two styles are not logical, but social.

From the standpoint of the individual researcher, the choice of problems in either style of work may be due to academic timidity, political disinterest, or even cowardice; but above all it is due to the institutional facts of the financial life of molecular research. Molecular work requires an organization of technicians and administrators, of equipment and money, and, as yet, of promoters. It can not proceed until agencies of research are sufficiently developed to provide detailed materials. It has arisen in definite institutional centers: in business, since the twenties among marketing agencies, and since the thirties, in the polling agencies; in academic life at two or three research bureaux; and in research branches of government. Since World War II the pattern has spread, but these are still the centers.

This institutionalization of the molecular style has involved the applied focus, which has typically been upon specific problems, presented so as to make clear alternatives of practical—which is to say, pecuniary and administrative—action. It is *not* true that only as general principles are discovered can social science offer "sound practical guidance"; often the administrator needs to know certain detailed facts and relations, and that is all he needs to know.

The sociologist in the applied focus no longer addresses "the public"; more usually he has specific clients with particular interests and perplexities. This shift from public to client, clearly destroys the idea of objectivity as aloofness, which perhaps meant responsiveness to vague, unfocused pressures, and thus rested more on the individual interests of the researcher. In applied research of the molecular style, the client's social operations and economic interests having often supplied the sometimes tacit but always present moral meaning and use to the problem and to its solution. This has meant that most molecular work of any scale has been socially guided by the concerns and worries set by practical government and business interests and has been responsible to them. Accordingly, there is little doubt that the applied focus has tended to lower the intellectual initiative and to heighten the opportunism of the researcher. However technically free he may be, his initiative and interest are in fact usually subordinate to those of the client, whether it be the selling of pulp magazines or the administration of an army's morale.

Very little except his own individual limitations has stood between the individual worker and macroscopic work of the highest order. But the rise of the molecular style means that the unattached man cannot pursue such research on any scale, for such work is dependent upon organization and money. If we would "solve" the problem raised by the coexistence of these two styles we must pay attention to the design of work that is possible for the unattached men who still comprise the bulk of those claiming membership in the sociological community.

The rise of applied molecular work, as it is now being organized, makes questions of moral and political policy of the social studies all the more urgent. As a bureaucratization of reflection, the molecular style is quite in line with dominant trends of modern social structure and its characteristic types of thought. I do not wish to consider these problems here except to say that they should not be confused with any differences of a logical character between the two styles of inquiry.

There are at least three relative differences of a logical sort between the macroscopic and the molecular styles of work as they are now practiced: the molecular is more objective; it is more open to cumulative development; and it is more open to statistical quantification.

Objectivity means that the work is so done and so presented that any other qualified person can repeat it, thus coming to the same results or showing that the results were mistaken. Subjectivity means the reverse, and thus that there is usually a persistent individual variation of procedure—and of solution. Under this difference lies the fact that when work is objective the procedures used are systematized or even codified and hence are available to any qualified operator; whereas in subjective work the procedures are often not systematized, much less standardized or codified.

This in turn means that in objective work there is a more distinct possibility of cumulation—or at least replication!—both in terms of empirical solutions and in terms of the procedures used. In the more subjective macroscopic work the sensitivity and talent of the individual worker weigh more heavily and although there may be those who "take up where he left off," this is usually a continuity of subject-matter, general ideas, and approach rather than an accumulation of procedure. It is possible within a few years to train competent persons to repeat a Sandusky job; [1] it is not so possible to train them to repeat a Middletown study. Another sample of soldiers in another war can be located on a morale scale and comparisons built up; Max Weber's analytic and historical essay on bureaucracy has not been repeated or checked in the same way, however much it has been criticized and "used." Macroscopic work has not experienced the sort of cumulative development that molecular work during the current generation of sociologists has.

It is descriptively true that the molecular style has been heavily statistical, whereas the macroscopic has not. This, again, is an aspect of the greater codification and the lower level of abstraction that molecular work entails. And it can be confidently supposed that as macroscopic work is made more systematic it will become more quantitative—at least as a general form of thought. For example, Darwin's *Origin* as well as many of Freud's theories are quantitative models of reflection.

Each of these three points is underpinned by the fact that molecular procedures can be, and have been, more explicitly codified than those of the macroscopic style; and by the fact that molecular terms are typically on a lower level of abstraction than most macroscopic conceptions.

Insofar as the logical differences between the two styles concern *procedures,* they are differences in the degree of systematic codification. Insofar as they involve *conceptions,* they are differences in level of abstraction.

When we say that molecular terms are on *lower* levels of abstraction we mean that they isolate from larger contexts a few precisely observed elements; in this sense they are of course quite abstract. When we say that macroscopic concepts are on *higher* levels of abstraction, we mean that they are more generalized, that the number of single variables which they cover are more numerous. The molecular term is narrow in scope, and specific in reference: it deals with a few discrete variables; the macroscopic researcher gains his broader scope by using concepts that cover, usually less specifically, a much larger number of variables.

There is no one clear-cut variable, the presence or absence of which allows application of the concept, "capitalism": under such concepts there is likely to be a pattern of interrelated variables. Thus, such concepts are not only high-level but their index structure is an elaborately compounded affair. Put technically, most big macroscopic concepts already have under them rather elaborate, and often unsystematic, cross-tabulations of several variables; most molecular terms stand for single variables useful for the stubs of such tables.

We can consider a term in its relation to some empirical items(s)—that is, its semantic dimension; and we can consider a term in its relation to other terms—that is, its syntactical dimension, or if you like, its conceptual implications.[2] It is characteristic of molecular terms that their semantic dimensions are pronounced, although syntactical relations may also be there. It is characteristic of macroscopic terms that their syntactical dimensions are pronounced, although semantical relations may also be available.

The higher macroscopic levels are more syntactically elaborate; semantically they involve a hierarchy of compounded indices pointing to whole gestalts of attributes. Macroscopic concepts are often sponge-like and unclarified in their semantic dimensions. Sometimes, in fact, they do not have any index structure that enables us to touch empirically observable facts or relations.[3] They have under them only a vague kind of many-dimensional indicator rather than an index. Yet, with all this, it may be that whether a statement is macroscopic or molecular is a matter of degree—a question of at what level we introduce our syntactical elaboration.

PETER L. BERGER

INVITATION TO SOCIOLOGY

HOW . . . are we to conceive of the sociologist? In discussing [above] the various images of him that abound in the popular mind we have already brought out certain elements that would have to go into our conception. We can now put them together. In doing so, we shall construct what sociologists themselves call an "ideal type." This means that what we delineate will not be found in reality in its pure form. Instead, one will find approximations to it and deviations from it, in varying degrees. Nor is it to be understood as an empirical average. We would not even claim that all individuals who now call themselves sociologists will recognize themselves without reservations in our conception, nor would we dispute the right of those who do not so recognize themselves to use the appellation. Our business is not excommunication. We would, however, contend that our "ideal type" corresponds to the self-conception of most sociologists in the mainstream of the discipline, both historically (at least in this century) and today.

The sociologist, then, is someone concerned with understanding society in a disciplined way. The nature of this discipline is scientific. This means that what the sociologist finds and says about the social phenomena he studies occurs within a certain rather strictly defined frame of reference. One of the main characteristics of this scientific frame of reference is that operations are bound by certain rules of evidence. As a scientist, the sociologist tries to be objective, to control his personal preferences and prejudices, to perceive clearly rather than to judge normatively. This restraint, of course, does not embrace the totality of the sociologist's existence as a human being, but is limited to his operations *qua* sociologist. Nor does the sociologist claim that his frame of reference is the only one within which society can be looked at. For that matter, very few scientists in any field would claim today that one should look at the world only scientifically. The botanist looking at a daffodil has no reason to dispute the right of the poet to look at the same object in a very different manner. There are many ways of playing. The point is not that one denies other people's games but that one is clear about the rules of one's own. The game of the sociologist, then, uses scientific rules. As a result, the sociologist must be clear in his own mind as to the meaning of these rules. That is, he must concern himself with methodological questions. Methodology

Reprinted from *Invitation to Sociology: A Humanistic Perspective* by Peter L. Berger (New York: Doubleday & Co., Anchor Books, 1963), pp. 12–16, by permission of the author, the publisher and Penguin Books Ltd.

does not constitute his goal. The latter, let us recall once more, is the attempt to understand society. Methodology helps in reaching this goal. In order to understand society, or that segment of it that he is studying at the moment, the sociologist will use a variety of means. Among these are statistical techniques. Statistics can be very useful in answering certain sociological questions. But statistics does not constitute sociology. As a scientist, the sociologist will have to be concerned with the exact significance of the terms he is using. That is, he will have to be careful about terminology. This does not have to mean that he must invent a new language of his own, but it does mean that he cannot naïvely use the language of everyday discourse. Finally, the interest of the sociologist is primarily theoretical. That is, he is interested in understanding for its own sake. He may be aware of or even concerned with the practical applicability and consequences of his findings, but at that point he leaves the sociological frame of reference as such and moves into realms of values, beliefs and ideas that he shares with other men who are not sociologists.

We daresay that this conception of the sociologist would meet with very wide consensus within the discipline today. But we would like to go a little bit further here and ask a somewhat more personal (and therefore, no doubt, more controversial) question. We would like to ask not only what it is that the sociologist is doing but also what it is that drives him to it. Or, to use the phrase Max Weber used in a similar connection, we want to inquire a little into the nature of the sociologist's demon. In doing so, we shall evoke an image that is not so much ideal-typical in the above sense but more confessional in the sense of personal commitment. Again, we are not interested in excommunicating anyone. The game of sociology goes on in a spacious playground. We are just describing a little more closely those we would like to tempt to join our game.

We would say then that the sociologist (that is, the one we would really like to invite to our game) is a person intensively, endlessly, shamelessly interested in the doings of men. His natural habitat is all the human gathering places of the world, wherever men come together. The sociologist may be interested in many other things. But his consuming interest remains in the world of men, their institutions, their history, their passions. And since he is interested in men, nothing that men do can be altogether tedious for him. He will naturally be interested in the events that engage men's ultimate beliefs, their moments of tragedy and grandeur and ecstasy. But he will also be fascinated by the commonplace, the everyday. He will know reverence, but this reverence will not prevent him from wanting to see and to understand. He may sometimes feel revulsion or contempt. But this also will not deter him from wanting to have his questions answered. The sociologist, in his quest for understanding, moves through the world of men without respect for the usual lines of demarcation. Nobility and degradation, power and obscurity, intelligence and folly—these are equally *interesting* to him, however unequal they may be in his personal values or tastes. Thus his questions may lead him to all possible levels of society, the best and the least known places, the most respected and the most despised. And, if he is a good sociologist, he will find himself in all these places because his own questions have so taken possession of him that he has little choice but to seek for answers.

It would be possible to say the same things in a lower key. We could say that the sociologist, but for the grace of his academic title, is the man who must listen to gossip despite himself, who is tempted to look through keyholes, to read other people's mail, to open closed cabinets. Before some otherwise unoccupied psychologist sets out now to construct an aptitude test for sociologists on the basis of sublimated voyeurism, let us quickly say that we are speaking merely by way of analogy. Perhaps some little boys consumed with curiosity to watch their maiden aunts in the bathroom later become inveterate sociologists. This is quite uninteresting. What interests us is the curiosity that grips any sociologist in front of a closed door behind which there are human voices. If he is a good sociologist, he will want to open that door, to understand these voices. Behind each closed door he will anticipate some new facet of human life not yet perceived and understood.

The sociologist will occupy himself with matters that others regard as too sacred or as too distasteful for dispassionate investigation. He will find rewarding the company of priests or of prostitutes, depending not on his personal preferences but on the questions he happens to be asking at the moment. He will also concern himself with matters that others may find much too boring. He will be interested in the human interaction that goes with warfare or with great intellectual discoveries, but also in the relations between people employed in a restaurant or between a group of little girls playing with their dolls. His main focus of attention is not the ultimate significance of what men do, but the action in itself, as another example of the infinite richness of human conduct. So much for the image of our playmate.

In these journeys through the world of men the sociologist will inevitably encounter other professional Peeping Toms. Sometimes these will resent his presence, feeling that he is poaching on their preserves. In some places the sociologist will meet up with the economist, in others with the political scientist, in yet others with the psychologist or the ethnologist. Yet chances are that the questions that have brought him to these same places are different from the ones that propelled his fellow-trespassers. The sociologist's questions always remain essentially the same: "What are people doing with each other here?" "What are their relationships to each other?" "How are these relationships organized in institutions?" "What are the collective ideas that move men and institutions?" In trying to answer these questions in specific instances, the sociologist will, of course, have to deal with economic or political matters, but he will do so in a way rather different from that of the economist or the political scientist. The scene that he contemplates is the same human scene that these other scientists concern themselves with. But the sociologist's angle of vision is different. When this is understood, it becomes clear that it makes little sense to try to stake out a special enclave within which the sociologist will carry on business in his own right. Like Wesley the sociologist will have to confess that his parish is the world. But unlike some latter-day Wesleyans he will gladly share this parish with others. There is, however, one traveler whose path the sociologist will cross more often than anyone else's on his journeys. This is the historian. Indeed, as soon as the sociologist turns from the present to the past, his preoccupations are very hard indeed to distinguish from those of the historian. However, we shall leave this relationship to a later part of our considerations. Suffice it to say here that the sociological journey will be much impoverished unless it is punctuated frequently by conversation with that other particular traveler.

Any intellectual activity derives excitement from the moment it becomes a trail of discovery. In some fields of learning this is the discovery of worlds previously unthought and unthinkable. This is the excitement of the astronomer or of the nuclear physicist on the antipodal boundaries of the realities that man is capable of conceiving. But it can also be the excitement of bacteriology or geology. In a different way it can be the excitement of the linguist discovering new realms of human expression or of the anthropologist exploring human customs in faraway countries. In such discovery, when undertaken with passion, a widening of awareness, sometimes a veritable transformation of consciousness, occurs. The universe turns out to be much more wonderful than one had ever dreamed. The excitement of sociology is usually of a different sort. Sometimes, it is true, the sociologist penetrates into worlds that had previously been quite unknown to him—for instance, the world of crime, or the world of some bizarre religious sect, or the world fashioned by the exclusive concerns of some group such as medical specialists or military leaders or advertising executives. However, much of the time the sociologist moves in sectors of experience that are familiar to him and to most people in his society. He investigates communities, institutions and activities that one can read about every day in the newspapers. Yet there is another excitement of discovery beckoning in his investigations. It is not the excitement of coming upon the totally unfamiliar, but rather the excitement of finding the familiar becoming transformed in its meaning. The fascination of sociology lies in the fact that its perspective makes us see in a new light the very world in which we have lived all our lives. This also constitutes a transformation of consciousness. Moreover, this transformation is more relevant existentially than that of many other intellectual disciplines, because it is more difficult to segregate in some special compartment of the mind. The astronomer does not live in the remote galaxies, and the nuclear physicist can, outside his laboratory, eat and laugh and marry and vote without thinking about the insides of the atom. The geologist looks at rocks only at appropriate times, and the linguist speaks English with his wife. The sociologist lives in society, on the job and off it. His own life, inevitably, is part of his subject matter. Men being what they are, sociologists too manage to segregate their professional insights from their everyday affairs. But it is a rather difficult feat to perform in good faith.

The sociologist moves in the common world of men, close to what most of them would call real. The categories he employs in his analyses are only refinements of the categories by which other men live—power, class, status, race, ethnicity. As a result, there is a deceptive simplicity and obviousness about some sociological investigations. One reads them, nods at the familiar scene, remarks that one has heard all this before and don't people have better things to do than to waste their time on truisms—until one is suddenly brought up against an insight that radically questions everything one had previously assumed about this familiar scene. This is the point at which one begins to sense the excitement of sociology.

Let us take a specific example. Imagine a sociology class in a Southern college where almost all the students are white Southerners. Imagine a lecture on the subject of the racial system of the South. The lecturer is talking here of matters that have been familiar to his students from the time of their infancy. Indeed, it may be that they are much more familiar with the minutiae of this system than he is. They are quite bored as a result. It seems to them that he is only using more pretentious words to describe what they already know. Thus he may use the term "caste," one commonly used now by American sociologists to describe the Southern racial system. But in explaining the term he shifts to traditional Hindu society, to make it clearer. He then goes on to analyze the magical beliefs inherent in caste tabus, the social dynamics of commensalism and connubium, the economic interests concealed within the system, the way in which religious beliefs relate to the tabus, the effects of the caste system upon the industrial development of the society and vice versa—all in India. But suddenly India is not very far away at all. The lecture then goes back to its Southern theme. The familiar now seems not quite so familiar any more. Questions are raised that are new, perhaps raised angrily, but raised all the same. And at least some of the students have begun to understand that there are functions involved in this business of race that they have not read about in the newspapers (at least not those in their hometowns) and that their parents have not told them—partly, at least, because neither the newspapers nor the parents knew about them.

It can be said that the first wisdom of sociology is this—things are not what they seem. This too is a deceptively simple statement. It ceases to be simple after a while. Social reality turns out to have many layers of meaning. The discovery of each new layer changes the perception of the whole.

Anthropologists use the term "culture shock" to describe the impact of a totally new culture upon a newcomer. In an extreme instance such shock will be experienced by the Western explorer who is told, halfway through dinner, that he is eating the nice old lady he had been chatting with the previous day——a shock with predictable physiological if not moral consequences. Most explorers no longer encounter cannibalism in their travels today. However, the first encounters with polygamy or with puberty rites or even with the way some nations drive their automobiles can be quite a shock to an American visitor. With the shock may go not only disapproval or disgust but a sense of excitement that things can *really* be that different from what they are at home. To some extent, at least, this is the excitement of any first travel abroad. The experience of sociological discovery could be described as "culture shock" minus geographical displacement. In other words, the sociologist travels at home—with shocking results. He is unlikely to find that he is eating a nice old lady for dinner. But the discovery, for instance, that his own church has considerable money invested in the missile industry or that a few blocks from his home there are people who engage in cultic orgies may not be drastically different in emotional impact. Yet we would not want to imply that sociological discoveries are always or even usually outrageous to moral sentiment. Not at all. What they have in common with exploration in distant lands, however, is the sudden illumunation of new and unsuspected facets of human existence in society. This is the excitement and, as we shall try to show later, the humanistic justification of sociology.

People who like to avoid shocking discoveries, who prefer to believe that society is just what they were taught in Sunday School, who like the safety of the rules and the maxims of what Alfred Schutz has called the "world-taken-for-granted," should stay away from sociology. People who feel no temptation before closed doors, who have no curiosity about human beings, who are content to admire scenery without wondering about the people who live in those houses on the other side of that river, should probably also stay away from sociology. They will find it unpleasant or, at any rate, unrewarding. People who are interested in human beings only if they can change, convert or reform them should also be warned, for they will find sociology much less useful than they hoped. And people whose interest is mainly in their own conceptual constructions will do just as well to turn to the study of little white mice. Sociology will be satisfying, in the long run, only to those who can think of nothing more entrancing than to watch men and to understand things human.

It may now be clear that we have, albeit deliberately, understated the case in the title of this chapter. To be sure, sociology is an individual pastime in the sense that it interests some men and bores others. Some like to observe human beings, others to experiment with mice. The world is big enough to hold all kinds and there is no logical priority for one interest as against another. But the word "pastime" is weak in describing what we mean. Sociology is more like a passion. The sociological perspective is more like a demon that possesses one, that drives one compellingly, again and again, to the questions that are its own. An introduction to sociology is, therefore, an invitation to a very special kind of passion. No passion is without its dangers. The sociologist who sells his wares should make sure that he clearly pronounces a *caveat emptor* quite early in the transaction.

Chapter **3**

BECOMING A MEMBER OF SOCIETY— SOCIALIZATION

In the biography of every person, the experience of society begins in childhood. In a very fundamental sense every child begins as a stranger in the world of adults. In the process commonly known as socialization every child is gradually initiated into the social structure with which he will probably live for the rest of his life. Processes of socialization vary widely, of course, from society to society: such fundamental matters as the schedule of weaning or the procedures of toilet-training differ drastically. Even within the same society the socialization pattern to which any particular person will be subjected differs by class, ethnic group, and other sociological variables.

Wherever socialization may take place it involves a process by which the attitudes, habits, and meanings of adults are inwardly absorbed by the growing child. This process is often called internalization. The social world, which the child first experiences as an external reality, is introduced into his own consciousness and becomes the major organizing principle of his own experience. This principle has both a cognitive and a moral dimension: the child learns what the world is supposed to be; he also learns how he ought properly to deal with it. Thus, childhood is a crucially important stage in every individual's relationship to society. And the way in which childhood is understood and arranged in any particular society is of crucial importance for the sociologist who wishes to understand how this society works. Even in the development of Western societies there have been far-reaching transformations in the patterning of childhood. A consequential transformation began at the onset of the modern era, when first in the bourgeoisie and then spreading to other classes, childhood was considered as a specially important biographical stage and children came to be given unprecedented attention. This great transformation has still not reached its conclusion. However important childhood is, socialization does not end with it. Every time an individual enters a new sector of society, new socialization processes start operating to initiate him into the patterns of conduct and thinking appropriate to that sector.

In the first selection, from Lillian Smith's *Killers of the Dream,* the author recounts her experiences as a white child in the American South. This is followed by a passage from Matthew Speier's "The Everyday World of the Child" describing a concrete instance in the everyday experience of a child and putting this data into a new theoretical context. A good deal of modern socialization theory is derived from George Herbert Mead, probably one of the leading figures of American social scientific thought. The selection from *Mind, Self and Society,* Mead's most influential book, discusses how children "learn" society through games. Finally, Martha Wolfenstein, in "Fun Morality: An Analysis of Recent American Child-training Literature," discusses recent patterns of child rearing in America.

LILLIAN SMITH

WHEN I WAS A CHILD

I was born and reared in a small Deep South town whose population was about equally Negro and white. There were nine of us who grew up freely in a rambling house of many rooms, surrounded by big lawn, back yard, gardens, fields, and barn. It was the kind of home that gathers memories like dust, a place filled with laughter and play and pain and hurt and ghosts and games. We were given such advantages of schooling, music, and art as were available in the South, and our world was not limited to the South, for travel to far places seemed a natural thing to us, and usually one of the family was in a remote part of the earth.

We knew we were a respected and important family of this small town but beyond this we gave little thought to status. Our father made money in lumber and naval stores for the excitement of making and losing it—not for what money can buy nor the security which it sometimes gives. I do not remember at any time wanting "to be rich" nor do I remember that thrift and saving were ideals which our parents considered important enough to urge upon us. In the family there was acceptance of risk, a mild delight in burning bridges, an expectant "What next?" We were not irresponsible; living according to the pleasure principle was by no means our way of life. On the contrary we were trained to think that each of us should do something of genuine usefulness, and the family thought it right to make sacrifices if necessary, to give each child preparation for such work. We were also trained to think learning important, and books; but "bad" books our mother burned. We valued music and art and craftsmanship but it was people and their welfare and religion that were the foci around which our lives seemed naturally to move. Above all else, the important thing was what we "planned to do." That each of us must do something was as inevitable as breathing for we owed a "debt to society which must be paid." This was a family commandment.

While many neighbors spent their energies in counting limbs on the family tree and grafting some on now and then to give symmetry to it, or in licking scars to cure their vague malaise, or in fighting each battle and turn of battle of the Civil War which has haunted the southern conscience so long, my father was pushing his nine children straight into the future. "You have your heritage," he used to say, "some of it good, some not so good; and as far as I know you had the usual number of grandmothers and grandfathers. Yes,

Reprinted from *Killers of the Dream* by Lillian Smith (New York: W. W. Norton & Co., 1949), pp. 152–56, by permission of the publisher.

there were slaves, too many of them in the family, but that was your grandfather's mistake, not yours. The past has been lived. It is gone. The future is yours. What are you going to do with it? He asked this question often and sometimes one knew it was but an echo of a question he had spent his life trying to answer for himself. For the future held my father's dreams; always there, not in the past, did he expect to find what he had spent his life searching for.

We lived the same segregated life as did other southerners but our parents talked in excessively Christian and democratic terms. We were told ten thousand times that status and money are unimportant (though we were well supplied with both); we were told that "all men are brothers," that we are a part of a democracy and must act like democrats. We were told that the teachings of Jesus are important and could be practiced if we tried. We were told that to be "radical" is bad, silly too; and that one must always conform to the "best behavior" of one's community and make it better if one can. We were taught that we were superior to hate and resentment, and that no member of the Smith family could stoop so low as to have an enemy. No matter what injury was done us, we must not injure ourselves further by retaliating. That was a family commandment.

We had family prayers once each day. All of us as children read the Bible in its entirety each year. We memorized hundreds of Bible verses and repeated them at breakfast, and said "sentence prayers" around the family table. God was not someone we met on Sunday but a permanent member of our household. It never occurred to me until I was fourteen or fifteen years old that He did not chalk up the daily score on eternity's tablets.

Despite the strain of living so intimately with God, the nine of us were strong, healthy, energetic youngsters who filled days with play and sports and music and books and managed to live most of the time on the careless level at which young lives should be lived. We had our times of anxiety of course, for there were hard lessons to be learned about the soul and "bad things" to be learned about sex. Sometimes I have wondered how we learned them with a mother so shy with words.

She was a wistful creature who loved beautiful things like lace and sunsets and flowers in a vague inarticulate way, and took good care of her children. We always knew this was not her world but one she accepted under duress. Her private world we rarely entered, though the shadow of it lay heavily on our hearts.

Our father owned large business interests, employed hundreds of colored and white laborers, paid them the prevailing low wages, worked them the prevailing long hours, built for them mill towns (Negro and white), built for each group a church, saw to it that religion was supplied free, saw to it that a commissary supplied commodities at a high price, and in general managed his affairs much as ten thousand other southern businessmen managed theirs. . . .

Against this backdrop the drama of the South was played out one day in my life:

A little white girl was found in the colored section of our town, living with a Negro family in a broken-down shack. This family had moved in a few weeks before and little was known of them. One of the ladies in my mother's club, while driving over to her washerwoman's, saw the child swinging on a gate. The shack, as she said, was hardly more than a pigsty and this white child was living with dirty and sick-looking colored folks. "They must have kidnapped her," she told her friends. Genuinely shocked, the clubwomen busied themselves in an attempt to do something, for the child was very white indeed. The strange Negroes were subjected to a grueling questioning and finally grew evasive and refused to talk at all. This only increased the suspicion of the white group. The next day the clubwomen, escorted by the town marshal, took the child from her adopted family despite their tears.

She was brought to our home. I do not know why my mother consented to this plan. Perhaps because she loved children and always showed concern for them. It was easy for one more to fit into our ample household and Janie was soon at home there. She roomed with me, sat next to me at the table; I found Bible verses for her to say at breakfast; she wore my clothes, played with my dolls, and followed me around from morning to night. She was dazed by her new comforts and by the interesting activities of this big lively family; and I was as happily dazed, for her adoration was a new thing to me; and as time passed a quick, childish, and deeply felt bond grew up between us.

But a day came when a telephone message was received from a colored orphanage. There was a meeting at our home. Many whispers. All afternoon the ladies went in and out of our house talking to Mother in tones too low for children to hear. As they passed us at play, they looked at Janie and quickly looked away again, though a few stopped and stared at her as if they could not tear their eyes from her face. When my father came home Mother closed her door against our young ears and talked a long time with him. I heard him laugh, heard Mother say, "But Papa, this is no laughing matter!" And then they were back in the living room with us and my mother was pale and my father was saying, "Well, work it out, Mame, as best you can. After all, now that you know, it is pretty simple."

In a little while my mother called my sister and me into her bedroom and told us that in the morning Janie would return to Colored Town. She said Janie was to have the dresses the ladies had given her and a few of my own, and the toys we had shared with her. She asked me if I would like to give Janie one of my dolls. She seemed hurried, though Janie was not to leave until next day. She said "Why not select it now?" And in dreamlike stiffness I brought in my dolls and chose one for Janie. And then I found it possible to say, "Why is she leaving? She likes us, she hardly knows them. She told me she had been with them only a month."

"Because," Mother said gently, "Janie is a little colored girl."

"But she's white!"

"We were mistaken. She is colored."

"But she looks—"

"She is colored. Please don't argue!"

"What does it mean?" I whispered.

"It means," Mother said slowly, "that she has to live in Colored Town with colored people."

"But why? She lived here three weeks and she doesn't belong to them, she told me so."

"She is a little colored girl."

"But you said yourself she has nice manners. You said that," I persisted.

"Yes, she is a nice child. But a colored child cannot live in our home."

"Why?"

"You know, dear! You have always known that white and colored people do not live together."

"Can she come to play?"

"No."

"I don't understand."

"I don't either," my young sister quavered.

"You're too young to understand. And don't ask me again, ever again, about this!" Mother's voice was sharp but her face was sad and there was no certainty left there. She hurried out and busied herself in the kitchen and I wandered through that room where I had been born, touching the old familiar things in it, looking at them, trying to find the answer to a question that moaned like a hurt thing. . . .

And then I went out to Janie, who was waiting, knowing things were happening that concerned her but waiting until they were spoken aloud.

I do not know quite how the words were said but I told her she was to return in the morning to the little place where she had lived because she was colored and colored children could not live with white children.

"Are you white?" she said.

"I'm white," I replied, "and my sister is white. And you're colored. And white and colored can't live together because my mother says so."

"Why?" Janie whispered.

"Because they can't," I said. But I knew, though I said it firmly, that something was wrong. I knew my mother and father whom I passionately admired had betrayed something which they held dear. And they could not help doing it. And I was shamed by their failure and frightened, for I felt they were no longer as powerful as I had thought. There was something Out There that was stronger than they and I could not bear to believe it. I could not confess that my father, who always solved the family dilemmas easily and with laughter, could not solve this. I knew that my mother who was so good to children did not believe in her heart that she was being good to this child. There was not a word in my mind that said it but my body knew and my glands, and I was filled with anxiety.

But I felt compelled to believe they were right. It was the only way my world could be held together. And, slowly, it began to seep through me: *I was white. She was colored. We must not be together. It was bad to be together. Though you ate with your nurse when you were little, it was bad to eat with any colored person after that. It was bad just as other things were bad that your mother had told you. It was bad that she was to sleep in the room with me that night. It was bad. . . .*

I was overcome with guilt. For three weeks I had done things that white children were not supposed to do. And now I knew these things had been wrong.

I went to the piano and began to play, as I had always done when I was in trouble. I tried to play my next lesson and as I stumbled through it, the little girl came over and sat on the bench with me. Feeling lost in the deep currents sweeping through our house that night, she crept closer and put her arms around me and I shrank away as if my body had been uncovered. I had not said a word, I did not say one, but she knew, and tears slowly rolled down her little white face. . . .

And then I forgot it. For more than thirty years the experience was wiped out of my memory. But that night, and the weeks it was tied to, worked its way like a splinter, bit by bit, down to the hurt places in my memory and festered there. And as I grew older, as more experiences collected around that faithless time, as memories of earlier, more profound hurts crept closer, drawn to that night as if to a magnet, I began to know that people who talked of love and children did not mean it. That is a hard thing for a child to learn. I still admired my parents, there was so much that was strong and vital and sane and good about them and I never forgot this; I stubbornly believed in their sincerity, as I do to this day, and I loved them. Yet in my heart they were under suspicion. Something was wrong.

Something was wrong with a world that tells you that love is good and people are important and then forces you to deny love and to humiliate people. I knew, though I would not for years confess it aloud, that in trying to shut the Negro race away from us, we have shut ourselves away from so many good, creative, honest, deeply human things in life. I began to understand slowly at first but more clearly as the years passed, that the warped, distorted frame we have put around every Negro child from birth is around every white child also. Each is on a different side of the frame but each is pinioned there. And I knew that what cruelly shapes and cripples the personality of one is as cruelly shaping and crippling the personality of the other. I began to see that though we may, as we acquire new knowledge, live through new experiences, examine old memories, gain the strength to tear the frame from us, yet we are stunted and warped and in our lifetime cannot grow straight again any more than can a tree, put in a steel-like twisting frame when young, grow tall and straight when the frame is torn away at maturity.

As I sit here writing, I can almost touch that little town, so close is the memory of it. There it lies, its main street lined with great oaks, heavy with matted moss that swings softly even now as I remember. A little white town rimmed with Negroes, making a deep shadow on the whiteness. There it lies, broken in two by one strange idea. Minds broken. Hearts broken. Conscience torn from acts. A culture split in a thousand pieces. That is segregation. I am remembering: a woman in a mental hospital walking four steps out, four steps in, unable to go further because she has drawn an invisible line around her small world and is terrified to take one step beyond it. . . . A man in a Disturbed Ward assigning "places" to the other patients and violently insisting that each stay in his place. . . . A Negro woman saying to me so quietly, "We cannot ride together on the bus, you know. It is not legal to be human down here."

Memory, walking the streets of one's childhood . . . of the town where one was born.

MATTHEW SPEIER

THE EVERYDAY WORLD OF THE CHILD

I. A NEW LOOK AT THE EMPIRICAL CONTENT OF CHILDHOOD SOCIALIZATION

Sociology considers the social life of the child as a basic area of study in so-called institutional analyses of family and school, for example. What is classically problematic about studying children is the fact of cultural induction, as I might refer to it. That is, sociologists (and this probably goes for anthropologists and psychologists) commonly treat childhood as a stage of life that builds preparatory mechanisms into the child's behavior so that he is gradually equipped with the competence to participate in the everyday activities of his cultural partners, and eventually as a bona fide adult member himself. This classical sociological problem has been subsumed under the major heading of socialization. In studying the organization of culture and society it seems quite natural to inquire into the process by which a new entrant acquires the status of a member in the eyes of others in his surrounding cultural milieu (unlike those entrants who arrive as adults, as "strangers" or immigrants, tourists and the like, the child enters upon the scene with a clean slate, because the process takes place without the underlay of previous cultural experience).

Reprinted from Jack D. Douglas, editor, *Understanding Everyday Life* (Chicago: Aldine Publishing Co., 1970), pp. 188–96; copyright © 1970 by Aldine Publishing Co. Reprinted by permission of the author and Aldine Publishing Company.

[Speier's notes deleted.—Ed.]

The classical formulation of the problem of socialization has centered on treatments of the child's entry and incorporation into culture as a *developmental process*. Its working paradigm has been to ask questions about child development, such as the general one: How does the child internalize the norms, values, attitudes, etc., of others in his society? Traditional anthropological ethnography has asked in addition: How does the child develop particular skills in social and economic ways of life? Psychologists have focused on maturational growth and on personality development. Lately, researchers from all these disciplines have become interested in the development of language skills.

I would like to propose an approach that differs sharply from the developmental one found in the classical formulations of socialization research. This approach sets aside questions of development yet retains the substantive interests of adult-child interaction central to the study of socialization.

I propose a simple definition of socialization that if acceptable to developmentally oriented research would imply the investigation of a hidden frame of analysis altogether: *socialization is the acquisition of interactional competences*. We can readily admit to the fact that children acquire "a sense of social structure," to use Cicourel's phrase. That is, for the child to develop from a newborn entrant to a participating member in social arrangements around him, he must undergo a learning process over the course of growing up through successive stages of life. However, to study this implicitly recognized acquisition process, presupposes a good knowledge of the features of interactional competences that are acquired. That is, what in fact are children developmentally acquiring? An investigation of the concrete features of competent interaction is nothing more or less than a study of what children normally and routinely do in their everyday activities, and as such it is not a study at all in the development of competence but a study in descriptive interactional analysis. It is my firm belief that no investigation of acquisition processes can effectively get underway until the concrete features of interactional competences are analyzed as topics in their own right. Without this preliminary step, which deliberately refrains from treating the topic of development, discussions of social competence must necessarily remain too vague and abstract to be of any direct use in empirical socialization research. . . .

In this reformulated context for studying children's everyday activity the notion of development takes on a new shape. The temporal scale is vastly reduced to interactional units of *occasioned and situated activity*. The focus now is upon developing sequences of interaction from one moment to the next, rather than upon stages of development in the child's life; on the way interactants build a social scene and build a conversation together, episodically, beginning with procedures for opening and entering into copresent interactions, next for sustaining them around practical purposes using conversational resources for so doing, and finally for terminating the interaction or shifting into new activities or situations along natural junctures. The notion of development, then, enters into the analysis of interactional sequences moving naturally through time as participants do and say things methodically together. Children presumably have sufficient competence to cooperate in interactional development over a great variety of social circumstances. As interactants they must be able to employ conversational procedures with those they routinely encounter in everyday life. What conversational resources are available to children and to their interlocutors when routine interactions arise and take shape? That is, how are such procedures for interacting employed resourcefully by participants as they go about their talking and acting and their everyday practical achievements?

In the discussion that follows I will attempt to demonstrate how a concrete instance of data in the child's everyday life can be analyzed so as to yield some key issues in the organization of childhood activity. These issues will form the basis for the general analysis in the remainder of the chapter.

II. TREATING AN INSTANCE AS AN OPENING GAMBIT

Outside his house he found Piglet, jumping up and down trying to reach the knocker.

"Hallo, Piglet," he said.

"Hallo, Pooh," said Piglet.

"What are *you* trying to do?"

"I was trying to reach the knocker," said Piglet. "I just came around—"

"Let me do it for you," said Pooh kindly. So he reached up and knocked at the door.

"I have just seen Eeyore," he began, (Pooh continues to explain his story about Eeyore when he interrupts himself)—

"What a long time whoever lives here is answering this door."

And he knocked again.

"But Pooh," said Piglet, "it's your own house!"

"Oh!" said Pooh. "So it is," he said. "Well let's go in." [1]

I will open with an analysis of a piece of data from the study of children's everyday activities. From it I hope to generate a few important issues that I will take up in more detail in the sections following this one. These issues will center on the analysis of interactional development as a sequence of conversation, and on the nature of some of the conversational resources used by participants in that sequence. I take it that the instance is typical of a mundane routine that children confront in daily life, namely, calling on their friends. Children's contacts often involve and sometimes require the intervention of adults. This is a point about the organization of childhood to keep in mind when examining the instance I am about to present.

The following complete event took place in an encounter between a neighborhood child and a household mother. I was a guest in this home and at the time of interaction was alone in a bedroom whose window was situated in a favorable position to overhear and record the entire sequence. It lasted for about one minute. The ecological arrangement is very important for an interpretation of the data. This house, on a street of private attached homes in San Francisco, has a front gate off the street that is always locked and that leads onto a tunnel or passage that has a staircase at the far end going up to the front door of the house. The staircase turns and a caller at the front gate at street level cannot see the front door one flight above. To gain entry a caller must always ring the bell at the gate first. This is a standard architectural arrangement for many homes in this city. It structures an interaction between caller and door-answer in such a way that neither party can see the other unless the answerer descends the staircase and turns the corner to look down the passageway at the caller. The entire passage and staircase are actually outdoors. The following transpired:

1. Caller: (Boy rings bell and waits for an answer to his ring.)
2. Mother: Who's there?
3. Caller: Can your son come out?
4. Mother: What?
5. Caller: Can your son come out?
6. Mother: What do you want?
7. Caller: Can your son come out?
8. Mother: (pause) Who is it?
9. Caller: Jerry. Can your son come out?
10. Mother: Oh— No he can't come right now. (Closes the front door.)
11. Caller: When do you think he could come out?
12. Mother: Silence, since mother has not heard, having closed the door before the boy spoke utterance 11.)
13. Caller: (leaves)

I want to make the following points about this piece:

1. One of the features of family arrangements consists of an ecological containment of members inside the confines of a physical setting, commonly thought of as a residence. Family members therefore carry on their household activities in a home territory. In this instance we find that home territories have entrance points, such as doors and front gates. Those who are bona fide residents of the territory have the right of free passage into their own homes (as Piglet points out to the forgetful Pooh) and in fact need not knock or ring to ask permission to be granted entry. In the case at hand, adults have keys, and others wishing entry, such as household children, ring, wait for voice identification, and without further question get passage by means of an electric button pressed by the door answerer. This does *not* constitute asking for and receiving permission where household members are concerned but merely requests clearance, a form of mechanical security to control the entry of *outsiders*. In other words it is an inconvenience to children of the house who may not own a key to the gate, the price paid for such security measures in big cities. But where nonhousehold members are concerned, entry does indeed involve the granting of permission to come in through the front door

(guests are treated as temporary household members and therefore the simple clearance pattern of ring-voice identification-entry applies).

2. A child wishing to call on another child must attend to this problem of entering another home territory (whether gates, locks, closed front doors, or some other physical arrangements exist). Now as far as conversational interaction or a state of talk goes, *the child must be prepared to identify himself as a caller and likewise the person on whom he is calling.* He must therefore have at his disposal the conversational resources to make such relevant identifications as necessary conditions, perhaps, for paying a visit to another child or getting him to come out to play.

3. As the data shows, the opening of the interaction is founded upon the principle of getting an adequate identification from the caller *as a second step in the sequence,* the first being the summoning of a household member by the bell-ring. The structural parallel is to that of the opening sequences of telephone calls, as analyzed by Schegloff. Where in telephoning activity the answerer speaks first after hearing the ring summon him, so too in the household entry situation in our data. But unlike the telephoning interaction, here the answerer provides a question that calls for explicit identification from the caller: "Who's there?" Unlike the telephone answerer, who cannot know where the caller is located, the door-bell answerer knows precisely his location and thereby presumably his most general intention: to speak to some member of the house and possibly to gain entry.

4. Now the boy caller in the data hears the mother's call for an identification, but rather than supplying it with an *identification term* he relies on voice recognition to do this identificational work. But it does not. His question, moreover, is his reason for being there. Instead of identifying himself in the terms of reference carried within his utterance, he offers an identification of the one he is calling on, the thirteen-year-old boy of the house: "Can your son come out?"

5. What can be gathered from this question? The caller could have made an identification by using the boy's first name (*FN*), but instead he has selected a term from an altogether different set of possible calling terms. He uses what I might call a *relational term* that is one of a number of such terms applying to members of a unit of social membership called "family." The selection of "your son" is interactionally viable because the caller has performed an analysis of his interlocutor, the mother of the house. Hearing (not seeing) her as the mother, because either he is familiar with her voice or he assumes any woman's voice will typically be that of the mother of the house, he transforms his own identification term for his friend, a *FN,* used to address him, into relational terms for the benefit of the answerer. In other words, when a child talks to another child's mother he can refer to his friend in terms of his relationship to her as her son. Now another consideration about this selection procedure suggests itself, namely, that when a neighborhood child doesn't in fact know the name of a boy on the block he has

played with in the past, he can formulate an identification using familial relationships in households. This is a conversational resource to accomplish the purpose of his call. Perhaps he just wants *someone* to play with.

6. The mother's failure to make voice identification leads her to ask the boy to repeat his first utterance. After he does, she still cannot identify him, so she then goes on to inquire about the purpose of his call: "What do you want?" The caller repeats himself for the third time, still waiting at the front gate and out of sight, and the mother calls once more for an identification of the caller. This is preceded by a brief pause in which she appears to be scrutinizing the voice for familiarity. It brings the boy's identification: "Jerry. Can your son come out?" He selects *fn,* but he doesn't employ a parallel term for the rest of his utterance, and continues to identify the boy of the house in family-relationship terms. I cannot prove it, but I would speculate that he does not know the *FN* of the boy of the house.

7. The mother then grasps fully the purpose of the call and what its interactional consequences might be, that her son is being asked to go outside and play. However, at that moment she knows that her son is playing with another child in the house, and also that he has not discerned the occasion of the caller's visit. This raises the next interesting point for our discussion. Instead of relaying the matter to her son, she instead tells the caller herself that he can't come out. Two aspects of this suggest themselves for consideration of the nature of adult interventions in children's contacts and on the nature of children's rights. On the one hand, this mother has the entitlement, as does presumably any mother, to make decisions about contacts between her child and other children. On the other hand, she does not feel obligated to inform her son of the event *before closing it off* on her own. Finally, she does not show obligation to the caller to provide for future contacts by saying "Come back later," for example. So, in no sense has she assigned herself the responsibility of being a go-between in a fully developed way. Her intervention then might be characterized in terms of the parental rights she typically exercises to *answer for or talk for* her child. In this way we see that a child has restricted rights as a speaker, given that we do indeed find in many different situations that parents enforce their entitlements to speak for their own children. Needless to say, the restriction of rights to speak is intimately connected to the restrictions on responsibilities, such as the child's presumed responsibility to take appropriate courses of action or to make suitable interactional decisions where other children are concerned. By talking for her son, a mother can practice interactional control over household activity.

8. Finally, the sequence terminates when the caller places a question designed to provide for future contact, and, recognizing that the answerer has retreated inside the house, leaving only silence, he takes leave of the front gate, never having gained entry.

G. H. MEAD

MIND, SELF AND SOCIETY

WE were speaking of the social conditions under which the self arises as an object. In addition to language we found two illustrations, one in play and the other in the game, and I wish to summarize and expand my account on these points. I have spoken of these from the point of view of children. We can, of course, refer also to the attitudes of more primitive people out of which our civilization has arisen. A striking illustration of play as distinct from the game is found in the myths and various of the plays which primitive people carry out, especially in religious pageants. The pure play attitude which we find in the case of little children may not be found here, since the participants are adults, and undoubtedly the relationship of these play processes to that which they interpret is more or less in the minds of even the most primitive people. In the process of interpretation of such rituals, there is an organization of play which perhaps might be compared to that which is taking place in the kindergarten in dealing with the plays of little children, where these are made into a set that will have a definite structure or relationship. At least something of the same sort is found in the play of primitive people. This type of activity belongs, of course, not to the everyday life of the people in their dealing with the objects about them—there we have a more or less definitely developed self-consciousness—but in their attitudes toward the forces about them, the nature upon which they depend; in their attitude toward this nature which is vague and uncertain, there we have a much more primitive response; and that response finds its expression in taking the rôle of the other, playing at the expression of their gods and their heroes, going through certain rites which are the representation of what these individuals are supposed to be doing. The process is one which develops, to be sure, into a more or less definite technique and is controlled; and yet we can say that it has arisen out of situations similar to those in which little children play at being a parent, at being a teacher—vague personalities that are about them and which affect them and on which they depend. These are personalities which they take, rôles they play, and in so far control the development of their own personality. This outcome is just what the kindergarten works toward. It takes the characters of these various vague beings and gets them into such an organized social relationship to each other that they build up the character of the little child.[1] The very introduction of organization from outside supposes a lack of organi-

Reprinted from *Mind, Self and Society* by G. H. Mead (Chicago: University of Chicago Press, 1934), pp. 152–64, by permission of the publisher.

zation at this period in the child's experience. Over against such a situation of the little child and primitive people, we have the game as such.

The fundamental difference between the game and play is that in the [former] the child must have the attitude of all the others involved in that game. The attitudes of the other players which the participant assumes organize into a sort of unit, and it is that organization which controls the response of the individual. The illustration used was of a person playing baseball. Each one of his own acts is determined by his assumption of the action of the others who are playing the game. What he does is controlled by his being everyone else on that team, at least in so far as those attitudes affect his own particular response. We get then an "other" which is an organization of the attitudes of those involved in the same process.

The organized community or social group which gives to the individual his unity of self may be called "the generalized other." The attitude of the generalized other is the attitude of the whole community.[2] Thus, for example, in the case of such a social group as a ball team, the team is the generalized other insofar as it enters—as an organized process or social activity—into the experience of any one of the individual members of it.

If the given human individual is to develop a self in the fullest sense, it is not sufficient for him merely to take the attitudes of other human individuals toward himself and toward one another within the human social process, and to bring that social process as a whole into his individual experience merely in these terms: he must also, in the same way that he takes the attitudes of other individuals toward himself and toward one another, take their attitudes toward the various phases or aspects of the common social activity or set of social undertakings in which, as members of an organized society or social group, they are all engaged; and he must then, by generalizing these individual attitudes of that organized society or social group itself, as a whole, act toward different social projects which at any given time it is carrying out, or toward the various larger phases of the general social process which constitutes its life and of which these projects are specific manifestations. This getting of the broad activities of any given social whole or organized society as such within the experiential field of any one of the individuals involved or included in that whole is, in other words, the essential basis and prerequisite of the fullest development of that individual's self: only insofar as he takes the attitudes of the organized social group to which he belongs toward the organized, cooperative social activity or set of such activities in which that group as such is engaged, does he develop a complete self or possess the sort of complete self he has developed. And on the other hand, the complex cooperative processes and activities and institutional functionings of organized human society are also possible only insofar as every individual involved in them or belonging to that society can take the general attitudes of all other such individuals with reference to these processes and activities and institutional functionings, and to the organized social whole of experiential relations and interactions thereby constituted—and can direct his own behavior accordingly.

It is in the form of the generalized other that the social process influences the behavior of the individuals involved in it and carrying it on, i.e., that the community exercises control over the conduct of its individual members; for it is in this form that the social process of community enters as a determining factor into the individual's thinking. In abstract thought the individual takes the attitude of the generalized other [3] toward himself, without reference to its expression in any particular other individuals; and in concrete thought he takes that attitude insofar as it is expressed in the attitudes toward his behavior of those other individuals with whom he is involved in the given social situation or act. But only by taking the attitude of the generalized other toward himself, in one or another of these ways, can he think at all; for only thus can thinking—or the internalized conversation of gestures which constitutes thinking—occur. And only through the taking by individuals of the attitude or attitudes of the generalized other toward themselves is the existence of a universe of discourse, as that system of common or social meanings which thinking presupposes at its context, rendered possible.

The self-conscious human individual, then, takes or assumes the organized social attitudes of the given social group or community (or of some one section thereof) to which he belongs, toward the social problems of various kinds which confront that group or community at any given time, and which arise in connection with the correspondingly different social projects or organized co-operative enterprises in which that group or community as such is engaged; and as an individual participant in these social projects or co-operative enterprises, he governs his own conduct accordingly. In politics, for example, the individual identifies himself with an entire political party and takes the organized attitudes of that entire party toward the rest of the given social community and toward the problems which confront the party within the given social situation; and he consequently reacts or responds in terms of the organized attitudes of the party as a whole. He thus enters into a special set of social relations with all the other individuals who belong to that political party; and in the same way he enters into various other special sets of social relations, with various other classes of individuals respectively, the individuals of each of these classes being the other members of some one of the particular organized subgroups (determined in socially functional terms) of which he himself is a member within the entire given society or social community. In the most highly developed, organized, and complicated human social communities—those evolved by civilized man—these various socially functional classes or subgroups of individuals to which any given individual belongs (and with the other individual members of which he thus enters into a special set of social relations) are of two kinds. Some of them are concrete social classes or subgroups, such as political parties, clubs, corporations, which are all actually functional social units, in terms of which their individual members are directly related to one another. The others are abstract social classes or subgroups, such as the class of debtors and the class of creditors, in terms of which their individual members are related to one another only more or less indirectly, and which only more or less indirectly function as social units, but which afford or represent unlimited possibilities for the widening and ramifying and enriching of the social relations among all the individual members of the given society as an organized and unified whole. The given individual's membership in several of these abstract social classes or subgroups makes possible his entrance into definite social relations (however indirect) with an almost infinite number of other individuals who also belong to or are included within one or another of these abstract social classes or subgroups cutting across functional lines of demarcation which divide different human social communities from one another, and including individual members from several (in some cases from all) such communities. Of these abstract social classes or subgroups of human individuals the one which is most inclusive and extensive is, of course, the one defined by the logical universe of discourse (or system of universally significant symbols) determined by the participation and communicative interaction of individuals; for of all such classes or subgroups, it is the one which claims the largest number of individual members, and which enables the largest conceivable number of human individuals to enter into some sort of social relation, however indirect or abstract it may be, with one another—a relation arising from the universal functioning of gestures as significant symbols in the general human social process of communication.

I have pointed out, then, that there are two general stages in the full development of the self. At the first of these stages, the individual's self is constituted simply by an organization of the particular attitudes of other individuals toward himself and toward one another in the specific social acts in which he participates with them. But at the second stage in the full development of the individual's self that self is constituted not only by an organization of these particular individual attitudes, but also by an organization of the social attitudes of the generalized other or the social group as a whole to which he belongs. These social or group attitudes are brought within the individual's field of direct experience, and are included as elements in the structure or constitution of his self, in the same way that the attitudes of particular other individuals are; and the individual arrives at them, or succeeds in taking them, by means of further organizing, and then generalizing, the attitudes of particular other individuals in terms of their organized social bearings and implications. So the self reaches its full development by organizing these individual attitudes of others into the organized social or group attitudes, and by thus becoming an individual reflection of the general systematic pattern of social or group behavior in which it and the others are all involved—a pattern which enters as a whole into the individual's experience in terms of these organized group attitudes which, through the mechanism of his central nervous system, he takes toward himself, just as he takes the individual attitudes of others.

The game has a logic, so that such an organization of the self is rendered possible: there is a definite end to be obtained; the actions of the different individuals are all related to each other with reference to that end so that they do not conflict; one is not in conflict with himself in the attitude of another man on the team. If one has the attitude of the person throwing the ball he can also have the response of catching the ball. The two are related so that they further the purpose of the game itself. They are interrelated in a unitary, organic fashion. There is a definite unity, then, which is introduced into the organization of other selves when we reach such a stage as that of the game, as over against the situation of play where there is a simple succession of one rôle after another, a situation which is, of course, characteristic of the child's own personality. The child is one thing at one time and another at another, and what he is at one moment does not determine what he is at another. That is both the charm of childhood as well as its inadequacy. You cannot count on the child; you cannot assume that all the things he does are going to determine what he will do at any moment. He is not organized into a whole. The child has no definite character, no definite personality.

The game is then an illustration of the situation out of which an organized personality arises. Insofar as the child does take the attitude of the other and allows that attitude of the other to determine the thing he is going to do with reference to a common end, he is becoming an organic member of society. He is taking over the morale of that society and is becoming an essential member of it. He belongs to it insofar as he does allow the attitude of the other that he takes to control his own immediate expression. What is involved here is some sort of an organized process. That which is expressed in terms of the game is, of course, being continually expressed in the social life of the child, but this wider process goes beyond the immediate experience of the child himself. The importance of the game is that it lies entirely inside of the child's own experience, and the importance of our modern type of education is that it is brought as far as possible within this realm. The different attitudes that a child assumes are so organized that they exercise a definite control over his response, as the attitudes in a game control his own immediate response. In the game we get an organized other, a generalized other, which is found in the nature of the child itself, and finds its expression in the immediate experience of the child. And it is that organized activity in the child's own nature controlling the particular response which gives unity, and which builds up his own self.

What goes on in the game goes on in the life of the child all the time. He is continually taking the attitudes of those about him, especially the rôles of those who in some sense control him and on whom he depends. He gets the function of the process in an abstract sort of a way at first. It goes over from the play into the game in a real sense. He has to play the game. The morale of the game takes hold of the child more than the larger morale of the whole community. The child passes into the game and the game expresses a social situation in which he can completely enter; its morale may have a greater hold on him than that of the family to which he belongs or the community in which he lives. There are all sorts of social organizations, some of which are fairly lasting, some temporary, into which the child is entering, and he is playing a sort of social game in them. It is a period in which he likes "to belong," and he gets into organizations which come into existence and pass out of existence. He becomes a something which can function in the organized whole, and thus tends to determine himself in his relationship with the group to which he belongs. That process is one which is a striking stage in the development of the child's morale. It constitutes him a self-conscious member of the community to which he belongs.

Such is the process by which a personality arises. I have spoken of this as a process in which a child takes the rôle of the other, and said that it takes place essentially through the use of language. Language is predominantly based on the vocal gesture by means of which co-operative activities in a community are carried out. Language in its significant sense is that vocal gesture which tends to arouse in the individual the attitude which it arouses in others, and it is this perfecting of the self by the gesture which mediates the social activities that gives rise to the process of taking the rôle of the other. The latter phrase is a little unfortunate because it suggests an actor's attitude which is actually more sophisticated than that which is involved in our own experience. To this degree it does not correctly describe that which I have in mind. We see the process most definitely in a primitive form in those situations where the child's play takes different rôles. Here the very fact that he is ready to pay out money, for instance, arouses the attitude of the person who receives money; the very process is calling out in him the corresponding activities of the other person involved. The individual stimulates himself to the response which he is calling out in the other person, and then acts in some degree in response to that situation. In play the child does definitely act out the rôle which he himself has aroused in himself. It is that which gives, as I have said, a definite content in the individual which answers to the stimulus that affects him as it affects somebody else. The content of the other that enters into one personality is the response in the individual which his gesture calls out in the other.

We may illustrate our basic concept by a reference to the notion of property. If we say "This is my property, I shall control it," that affirmation calls out a certain set of responses which must be the same in any community in which property exists. It involves an organized attitude with reference to property which is common to all the members of the community. One must have a definite attitude of control of his own property and respect for the property of others. Those attitudes (as organized sets of responses) must be there on the part of all, so that when one says such a thing he calls out in himself the response of the others. He is calling out the response of what I have called a generalized other. That which makes society possible is such common responses, such organized attitudes, with reference to what we term property, the cults of religion, the process of education, and the relations of the family. Of course, the wider the society the more definitely universal these objects must be. In any case there must be a definite set of responses, which we may speak of as abstract, and which can belong to a very large group. Property is in itself a very abstract concept. It is that which the individual himself can control and nobody else can control. The attitude is different from that of a dog toward a bone. A dog will fight any other dog trying to take the bone. The dog is not taking the attitude of the other dog. A man who says "This is my property" is taking an attitude of the other person. The man is appealing to his rights because he is able to take the attitude which everybody else in the group has with reference to property, thus arousing in himself the attitude of others.

What goes to make up the organized self is the organization of the attitudes which are common to the group. A person is a personality because he belongs to a community, because he takes over the institutions of that community into his own conduct. He takes its language as a medium by which he gets his personality, and then through a process of taking the different rôles that all the others furnish he comes to get the attitude of the members of the community. Such, in a certain sense, is the structure of a man's personality. There are certain common responses which each individual has toward certain common things, and insofar as those common responses are awakened in the individual when he is affecting other persons he arouses his own self. The structure, then, on which the self is built is this response which is common to all, for one has to be a member of a community to be a self. Such responses are abstract attitudes, but they constitute just what we term a man's character. They give him what we term his principles, the acknowledged attitudes of all members of the community toward what are the values of that community. He is putting himself in the place of the generalized other, which represents the organized responses of all the members of the group. It is that which guides conduct controlled by principles, and a person who has such an organized group of responses is a man whom we say has character, in the moral sense.

It is a structure of attitudes, then, which goes to make up a self, as distinct from a group of habits. We all of us have, for example, certain groups of habits, such as the particular intonations which a person uses in his speech. This is a set of habits of vocal expression which one has but which one does not know about. The sets of habits which we have of that sort mean nothing to us; we do not hear the intonations of our speech that others hear unless we are paying particular attention to them. The habits of emotional expression which belong to our speech are of the same sort. We may know that we have expressed ourselves in a joyous fashion but the detailed process is one which does not come back to our conscious selves. There are whole bundles of such habits which do not enter into a conscious self, but which help to make up what is termed the unconscious self.

After all, what we mean by self-consciousness is an awakening in ourselves of the group of attitudes which we are arousing in others, especially when it is an important set of responses which go to make up the members of the community. It is unfortunate to fuse or mix up consciousness, as we ordinarily use that term, and self-consciousness. Consciousness, as frequently used, simply has reference to the field of experience, but self-consciousness refers to the ability to call out in ourselves a set of definite responses which belong to the others of the group. Consciousness and self-consciousness are not on the same level. A man alone has, fortunately or unfortunately, access to his own toothache, but that is not what we mean by self-consciousness.

I have so far emphasized what I have called the structures upon which the self is constructed, the framework of the self, as it were. Of course we are not only what is common to all: each one of the selves is different from everyone else; but there has to be such a common structure as I have sketched in order that we may be members of a community at all. We cannot be ourselves unless we are also members in whom there is a community of attitudes which control the attitudes of all. We cannot have rights unless we have common attitudes. That which we have acquired as self-conscious persons makes us such members of society and gives us selves. Selves can only exist in definite relationships to other selves. No hard-and-fast line can be drawn between our own selves and the selves of others, since our own selves exist and enter as such into our experience only insofar as the selves of others exist and enter as such into our experience also. The individual possesses a self only in relation to the selves of the other members of his social group; and the structure of his self expresses or reflects the general behavior pattern of this social group to which he belongs, just as does the structure of the self of every other individual belonging to this social group.

MARTHA WOLFENSTEIN

FUN MORALITY: AN ANALYSIS OF RECENT AMERICAN CHILD-TRAINING LITERATURE

A recent development in American culture is the emergence of what we may call "fun morality." Here fun, from having been suspect, if not taboo, has tended to become obligatory. Instead of feeling guilty for having too much fun, one is inclined to feel ashamed if one does not have enough. Boundaries formerly maintained between play and work break down. Amusements infiltrate into the sphere of work, while, in play, self-estimates of achievement become prominent. This development appears to be at marked variance with an older, Puritan ethic, although, as we shall see, the two are related.

The emergence of fun morality may be observed in the ideas about child training of the last forty years. In these one finds a changing conception of human impulses and an altered evaluation of play and fun which express the transformation of moral outlook. These changing ideas about child training may be regarded as part of a larger set of adult attitudes current in contemporary American culture. Thus I shall interpret the development which appears in the child-training literature as exemplifying a significant moral trend of our times.

The ideas on child training which I shall present are taken from the publications of the United States Department of Labor Children's Bureau. These publications probably express at any given time a major body of specialized opinion in the field, though how far they are representative would have to be determined by further study of other publications. In taking these publications as indicative of certain changing attitudes, I leave undetermined to what extent these attitudes are diffused among parents and also to what extent parents' actual behavior with their children conforms to these ideas. Both these topics would require further research.

The innovations in child-training ideas of the past few decades may readily be related to developments in psychological research and theory (notably behaviorism, Gesell's norms of motor development, and psychoanalysis). However, the occurrence and particularly the diffusion of certain psychological

Reprinted from *Childhood in Contemporary Cultures,* ed. Margaret Mead and Martha Wolfenstein (Chicago: University of Chicago Press, 1955), pp. 168–74, by permission of the author and the publisher.

ideas at certain periods are probably related to the larger cultural context. A careful study of the ways in which psychological theories have been adapted for parent guidance and other pedagogical purposes would show that a decided selection is made from among the range of available theories, some points being overstressed, others omitted, and so on.

The *Infant Care* bulletin of the Children's Bureau, the changing contents of which I shall analyze, was first issued in 1914. The various editions fall into three main groupings: 1914 and 1921, 1929 and 1938, 1942 and 1945 (i.e., the most drastic revisions occurred in 1929 and 1942).[1] For the present purpose I shall mainly contrast the two ends of the series, comparing the 1914 edition with those of 1942 and 1945 (the two latter are practically identical) and skipping over the middle period. Thus I shall attempt to highlight the extent of the change rather than to detail the intermediate stages (which in any case show some complicated discontinuities).

As the infant embodies unmodified impulses, the conception of his nature is a useful index of the way in which the impulsive side of human nature generally is regarded. The conception of the child's basic impulses has undergone an extreme transformation from 1914 to the 1940's. At the earlier date, the infant appeared to be endowed with strong and dangerous impulses. These were notably autoerotic, masturbatory, and thumb-sucking. The child is described as "rebelling fiercely" if these impulses are interfered with.[2] The impulses "easily grow beyond control" [3] and are harmful in the extreme: "children are sometimes wrecked for life." [4] The baby may achieve the dangerous pleasures to which his nature disposes him by his own movements or may be seduced into them by being given pacifiers to suck or having his genitals stroked by the nurse.[5] The mother must be ceaselessly vigilant; she must wage a relentless battle against the child's sinful nature. She is told that masturbation "must be eradicated . . . treatment consists in mechanical restraints." The child should have his feet tied to opposite sides of the crib so that he cannot rub his thighs together; his nightgown sleeves should be pinned to the bed so that he cannot touch himself.[6] Similarly for thumb-sucking, "the sleeve may be pinned or sewed down over the fingers of the offending hand for several days and nights," or a patent cuff may be used which holds the elbow stiff.[7] The mother's zeal against thumb-sucking is assumed to be so great that she is reminded to allow the child to have his hands free some of the time so that he may develop legitimate manual skills; "but with the approach of sleeping time the hand must be covered." [8] The image of the child at this period is that he is centripetal, tending to get pleasure from his own body. Thus he must be bound down with arms and legs spread out to prevent self-stimulation.

In contrast to this we find in 1942–45 that the baby has been transformed into almost complete harmlessness. The intense and concentrated impulses of the past have disappeared. Drives toward erotic pleasure (and also toward domination, which was stressed in 1929–38) have become weak and incidental. Instead, we find impulses of a much more diffuse and moderate character. The baby is interested in exploring his world. If he happens to put his thumb in his mouth or to touch his genitals, these are merely incidents, and unimportant ones at that, in his over-all exploratory progress. The erogenous zones do not have the focal attraction which they did in 1914, and the baby easily passes beyond them to other areas of presumably equal interest. "The baby will not spend much time handling his genitals if he has other interesting things to do." [9] This infant explorer is centrifugal as the earlier erotic infant was centripetal. Everything amuses him, nothing is excessively exciting.

The mother in this recent period is told how to regard autoerotic incidents: "Babies want to handle and investigate everything that they can see and reach. When a baby discovers his genital organs he will play with them. . . . A wise mother will not be concerned about this." [10] As against the older method of tying the child hand and foot, the mother is now told: "See that he has a toy to play with and he will not need to use his body as a plaything." [11] The genitals are merely a resource which the child is thrown back on if he does not have a toy. Similarly with thumb-sucking: "A baby explores everything within his reach. He looks at a new object, feels it, squeezes it, and almost always puts it in his mouth." [12] Thus again what was formerly a "fierce" pleasure has become an unimportant incident in the exploration of the world. Where formerly the mother was to exercise a ceaseless vigilance, removing the thumb from the child's mouth as often as he put it in, now she is told not to make a fuss. "As he grows older, other interests will take the place of sucking." [13] (Incidentally, this unconcerned attitude toward thumb-sucking is a relatively late development. The 1938 edition still had an illustration of a stiff cuff which could be put on the infant at night to prevent his bending his elbow to get his fingers to his mouth. The attitude toward masturbation relaxed earlier, diversion having already been substituted for mechanical restraints in 1929.)

This changing conception of the nature of impulses bears on the question: Is what the baby likes good for him? The opposition between the pleasant and the good is deeply grounded in older American morals (as in many other ascetic moral codes). There are strong doubts as to whether what is enjoyable is not wicked or deleterious. In recent years, however, there has been a marked effort to overcome this dichotomy, to say that what is pleasant is also good for you. The writers on child training reflect the changing ideas on this issue.

In the early period there is a clear-cut distinction between what the baby "needs," his legitimate requirements, whatever is essential to his health and well-being, on the one hand, and what the baby "wants," his illegitimate pleasure strivings, on the other. This is illustrated, for instance, in the question of whether to pick the baby up when he cries. In 1914 it was essential to determine whether he really needed something or whether he only wanted something. Crying is listed as a bad habit. This is qualified with the remark that the baby has no other way of expressing his "needs"; if he is expressing a need, the mother should respond. "But when the baby cries simply because he has learned from experience that this brings him what he wants, it is one of the worst habits he can learn." If the baby cries, "the mother may suspect illness, pain, hunger or thirst." These represent needs. If checking on all these shows they are not present, "the baby probably wants to be taken up, walked with, played with," etc. "After the baby's needs have been fully satisfied, he should be put down and allowed to cry." [14] (This position remained substantially unchanged up to 1942.)

In 1942–45, wants and needs are explicitly equated. "A baby sometimes cries because he wants a little more attention. He probably needs a little extra attention under some circumstances just as he sometimes needs a little extra food and water. Babies want attention; they probably need plenty of it." [15] What the baby wants for pleasure has thus become as legitimate a demand as what he needs for his physical well-being and is to be treated in the same way.[16]

The question of whether the baby wants things which are not good for him also occurs in connection with feeding. The baby's appetite was very little relied on to regulate the quantity of food he took in the early period. Overfeeding was regarded as a constant danger; the baby would never know when he had enough. This is in keeping with the general image of the baby at this time as a creature of insatiable impulses. In contrast to this, we find in the recent period that "the baby's appetite usually regulates successfully the amount of food he takes." [17] Thus again impulses appear as benevolent rather than dangerous.

Formerly, giving in to impulse was the way to encourage its growing beyond control. The baby who was picked up when he cried, held and rocked when he wanted it, soon grew into a tyrant.[18] This has now been strikingly reversed. Adequate early indulgence is seen as the way to make the baby less demanding as he grows older.[19] Thus we get the opposite of the old maxim, "Give the devil the little finger, and he'll take the whole hand." It is now "Give him the whole hand, and he'll take only the little finger."

The attitude toward play is related to the conception of impulses and the belief about the good and the pleasant. Where impulses are dangerous and the good and pleasant are opposed, play is suspect. Thus in 1914, playing with the baby was regarded as dangerous; it produced unwholesome pleasure and ruined the baby's nerves. Any playful handling of the baby was titillating, excessively exciting, deleterious. Play carried the overtones of feared erotic excitement. As we noted, this was the period of an intensive masturbation taboo, and there were explicit apprehensions that the baby might be seduced into masturbation by an immoral nurse who might play with his genitals.

The mother of 1914 was told: "The rule that parents should not play with the baby may seem hard, but it is without doubt a safe one. A young, delicate and nervous baby needs rest and quiet, and however robust the child much of the play that is indulged in is more or less harmful. It is a great pleasure to hear the baby laugh and crow in apparent delight, but often the means used to produce the laughter, such as tickling, punching, or tossing, makes him irritable and restless. It is a regrettable fact that the few minutes' play that the father has when he gets home at night . . . may result in nervous disturbance of the baby and upset his regular habits." [20] It is relevant to note that at this time "playthings . . . such as rocking horses, swings, teeter boards, and the like" are cited in connection with masturbation, as means by which "this habit is learned." [21] The dangerousness of play is related to that of the ever present sensual impulses which must be constantly guarded against. (In 1929–38, play becomes less taboo, but must be strictly confined to certain times of the day. In this period the impulse to dominate replaces erotic impulses as the main hazard in the child's nature, and the corresponding danger is that he may get the mother to play with him whenever he likes.)

In the recent period, play becomes associated with harmless and healthful motor and exploratory activities. It assumes the aspect of diffuse innocuousness which the child's impulse life now presents. Play is derived from the baby's developing motor activities, which are now increasingly stressed. "A baby needs to be able to move all parts of his body. He needs to exercise. . . . At a very early age the baby moves his arms and legs aimlessly. . . . As he gets older and stronger and his movements become more vigorous and he is better able to control them he begins to play." [22] Thus play has been successfully dissociated from unhealthy excitement and nervous debilitation and has become associated with muscular development, necessary exercise, strength, and control. This is in keeping with the changed conception of the baby, in which motor activities rather than libidinal urges are stressed. For the baby who is concerned with exploring his world rather than with sucking and masturbating, play becomes safe and good.

Play is now to be fused with all the activities of life. "Play and singing make both mother and baby enjoy the routine of life." [23] This mingling of play with necessary routines is consonant with the view that the good and pleasant coincide. Also, as the mother is urged to make play an aspect of every activity, play assumes a new obligatory quality. Mothers are told that "a mother usually enjoys entering into her baby's play. Both of them enjoy the little games that mothers and babies have always played from time immemorial." (This harking back to time immemorial is a way of skipping over the more recent past.) "Daily tasks can be done with a little play and singing thrown in." [24] Thus it is now not adequate for the mother to perform efficiently the necessary routines for her baby; she must also see that these are fun for both of them. It seems difficult here for anything to become permissible without becoming compulsory. Play, having ceased to be wicked, having become harmless and good, now becomes a new duty.

In keeping with the changed evaluation of impulses and play, the conception of parenthood has altered. In the earlier period the mother's character was one of strong moral devotion. There were frequent references to her "self-control," "wisdom," "strength," "persistence," and "unlimited patience." The mothers who read these bulletins might either take pride in having such virtues or feel called upon to aspire to them. The writers supposed that some mothers might even go to excess in their devoted self-denial. Thus the mothers were told that, for their own health and thus for the baby's good, they should not stay bound to the crib-side without respite, but should have some pleasant, although not too exhausting, recreation.[25] The mother at this time is pictured as denying her own impulses just as severely as she does those of her child. Just as she had to be told to let the baby's hands free occasionally (not to overdo the fight against thumb-sucking), so she must be counseled to allow herself an intermission from duty. (In the 1929–38 period parenthood became predominantly a matter of knowhow. The parents had to use the right technique to impose routines and to keep the child from dominating them.)

In the most recent period parenthood becomes a major source of enjoyment for both parents (the father having come much more into the picture than he was earlier). The parents are promised that having children will keep them together, keep them young, and give them fun and happiness. As we have seen, enjoyment, fun, and play now permeate all activities with the child. "Babies—and usually their mothers—enjoy breast feeding"; nursing brings "joy and happiness" to the mother. At bath time the baby "delights" his parents, and so on.[26]

The characterization of parenthood in terms of fun and enjoyment may be intended as an inducement to parents in whose scheme of values these are presumed to be priorities. But also it may express a new imperative: You ought to enjoy your child. When a mother is told that most mothers enjoy nursing, she may wonder what is wrong with her in case she does not. Her self-evaluation can no longer be based entirely on whether she is doing the right and necessary things but becomes involved with nuances of feeling which are not under voluntary control. Fun has become not only permissible but required, and this requirement has a special quality different from the obligations of the older morality.

Chapter **4**

WHAT IS AN INSTITUTION? THE CASE OF LANGUAGE

EMILE Durkheim provided a classical description of social institutions (or, as he called them, "social facts"). Institutions are external to the individual; they are *there* whether the individual wishes it or not. Institutions have the quality of objectivity; the individual cannot determine their characteristics in accordance with his subjective preferences or idiosyncratic perception, but he must come to terms with the way in which any particular institution is collectively understood and "acted out." Significantly, institutions have coercive power: if the individual decides to ignore or resist a genuine institution, sanctions are available against him—any number of more or less unpleasant things may happen to him. Finally, institutions have moral authority and historicity: an institution not only exists as a fact but almost invariably is justified as having a right to exist; therefore, the individual is supposed to adhere to institutionally appropriate conduct not only because he fears the sanctions but also because he acknowledges the moral authority. And every genuine institution is experienced by the individual as a historical phenomenon: it was there before he was born, and it will outlast his own life.

Every one of these characteristics is exemplified by language. It is not only one of the basic institutions of all institutions, but it is also the primary medium of social interaction on which all other institutions of society depend in maintaining themselves as reality.

Through language the objective and constraining reality of society is first experienced by the child and continues to be experienced by the adult throughout his life. The first selection in this chapter is from a work by Iona and Peter Opie, *The Lore and Language of Schoolchildren,* in which the authors show how British schoolchildren reflect the experience of society as a strange phenomenon and how they deal with it linguistically. There follows another selection from Matthew Speier's "The Everyday World of the Child," which describes the way in which the categories of society are imposed on a child through language.

Helen Keller, blind and deaf-mute from early childhood, is one of the most moving cases of an individual who became an influential figure in society despite overwhelming handicaps. The account of her acquisition of language by her extraordinary teacher Anne Mansfield Sullivan, taken from Miss Keller's *Story of My Life,* offers dramatic testimony of the power of language in relating the individual to the social world.

Jean Piaget is generally considered the foremost child psychologist of this century. Much of his early work deals with language development. The passages selected from *The Language and Thought of the Child* describe how language leads the child from a self-enclosed world of his egocentric experiences to the social world shared with others. This selection, together with the three preceding, illustrates how any language both constrains and liberates the individual. However, just as society is differentiated in terms of groups (such as social classes), so it is differentiated linquistically. The final selection —Russell Lynes's commentary on Nancy Mitford's well-known book on the language of the English upper class, *Noblesse Oblige*—discusses class differences in language both in England and in the United States.

IONA AND PETER OPIE

THE LORE AND LANGUAGE OF SCHOOLCHILDREN

THE scraps of lore which children learn from each other are at once more real, more immediately serviceable, and more vastly entertaining to them than anything which they learn from grown-ups. To a child it can be a 'known fact' that the Lord's Prayer said backwards raises the devil, that a small knife-wound between the thumb and forefinger gives a person lock-jaw, that a hair from the head placed on the palm will split the master's cane. It can be a useful piece of knowledge that the reply to 'A pinch and a punch for the first of the month' is 'A pinch and a kick for being so quick'. And a verse a child hears the others saying,

> Mister Fatty Belly, how is your wife?
> Very ill, very ill, up all night,
> Can't eat a bit of fish
> Nor a bit of liquorice.
> O-U-T spells out and out you must go
> With a jolly good clout upon your ear hole spout,

may seem the most exciting piece of poetry in the language.

Such a verse, recited by 8-year-olds in Birmingham, can be as traditional and as well known to children as a nursery rhyme; yet no one would mistake it for one of Mother Goose's compositions. It is not merely that there is a difference in cadence and subject-matter, the manner of its transmission is different. While a nursery rhyme passes from a mother or other adult to the small child on her knee, the school rhyme circulates simply from child to child, usually outside the home, and beyond the influence of the family circle. By its nature a nursery rhyme is a jingle preserved and propagated not by children but by adults, and in this sense it is an 'adult' rhyme. It is a rhyme which is adult approved. The schoolchild's verses are not intended for adult ears. In fact part of their fun is the thought, usually correct, that adults know nothing about them. Grownups have outgrown the schoolchild's lore. If made aware of it they tend to deride it; and they actively seek to suppress its livelier manifestations. Certainly they do nothing to encourage it. And the folklorist and anthropologist can, without travelling a mile from his door, examine a thriving unselfconscious culture (the word 'culture' is used here deliberately) which is as unnoticed by the sophisticated world, and quite as little affected by it, as is the culture of some dwindling aboriginal tribe living out its helpless existence in the hinterland of a native reserve. Perhaps, indeed, the subject is worthy of a more formidable study than is accorded it here. As Douglas Newton has pointed out: 'The world-wide fraternity of children is the greatest of savage tribes, and the only one which shows no sign of dying out.'

No matter how uncouth schoolchildren may outwardly appear, they remain tradition's warmest friends. Like the savage, they are respecters, even venerators, of custom; and in their self-contained community their basic lore and language seems scarcely to alter from generation to generation. Boys continue to crack jokes that Swift collected from his friends in Queen Anne's time; they play tricks which lads used to play on each other in the heyday of Beau Brummel; they ask riddles which were posed when Henry VIII was a boy. Young girls continue to perform a magic feat (levitation) of which Pepys heard tell ('One of the strangest things I ever heard'); they hoard bus tickets and milk-bottle tops in distant memory of a love-lorn girl held to ransom by a tyrannical father; they learn to cure warts (and are successful in curing them) after the manner which Francis Bacon learnt when he was young. They call after the tearful the same jeer Charles Lamb recollected; they cry 'Halves!' for something found as Stuart children were accustomed to do; and they rebuke one of their number who seeks back a gift with a couplet used in Shakespeare's day. They attempt, too, to learn their fortune from snails, nuts, and apple-parings—divinations which the poet Gay described nearly two and a half centuries ago; they span wrists to know if someone loves them in the way that Southey used at school to tell if a boy was a bastard; and when they confide to each other that the Lord's Prayer said backwards will make Lucifer appear, they are perpetuating a story which was gossip in Elizabethan times.[1]

The oral rhymes which children inherit, almost automatically, after about seven years' residence in this island, may be divided into two classes. There are those which are essential to the regulation of their games and their relationships with each other; and there are those, seemingly almost as necessary to them, which are mere expressions of exuberance: a discordant symphony of jingles, slogans, nonsense verses, tongue-twisters, macabre rhymes, popular songs, parodies, joke rhymes, and improper verses, epitomized in the nonsensical couplet,

Oh my finger, oh my thumb,
Oh my belly, oh my bum,

which is repeated for no more reason than that they heard someone else say it, that they like the sound of the rhyme *thumb* and *bum,* that it is a bit naughty, and that for the time being, in the playground or in the gang, it is considered the latest and smartest thing to say—for they are not to know that the couplet was already old when their parents were youngsters.[2]

Rhyme seems to appeal to a child as something funny and remarkable in itself, there need be neither wit nor reason to support it.

Mrs. White had a fright
In the middle of the night,
She saw a ghost eating toast
Half-way up the lamp post.

I think what's so clever about this', says a 9-year-old, 'is the way it all rhymes.' Hence, apparently, the popularity of the rather horrid and otherwise pointless jests they have: 'Do you want an apple?' 'Yes.' 'Wipe your nose and go to chapel.' 'Do you want a sweet?' 'Yes.' 'Suck your feet.' 'Do you want jelly?'—'Rub your belly.' 'Do you want treacle?'—'You're a big fat beetle.' And hence the way lines of current dance songs become catch phrases: 'O Nicholas, don't be so ridiculous; and 'See you later, alligator'—'In a while, crocodile', repeated *ad nauseam* in 1956.

Listening to children as they 'tumble and rhyme' out of school (as Dylan Thomas described them) they seem to have a chant on their lips as constantly as they have a comic in their hands, or a sweet in their mouth. A group of small, round-faced toughs come step-hopping out of school, chanting over and over again, 'Roly poly barley sugar; roly poly barley sugar', and laughing at the littlest one who cannot get his step-hop in time with the rest. A party of girls sauntering along the pavement, arms intertwined in friendship, croon a happy song of little meaning:

I'm a knock-kneed chicken, I'm a bow-legged sparrow,
Missed my bus so I went by barrow.
I went to the café for my dinner and my tea,
Too many radishes—Hick! Pardon me.

And in the school forecourt twenty or thirty children, those who live in the outlying villages, are killing time, while they wait for the school bus 'which is always late', chorusing at the full extent of their voices:

Never let your braces dangle,
Never let your braces dangle.
Poor old sport
He got caught
And went right through the mangle;
Went through the mangle he did, by gum,
Came out like linoleum,
Now he sings in kingdom-come:
Never let your braces dangle, chum.

These rhymes are more than playthings to children. They seem to be one of their means of communication with each other. Language is still new to them, and they find difficulty in expressing themselves. When on their own they burst into rhyme, of no recognizable relevancy, as a cover in unexpected situations, to pass off an awkward meeting, to fill a silence, to hide a deeply felt emotion, or in a gasp of excitement. And through these quaint ready-made formulas the ridiculousness of life is underlined, the absurdity of the adult world and their teachers proclaimed, danger and death mocked, and the curiosity of language itself is savoured.

MATTHEW SPEIER

SELECTING AN IDENTIFICATION BY MEANS OF SOCIAL CATEGORIES

EARLY uses of referential procedures involve the application of social category terms for addressing, for referring *in absentia,* and for distinguishing what persons appropriately can be called. The child at a very early stage in life is confronted with one very basic problem of reference, practically speaking—how to go about selecting an identification for another. The following material illustrates the very young child's first attempts at using referential procedures. It is from the notes of a mother reported by Church. . . .

a. at 11 months: Whenever Ruth sees an infant or a child in a carriage, or even a toddler, she says "Hi, babee."
b. at 11 months: If I scold Ruth and slap her hand for doing something she shouldn't, she either hits her own hand or mine and says in a scolding tone, "Da-da-da", which means "bad", I think, in addition to "Daddy" and some other things. But while "da-da" means several things, "Ma-ma" now means only "mother".
c. at 15–16½ months: When Ruth sees any of our friends who have babies, she says, "Baby, baby", whether or not the baby is present.
d. at 17–18 months: It isn't confusing yet to Ruth that I am "Mother", "Ellen", and "Mommy" or that my husband is "Karl", "Daddy", or "Father". . . . Lately she calls me "Mommy Ellen".
e. at 19 months: Ruthie connects couples. Whenever she hears someone's name, she adds the name of the spouse and the baby. Example: "Herb-Elaine and baby Janet".

Reprinted from Jack D. Douglas, editor, *Understanding Everyday Life* (Chicago: Aldine Publishing Co., 1970), pp. 199–202;

[Speier's cross-references and notes deleted.—Ed.]

Geertz . . . reports about a Javanese family household in which a mother, separated from her husband, returned with her small son to her parent's home:

> Tuti's son Ta, aged one and a half, was the pet of his grandmother, who still had children of her own aged about 10 and older. Ta called his grandmother "mother" and his grandfather, "father", partly following the usage of his own mother toward them and partly because he was taught so by the affectionate grandmother. His own mother he called by her name, a familiar practice usually only permissible toward kin younger than oneself.

This child appeared, then, to have solved the task of selecting an identification for other household members by using referential terms for them interactionally mobilized by those around him; that is, he called his grandparents what his mother called them and his grandmother what his grandmother instructed him to call her. From this illustration we find that referential procedures and terms are not found by the analyst by looking for kin terms in the family, for example, that are used only in accordance with consanguinal rules of address, as anthropologists refer to them. The Javanese child called his biological mother by her proper name and used the biological kin terms normally reserved for the parent-child relationship for nonparents. A similar case would be in our own culture when a husband addresses his wife with "mother" in the presence of his children. He is calling *their* mother what he knows is a term of reference used by *them*.

The child's problem in using referential procedures in the family household appears quite complicated from the start because of the wide range of alternative selections of social categories that he can make in given interactional circumstances; that is, who can be called what in the presence of relevant others, when speaking either to or about them. It is important to distinguish between situations where referential procedures are used for those present or absent to the speaker. One way of conceptualizing this distinction is to say that the child learns to use different procedures when he talks *to* others as compared to when he talks *about* others. A four-year-old boy expressed the relevance of this distinction when his father overheard him speaking to his mother, using his father's first name, which was not common practice for him to do. The father asked the child when he called him by his name and when he called him "daddy." The child replied: "I call you daddy when I'm telling you something." His father then inquired further into when he called him by his first name and the child replied: "When I'm talking about you." The point I want to stress here is that probably the child very rarely used this *in absentia* procedure as an optional way of referring to his father, but that it was part of a real repertoire of possibilities apparently governed by the child's perceived social constraint upon proper forms of address for his father.

Schneider . . . has spelled out the range of normal alternatives for referential terms in American kinship. It should dispel any idea that socialization to referential procedures is based upon the acquisition of only a few basic terms for identifying and calling others in the household. He says:

> The first point which must be made is that there is a wide variety of *alternate* terms and usages applicable to any particular kind of person as a relative. To put it another way, there are far more kinship terms and terms for kinsmen than there are kinds of kinsmen, or categories of kinsmen.
>
> Mother may be called "mother", "mom", "ma", "mummy", "mama", "old woman", or by her first name, nickname, a diminutive, or a variety of other designations, including unique or idiosyncratic appellations, sometimes related to baby-talk. Father may be called "father", "pop", "pa", "dad", "daddy", "old man", "boss", or by his first name, nickname, a diminutive, or a variety of less commonly used designations, including unique or idiosyncratic appellations, sometimes related to baby-talk. Uncles may be addressed or referred to as uncle-plus-first-name, first name alone, or uncle alone. And so, too, aunts. Grandparents may be called "grandma", "grandpa", "gramma", "grapa", "nana"; last names may be added to distinguish "Gramma Jones" from "Gramma Smith". Cousins are addressed by their first name, nickname, a diminutive, or other personal forms of designation, or as cousin-plus-first-name ("Cousin Jill"). Son may be called "son", "sonny", "kid", "kiddy", "boy", or by his first name, nickname, or diminutive, or other forms of personal designation. And daughter may be called "girl", "sister", "daughter", by her first name, nickname, a diminutive, or sister-plus-first-name ("Sister Jane"), as well as idiosyncratic and personal forms. "Kid" as a form for child does not distinguish son from daughter. Brother may be "brother", brother-plus-first-name, first name alone, nickname, diminutives, or personal forms. Sister may be "sister", sister-plus-first-name, first name alone, nick-name, diminutives, or personal forms.[1]

What I want to stress about this list of possible referential alternatives is not that the researcher can simply make a list of them, as Schneider has done, but that he can also search out their *systematic procedural basis* in family interaction. That is, by examining their conversational use he can attempt to describe the child's practices of selecting identifications for others in the family household.

A further procedure that a child can use involves his knowing an identification term for all the family members as *a unit of membership*. That is, he can refer to his family as "my family" or "our family," or his family's residence as "my house" or "our house," which adds the locational or territorial aspect to the referential procedure. Schneider points out that "members of the family" can be differentiated from "the family as a cultural unit." That is, in some sense the child learns that all of his relatives (another identification term for a whole collection of familially related persons), are referentially distinguishable with respect to the unit he calls "family." The latter constitutes for the child a stable set of categories used in everyday interaction with household members. In cases where extended relatives, such as grandparents or uncles, are permanent household residents, the child presumably acquires suitable methods of reference that take on additional categorials. In other words, the composition of the unit he calls his family consists of various stable identifiable members whose daily presence establishes interactional requirements for the child's use of referential procedures. In other words, referential demands are placed upon his speech interaction relative to each and every member.

To summarize the discussion so far, one major problem in the child's socialization to family life is how he competently used referential terms for relevant others in the family membership unit. The correct use of referential procedures for naming others comprises one part of the child's cultural apparatus pertaining to typical family arrangements and to the structure of social relationships these arrangements encompass.

ANNE MANSFIELD SULLIVAN

HELEN KELLER ACQUIRES LANGUAGE

March 20, 1887.

My heart is singing for joy this morning. A miracle has happened! The light of understanding has shone upon my little pupil's mind, and behold, all things are changed!

The wild little creature of two weeks ago has been transformed into a gentle child. She is sitting by me as I write, her face serene and happy, crocheting a long red chain of Scotch wool. She learned the stitch this week, and is very proud of the achievement. When she succeeded in making a chain that would reach across the room, she patted herself on the arm and put the first work of her hands lovingly against her cheek. She lets me kiss her now, and when she is in a particularly gentle mood, she will sit in my lap for a minute or two; but she does not return my caresses. The great step—the step that counts—has been taken. The little savage has learned her first lesson in obedience, and finds the yoke easy. It now remains my pleasant task to direct and mould the beautiful intelligence that is beginning to stir in the child-soul. Already people remark the change in Helen. Her father looks in at us morning and evening as he goes to and from his office, and sees her contentedly stringing her beads or making horizontal lines on her sewing-card, and exclaims, "How quiet she is!" When I came, her movements were so insistent that one always felt there was something unnatural and almost weird about

Reprinted from *The Story of My Life* by Helen Keller, "Letters by Anne Mansfield Sullivan" (New York: Doubleday & Co., Anchor Books, 1954), pp. 268–76, by permission of the publisher.

her. I have noticed also that she eats much less, a fact which troubles her father so much that he is anxious to get her home. He says she is homesick. I don't agree with him; but I suppose we shall have to leave our little bower very soon.

Helen has learned several nouns this week. "M-u-g" and "m-i-l-k," have given her more trouble than other words. When she spells "milk," she points to the mug, and when she spells "mug," she makes the sign for pouring or drinking, which shows that she has confused the words. She has no idea yet that everything has a name.

Yesterday I had the little Negro boy come in when Helen was having her lesson, and learn the letters, too. This pleased her very much and stimulated her ambition to excel Percy. She was delighted if he made a mistake, and made him form the letter over several times. When he succeeded in forming it to suit her, she patted him on his woolly head so vigorously that I thought some of his slips were intentional.

One day this week Captain Keller brought Belle, a setter of which he is very proud, to see us. He wondered if Helen would recognize her old playmate. Helen was giving Nancy a bath, and didn't notice the dog at first. She usually feels the softest step and throws out her arms to ascertain if anyone is near her. Belle didn't seem very anxious to attract her attention. I imagine she has been rather roughly handled sometimes by her little mistress. The dog hadn't been in the room more than half a minute, however, before Helen began to sniff, and dumped the doll into the wash-bowl and felt about the room. She stumbled upon Belle, who was crouching near the window where Captain Keller was standing. It was evident that she recognized the dog; for she put her arms round her neck and squeezed her. Then Helen sat down by her and began to manipulate her claws. We couldn't think for a second what she was doing; but when we saw her making the letters "d-o-l-l" on her own fingers, we knew that she was trying to teach Belle to spell.

March 28, 1887.

Helen and I came home yesterday. I am sorry they wouldn't let us stay another week; but I think I have made the most I could of the opportunities that were mine the past two weeks, and I don't expect that I shall have any serious trouble with Helen in the future. The back of the greatest obstacle in the path of progress is broken. I think "no" and "yes," conveyed by a shake or nod of my head, have become facts as apparent to her as hot and cold or as the difference between pain and pleasure. And I don't intend that the lesson she has learned at the cost of so much pain and trouble shall be unlearned. I shall stand between her and the over-indulgence of her parents. I have told Captain and Mrs. Keller that they must not interfere with me in any way. I have done my best to make them see the terrible injustice to Helen of allowing her to have her way in everything, and I have pointed out that the processes of teaching the child that everything cannot be as he wills it, are apt to be painful both to him and to his teacher. They have promised to let me have a free hand and help me as much as possible. The improvement they cannot help seeing in their child has given them more confidence in me. Of course, it is hard for them. I realize that it hurts to see their afflicted little child punished and made to do things against her will. Only a few hours after my talk with Captain and Mrs. Keller (and they had agreed to everything), Helen took a notion that she wouldn't use her napkin at table. I think she wanted to see what would happen. I attempted several times to put the napkin round her neck; but each time she tore it off and threw it on the floor and finally began to kick the table. I took her plate away and started to take her out of the room. Her father objected and said that no child of his should be deprived of his food on any account.

Helen didn't come up to my room after supper, and I didn't see her again until breakfast-time. She was at her place when I came down. She had put the napkin under her chin, instead of pinning it at the back, as was her custom. She called my attention to the new arrangement, and when I did not object she seemed pleased and patted herself. When she left the dining-room, she took my hand and patted it. I wondered if she was trying to "make up." I thought I would try the effect of a little belated discipline. I went back to the dining-room and got a napkin. When Helen came upstairs for her lesson, I arranged the objects on the table as usual, except that the cake, which I always give her in bits as a reward when she spells a word quickly and correctly, was not there. She noticed this at once and made the sign for it. I showed her the napkin and pinned it around her neck, then tore it off and threw it on the floor and shook my head. I repeated this performance several times. I think she understood perfectly well; for she slapped her hand two or three times and shook her head. We began the lesson as usual. I gave her an object, and she spelled the name (she knows twelve now). After spelling half the words, she stopped suddenly, as if a thought had flashed into her mind, and felt for the napkin. She pinned it round her neck and made the sign for cake (it didn't occur to her to spell the word, you see). I took this for a promise that if I gave her some cake she would be a good girl. I gave her a larger piece than usual, and she chuckled and patted herself.

April 3, 1887.

We almost live in the garden, where everything is growing and blooming and glowing. After breakfast we go out and watch the men at work. Helen loves to dig and play in the dirt like any other child. This morning she planted her doll and showed me that she expected her to grow as tall as I. You must see that she is very bright, but you have no idea of how cunning she is.

At ten we come in and string beads for a few minutes. She can make a great many combinations now, and often invents new ones herself. Then I let her decide whether she will sew or knit or crochet. She learned to knit very quickly, and is making a wash-cloth for her mother. Last week she made her doll an apron, and it was done as well as any child her age could do it. But I am always glad when this work is over for the day. Sewing and crocheting are inventions of the devil, I think. I'd rather break stones on the king's highway than hem a handkerchief. At eleven we have gymnastics. She knows all the free-hand movements and the "Anvil Chorus" with the dumb-bells. Her father says he is going to fit up a gymnasium for her in the pump-house; but we both like a good romp better than set exercises. The hour from twelve to one is devoted to the learning of new words. *But you mustn't think this is the only time I spell to Helen; for I spell in her hand everything we do all day long, although she has no idea as yet what the spelling means.* After dinner I rest for an hour, and Helen plays with her dolls or frolics in the yard with the little darkies, who were her constant companions before I came. Later I join them, and we make the rounds of the outhouses. We visit the horses and mules in their stalls and hunt for eggs and feed the turkeys. Often, when the weather is fine, we drive from four to six, or go to see her aunt at Ivy Green or her cousins in the town. Helen's instincts are decidedly social; she likes to have people about her and to visit her friends, partly, I think, because they always have things she likes to eat. After supper we go to my room and do all sorts of things until eight, when I undress the little woman and put her to bed. She sleeps with me now. Mrs. Keller wanted to get a nurse for her; but I concluded I'd rather be her nurse than look after a stupid, lazy negress. Besides, I like to have Helen depend on me for everything, *and I find it much easier to teach her things at odd moments than at set times.*

On March 31st I found Helen knew eighteen nouns and three verbs. Here is a list of the words. Those with a cross after them are words she asked for herself: *Doll, mug, pin, key, dog, hat, cup, box, water, milk, candy, eye* (*x*), *finger* (*x*), *toe* (*x*), *head* (*x*), *cake, baby, mother, sit, stand, walk.* On April 1st she learned the nouns *knife, fork, spoon, saucer, tea, papa, bed,* and the verb *run.*

April 5, 1887.

I must write you a line this morning because something very important has happened. Helen has taken the second great step in her education. She has learned that *everything has a name, and that the manual alphabet is the key to everything she wants to know.*

In a previous letter I think I wrote you that "mug" and "milk" had given Helen more trouble than all the rest. She confused the nouns with the verb "drink." She didn't know the word for "drink," but went through the pantomime of drinking whenever she spelled "mug" or "milk." This morning, while she was washing, she wanted to know the name for "water." When she wants to know the name of anything, she points to it and pats my hand. I spelled "w-a-t-e-r" and thought no more about it until after breakfast. Then it occurred to me that with the help of this new word I might succeed in straightening out the "mug-milk" difficulty. We went out to the pump-house, and I made Helen hold her mug under the spout while I pumped. As the cold water gushed forth, filling the mug, I spelled "w-a-t-e-r" in Helen's free hand. The word coming so close upon the sensation of cold water rushing over her hand seemed to startle her. She dropped the mug and stood as one transfixed. A new light came into her face. She spelled "water" several times. Then she dropped on the ground and asked for its name and pointed to the pump and the trellis, and suddenly turning round she asked for my name. I spelled "Teacher." Just then the nurse brought Helen's little sister into the pump-house, and Helen spelled "baby" and pointed to the nurse. All the way back to the house she was highly excited, and learned the name of every object she touched, so that in a few hours she had added thirty new words to her vocabulary. Here are some of them: *Door, open, shut, give, go, come,* and a great many more.

P.S.—I didn't finish my letter in time to get it posted last night; so I shall add a line. Helen got up this morning like a radiant fairy. She has flitted from object to object, asking the name of everything and kissing me for very gladness. Last night when I got in bed, she stole into my arms of her own accord and kissed me for the first time, and I thought my heart would burst, so full was it of joy.

April 10, 1887.

I see an improvement in Helen day to day, almost from hour to hour. Everything must have a name now. Wherever we go, she asks eagerly for the names of things she has not learned at home. She is anxious for her friends to spell, and eager to teach the letters to every one she meets. She drops the signs and pantomine she used before, as soon as she has words to supply their place, and the acquirement of a new word affords her the liveliest pleasure. And we notice that her face grows more expressive each day.

I have decided not to try to have regular lessons for the present. I am going to treat Helen exactly like a two-year-old child. It occurred to me the other day that it is absurd to require a child to come to a certain place at a certain time and recite certain lessons, when he has not yet acquired a working vocabulary. I sent Helen away and sat down to think. I asked myself, "*How does a normal child learn language?*" The answer was simple, "By imitation." The child comes into the world with the ability to learn, and he learns of himself, provided he is supplied with sufficient outward stimulus. He sees people do things, and he tries to do them. He hears others speak, and he tries to speak. *But long before he utters his first word, he understands what is said to him.* I have been observing Helen's little cousin lately. She is about fifteen months old, and already understands a great deal. In response to questions she points out prettily her nose, mouth, eye, chin, cheek, ear. If I say, "Where is baby's other ear?" she points it out correctly. If I hand her a flower, and say, "Give it to mamma," she takes it to her mother. If I say, "Where is the little rogue?" she hides behind her mother's chair, or covers her face with her hands and peeps out at me with an expression of genuine roguishness. She obeys many commands like these: "Come," "Kiss," "Go to papa," "Shut the door," "Give me the biscuit." But I have not heard her try to say any of these words, although they have been repeated hundreds of times in her hearing, and it is perfectly evident that she understands them. These observations have given me a clue to the method to be followed in teaching Helen language. *I shall talk into her hand as we talk into the baby's ears.* I shall assume that she has the normal child's capacity of assimilation and imitation. *I shall use complete sentences in talking to her,* and fill out the meaning with gestures and her descriptive signs when necessity requires it; but I shall not try to keep her mind fixed on any one thing. I shall do all I can to interest and stimulate it, and wait for results.

April 24, 1887.

The new scheme works splendidly. Helen knows the meaning of more than a hundred words now, and learns new ones daily without the slightest suspicion that she is performing a most difficult feat. She learns because she can't help it, just as the bird learns to fly. But don't imagine that she "talks fluently." Like her baby cousin, she expresses whole sentences by single words. "Milk," with a gesture means, "Give me more milk." "Mother," accompanied by an inquiring look, means, "Where is mother?" "Go" means, "I want to go out." But when I spell into her hand, "Give me some bread," she hands me the bread, or if I say, "Get your hat and we will go to walk," she obeys instantly. The two words, "hat" and "walk" would have the same effect; *but the whole sentence, repeated many times during the day, must in time impress itself upon the brain, and by and by she will use it herself.*

We play a little game which I find most useful in developing the intellect, and which incidentally answers the purpose of a language lesson. It is an adaptation of hide-the-thimble. I hide something, a ball or a spool, and we hunt for it. When we first played this game two or three days ago, she showed no ingenuity at all in finding the object. She looked in places where it would have been impossible to put the ball or the spool. For instance, when I hid the ball, she looked under her writing-board. Again, when I hid the spool, she looked for it in a little box not more than an inch long; and she very soon gave up the search. Now I can keep her interest in the game for an hour or longer, and she shows much more intelligence, and often great ingenuity in the search. This morning I hid a cracker. She looked everywhere she could think of without success, and was evidently in despair when suddenly a thought struck her, and she came running to me and made me open my mouth very wide, while she gave it a thorough investigation. Finding no trace of the cracker there, she pointed to my stomach and spelled "eat," meaning, "Did you eat it?"

Friday we went down town and met a gentleman who gave Helen some candy, which she ate, except one small piece which she put in her apron pocket. When we reached home, she found her mother, and of her own accord said, "Give baby candy." Mrs. Keller spelled, "No—baby eat—no." Helen went to the cradle and felt of Mildred's mouth and pointed to her own teeth. Mrs. Keller spelled "teeth." Helen shook her head and spelled "Baby teeth—no, baby eat—no," meaning of course, "Baby cannot eat because she has no teeth."

JEAN PIAGET

THE LANGUAGE AND THOUGHT OF THE CHILD

WHAT are the conclusions we can draw from these facts? It would seem that up to a certain age we may safely admit that children think and act more ego-centrically than adults, that they share each other's intellectual life less than we do. True, when they are together they seem to talk to each other a great deal more than we do about what they are doing, but for the most part they are only talking to themselves. We, on the contrary, keep silent far longer about our action, but our talk is almost always socialized.

Such assertions may seem paradoxical. In observing children between the ages of 4 and 7 at work together in the classes of the *Maison des Petits,* one is certainly struck by silences, which are, we repeat, in no way imposed nor even suggested by the adults. One would expect, not indeed the formation of working groups, since children are slow to awake to social life, but a hubbub caused by all the children talking at once. This is not what happens. All the same, it is obvious that a child between the ages of 4 and 7, placed in the conditions of spontaneous work provided by the educational games of the *Maison des Petits,* breaks silence far oftener than does the adult at work, and seems at first sight to be continuously communicating his thoughts to those around him.

Ego-centrism must not be confused with secrecy. Reflexion in the child does not admit of privacy. Apart from thinking by images or autistic symbols which cannot be directly communicated, the child up to an age as yet unde-

Reprinted from *The Language and Thought of the Child* by Jean Piaget (New York: Humanities Press, Inc. Meridian Book, 1971), pp. 58–68, by permission of the publisher.

[Piaget's notes deleted.—Ed.]

termined, but probably somewhere about seven, is incapable of keeping to himself the thoughts which enter his mind. He says everything. He has no verbal continence. Does this mean that he socializes his thought more than we do? That is the whole question, and it is for us to see to whom the child really speaks. It may be to others. We think on the contrary that, as the preceding study shows, it is first and foremost to himself, and that speech, before it can be used to socialize thought, serves to accompany and reinforce individual activity. Let us try to examine more closely the difference between thought which is socialized but capable of secrecy, and infantile thought which is ego-centric but incapable of secrecy.

The adult, even in his most personal and private occupation, even when he is engaged on an enquiry which is incomprehensible to his fellow-beings, thinks socially, has continually in his mind's eye his collaborators or opponents, actual or eventual, at any rate members of his own profession to whom sooner or later he will announce the result of his labours. This mental picture pursues him throughout his task. The task itself is henceforth socialized at almost every stage of its development. Invention eludes this process, but the need for checking and demonstrating calls into being an inner speech addressed throughout to a hypothetical opponent, whom the imagination often pictures as one of flesh and blood. When, therefore, the adult is brought face to face with his fellow-beings, what he announces to them is something already socially elaborated and therefore roughly adapted to his audience, *i.e.,* it is comprehensible. Indeed, the further a man has advanced in his own line of thought, the better able is he to see things from the point of view of others and to make himself understood by them.

The child, on the other hand, placed in the conditions which we have described, seems to talk far more than the adult. Almost everything he does is to the tune of remarks such as "I'm drawing a hat," "I'm doing it better than you," etc. Child thought, therefore, seems more social, less capable of sustained and solitary research. This is so only in appearance. The child has less verbal continence simply because he does not know what it is to keep a thing to himself. Although he talks almost incessantly to his neighbors, he rarely places himself at their point of view. He speaks to them for the most part as if he were alone, and as if he were thinking aloud. He speaks, therefore, in a language which disregards the precise shade of meaning in things and ignores the particular angle from which they are viewed, and which above all is always making assertions, even in argument, instead of justifying them. Nothing could be harder to understand than the note-books which we have filled with the conversation of Pie and Lev. Without full commentaries, taken down at the same time as the children's remarks, they would be incomprehensible. Everything is indicated by allusion, by pronouns and demonstrative articles —"he, she, the, mine, him," etc.—which can mean anything in turn, regardless of the demands of clarity or even of intelligibility. (The examination of this style must not detain us now; it will appear again . . . in connexion with verbal explanation between one child and another.) In a word, the child hardly ever even asks himself whether he has been understood. For him, that goes without saying, for he does not think about others when he talks. He utters a "collective monologue." His language only begins to resemble that of adults when he is directly interested in making himself understood; when he gives orders or asks questions. To put it quite simply, we may say that the adult thinks socially, even when he is alone, and that the child under 7 thinks ego-centrically, even in the society of others.

What is the reason for this? It is, in our opinion, twofold. It is due, in the first place, to the absence of any sustained social intercourse among the children of less than 7 or 8, and in the second place to the fact that the language used in the fundamental activity of the child—play—is one of gestures, movement and mimicry as much as of words. There is, as we have said, no real social life among children of less than 7 or 8 years. The type of children's society represented in a classroom of the *Maison des Petits* is obviously of a fragmentary character, in which consequently there is neither division of work, centralization of effort, nor unity of conversation. We may go further, and say that it is a society in which, strictly speaking, individual and social life are not differentiated. An adult is at once far more highly individualized and far more highly socialized than a child forming part of such a society. He is more individualized, since he can work in private without perpetually announcing what he is doing, and without imitating his neighbours. He is more socialized for the reasons which have just been given. The child is neither individualized, since he cannot keep a single thought secret, and since everything done by one member of the group is repeated through a sort of imitative repercussion by almost every other member, nor is he socialized, since this imitation is not accompanied by what may properly be called an interchange of thought, about half the remarks made by children being ego-centric in character. If, as Baldwin and Janet maintain, imitation is accompanied by a sort of confusion between one's own action and that of others, then we may find in this fragmentary type of society based on imitation some sort of explanation of the paradoxical character of the conversation of children who, while they are continually announcing their doings, yet talk only for themselves, without listening to anyone else.

Social life at the *Maison des Petits* passes, according to the observations of Mlles. Audemars and Lafendel, through three stages. Up till the age of about 5, the child almost always works alone. From 5 to about 7½, little groups of two are formed, like that of Pie and Ez (*cf.* the remarks taken down under the heading "adapted information"). These groups are transitory and irregular. Finally, between 7 and 8 the desire manifests itself to work with others. Now it is in our opinion just at this age that ego-centric talk loses some of its importance, and it is at this age . . . that we shall place the higher stages of conversation properly so-called as it takes place between children. It is also at this age . . . that children begin to understand each other in spoken explanations, as opposed to explanations in which gestures play as important a part as words.

A simple way of verifying these hypotheses is to re-examine children between 7 and 8 whose ego-centrism at an earlier stage has been ascertained. This is the task which Mlle. Berguer undertook with Lev. She took down under the same conditions as previously some 600 remarks made by Lev at the age of 7 and a few months. The co-efficient of ego-centrism was reduced to 0.27.

These stages of social development naturally concern only the child's intellectual activity (drawings, constructive games, arithmetic, etc.). It goes without saying that in outdoor games the problem is a completely different one; but these games touch only on a tiny portion of the thought and language of the child.

If language in the child of about 6½ is still so far from being socialized, and if the part played in it by the ego-centric forms is so considerable in comparison to information and dialogue etc., the reason for this lies in the fact that childish language includes two distinct varieties, one made up of gestures, movements, mimicry etc., which accompany or even completely supplant the use of words, and the other consisting solely of the spoken word. Now, gesture cannot express everything. Intellectual processes, therefore, will remain ego-centric, whereas commands etc., all the language that is bound up with action, with handicraft, and especially with play, will tend to become more socialized. We shall come across this essential distinction again. . . . It will then be seen that verbal understanding between children is less adequate than between adults, but this does not mean that in their games and in their manual occupations they do not understand each other fairly well; this understanding, however, is not yet altogether verbal.

RUSSELL LYNES

INTRODUCTION TO NANCY MITFORD'S *NOBLESSE OBLIGE*

"NOTHING stirs us to such a frenzy of shame-faced excitement," Philip Toynbee wrote in the London *Observer* last February, "as a public issue which involves class distinctions. Miss Nancy Mitford's article 'The English-Aristocracy,' was published in *Encounter* only four months ago, but the terms 'U' (upper-class speech) and 'Non-U' are already part of the current literary vocabulary."

There were just a few paragraphs in Miss Mitford's article that started the frenzy. Ostensibly her essay was an attempt by one of the most gifted comic writers of our day to characterize, justify, and define her peers. Miss Mitford is the daughter of a baron, and therefore a "Hon." [1] In order to demonstrate, as she said, that "the upper middle class [2] does not merge imperceptibly into the middle class" she used the researches of Professor Alan S. C. Ross of Birmingham University [3] into the differences of speech which distinguish the members of one social class in England from another. Professor Ross, in whose mind there seems to be little doubt that upper-middle-class usage is superior to middle-class usage, first published his findings in a Finnish philological journal where, presumably, they didn't cause the flutter of a single Finn.

The flutter in the Hon-coop [4] as a result of Miss Mitford's quotations from Professor Ross's learned findings was, however, exceedingly noisy and somewhat acrimonious. The distinctions between "U" and "non-U" language and behavior became something of a national parlor game, a sort of linguistic "How to Tell Your Friends from the Apes." The demand for copies of Miss Mitford's article was so great that *Encounter,* which cannot be called a popular magazine (it is published in London by the Committee on Cultural Freedom, and its literary editor is Stephen Spender), sold out the edition "immediately after publication."

There is little to be gained by adumbrating in this introduction the distinctions between "U" and "non-U" which you will find explored and exploded in the essays in this volume. But the extent and nature of the response which they evoked are entertaining and a hint of the meaning of the terms is necessary to introduce them. It is "U," for example, to say *lavatory paper,* but "non-U" to say *toilet paper* (which makes all Americans non-U), but *writing paper* is "U" and *note paper* is "non-U"; *wealthy* is "non-U" and *rich* is "U." According to Miss Mitford it is profoundly and inexcusably "non-U" to say "Cheers" before drinking, and it is sufficient cause to tear up a letter without reading it if the salutation reads "Dear Nancy Mitford"; Dear Miss Mitford, or Dear Nancy, but Dear Nancy Mitford, never!

Some of the most entertaining and infuriated responses to Miss Mitford's essay are contained in this book, though it is only fair to John Betjeman to note that his verses, "How to Get On in Society," appeared in *Time and Tide* many months before Miss Mitford's piece was published. But there was a great deal that appeared elsewhere, and I would like to quote some of it to justify Mr. Philip Toynbee's use of the word "frenzy."

Graham Greene in a letter to the *Observer* wrote:

Sir,—It is sad to find that by Miss Mitford's exacting standard Henry James was often Non-U in his correspondence. Frequently he followed the "unspeakable usage" of writing to someone as "Dear X X."

Many examples will be found in the last edition of his letters: "Dear Walter Besant," "Dear Auguste Monod," and surely most shocking of all to Miss Mitford, "Dear Margot Asquith."

To this Miss Mitford replied in a letter that also appeared in the *Observer:* ". . . As for Henry James writing 'dear Margot Asquith,' he was an American." Which seems to take care of that, and of us.

Another correspondent to the *Observer* P. B. S. Andrews (how many initials, I wonder, does it take to make one "U"?) wrote:

Sir—Mr. Philip Toynbee has been hunting through a dictionary of quotations in order to rebut Miss Mitford's "extraordinary claim that the U circumlocution, 'looking glass' should be preferred to the apparently non-U 'Mirror.' " There is surely no need to look far for the perfect comment on the difference between Miss Mitford's pretentiousness and what used to be called the King's English:—

King Richard:
An if my word be sterling yet in England,
Let it command a mirror hither straight. . . .
Bolingbroke:
Go some of you and fetch a looking-glass.

[Shakespeare, *Richard II*]

To this Miss Mitford replied: "It is probable that Richard II, like many monarchs, was Non-U."

Malcolm Muggeridge, the editor of *Punch,* who knows a band wagon when he sees one, devoted almost an entire issue of his magazine to Miss Mitford's little tempest. Its cover bore a mock coat of arms and the device "Snoblesse Oblige," and there was a coronet printed at the top of each page of the issue. The book review section (except for a piece on Saint-Simon, who "staggered even Louis XIV by the emphasis he put on rank and pedigree," and one on Lady Waldegrave) was devoted entirely to books by peers reviewed by other peers. In a piece called "Aunt Nancy's Casebook (*Correspondents requiring a private reply should enclose a stamped, embossed envelope*)" were such delightful questions and answers as:

My mother forbids me to use the word "Tuesday"; she says it is common. What can I say instead?

"Morbid," Wylye Valley

Your mother is quite right. Tuesday is a very Non-U word,[5] indicating the day people who stay on after a Friday to Monday fail to leave.

Edna St. Vincent Millay, once a U poet whose reputation has become rather non-U in recent years, wrote inaccurately that there are no islands any more. The charm of the essays in this book is at least partly their insularity. This is a family joke told in family language, but like most good jokes it is easily translated into one's own experience.

Direct translation is, of course, impossible. A boot does not fit a foot that is used to a shoe, or a bonnet a car that is used to a hood. A society that perpetuates an inherited aristocracy, even though it often makes fun of its behavior, does not joke about its ruling class in at all the same way that we joke about our tycoons and politicians and F.F.V's. The tone of voice is as different as the substance of the joke. But the nature of English snobbery on which this book is based is less different from our own than one might think. We live, as these essays demonstrate beyond the shadow of a doubt, in an age of Reverse Snobbism.

There is an anecdote that demonstrates what I mean. Dame May Whitty, the distinguished actress, was shopping in London shortly after the last war and the salesgirl who waited on her was annoyingly indifferent. Dame May said to her: "I suppose you know who I am." The girl said she did. "I suppose you think you're as good as I am," said the actress. "I certainly do," replied the girl. "Then why," asked Dame May, "can't you be civil to your equals?"

We are somewhat more direct in our anti-snobbism, and lest we seem to be snobbish, we tend to affect the language and manners of the people we are with. The chairman of the board, for example, is almost, but distinctly not quite, as slangy and back-slapping as his salesmen when he is at a sales conference, though his manner and his language with his board of directors is dignified, proper, and filled with the U phrases of top management. The reverse is likely to be true of his wife, whose language at a company outing is apt to be rather more U than when she is with her intimates; at such moments she is very conscious of being the wife of the chairman and feels the responsibilities of *noblesse oblige*.

Class mannerisms of speech in America are difficult to pin down; regional differences are so much more important. A Bostonian can get away with mannerisms of speech that would be intolerable in a New Yorker or a Charlestonian. If there are usages that betray class distinctions in America, they are likely to be of the very same kind that Miss Mitford has singled out—the use of a fancy phrase when a simple one does as well.

The analogy is far from exact, but in some respects our Nancy Mitford is Emily Post, who has for many years occupied a kind of quasi-official position as arbiter of U behavior in America. In the latest edition of *Etiquette, the Blue Book of Social Usage,* Mrs. Post in a tone of voice not much different from Miss Mitford's devotes a chapter to U and non-U (under another name) words and phrases. It is called "The Words We Use and How to Choose Them." Here is a sample:

Words and phrases to watch out for

Never say	Say instead
I desire to purchase	I would like to buy
I presume	I suppose
Tendered him a banquet	Gave a dinner for him
Mansion	Big house

Unintentional vulgarities

Lovely food	Good food, or delicious food
Elegant home	Beautiful house or place

In very bad taste	
Formals	Formal clothes
Boy (*when over twenty-one years of age*) [6]	Man
Drapes—*this word is an inexcusable vulgarism*	Curtains *are hung at a window;* hangings *as decoration for walls. It is true,* draperies *would be correct for many loopings or shirrings or pleatings, especially on a woman's dress.*
Corsage	*A word cherished by many, but distasteful to the fastidious who prefer the phrase,* flowers to wear.
Going steady with	*There is no proper equivalent for the phrase because according to etiquette the situation does not exist; no man is given the exclusive right to be devoted to any girl unless engaged to her.*

It is obvious that Mrs. Post plays by the same rules as Miss Mitford. But it is unlikely that anyone who is U in America would think of quoting Mrs. Post as an authority on manners of speech, or, even worse, of deigning to argue with her. "In best—meaning most distinguished—society no one arises, or retires, or resides in a residence," according to Mrs. Post. "One gets up, takes a bath,[7] goes to bed, and lives in a house." It may be a rather circumscribed formula for daily existence, but, as Mrs. Post adds, "Everything that is simple and direct is better form than the cumbersome and pretentious."

As you will see, there is trenchant disagreement among the U's who have contributed to this book, as there will be equally sharp controversy about what is U and non-U in America. Recently a friend in London sent me the clippings from which I have quoted above, and he wrote: "Altogether I think it is fair to say the articles have caused a great deal of light-hearted controversy. I rather think that when they come out in a book the tone of the comment will change. . . . These shibboleths will come to the attention of thousands of people who have been happily talking about serviettes and toilets all their lives without realizing that they were writing themselves off, in certain eyes, as socially benighted, beyond the pale, outcasts. This may make them, justifiably, sore. For my own part," he added, "I think there could be nothing more hopelessly Non-U than to write a book about it."

Chapter 5

THE FAMILY

THE family has always been and continues to be the most important institutional "launching pad" for the individual's journey through society. This has always been so. Cultural anthropologists and other social scientists have therefore had good grounds for focusing attention on the family as a fundamental institution. In recent Western society, however, the family has undergone fundamental transformations. Most of these, directly or indirectly, are the result of the industrial revolution. Probably the most important transformation has been the separation of the family from the processes of economic production and its segregation in that peculiarly modern sphere we call private life.

In the contemporary family the individual withdraws from his "outside" involvements into a protective circle within which, it is assumed, he will find his basic satisfaction as a human being. This transformation has given the family unprecedented meaning, but it has also subjected it to unprecedented stress—all the more so because the modern family has greatly shrunk in size and, for most people, is now effectively reduced to what is commonly called the conjugal nuclear family, that is, the isolated married couple and their (usually few) offspring. Thus, three or four people are supposed to provide unprecedented meaning for each other, a social demand that not surprisingly has been difficult to meet in many cases. The stress has been increased by the separation of the family from economic production. Economically speaking, most families produce nothing together—except expenditures. Many of the instabilities in the modern family (such as divorce rates, the "generation gap," and the like) can be traced to these facts. Within American society there continue to be important differences in family patterns between different classes and other social groupings. In all of them, however, the institution of the family is facing serious problems. In consequence there have recently been not only strong attacks on the family but attempts to innovate it or to find alternatives to it.

Many sociologists—and, of course, many writers of all sorts interested in social mores—have dealt with the different family patterns in different groups and classes in American society. The six selections that follow illustrate some of them. Herbert J. Gans, in his *Urban Villagers,* discusses different family types and then relates the group he was studying, the working class Italian ethnics in Boston, to this typology. J. R. Seeley, R. A. Sim, and E. W. Lossley, in a selection from one of the most influential recent studies of the upper-middle-class family, *Crestwood Heights,* delineate the patterns of that group. The sharp controversy in the late 1960s about the black family in America, triggered by the work on this subject by Daniel P. Moynihan, occasioned Frank Riessman's "In Defense of the Negro Family," a spirited defense of the socializing capacity of the black family. Recent numerous experiments in communal and other countercultural "alternatives" to the family are the subject of Bennett M. Berger, Bruce M. Hackett, and R. Mervyn Millar in *Child-Rearing Practices in the Communal Family,* a report on the new emerging models. Peter L. Berger and Hansfried Kellner in "Marriage and the Construction of Reality" discuss the role of the family and especially the married couple in constructing a world of common meaning for the participants. This selection is followed by a passage from John Updike's novel, *Couples,* which makes a similar point in an entirely different way.

HERBERT J. GANS

FAMILY TYPES AND THE URBAN VILLAGERS

SOCIOLOGISTS and anthropologists generally distinguish between the *nuclear* family, made up of husband, wife, and children but separated from other relatives; and the *extended* family, in which a group of nuclear families and related individuals from several generations act together as a virtual unit. The extended family is found most often in agricultural or hunting societies, where such groups are functioning economic units. The nuclear family is associated with the urban-industrial society in which family members cannot be employed together, and in which, because of rapid social change, cultural differences between the generations and the resulting conflict between young and old make life together difficult.

West Enders fall squarely between the two ideal types. The nature of the family, however, can best be understood if one can distinguish between households and families. West End households are nuclear, with two qualifications. Married daughters often retain close ties with their mothers and try to settle near them. They do not share the same apartment, because however close the ties, there are differences between the generations—or at least between husband and mother-in-law—that are likely to create conflict. Some households take in close relatives who would otherwise be alone, especially unmarried brothers, sisters, or even cousins, because of feelings of obligation, love, and the desire to reduce the loneliness of the single person. Pitkin, observing a similar pattern in his study of a Southern Italian village, described this family as *expanded.*[1] Since rents were low in the West End, unmarried siblings often had their own apartments. Much of their spare time, however, was spent with married brothers or sisters, and they often participated in child-rearing as quasi-parental aunts and uncles.

Reprinted from *The Urban Villagers* by Herbert J. Gans (New York: Free Press, 1962), pp. 45–57, by permission of The Macmillan Company.
[Some of Gans's notes deleted.—Ed.]

But although households are nuclear or expanded, the family itself is still closer to the extended type. It is not an economic unit, however, for there are few opportunities for people to work together in commercial or manufacturing activities. The extended family actually functions best as a social circle, in which relatives who share the same interests, and who are otherwise compatible, enjoy each other's company. Members of the family circle also offer advice and other help on everyday problems. There are some limits to this aid, however, especially if the individual who is being helped does not reciprocate. For example, one family I met in the West End had a member who suffered from spells of deep depression. The family circle visited him frequently to cheer him up, to give advice, and to urge him to join in family activity, but when he failed to accept their ministrations, his relatives became impatient. They continued to visit him, but did so grudgingly. As one of his relatives put it: "He has no interests, why should anyone care about him?"

The extended family system is limited generationally, for relationships between adults and their parents—the immigrant generation—are fewer and less intimate than those between adults of the same generation.[2] Visits with parents are exchanged, but parents are generally not part of the continuing social life of the family circle. Widowed parents do not live with their children if other alternatives are available. While old people are allowed to function as grandparents, they are freely criticized for spoiling their grandchildren, or for insisting on outmoded ideas. Compared with the middle class, in fact, the older generation receives little respect or care. Social workers in the West End told of families who sent old people to welfare agencies even when they could afford to support them, although this is not typical. The lack of respect toward the older generation is especially noticeable among children, who tease and insult old people behind their backs, including their own grandparents.

The only exception to this pattern is the previously mentioned tie between mother and married daughter, and a more infrequent one between mother and unmarried son. Even so, mothers tend to assist rather than guide their married daughters. They help out in the household and in the rearing of children, but they have neither the power nor authority of the "Mum," the ruling matriarch of the English working-class family.[3]

The expanded family that I have described is common to both the routine-seekers and the action-seekers. The analysis of family life in the following pages will deal principally with male-female and parent-child relationships among the routine-seekers. As already indicated, most West Enders are routine-seekers—or become so when they marry—and families in which the husband is an action-seeker are relative few in number.

MALE-FEMALE RELATIONSHIPS

I pointed out [earlier] that West Enders socialize primarily with people of their own age and sex, and are much less adept than middle-class people at heterosexual relationships. In many working-class cultures, the man is away from the house even after work, taking his leisure in the corner taverns that function as men's clubs. But, since the Italian culture is not a drinking one, this is less frequent among West Enders. Consequently, much of their segregation of leisure takes place within the home: the women sit together in one room, the men in another. Even when everyone gathers around the kitchen table, the men group together at one end, the women at the other, and few words are exchanged between them. Men are distinctly uncomfortable in the company of women, and vice versa, but the men find it harder to interact with the women than the women with the men. At social gatherings I attended, whenever women initiated conversations with men, the men would escape as quickly as possible and return to their own group. They explained that they could not keep up with the women, that the women talked faster and more readily, turning the conversation to their own feminine interests and that they tried to dominate the men. The men defended themselves either by becoming hostile or by retreating. Usually, they retreated.

The men's inability to compete conversationally with women is traditional. Second-generation Italians grew up in a patriarchal authority system with a strictly enforced double standard of behavior for boys and girls. The boys were freer to indulge their gratifications than the girls. In order to be able to do what they wanted, the girls thus had to learn early how to subvert the male authority by verbal means—"how to get around the men"—and what they did not learn elsewhere, they learned from the mother's wile in getting her way with her husband. As will be noted subsequently, the father enforces discipline and administers punishment; he does not need to talk. The mother can influence her husband only by talking to him, reinterpreting the child's deeds so that he will not punish the child any more than she feels desirable. Talk is the woman's weapon for reducing inequities in power between male and female.

With unrelated women, the male reaction sometimes resembles fear. The men are afraid that the women will overpower them through their greater verbal skill, and thus overturn the nominal dominance of the man over the woman. In a culture that puts great stress on what David Riesman has called "male vanity," placing a man in an inferior position is thought to impugn his masculinity. In other situations, his fear is based on an opposite motive, that undue contact with a woman may produce sexual desire that cannot be satisfied. Among West End men, the unrelated woman is conceived mainly as a sexual object. At the same time, the strict double standard makes her sexually inaccessible. Consequently, while men are freely aggressive, both sexually and verbally, with a "bad" girl, they must control themselves with an inaccessible "good" girl. Among unmarried people, for example, when a "good" girl enters an all-male group, profanity and sexual talk are immediately halted, and the men seem momentarily paralyzed before they can shift conversational gears.

What the men fear is their own ability at self-control. This attitude, strongest among young, unmarried people, often carries over into adulthood. The traditional Italian belief—that sexual intercourse is unavoidable when a man and a woman are by themselves—is maintained intact among second-generation West Enders, and continues even when sexual interest itself is on the wane. For example, I was told of an older woman whose apartment was adjacent to that of an unmarried male relative. Although they had lived in the same building for almost twenty years and saw each other almost every day, she had never once been in his apartment because of this belief.

As a result, the barriers between the sexes are high, and they are crossed mainly by deviant types. The only men who carry on a consistent social relationship with women are "ladies' men," who are in varying degrees effeminate. Likewise, the only women who carry on such a relationship with men are likely to be those with strong masculine tendencies. Some of my neighbors used to anger their wives by sharing sexually connotative jokes and indulging in sexual banter with a young woman who appeared to be masculine in some of her ways. They were able to do this because the girl did not represent a potential sexual object. Although she still saw herself as a potential bride, and expressed great, though false, embarrassment at the men's behavior, she was a safe target for the expression of the sexual hostility of the males toward the women. At the same time, she never discouraged these attacks because they were the only kinds of advances she was likely to get from men, and perhaps because she was masculine enough to be able to enjoy the joking. West End women indulged in sexual banter, too, but only among themselves.

The male fear of "good" girls was vividly portrayed one evening when a group of men in their twenties were pursuing another man who had slashed one of the group in a tavern brawl. The man ran into his apartment building, leaving the entrance blocked by his mother and his sister. Armed with sticks, the men pushed the mother—a woman in her sixties—out of the way. The girl, however, was able to stop them from coming into the building. While they did attack her verbally, they did not touch her, and then, promising to carry out justice at some other time, they eventually withdrew.

HUSBAND-WIFE RELATIONSHIPS

The general pattern of male-female interaction carries over into the relationship between husbands and wives. The barriers between male and female are translated into a marital relationship that can be best described as *segregated,* as distinguished from the *joint* relationship that characterizes the middle-class family.[4] Bott's description of this phenomenon among English families applies to the West Enders as well:

> Husband and wife have a clear differentiation of tasks and a considerable number of separate interests and activities. They have a clearly defined division of labor into male tasks and female tasks. They expect to have different leisure pursuits, and the husband has his friends . . . the wife hers.[5]

While the husband's main role is breadwinning, the wife is responsible for all functions concerning home and child, even the finding of an apartment. Women speak of the family apartment as "my rooms"; husbands speak to wives about "your son." Responsible for overseeing the rearing of the child, the mother may even administer discipline, although this is usually left to the father when he comes home from work.

On the surface, this pattern differs little from the middle-class one. In middle-class society, as in most societies, most of the tasks connected with home and child are also the mother's duty. In the West End, however, the boundary between tasks is quite rigid. As one West End housewife put it, "when my husband comes home with the pay, I can't ask him to help in the house." Whereas the middle-class husband expects to help out in the household, and to share the responsibilities of child-rearing, the West End husband does not expect to do so, and will help out only in unusual situations. It is not that he rejects the possibility of joint action; it is simply something outside of his experience.

The segregation of functions is more clearly visible in the emotional aspects of the husband-wife relationship. Although young West Enders are as much concerned with romantic love as other Americans, and although couples do marry on the basis of love, the marital relationship is qualitatively different from that of the middle class. Not only is there less communication and conversation between husband and wife, but there is also much less gratification of the needs of one spouse by the other. Husbands and wives come together for procreation and sexual gratification, but less so for the mutual satisfaction of emotional needs or problem solving. Among my neighbors was a bachelor. When I asked one of his relatives whether he would ever marry, it was explained that he would probably not, since his work brought him into contact with women who satisfied his sexual needs. In addition, his frequent visits to his married sister's household provided the opportunity for the little relationship with children expected of the man.

Thus the marriage partners are much less "close" than those in the middle class. They take their troubles less to each other than to brothers, sisters, other relatives, or friends. Men talk things over with brothers, women with sisters and mothers; each thus remains on his side of the sexual barrier.

I can best illustrate the nature of the marital relationship anecdotally. Most of the small stores in the West End were family enterprises. Two of the Italian stores that I frequented were each run by a man and a woman who I knew to be related, and who I thought were either brother and sister, or cousins. In both cases, my assumption was based on the man's lack of interest in the woman's children when they were in the store, as well as her total lack of interest in his business. One day, when I raised the question of relocation plans in one of the stores, the woman replied curtly, "I don't care about the store, it's his; it's his business to make a living, not mine." This was said matter of factly, without a trace of anger or malice. If the woman did interfere in the man's activities, especially in one of the many extracurricular ones that commonly took place in small stores, she was rebuked and told to mind her own business. In both stores, the lack of communication convinced me that the relationship was that of two individuals who were brought together by economic necessity and by kinship ties, but who otherwise were not close. My assumption that they were siblings or cousins had been based on my middle-class expectations, and I was much surprised to discover that, in both places, they were husband and wife.

The segregated conjugal pattern is closely associated with the extended family, for the functions that are not performed by husband and wife for each other are handled primarily by other members of the extended family. In a society where male and female roles are sharply distinguished, the man quickly learns that, on many occasions, his brother is a better source of advice and counsel than his wife. The recruitment of the family circle on the basis of compatibility enhances this pattern, for those relatives who provide helpful advice are also likely to be compatible in other ways, and thus to be part of the circle.

Although the middle-class observer may find it hard to imagine the absence of the marital closeness that exists in his own culture, this pattern has been functional for people like the West Enders. Until recently, they have lived under conditions in which one of the spouses could easily be removed from the household by mental or physical breakdown, or premature death. In years past more than today, job insecurity, occupational hazards, and poor living conditions meant that every wife might have to reckon with the incapacitation or removal of the breadwinner—even though male desertion was and is rare. Likewise, illness or death in childbirth might remove the woman before her time. The lack of closeness, however, makes it easier for the remaining parent to maintain the household, raise the children—usually with the help of members of the extended family—and to overcome the emotional loss of the spouse. For example, a West Ender I met had lost her husband, with whom she was said to have been exceptionally close, a few months earlier. With her children married, however, she began to think of herself as a single woman again, and participated in social activities with a number of unmarried women of whom she said—only half in jest—she would join in "man-catching" endeavors. At the same time, however, she was able to talk about events which she had shared with her husband as if he were still alive. She did not have to shut him out of her mind in order to overcome the pain of her loss.

Although the segregated conjugal pattern is clearly dominant among the West Enders, signs of its eventual disappearance are making themselves felt. Between the first and second generation, the major change has been that of bringing the men into the house for their evening activities. While Italians have never been frequenters of neighborhood taverns, the immigrant generation did set up club houses for card playing and male sociability that kept some men away from the house after work. These have disappeared, however, and, as already noted, second-generation men now segregate themselves from the women inside the home, and spend only one or two evenings a week in activities "with the boys." The women also have begun to conceive of their husbands as helping them in the home, although they are not yet ready to insist or even to ask for their aid. The move to the suburbs is probably one indication of the ascendancy of the wife to greater equality, for in these areas, where the joint conjugal pattern is dominant, it is somewhat harder for the man to maintain the old pattern. West Enders occasionally mentioned couples in which the wife, shortly after marriage, had persuaded the husband to move out of the West End in order to "get him away from the boys." But this does not always work; some men, even after twenty years of living elsewhere, return to the West End for evening visits to male friends. One man I knew, who used to come back for male companionship, was kidded about being dominated by his wife, and this has driven him further away from his old peers.

CHILD-REARING

In the West End, children come because marriage and God bring them. This does not mean that West Enders believe children to be caused by God, but that the Catholic church opposes birth control, and that this is God's wish. There is some planning of conception, either through the use of the church-approved rhythm method, or, more rarely, through contraception. But while the sale of contraceptives is illegal in Massachusetts, this does not prevent their acquisition. West Enders, however, do reject their use—or at least talking about their use—on religious grounds. The major method of family planning seems to be ex post facto. Should the wife become pregnant after a couple has had what they deem to be enough children, she may attempt to abort herself, using traditional methods that she has learned from other women. If the attempt fails, as it probably does in many cases, the new child is accepted fatalistically—and usually happily—as yet another manifestation of the will of God. Even so, families are smaller among second-generation Italians than among their parents. The couple with six to eight children, which seems to have been prevalent among the first generation, now has become a rarity. A large family is still respected, however, because children themselves are still highly valued.

The fact that children are not planned affects the way in which parents relate to them, and the methods by which they bring them up. Indeed, American society today is characterized by three types of families: the *adult-centered*—prevalent in working-class groups—run by adults for adults, where the role of the children is to behave as much as possible like miniature adults; the *child-centered*—found among families who plan their children, notably in the lower middle class—in which parents subordinate adult pleasures to give the child what they think he needs or demands; and the *adult-directed*—an upper-middle-class pattern—in which parents also place lower priorities on their own needs, in order to guide the children toward a way of life the parents consider desirable.[6]

In the lower middle class of the present generation, husband and wife are likely to have finished high school, perhaps even the same one. This shared background helps them to communicate with each other, and creates some common interests, although much spare time still is spent with peers of the same sex. The most easily shared interest is the children, and the parents communicate best with each other through joint child-rearing. As a result, this family is child-centered. Parents play with their children—which is rare in the working class—rear them with some degree of self-consciousness, and give up some of their adult pleasures for them. Family size is strongly influenced by educational aspirations. If the parents are satisfied with their own occupational and social status, and feel no great urgency to send their children to college, they may have as many children as possible. For each child adds to their shared enjoyment and to family unity—at least while the children are young. Sometimes, the child will dominate his parents unmercifully, although child-centered parents are not necessarily permissive in their child-rearing. Rather, they want the child to have a happier childhood than they experienced, and will give him what they believe is necessary for making it so. One of their child-centered acts is the move to the suburb, made not only for the child's benefit, but also to make their child-rearing easier for themselves, and to reduce some of the burdens of child-centeredness. They give the child freely over to the care of the school, and to organizations like the Scouts or Little League, because these are all child-centered institutions.

Among college-educated parents, education and educational aspirations shape family life. College education adds immeasurably to the number of common interests between husband and wife, including activities other than child-rearing. Consequently, these parents know what they want for their children much more clearly than does the child-centered family, and their relationship to the children is adult-directed. Child-rearing is based on a model of an upper-middle-class adulthood characterized by individual achievement and social service for which parents want the child to aim. As a result, the child's wants are of less importance. Such parents devote much time and effort to assuring that the child receives the education which will help him to become a proper adult. For this purpose, they may limit the size of their families; they will choose their place of residence by the quality of the school system; they will ride herd on the school authorities to meet their standards; and, of course, they will exert considerable pressure on the children to do well in school.[7]

The West End family is an adult-centered one. Since children are not planned, but come naturally and regularly, they are not at the center of family life. Rather, they are raised in a household that is run to satisfy adult wishes first. As soon as they are weaned and toilet-trained, they are expected to behave themselves in ways pleasing to adults. When they are with adults, they must act as the adults want them to act: to play quietly in a corner, or to show themselves off to other adults to demonstrate the physical and psychological virtues of their parents. Parents talk to them in an adult tone as soon as possible, and, once they have passed the stage of babyhood, will cease to play with them. When girls reach the age of seven or eight, they start assisting the mother, and become miniature mothers. Boys are given more freedom to roam, and, in that sense, are treated just like their fathers.

But while children are expected to behave like adults at home, they are able to act their age when they are with their peers. Thus, once children have moved into their own peer group, they have considerable freedom to act as they wish, as long as they do not get into trouble. The children's world is their own, and only within it can they really behave like children. Parents are not expected to supervise, guide, or take part in it. In fact, parent-child relationships are segregated almost as much as male-female ones. The child will report on his peer group activities at home, but they are of relatively little interest to parents in an adult-centered family. If the child performs well at school or at play, parents will praise him for it. But they are unlikely to attend his performance in a school program or a baseball game in person. This is his life, not theirs.

Schoolteachers and social workers who dealt with West End children often interpreted the family segregation patterns from a more child-centered perspective, and assumed that the parents had lost interest in their children or were ignoring them. But this is not the case. At home, they are still part of the family circle, and continue to play their assigned roles. In fact, West End children continue to attend family gatherings at ages at which middle-class children are usually excused from them. They also sit in on social gatherings from which middle-class children might be excluded altogether. But then West Enders do not make the same distinction between family and social gatherings, since they usually involve the same people.

J. R. SEELEY, R. A. SIM, AND E. W. LOOSLEY

FUNCTIONS OF THE CRESTWOOD HEIGHTS FAMILY

THE Crestwood Heights family, although it varies in organization, composition, and orientation from families in other cultures and from the family of

Reprinted from *Crestwood Heights* by J. R. Seeley, R. A. Sim, and E. W. Loosley (New York: Basic Books, 1956), pp. 162–64, by permission of Basic Books, University of Toronto Press, and Constable & Co., Ltd.

other periods in the history of Western civilization, regulates, as the family always has, sexual and affectional expression in the patterns approved by the society. The life-long relationship between one man and one woman for the procreation of children is the basis of the family in Crestwood Heights. Given the highly mobile nature of the population, the relative lack of secondary institutions capable of absorbing unmarried individuals, and the absence of kin, the tie between husband and wife indeed becomes pre-eminently important, providing, as it does, the one enduring human relationship in the society. Even child-rearing, central as it is in Crestwood, is not so lasting a commitment, since, given the lengthened life-span of the individual, it now occupies relatively few years.

Child-rearing is still considered as essentially the responsibility of the family, and it is the father and mother who must legally assume the child's economic support. And no matter to what degree one parent or both may rely upon institutions and people outside the immediate family for guidance in the child-rearing process itself, it is the parents alone who must decide whether to accept or to reject the proffered services. Strong pressures will, of course, be brought to bear to ensure that they make the "right" decision, but technically it is they who decide.

The family, too, remains the primary social unit in which individuals ideally relate to each other, first of all as human beings, free to express feelings of love and affection, anger or hostility. It is the family which ideally gives emotional security, enabling the individuals comprising it to play their cultural roles in the larger society, and which provides, if they are provided, the models for identification without which the psycho-sexual development of the child cannot be assured. Thus a family is both the biological and psychological nexus of the society.

Most Crestwooders value highly the privilege of functioning individually in the institutions of the culture. It is the economic and social status of the family which allows them so to function. Economic and social status is closely tied in with the ownership of property. The house, then, as has been already pointed out, in addition to being a powerful symbol of material prosperity, stands for family solidarity and strength; from this base of home and male career, which are almost inseparable, the family can articulate with the community. The family, together with the male career, provides the rationale and the validation of the man's role; and the family provides the chief content of the woman's role.

Finally, the family in Crestwood Heights, since it is ultimately responsible for child-rearing, must, in some sense, "transmit the culture." This transmission of the culture (or, rather, the cultures), as in the case of another important institution, the school, is no longer to any great degree a process of passing on fixed items of tradition; where the family is concerned, the process seems rather to result in the freeing to a considerable extent of individual members to create new cultural and family patterns, independent of those of the family of orientation, and often radically different. Parents and children alike must learn continually to accept forms of social behavior unlike those previously known and practised.

One highly intelligent and understanding Crestwood mother, when asked what were the professional or business aspirations of her teen-age sons, at first said she didn't know, and only remembered, under questioning, that the boys *had* expressed certain wishes. She took a day to consider the matter; then added further details after "checking" with her husband and sons. The family owns a business which could be inherited by the children. The father had entered the firm (owned by his family) during the Depression, renouncing the higher academic career for which he had been preparing.

This mother's ignorance of her children's ambitions and the lack of pressure on the children to enter the family business suggest the typical "freeing" of the children to develop new patterns of their own—"within limits, of course." Freedom for the children, as in this instance, can also mean freedom for the parents, who need not now feel constrained to keep the business after the father's retirement. Thus a relatively inexpensive investment in a university education for both sons would allow father and mother later to liquidate their assets for their own old age. New patterns of living would then be created for all family members.

The Crestwood family, as we have seen in other contexts, is definitely oriented to the future, which, apart from its likely disparity with the past, is felt as unknown and unpredictable. The past tends otherwise to be obliterated from the collective thinking of the family. Consciously, the future is optimistically viewed; and the task of the family is to equip the child as effectively as possible in the present with all available means for his later solitary climb to better and more prosperous worlds lying far ahead in time. This passionate optimism tends to be of an essentially catastrophic type—useless as belief and motive force unless *all* its conditions are realized. That the chances for a total realization are slight indeed, is a fact which many Crestwooders cannot easily accept. But the future nevertheless beckons with sufficient force that the parental generation, if it seriously hinders the child's "upward" progress, must be virtually abandoned; this is well understood by both parents and children. Only the promise of continuous upward social mobility (or, at the very least, continually validated status at the present level) can nerve the Crestwood Heights family to its obligations; and for its members to feel the full poignancy of the separation to which it, as a family, is dedicated, might well wreck the whole precarious structure. Such a family is peculiarly vulnerable to outside circumstance. A severe economic depression, for example, would have immediate and serious repercussions which the Crestwood family, given the primary functions outlined, has, it would seem, but few inner resources to meet.

FRANK RIESSMAN

IN DEFENSE OF THE NEGRO FAMILY

I am convinced that Daniel Moynihan is sincerely opposed to discrimination and his report is intended to document its negative consequences for the Negro. From my point of view, however, the report represents a highly inappropriate approach to the development of programs and policy to fulfill the rights of the Negro. —F.R.

The Moynihan Report employs supposed inadequacies of the Negro family as an explanatory tool for understanding why the Negro has not taken his place fully in the economic structure of our society. Presumed family pathology, particularly illegitimacy, is said to be an important cause for the underemployment and inadequate education of the Negro.

The questions raised by Moynihan and many others at the White House Conference on Civil Rights are crucial ones: Would the removal of all forms of discrimination enable the Negro to take his place economically in American society? Has the Negro been so damaged by discrimination that even if these barriers were removed, he could not be adequately employed and educated? Perhaps full employment and the removal of discrimination would bring the Negro into the mainstream far more rapidly than is assumed. But one thing seems clear: if Negroes are offered more of the same in the way of education, training and employment, they will be less than responsive. They are not terribly interested in training which does not lead to jobs or which leads at best to dead-end menial jobs. They are not attracted to the summer programs of the busy-work, anti-riot type. And they have not been receptive to "compensatory education" which constantly stresses their deficits.

The basic defect in the Moynihan thesis is a *one-sided presentation of the consequences of segregation and discrimination*. That damage has been done to the Negro as a result of discrimination cannot and should not be denied. But the Negro has responded to his oppressive conditions by many powerful coping endeavors. He has developed many ways of fighting the system, protecting himself, providing self-help and even joy. One of the most significant forms of his adaptation has been the extended, female-based family.

To overlook this adaptation and instead to emphasize one-sidedly the limiting aspects and presumed pathology of this family is to do the Negro a deep injustice. And it is most inappropriate to attempt to involve people in change by emphasizing some alleged weakness in their make-up. People are much

Reprinted from *Dissent* (March–April 1966): 266–67 by permission of the author and the publishers.

more readily moved through an emphasis on their strengths, their positives, their coping abilities. Thus, Dr. John Spiegel, in providing psychotherapy for low-income populations, is most concerned to work with the extended family rather than ignore it or stress its difficulties.

Conceptualizations developed in psychiatry have important bearing on this issue of strength and weakness. For example, it has come to be recognized that *mental health and mental illness can co-exist in the same individual—that a person may have considerable pathology and at the same time have considerable strength;* one is not the inverse of the other. In other words, it is not accurate to assume that because an individual has more pathology, he has less strength or health. Health and pathology are two continuums, albeit overlapping ones. This concept has enormous implications for social action because it draws attention to the need for concentrating on the health-producing aspects of an organism or group.

Moynihan's stress on pathology and deficit is not unique. The HARYOU Report similarly stresses various indices of pathology in Harlem and fails to accent the strengths, the cohesion, etc. It is thus no accident that the Haryou Report misinterprets the low rate of suicide among the Negro population. If it followed the classic Durkheim theory, it would note that a low rate of suicide is generally an index of the *cohesion* of a group.

THE LIMITATIONS OF "COMPENSATORY EDUCATION"

In the field of education there is a widely heralded, although essentially unsuccessful approach, which, like Moynihan's one-sidedly attempts to develop a theory of action based on deficits. This is the "compensatory education" thesis. It does not build on the action style, the cooperative (team) learning potential, or the hip language of the poor. Instead, "compensatory education" stresses deficits and attempts to build an entire program on overcoming these deficits.[1] Like the Moynihan family approach, it is one-sided in principle and doomed to failure in practice.

As long ago as 1955 I reported that Negro families in large numbers stated that education was what *they had missed most in life and what they would like their children to have.* These matriarchal Negro families are very pro-education, although they have long criticized I.Q. tests, all-white readers, condescending PTA's and together with their children have indicated a strong desire for a livelier, more vital school. Their surprisingly tough, masculine youngsters would like a more male-oriented school. This can be achieved by hiring large numbers of males as non-professional teacher aides, recreation aides, parent education coordinators. Since much of the learning in the family comes less from the parents and more from the brothers and sisters and friends on the street, this method of peer learning might be well adapted in the school itself. (Lippitt has demonstrated how disadvantaged 6th grade youngsters helped 4th grade youngsters for a short time during the week and both groups dramatically improved in their performance!)

If one wants to improve the educability of the Negro, it would seem much more relevant to stress changes in school practice and to develop this practice so that it is more attuned to the style and strengths of the population in question rather than to emphasize the reorganization of the family. The latter emphasis seems much more indirect and less likely to produce the desired result. Similarly in the employment area, it would appear more appropriate to develop meaningful jobs for large numbers of low-income Negroes—e.g., non-professional jobs, where the job is provided first and the training built-in, rather than being concerned about the family. In the Howard University Community Apprentice Experiment, it was demonstrated that highly delinquent functionally illiterate Negro youngsters were able to function in non-professional jobs when the training was built into the job. Their life patterns and work patterns were markedly changed as a result of this new non-professional job experience even though there were no changes in their family pattern.

The emphasis on the deficits and damages of the Negro people kept sociologists from predicting the powerful Negro upheaval that developed into the Civil Rights Movement. This behavior could not be accounted for by an emphasis on the weaknesses, deficiencies and supposed lack of bootstraps. The uprising evolved from the Negro's *strength,* his protest, his anger, and when the conditions were ripe, these traits produced a powerful movement.

In response to the deficiencies of the system the Negro has developed his own informal system and traditions in order to cope and survive. Storefront churches, the extended family, the use of the street as a playground, the block party, the mutual help of siblings, the informal know-how and self-help of the neighborhood, the use of peer learning, hip language, the rent strike and other forms of direct social action are just a few illustrations.

The community action phase of the anti-poverty program is attempting to build on these positive traditions; thus the use of neighborhood service-center storefronts and non-professional neighborhood helpers are strategic features of the new community action program.

Only by calling upon these traditions can the Negro move into the mainstream of our society and, not incidentally, can our society benefit enormously from incorporation of some of these traditions. It is the Negro who is basically challenging our educational system and producing the demand for changes in educational technology and organization that will be of benefit to everyone.

There is increasing evidence that the integrated education drive is actually powering educational benefits for *all* children, not only Negro children. For example, the great hue and cry that arose regarding the "segregated," white face, white theme "Dick and Jane" readers, has led to the development of a variety of new "urban" readers that appear not only to improve the reading ability of the Negro children but also the white children's reading. Even more dramatic, the *Wall Street Journal* (Jan 20, 1965) reports that in approximately 10% of *Southern* schools where some desegregation has taken place, not only do the Negro pupils improve rapidly but the white youngsters appear to advance also.

It is only through full recognition of strengths that the condescension of welfarism can be avoided. Because if the have-nots have nothing—no culture, no strength, no anger, no organization, no cooperativeness, no inventiveness, no vitality—if they are only depressed, apathetic, fatalistic and pathological, then where is the force for their liberation to come from?

BENNETT M. BERGER, BRUCE M. HACKETT,
AND R. MERVYN MILLAR

FAMILY STRUCTURE AND SEXUAL RELATIONS IN THE COMMUNAL FAMILY

EVERYTHING we have said about the children of the communes occurs in the context of hippie relationships and family structures, and it is important to understand these, not only because they are the most palpably real aspect of the research scene but because they contain the seeds of the potential futures of the commune movement.

The most important single feature of hip relationships is their fragility. We mean by this not that many of the relationships don't last; quite the contrary. In several of our more stable communes couples have been "together" as long as the commune has existed (two to three years), and sometimes longer. We mean, rather, that there tend to be few if any cultural constraints or structural underpinnings to sustain relationships when and if they become tension-ridden or otherwise unsatisfying. The uncertainty of futures hovers over hip relationships like a probation officer, reminding the parties of the necessary tentativeness of their commitments to each other.

Very few nuclear units, for example, are legally married; neither the men nor the women have the kinds of jobs that bind them to a community; in

Reprinted from *Child-Rearing Practices in the Communal Family,* Progress Report on Communal Child Rearing, Department of Sociology, University of California, Davis, 6971, by Bennett M. Berger, Bruce M. Hackett, and R. Mervyn Millar, pp. 17–23 by permission of the authors. This study was supported by grant #MN-16579-03 from the National Institute of Mental Health, Education, and Welfare.

other respects their investments in the environmental locale or its institutions are minimal. Like many of their parents (whom theorists have suggested have been highly mobile—a hypothesis which we will test in our interviewing), they move around a great deal, getting into and out of "intimate" relations rather quickly through such techniques as spontaneous "encounter" and other forms of "upfrontness." And above and beyond these, there is a very heavy emphasis on "present orientation"—a refusal to *count on* futures as a continuation of present arrangements—and a diffuse desire to remain "kids" themselves in the sense of unencumberedness, a freedom *from* the social ties that constrain one toward instrumental action.

Yet despite the fact of (and the attitudinal adjustment to) the fragility of relationships, there are romantic images also superimposed. Although the fragility of old man—old lady relationships is a fact, communards of all sorts are generally reluctant to believe in a future of serial monogamy. Many communards, particularly the women, hope for an ideal lover or a permanent mate but tend to have not much real expectation that it will happen. Instead, compensatory satisfactions are found in the *image* of the communal family and household, always full of people, where a group of brothers and sisters, friends as kin, spend all or most of their time with each other, working, playing, loving, rapping, "hanging out"—where wedding bells, far from breaking up the old gang, are themselves so rare that they are occasions for regional celebrations of solidarity when they do ring out.

Where it exists, it is the fact of communal solidarity which functions as the strongest support for fragile relations among couples. For when the communal scene is a wholesome and attractive one, as it sometimes is, couples whose relationship is very unstable may elect to stay together in order to share those benefits rather than threaten them by breaking up.

But in spite of the fragility of relationships in a system which defines futures as uncertain and in an ideology emphasizing spontaneity and freedom, heterosexual couples are the backbone of most communes, urban or rural, creedal or not. They seem more stable and dependable as members than single people do, if only because their search for partners is ended, even if that ending is temporary. The temporary character of the relationships is more pronounced in urban communes, both, we believe, because the very presence of couples in rural communes is itself generally evidence of more stable commitment, and because of the higher probability in urban scenes of meeting another man or woman who is ready and willing to enter into a close relationship at little more than a moment's notice.

When a couple has a child, their mobility is reduced somewhat, of course, even when the child is the product of a previous union of either the female or male. But only "somewhat," because of the importance of what we call the "splitting" phenomenon, particularly as it applies to men. We mentioned previously that children (especially very young ones) "belong" to their mothers, and that norms *requiring* paternal solicitude for children are largely absent.

What this means is that fathers are "free"—at the very least free to split whenever they are so moved. Since they are not "legally" fathers (even if they biologically are) they have no claims on the child, and since there is generally a strong communal norm *against* invoking the legal constraints of straight society (i.e., calling the police), fathers have no obligation to the child that anyone is willing to enforce. Moreover, no norm takes priority over the individual's (particularly the male's) search for himself, or meaning, or transcendence, and if this search requires father's wandering elsewhere "for a while," there is nothing to prevent it.

One consequence of this family pattern is the frequency of woman-with-child (and without old man) in many of the communes we have studied—although this occurs as often as a result of the woman-with-child arriving on the commune scene that way as it does as a result of her partner "splitting." A situation like this does not typically last a long time in any commune we have studied, although it was present in almost all of them. Even when the women involved say they prefer celibacy, there is some doubt that they actually do. One afternoon in a tepee, three young women (without men) with infants on the breast agreed that they welcomed a respite from men, what with their bodies devoted almost full time to the nursing of infants. Within a week, two of them had new old men and the third had gone back to her old one. Celibacy or near celibacy occurs only in those creedal communes whose doctrines define sexual activity as impure or as a drain on one's physical and spiritual resources for transcendence.

But although celibacy is rare and although couple relations are fragile, this should not be taken to mean that sex is either promiscuous or disordered. At any given time, monogamous coupling is the norm in all the communes we studied closely; in this respect hippies tend to be more traditional than the "swingers" and wife-swappers one reads about in the middle class. Although there are communes whose creed requires group marriage (in the sense that all the adults are regarded as married to all the others, and expected to have sexual relations with each other), we have not studied any of these at first hand. But even in communes where coupling is the norm, there seems to be evidence of a natural drift toward group marriage—although it may still be ideologically disavowed. For one thing, when couples break up in rural communes, it is as likely as not that each will remain on the land; and this occurs frequently in urban communes too. Without a drift toward group marriage, situations like this could and do cause great communal tensions which threaten the survival of the group. Whereas, on the other hand, a not uncommon feature of communes is a situation in which over a long period of time, many of the adults have had sexual relations with each other at one or another point between the lapses of "permanent" coupling. Under these conditions, group marriage can seem like a "natural" emergence rather than unnaturally "forced" by a creed—a natural emergence which, by gradually being made an item of affirmed faith, can conceivably solve some of the problems and ease some of the tensions generated by the fragility of couple relations and the break-ups which are a predictable result of them. Broken-up couples may still "love" each other as kin, under these conditions—even if they find themselves incapable of permanently sharing the same tent, cabin, or bed, an incapacity more likely to be explained astrologically than any other way. (Astrology is used to explain "problems" with respect to children and intimate relations between couples.) [1]

But the widespread presence of women-with-children as nuclear units in the communes is not merely an artifact of the splitting of men or an expression of the belief of hip parents in the unwisdom of staying together "for the sake of the child." The readiness of hip women to bear the child even of a "one-night stand" is supported by social structures which indicate its "logic." Unlike middle-class women, for example, a hippie female's social status does not depend upon her old man's occupation; she doesn't need him for that. The state is a much better provider than most men who are available to her. Having a baby, moreover, helps solve an identity problem by giving her something to do. An infant to care for provides more meaning and security in her life than most men could. And in addition, these women are often very acceptable to communes as new members. They are likely to be seen as potentially less disruptive to ongoing commune life than a single man; they are likely to be seen as more dependable and stable than a single man; and these women provide a fairly stable source of communal income through the welfare payments that most of them receive. From the point of view of the hip mothers, commune living is a logical choice; it solves some of the problems of loneliness—there are always others around; it provides plenty of opportunities for interaction with men—even if they aren't always immediately "available"; instead of having to go out to be picked up, a hip mother can rely on a fairly large number of male visitors passing through the commune, with whom she may establish a liaison. And if she does want to go out, there are usually other members of the family present to look after her child, and other males to act as surrogate fathers.

If these descriptions sound as if they bear some similarity to working-class or lower-class patterns in extended-kin groups, the similarity is not inadvertent, although the correspondence is far from perfect. Communal life tends to be very dense, although most communes do have clearly marked areas of privacy. Most communes of all kinds are typically divided into public or communal areas and private areas. In rural communes, there is usually a communal house where people cook, eat, and engage in other collective activities such as meetings, musicales, entertainment of visitors, and so on. In addition there may be a library, sewing rooms, room for spare clothing and other needs for whose satisfactions collective solutions are made. But rural communes tend to discourage "living" (i.e., sleeping) in the communal house, except when the commune is crowded with visitors, guests, or new prospective members. Sleeping quarters are private, and one of the first expressive commitments of a new member in a rural commune is building his own house (containing usually a single room)—a tepee, an A frame, a dome, a shack or lean-to—out of available local materials, and ideally out of sight of the nearest other private dwelling.

In urban communes, the kitchen and living room-dining room generally serve as communal areas, whereas the bedrooms are private and "belong to" the couples who sleep in them. Privacy, of course, is more difficult to sustain in urban communes than in rural ones, even though knocking on closed or almost closed bedroom doors before entering is an item of communal good manners.

In urban and rural communes, children tend to sleep in the same room as their parents (or mother), although if space is available older children may sleep in a room of their own or, as in one rural commune, in a separate house. Although a typical item of commune architecture is the use of sleeping lofts both to increase privacy and to make use of unused space above the head but below the roof, children are regularly exposed to sexual activities—as is true in any community where people cannot afford a lot of space. But the less than perfect privacy for sexual and excretory functions—particularly when the commune is crowded with visitors or crashers—although sometimes a source of tension, is not typically a major problem because of the latent communal belief in most places that no normal and honorable functions *need* to be hidden from public view. The high value of upfrontness, the commonness of nudity, the glass on bathroom or outhouse doors (or no doors at all) and the general belief that people are and should be perfectly transparent to each other is not always enough to overcome years of training in shyness, modesty, etc., regarding sexual and excretory functions, but it generally is enough to at least constrain people to regard their remaining shynesses as hang-ups which they should try to overcome in the name of communal sharing of as much as can conceivably be shared.

Nevertheless, even under crowded conditions, communards develop ways of creating private spaces for the activities, such as sex, for which they still require privacy. Thus tapestries will often be tacked up between one mattress and the next or music will constantly be coming from a radio or record player to cover sounds of love-making or private conversations. People sometimes forgo sexual activity when conditions are crowded, but we have also seen strong compensatory satisfactions taken from the simple fact of a lot of people just sleeping together.

In the report for 1970 we mentioned that the women's liberation movement would probably not approve of the position of women in most communes. Although this is still largely true, it requires some explication. The fact is that in most communes of all types women tend to do traditional women's work: most of the cooking and cleaning (they are more concerned with tidiness than most men), and, in the rural communes, much of the traditional female farm roles in addition. But it is also true that women share in the general ethos of equalitarianism of most communes. With the exception of those religious communes which have an explicitly "sexist" creed, women can be found doing any but the most physically arduous labors, and in several communes we have studied closely, women do play important leadership roles. But on the whole they are less ideologically forceful than men, and express themselves with generally less authority—although we have encountered important exceptions to this tendency.

Concern over the status of women is more common in urban communes than rural ones (this is true in general of political matters), and female liberation has been a heavy topic of conversation in two urban communes we have studied (along with the male liberation which female liberation is said to bring in its wake). And in one of these communes, there is a distinctly "funky" working-class atmosphere, combining a lot of roughhouse play (ass- and crotch-grabbing, mock-rape, etc.) by both the men and the women, with a fairly equal sexual division of labor.

PETER L. BERGER AND HANSFRIED KELLNER

MARRIAGE AND THE CONSTRUCTION OF REALITY

THE ethnologists keep reminding us that the family in our society is of the conjugal type and that the central relationship in this whole area is the marital one. It is on the basis of marriage that, for most adults in our society, existence in the private sphere is built up. It will be clear that this is not at all a universal or even cross culturally wide function of marriage. Rather has marriage in our society taken on a very peculiar character and functionality. It has been pointed out that marriage in contemporary society has lost some of its older functions and taken on new ones instead.[1] This is certainly correct, but we would prefer to state the matter a little differently. Marriage and the family used to be firmly embedded in a matrix of wider community relationships, serving as extensions and particularizations of the latter's social controls. There were few separating barriers between the world of the individual family and the wider community, a fact even to be seen in the physical conditions under which the family lived before the industrial revolution.[2] The same social life pulsated through the house, the street and the community. In our

Reprinted from *Diogenes* 46: 1–13 by permission of the authors and the publisher.

terms, the family and within it the marital relationship were part and parcel of a considerably larger area of conversation. In our contemporary society, by contrast, each family constitutes its own segregated sub-world, with its own controls and its own closed conversation.

This fact requires a much greater effort on the part of the marriage partners. Unlike an earlier situation in which the establishment of the new marriage simply added to the differentiation and complexity of an already existing social world, the marriage partners now are embarked on the often difficult task of constructing for themselves the little world in which they will live. To be sure, the larger society provides them with certain standard instructions as to how they should go about this task, but this does not change the fact that considerable effort of their own is required for its realization. The monogamous character of marriage enforces both the dramatic and the precarious nature of this undertaking. Success or failure hinges on the present idiosyncracies and the fairly unpredictable future development of those idiosyncracies of only two individuals (who, moreover, do not have a shared past)—as Simmel has shown, the most unstable of all possible social relationships.[3] Not surprisingly, the decision to embark on this undertaking has a critical, even cataclysmic connotation in the popular imagination, which is underlined as well as psychologically assuaged by the ceremonialism that surrounds the event.

Every social relationship requires objectivation, that is, requires a process by which subjectively experienced meanings become objective to the individual and, in interaction with others, become common property and thereby massively objective.[4] The degree of objectivation will depend on the number and the intensity of the social relationships that are its carriers. A relationship that consists of only two individuals called upon to sustain, by their own efforts, an ongoing social world will have to make up in intensity for the numerical poverty of the arrangement. This, in turn, accentuates the drama and the precariousness. The later addition of children will add to the, as it were, density of objectivation taking place within the nuclear family, thus rendering the latter a good deal less precarious. It remains true that the establishment and maintenance of such a social world makes extremely high demands on the principal participants.

The attempt can now be made to outline the ideal-typical process that takes place as marriage functions as an instrumentality for the social construction of reality. The chief protagonists of the drama are two individuals, each with a biographically accumulated and available stock of experience.[5] As members of a highly mobile society, these individuals have already internalized a degree of readiness to re-define themselves and to modify their stock of experience, thus beinging with them considerable psychological capacity for entering new relationships with others.[6] Also, coming from broadly similar sectors of the larger society (in terms of region, class, ethnic, and religious affiliations), the two individuals will have organized their stock of experience in similar fashion. In other words, the two individuals have internalized the same overall world, including the general definitions and expectations of the marriage relationship itself. Their society has provided them with a taken-for-granted image of marriage and has socialized them into an anticipation of stepping into the taken-for-granted roles of marriage. All the same, these relatively empty projections now have to be actualized, lived through and filled with experiential content by the protagonists. This will require a dramatic change in their definitions of reality and of themselves.

As of the marriage, most of each partner's actions must now be projected in conjunction with those of the other. Each partner's definitions of reality must be continually correlated with the definitions of the other. The other is present in nearly all horizons of everyday conduct. Furthermore, the identity of each now takes on a new character, having to be constantly matched with that of the other, indeed being typically perceived by people at large as being symbiotically conjoined with the identity of the other. In each partner's psychological economy of significant others, the marriage partner becomes the other *par excellence,* the nearest and most decisive co-inhabitant of the world. Indeed, all other significant relationships have to be almost automatically re-perceived and re-grouped in accordance with this drastic shift.

In other words, from the beginning of the marriage each partner has new modes in his meaningful experience of the world in general, of other people and of himself. By definition, then, marriage constitutes a nomic rupture. In terms of each partner's biography, the event of marriage initiates a new nomic process. Now, the full implications of this fact are rarely apprehended by the protagonists with any degree of clarity. There rather is to be found the notion that one's world, one's other-relationships and, above all, oneself have remained what they were before—only, of course, that world, others and self will now be shared with the marriage partner. It should be clear by now that this notion is a grave misapprehension. Just because of this fact, marriage now propels the individual into an unintended and unarticulated development, in the course of which the nomic transformation takes place. What typically *is* apprehended are certain objective and concrete problems arising out of the marriage—such as tensions with in-laws, or with former friends, or religious differences between the partners, as well as immediate tensions between them. These are apprehended as external, situational and practical difficulties. What is *not* apprehended is the subjective side of these difficulties, namely, the transformation of *nomos* and identity that has occurred and that continues to go on, so that all problems and relationships are experienced in a quite new way, that is, experienced within a new and ever-changing reality.

Take a simple and frequent illustration—the male partner's relationships with male friends before and after the marriage. It is a common observation that such relationships, especially if the extra-marital partners are single, rarely survive the marriage, or, if they do, are drastically re-defined after it. This is typically the result of neither a deliberate decision by the husband nor deliberate sabotage by the wife. What rather happens, very simply, is a slow process in which the husband's image of his friend is transformed as he keeps talking about this friend with his wife. Even if no actual talking goes on, the mere presence of the wife forces him to see his friend differently. This need not mean that he adopts a negative image held by the wife. Regardless of what image she holds or is believed by him to hold, it will be different from that held by the husband. This difference will enter into the joint image that now must needs be fabricated in the course of the ongoing conversation between the marriage partners—and, in due course, must act powerfully on the image previously held by the husband. Again, typically, this process is rarely apprehended with any degree of lucidity. The old friend is more likely to fade out of the picture by slow degrees, as new kinds of friends take his place. The process, if commented upon at all within the marital conversation, can always be explained by socially available formulas about "people changing," "friends disappearing" or oneself "having become more mature." This process of conversational liquidation is especially powerful because it is onesided—the husband typically talks with his wife about his friend, but *not* with his friend about his wife. Thus the friend is deprived of the defense of, as it were, counter-defining the relationship. This dominance of the marital conversation over all others is one of its most important characteristics. It may be mitigated by a certain amount of protective segregation of some non-marital relationships (say, "Tuesday night out with the boys," or "Saturday lunch with mother"), but even then there are powerful emotional barriers against the sort of conversation (conversation *about* the marital relationship, that is) that would serve by way of counter-definition.

Marriage thus posits a new reality. The individual's relationship with this new reality, however, is a dialectical one—he acts upon it, in collusion with the marriage partner, and it acts back upon both him and the partner, welding together their reality. Since, as we have argued before, the objectivation that constitutes this reality is precarious, the groups with which the couple associates are called upon to assist in co-defining the new reality. The couple is pushed towards groups that strengthen their new definition of themselves and the world, avoids those that weaken this definition. This, in turn, releases the commonly known pressures of group association, again acting upon the marriage partners to change their definitions of the world and of themselves. Thus the new reality is not posited once and for all, but goes on being re-defined not only in the marital interaction itself but in the various maritally based group relationships into which the couple enters.

In the individual's biography marriage, then, brings about a decisive phase of socialization that can be compared with the phases of childhood and adolescence. This phase has a rather different structure from the earlier ones. There the individual was in the main socialized into already existing patterns. Here he actively collaborates rather than passively accommodates himself. Also, in the previous phases of socialization, there was an apprehension of entering into a new world and being changed in the course of this. In marriage there is little apprehension of such a process, but rather the notion that the world has remained the same, with only its emotional and pragmatic connotations having changed. This notion, as we have tried to show, is illusionary.

The re-construction of the world in marriage occurs principally in the course of conversation, as we have suggested. The implicit problem of this conversation is how to match two individual definitions of reality. By the very logic of the relationship, a common overall definition must be arrived at—otherwise the conversation will become impossible and, *ipso facto,* the relationship will be endangered. Now, this conversation may be understood as the working away of an ordering and typifying apparatus—if one prefers, an objectivating apparatus. Each partner ongoingly contributes his conceptions of reality, which are then *"talked through,"* usually not once but many times, and in the process become objectivated by the conversational apparatus. The longer this conversation goes on, the more massively real do the objectivations become to the partners. In the marital conversation a world is not only built, but it is also kept in a state of repair and ongoingly refurnished. The subjective reality of this world for the two partners is sustained by the same conversation. The nomic instrumentality of marriage is concretized over and over again, from bed to breakfast table, as the partners carry on the endless conversation that feeds on nearly all they individually or jointly experience. Indeed, it may happen eventually that no experience is fully real unless and until it has been thus "talked through."

JOHN UPDIKE

AFTER THE PARTY

NOW, thinking of this house from whose purchase he had escaped and from whose sale he had realized a partner's share of profit, Piet conservatively rejoiced in the house he had held. He felt its lightly supporting symmetry all around him. He pictured his two round-faced daughters asleep in its shelter. He gloated upon the sight of his wife's body, her fine ripeness.

Having unclasped her party pearls, Angela pulled her dress, the black décolleté knit, over her head. Its soft wool caught in her hairpins. As she struggled, lamplight struck zigzag fire from her slip and static electricity made its nylon adhere to her flank. The slip lifted, exposing stockingtops and garters. Without her head she was all full form, sweet, solid.

Pricked by love, he accused her: "You're not happy with me."

She disentangled the bunched cloth and obliquely faced him. The lamplight, from a bureau lamp with a pleated linen shade, cut shadows into the line of her jaw. She was aging. A year ago, she would have denied the accusation. "How can I be" she asked, "when you flirt with every woman in sight?"

"In sight? Do I?"

"Of course you do. You know you do. Big or little, old or young, you eat them up. Even the yellow ones, Bernadette Ong. Even poor little soused Bea Guerin, who has enough troubles."

"You seemed happy enough, conferring all night with Freddy Thorne."

"Piet, we can't keep going to parties back to back. I come home feeling dirty. I hate it, this way we live."

"You'd rather we went belly to belly? Tell me"—he had stripped to his waist, and she shied from that shieldlike breadth of taut bare skin with its cruciform blazon of amber hair—"what do you and Freddy find to talk about for hours on end? You huddle in the corner like children playing jacks." He took a step forward, his eyes narrowed and pink, party-chafed. She resisted the urge to step backwards, knowing that this threatening mood of his was supposed to end in sex, was a plea.

Instead she reached under her slip to unfasten her garters. The gesture, so vulnerable, disarmed him; Piet halted before the fireplace, his bare feet chilled by the hearth's smooth bricks.

"He's a jerk," she said carelessly, of Freddy Thorne. Her voice was lowered by the pressure of her chin against her chest; the downward reaching of her arms gathered her breasts to a dark crease. "But he talks about things that interest women. Food. Psychology. Children's teeth."

"What does he say psychological?"

"He was talking tonight about what we all see in each other."

"Who?"

"You know. Us. The couples."

"What Freddy Thorne sees in me is a free drink. What he sees in you is a gorgeous fat ass."

She deflected the compliment. "He thinks we're a circle. A magic circle of heads to keep the night out. He told me he gets frightened if he doesn't see us over a weekend. He thinks we've made a church of each other."

"That's because he doesn't go to a real church."

"Well Piet, you're the only one who does. Not counting the Catholics." The Catholics they knew socially were the Gallaghers and Bernadette Ong. The Constantines had lapsed.

"It's the source," Piet said, "of my amazing virility. A stiffening sense of sin." And in his chalkstripe suit pants he abruptly dove forward, planted his weight on his splayed raw-knuckled hands, and stood upside down. His tensed toes reached for the tip of his conical shadow on the ceiling; the veins in his throat and forearms bulged. Angela looked away. She had seen this too often before. He neatly flipped back to his feet; his wife's silence embarrassed him. "Christ be praised," he said, and clapped, applauding himself.

"Shh. You'll wake the children."

"Why the hell shouldn't I, they're always waking me, the little blood-suckers." He went down on his knees and toddled to the edge of the bed. "Dadda, Dadda, wake up-up, Dadda. The Sunnay paper's here, guess what? Jackie Kennedy's having a *ba*by!"

"You're so cruel," Angela said, continuing her careful undressing, parting vague obstacles with her hands. She opened her closet door so that from her husband's angle her body was hidden. Her voice floated free. "Another thing Freddy thinks, he thinks the children are suffering because of it."

"Because of what?"

"Our social life."

"Well, I have to have a social life if you won't give me a sex life."

"If you think *that* approach is the way to a lady's heart, you have a lot to learn." He hated her tone; it reminded him of the years before him, when she had instructed children.

He asked her, "Why shouldn't children suffer? They're supposed to suffer. How else can they learn to be good?" For he felt that if only in the matter of suffering he knew more than she, and that without him she would raise their daughters as she had been raised, to live in a world that didn't exist.

She was determined to answer him seriously, until her patience dulled his pricking mood. "That's positive suffering," she said. "What we give them is neglect so subtle they don't even notice it. We aren't abusive, we're just evasive. For instance, Frankie Appleby is a bright child, but he's just going to waste, he's just Jonathan Little-Smith's punching bag because their parents are always together.

"Hell. Half the reason we all live in this silly hick town is for the sake of the children."

"But we're the ones who have the fun. The children just get yanked along. They didn't enjoy all those skiing trips last winter, standing in the T-bar line shivering and miserable. The girls wanted all winter to go some Sunday to a museum, a nice warm museum with stuffed birds in it, but we wouldn't take them because we would have had to go as a family and our friends might do something exciting or ghastly without us. Irene Saltz finally took them, bless her, or they'd never have gone. I like Irene; she's the only one of us who has somehow kept her freedom. Her freedom from crap."

"How much did you drink tonight?"

"It's just that Freddy didn't let me talk enough."

"He's a jerk," Piet said and, suffocated by an obscure sense of exclusion, seeking to obtain at least the negotiable asset of a firm rejection, he hopped across the hearth-bricks worn like a passageway in Delft and sharply kicked shut Angela's closet door, nearly striking her. She was naked.

He too was naked. Piet's hands, feet, head, and genitals were those of a larger man, as if his maker, seeing that the cooling body had been left too small, had injected a final surge of plasma which at these extremities had ponderously clotted. Physically, he held himself, his tool-toughened palms curved and his acrobat's back a bit bent, as if conscious of a potent burden.

Angela had flinched and now froze, one arm protecting her breasts. A luminous polleny pallor, the shadow of last summer's bathing suit, set off her surprisingly luxuriant pudenda. The slack forward cant of her belly remembered her pregnancies. Her thick-thighed legs were varicose. But her tipped arms seemed, simple and symmetrical, a maiden's; her white feet were high-arched and neither little toe touched the floor. Her throat, wrists and triangular bush appeared the pivots for some undeniable effort of flight, but like Eve on a portal she crouched in shame, stone. She held rigid. Her blue irises cupped light catlike, shallowly. Her skin breathed hate. He did not dare touch her, though her fairness gathered so close dried his tongue. Their bodies hung upon them as clothes too gaudy. Piet felt the fireplace draft on his ankles and became sensitive to the night beyond her hunched shoulders, an extensiveness pressed tight against the bubbled old panes and the frail mullions, a blackness charged with the ache of first growth and the suspended skeletons of Virgo and Leo and Gemini.

She said, "Bully."

He said, "You're lovely."

"That's too bad. I'm going to put on my nightie."

Sighing, immersed in a clamor of light and paint, the Hanemas dressed and crept to bed, exhausted.

Chapter 6

THE COMMUNITY

IN the individual's biography, the first steps (literally) taken beyond the confines of the family bring him into the wider world of the local community. What this involves in terms of social experiences depends very largely on the type of community it is. The most basic differentiation between communities is that between urban and rural. Indeed, in the 1920s and into the 1930s much of American sociology was divided between the students of urban and rural communities. Since urbanization has been one of the great recent forces of change, its study has been a perennial concern of sociologists. The so-called Chicago School in this country produced some classic studies of modern urban life. Following World War II the vast growth of suburbs considerably complicated the picture. The categories of "urban" and "rural" became increasingly less distinct, and sociologists began to refer to a continuum in which one type of community blends over into another.

In the traditional rural community or small town the individual experienced social life mainly in encounters with others well known to him and sharing most basic values with him. Urbanization has brought about fundamental changes on this level of elementary face-to-face experiences. In a big city most social contacts are with strangers. Also, in a big city people with widely differing values and styles of life live in close physical proximity. The large city has, therefore, produced social problems and tensions of a very peculiar kind. Suburbanization seems to have bridged this classic dichotomy. Although the growth of suburbs can be interpreted in demographic and economic terms, one of its essential characteristics is a quest for community that will, in some way, restore to the individual the experiences of belonging and personal security that used to be associated with life in small towns. A contemporary suburb, however, is quite different from an old-fashioned rural community. Sociologists have tried to understand its curious blend of smallness and bigness, personalism and anonymity.

The Chicago School, which flourished from the 1920s on, paid particular attention to the enormous variety of urban community life. The selection from Harvey Warren Zorbaugh's *The Gold Coast and the Slum* describes some sharply distinct communities with the city of Chicago at the end of the 1920s. Herbert J. Gans, in "Urbanism and Suburbanism as Ways of Life: A Re-evaluation of Definitions," summarizes the differences between urban and suburban communities that arose in America in the years following World War II. Since anonymity has become a key feature of social experience, I have included passages by George Simmel, an important German sociologist of the "classical" period who wrote a definitive essay on the psychological consequences of urban life, "The Metropolis and Mental Life." Segregation and conflict between ethnic and racial groups have, too, been universal features of urban life in America. In the final selection, "Ethnogenesis and Negro-Americans Today," Lester Singer deals with the emergence of a distinctive community in recent years.

HARVEY WARREN ZORBAUGH

THE GOLD COAST AND THE SLUM

THE Chicago River, its waters stained by industry, flows back upon itself, branching to divide the city into the South Side, the North Side, and "the great West Side." In the river's southward bend lies the Loop, its skyline looming toward Lake Michigan. The Loop is the heart of Chicago, the knot in the steel arteries of elevated structure which pump in a ceaseless stream the three millions of population of the city into and out of its central business district. The canyon-like streets of the Loop rumble with the traffic of commerce. On its sidewalks throng people of every nation, pushing unseeingly past one another, into and out of office buildings, shops, theaters, hotels, and ultimately back to the north, south, and west "sides" from which they came. For miles over what once was prairie now sprawls in endless blocks the city.

The city's conquest of the prairie has proceeded stride for stride with the development of transportation. The outskirts of the city have always been about forty-five minutes from the heart of the Loop. In the days of the horse-drawn car they were not beyond Twenty-second Street on the South Side. With the coming of the cable car they were extended to the vicinity of Thirty-sixth Street. The electric car—surface and elevated—again extended the city's outskirts, this time well past Seventieth Street. How far "rapid transit" will take them, no one can predict.

Apace with the expansion of the city has gone the ascendancy of the Loop. Every development in transportation, drawing increasing throngs of people into the central business district, has tended to centralize there not only commerce and finance, but all the vital activities of the city's life. The development of communication has further tightened the Loop's grip on the life of the city. The telephone has at once enormously increased the area over which the central business district can exert control and centralized that control. The newspaper, through the medium of advertising, has firmly established the supremacy of the Loop and, through the news, focused the attention of the city upon the Loop. The skyscraper is the visible symbol of the Loop's domination of the city's life. The central business district of the old city—like that of modern London—with its six- and eight-story buildings, sprawled over an unwieldy area. But the skyscraper, thrusting the Loop skyward thirty, forty, fifty stories, has made possible an extraordinary centralization and articulation of the central business district of the modern city. Drawing thousands

Reprinted from *The Gold Coast and the Slum* by Harvey Warren Zorbaugh (Chicago: University of Chicago Press, 1929), pp. 14–19, by permission of the publisher.

daily into the heart of the city, where the old type of building drew hundreds, the cluster of skyscrapers within the Loop has become the city's vortex.

As the Loop expands it literally submerges the areas about it with the traffic of its commerce. Business and industry encroach upon residential neighborhoods. As the roar of traffic swells, and the smoke of industry begrimes buildings, land values rise. The old population moves slowly out, to be replaced by a mobile, shifting, anonymous population bringing with it transitional forms of social life. Within the looming shadow of the skyscraper, in Chicago as in every great city, is found a zone of instability and change—the tidelands of city life.

In a part of these tidelands, within ten minutes' walk of the Loop and the central business district, within five minutes by street car or bus, just across the Chicago River, lies the Near North Side, sometimes called "North Town." Within this area, a mile and a half long and scarcely a mile wide, bounded by Lake Michigan on the east and by the Chicago River on the south and west, under the shadow of the Tribune Tower, a part of the inner city, live ninety thousand people, a population representing all the types and contrasts that lend to the great city its glamor and romance.

The first settlers of Chicago built upon the north bank of the Chicago River, and Chicago's first business house and first railroad were on Kinzie Street. But early in Chicago's history destiny took its great commercial and industrial development southward, and for several decades the North Side was a residential district, well-to-do and fashionable. The story of early Chicago society centers about homes on Ohio, Erie, Cass, and Rush streets; and street after street of old stone fronts, curious streets some of them, still breathe an air of respectability reminiscent of earlier and better days and belying the slow conquest of the slum.

Here change has followed fast upon change. With the growth of the city commerce has encroached upon residential property, relentlessly pushing it northward or crowding it along the lake shore, until now the Near North Side is chequered with business streets. Into this area, where commerce is completing the conquest of the community, has crept the slum. Meantime great industries have sprung up along the river, and peoples speaking foreign tongues have come to labor in them. The slum has offered these alien peoples a place to live cheaply and to themselves; and wave upon wave of immigrants has swept over the area—Irish, Swedish, German, Italian, Persian, Greek, and Negro—forming colonies, staying for a while, then giving way to others. But each has left its impress and its stragglers, and today there live on the Near North Side twenty-nine or more nationalities, many of them with their Old World tongues and customs.

The city's streets can be read as can the geological record in the rock. The old stone fronts of the houses on the side streets; old residences along lower Rush and State, crowded between new business blocks, or with shops built along the street in front of them; a garage with "Riding Academy" in faded letters above its doors; the many old churches along La Salle and Dearborn streets; an office building growing out of a block of rooming-houses; "Deutsche Apotheke" on the window of a store in a neighborhood long since Italian—these are signs that record the changes brought about by the passing decades, changes still taking place today.

The Near North Side is an area of high light and shadow, of vivid contrasts—contrasts not only between the old and the new, between the native and the foreign, but between wealth and poverty, vice and respectability, the conventional and the bohemian, luxury and toil.

Variety is the spice of life, as depicted in the books of the Board of Assessors; autocracy and democracy mingle on the same pages; aphorisms are borne out; and "art for art's sake" remains the slogan of the twentieth century.

On one page of North District Book 18, the record of the worldly holdings of James C. Ewell, artist, 4 Ohio Street, is set down as "Total personal property, $19." So-and-so, artists, are reported thruout the district with this notation. "Attic room, ill-furnished, many paintings: unable to estimate."

The art colony is located in this section, as is the colony of the rich and the nearly rich. And on the same page are the following three entries which span the stream of life:

Cyrus H. McCormick, 50 E. Huron St., $895,000; taxable assessment, $447,500.
Mary V. McCormick, 678 Rush St., $480,000; taxable assessment, $240,000.

And then—as another contrast—the following entry appears on record:

United States Senator Medill McCormick, guest at the Drake Hotel, $——,000,000,000.[1]

At the corner of Division Street and the Lake Shore Drive stands a tall apartment building in which seventeen-room apartments rent at one thousand dollars a month. One mile west, near Division Street and the river, Italian families are living in squalid basement rooms for which they pay six dollars a month. The greatest wealth in Chicago is concentrated along the Lake Shore Drive, in what is called the "Gold Coast." Almost at its back door, in "Little Hell," is the greatest concentration of poverty in Chicago. Respectability, it would seem, is measured by rentals and land values.[2]

The Near North Side is not merely an area of contrasts; it is an area of extremes. All the phenomena characteristic of the city are clearly segregated and appear in exaggerated form. Not only are there extremes of wealth and poverty. The Near North Side has the highest residential land values in the city, and among the lowest; it has more professional men, more politicians, more suicides, more persons in *Who's Who,* than any other "community" in Chicago.[3]

The turgid stream of the Chicago River, which bounds the Near North Side on the south and the west, has played a prominent part in its history. A great deal of shipping once went up the river, and tugs, coal barges, tramp freighters, and occasional ore boats still whistle at its bridges and steam slowly around its bends. This shipping caused commerce and industry to locate along the river, and today wharves, lumber and coal yards, iron works, gas works, sheet metal works, light manufacturing plants and storage plants, wholesale houses for spices, furs, groceries, butter, and imported oils line both sides of the river for miles, and with the noise and smoke of the railroads make a great barrier that half encircles the Near North Side, renders the part of it along the river undesirable to live in, and slowly encroaches northward and eastward.

"North Town" is divided into east and west by State Street. East of State Street lies the Gold Coast, Chicago's most exclusive residential district, turning its face to the lake and its back upon what may lie west toward the river. West of State Street lies a nondescript area of furnished rooms: Clark Street, the Rialto of the half-world; "Little Sicily," the slum.

The Lake Shore Drive is the Mayfair of the Gold Coast. It runs north and south along Lake Michigan, with a wide parkway, bridle path, and promenade. On its western side rise the imposing stone mansions, with their green lawns and wrought-iron-grilled doorways, of Chicago's wealthy aristocracy and her industrial and financial kings. South of these is Streeterville, a "restricted" district of tall apartments and hotels. Here are the Drake Hotel and the Lake Shore Drive Hotel, Chicago's most exclusive. And here apartments rent for from three hundred fifty to a thousand dollars a month. Indeed, the Lake Shore Drive is a street more of wealth than of aristocracy; for in this midwest metropolis money counts for more than does family, and the aristocracy is largely that of the financially successful.

South of Oak Street the Lake Shore Drive, as it turns, becomes North Michigan Avenue, an avenue of fashionable hotels and restaurants, of smart clubs and shops. North Michigan Avenue is the Fifth Avenue of the middle West; and already it looks forward to the day when Fifth Avenue will be the North Michigan Avenue of the East.

On a warm spring Sunday "Vanity Fair" glides along "the Drive" in motor cars of expensive mark, makes colorful the bridlepaths, or saunters up the promenade between "the Drake" and Lincoln Park. The tops of the tan motor busses are crowded with those who live farther out, going home from church—those of a different world who look at "Vanity Fair" with curious or envious eyes. Even here the element of contrast is not lacking, for a mother from back west, with a shawl over her head, waits for a pause in the stream of motors to lead her eager child across to the beach, while beside her stand a collarless man in a brown derby and his girl in Sunday gingham, from some roominghouse back on La Salle Street.

For a few blocks back of "the Drive"—on Belleview Place, East Division Street, Stone, Astor, Banks, and North State Parkway, streets less pretentious but equally aristocratic—live more than a third of the people in Chicago's social register, "of good family and not employed." Here are the families that lived on the once fashionable Prairie Avenue, and later Ashland Boulevard, on the South and West sides. These streets, with the Lake Shore Drive, constitute Chicago's much vaunted Gold Coast, a little world to itself, which the city, failing to dislodge, has grown around and passed by.

At the back door of the Gold Coast, on Dearborn, Clark, and La Salle streets, and on the side streets extending south to the business and industrial area, is a strange world, painfully plain by contrast, a world that lives in houses with neatly lettered cards in the window: "Furnished Rooms." In these houses, from midnight to dawn, sleep some twenty-five thousand people. But by day houses and streets are practically deserted. For early in the morning this population hurries from its houses and down its streets, boarding cars and busses, to work in the Loop. It is a childless area, an area of young men and young women, most of whom are single, though some are married, and others are living together unmarried. It is a world of constant comings and goings, of dull routine and little romance, a world of unsatisfied longings.

The Near North Side shades from light to shadow, and from shadow to dark. The Gold Coast gives way to the world of furnished rooms; and the rooming-house area, to the west again, imperceptibly becomes the slum. The common denominator of the slum is its submerged aspect and its detachment from the city as a whole. The slum is a bleak area of segregation of the sediment of society; an area of extreme poverty, tenements, ramshackle buildings, of evictions and evaded rents; an area of working mothers and children, of high rates of birth, infant mortality, illegitimacy, and death; an area of pawnshops and second-hand stores, of gangs, of "flops" where every bed is a vote. As distinguished from the vice area, the disintegrating neighborhood, the slum is an area which has reached the limit decay and is on the verge of reorganization as missions, settlements, playparks, and business come in.

The Near North Side, west of Clark Street from North Avenue to the river, and east of Clark Street from Chicago Avenue to the river, we may describe as a slum, without fear of contradiction. For this area, cut off by the barrier of river and industry, and for years without adequate transportation, has long been a backwater in the life of the city. This slum district is drab and mean. In ten months the United Charities here had 460 relief cases. Poverty is extreme. Many families are living in one or two basement rooms for which they pay less than ten dollars a month. These rooms are stove heated, and wood is sold on the streets in bundles, and coal in small sacks. The majority of houses, back toward the river, are of wood, and not a few have windows broken out. Smoke, the odor from the gas works, and the smell of dirty alleys is in the air. Both rooms and lots are overcrowded. Back tenements, especially north of Division Street, are common.[4]

Life in the slum is strenuous and precarious. One reads in the paper of a mother on North Avenue giving away her baby that the rest of her children may live. Frequently babies are found in alleyways. A nurse at the Passavant Hospital on North La Salle tells of a dirty little gamin, brought in from Wells Street, whose toe had been bitten off by a rat while he slept. Many women from this neighborhood are in the maternity ward four times in three years. A girl, a waitress, living at the Albany Hotel on lower Rush Street, recently committed suicide leaving the brief note, "I am tired of everything. I have seen too much. That is all." [5]

Clark Street is the Rialto of the slum. Deteriorated store buildings, cheap dance halls and movies, cabarets and doubtful hotels, missions, "flops," pawnshops and second-hand stores, innumerable restaurants, soft-drink parlors and "fellowship" saloons, where men sit about and talk, and which are hangouts for criminal gangs that live back in the slum, fence at the pawnshops, and consort with the transient prostitutes so characteristic of the North Side—such is "the Street." It is an all-night street, a street upon which one meets all the varied types that go to make up the slum.

The slum harbors many sorts of people: the criminal, the radical, the bohemian, the migratory worker, the immigrant, the unsuccessful, the queer and unadjusted. The migratory worker is attracted by the cheap hotels on State, Clark, Wells, and the streets along the river. The criminal and underworld find anonymity in the transient life of the cheaper rooming-houses such as exist on North La Salle Street. The bohemian and the unsuccessful are attracted by cheap attic or basement rooms. The radical is sure of a sympathetic audience in Washington Square. The foreign colony, on the other hand, is found in the slum, not because the immigrant seeks the slum, nor because he makes a slum of the area in which he settles, but merely because he finds there cheap quarters in which to live, and relatively little opposition to his coming. From Sedgwick Street west to the river is a colony of some fifteen thousand Italians, familiarly known as "Little Hell." Here the immigrant has settled blocks by villages, bringing with him his language, his customs, and his traditions, many of which persist.

Other foreign groups have come into this area. North of "Little Sicily," between Wells and Milton streets, there is a large admixture of Poles with Americans, Irish, and Slavs. The Negro, too, is moving into this area and pushing on into "Little Hell." There is a small colony of Greeks grouped about West Chicago Avenue, with its picturesque coffee houses on Clark Street. Finally, there has come in within the past few years a considerable colony of Persians, which has also settled in the vicinity of Chicago Avenue. The slum on the Near North Side is truly cosmopolitan.

In the slum, but not of it, is "Towertown," or "the village." South of Chicago Avenue, along east Erie, Ohio, Huron, and Superior streets, is a considerable colony of artists and of would-be artists. The artists have located here because old buildings can be cheaply converted into studios. The would-be artists have followed the artists. And the hangers-on of bohemia have come for atmosphere, and because the old residences in the district have stables.

"The village" is full of picturesque people and resorts—tearooms with such names as the Wind Blew Inn, the Blue Mouse, and the Green Mask. And many interesting art stores, antique shops, and stalls with rare books are tucked away among the old buildings. All in all, the picturesque and unconventional life of "the village" is again in striking contrast to the formal and conventional life of the Gold Coast, a few short blocks to the north.

One has but to walk the streets of the Near North Side to sense the cultural isolation beneath these contrasts. Indeed, the color and picturesqueness of the city exists in the intimations of what lies behind the superficial contrasts of its life. How various are the thoughts of the individuals who throng up Michigan Avenue from the Loop at the close of the day—artists, shop girls, immigrants, inventors, men of affairs, women of fashion, waitresses, clerks, entertainers. How many are their vocational interests; how different are their ambitions. How vastly multiplied are the chances of life in a great city, as compared with those of the American towns and European peasant villages from which most of these individuals have come. What plans, plots, conspiracies, and dreams for taking advantage of these chances different individuals must harbor under their hats. Yet they have little in common beyond the fact that they jostle one another on the same street. Experience has taught them different languages. How far they are from understanding one another, or from being able to communicate save upon the most obvious material matters!

As one walks from the Drake Hotel and the Lake Shore Drive west along Oak Street, through the world of rooming-houses, into the slum and the streets of the Italian Colony one has a sense of distance as between the Gold Coast and Little Hell—distance that is not geographical but social. There are distances of language and custom. There are distances represented by wealth and the luster it adds to human existence. There are distances of horizon—the Gold Coast living throughout the world while Little Hell is still only slowly emerging out of its old Sicilian villages. There are distances represented by the Gold Coast's absorbing professional interests. It is one world that revolves about the Lake Shore Drive, with its mansions, clubs, and motors, its benefits and assemblies. It is another world that revolves about the Dill Pickle Club, the soap boxes of Washington Square, or the shop of Romano the Barber. And each little world is absorbed in its own affairs. . . .

HERBERT J. GANS

URBANISM AND SUBURBANISM AS WAYS OF LIFE: A RE-EVALUATION OF DEFINITIONS

THE contemporary sociological conception of cities and of urban life is based largely on the work of the Chicago School, and its summary statement in Louis Wirth's essay, "Urbanism as a Way of Life." [1] In that paper, Wirth developed a "minimum sociological definition of the city" as "a relatively large, dense and permanent settlement of socially heterogeneous individuals." [2] From these prerequisites, he then deduced the major outlines of the urban way of life. As he saw it, number, density, and heterogeneity created a social structure in which primary-group relationships were inevitably replaced by secondary contacts that were impersonal, segmental, superficial, transitory, and often predatory in nature. As a result, the city dweller became anonymous, isolated, secular, relativistic, rational, and sophisticated. In order to function in the urban society, he was forced to combine with others to organize corporations, voluntary associations, representative forms of government, and the impersonal mass media of communications.[3] These replaced the primary groups and the integrated way of life found in rural and other pre-industrial settlements.

Wirth's paper has become a classic in urban sociology, and most texts have followed his definition and description faithfully.[4] In recent years, however, a considerable number of studies and essays have questioned his formulations.[5] In addition, a number of changes have taken place in cities since the article was published in 1938, notably the exodus of white residents to low- and medium-priced houses in the suburbs, and the decentralization of industry. The evidence from these studies and the changes in American cities suggest that Wirth's statement must be revised.

There is yet another, and more important reason for such a revision. Despite its title and intent, Wirth's paper deals with urban-industrial society, rather than with the city. This is evident from his approach. Like other urban sociologists, Wirth based his analysis on a comparison of settlement types, but unlike his colleagues, who pursued urban-rural comparisons, Wirth con-

Reprinted from *Human Behavior and Social Processes: An Interactionist Approach* (ed. Arnold Rose). (Boston: Houghton Mifflin, 1962), pp. 625–39 and 644–48. Reprinted by permission of the publishers.

trasted the city to the folk society. Thus, he compared settlement types of pre-industrial and industrial society. This allowed him to include in his theory of urbanism the entire range of modern institutions which are not found in the folk society, even though many such groups (e.g., voluntary associations) are by no means exclusively urban. Moreover, Wirth's conception of the city dweller as depersonalized, atomized, and susceptible to mass movements suggests that his paper is based on, and contributes to, the theory of the mass society.

Many of Wirth's conclusions may be relevant to the understanding of ways of life in modern society. However, since the theory argues that all of society is now urban, *his analysis does not distinguish ways of life in the city from those in other settlements within modern society*. In Wirth's time, the comparison of urban and pre-urban settlement types was still fruitful, but today, the primary task for urban (or community) sociology seems to me to be the analysis of the similarities and differences between contemporary settlement types.

This paper is an attempt at such an analysis; it limits itself to distinguishing ways of life in the modern city and the modern suburb. A re-analysis of Wirth's conclusions from this perspective suggests that his characterization of the urban way of life applies only—and not too accurately—to the residents of the inner city. The remaining city dwellers, as well as most suburbanites, pursue a different way of life, which I shall call "quasi-primary." This proposition raises some doubt about the mutual exclusiveness of the concepts of city and suburbs and leads to a yet broader question: whether settlement concepts and other ecological concepts are useful for explaining ways of life.

THE INNER CITY

Wirth argued that number, density, and heterogeneity had two social consequences which explain the major features of urban life. On the one hand, the crowding of diverse types of people into a small area led to the segregation of homogeneous types of people into separate neighborhoods.[6] On the other hand, the lack of physical distance between city dwellers resulted in social contact between them, which broke down existing social and cultural patterns and encouraged assimilation as well as acculturation—the melting pot effect.[7] Wirth implied that the melting pot effect was far more powerful than the tendency toward segregation and concluded that, sooner or later, the pressures engendered by the dominant social, economic, and political institutions of the city would destroy the remaining pockets of primary-group relationships.[8] Eventually, the social system of the city would resemble Toennies' *Gesellschaft*—a way of life which Wirth considered undesirable.

Because Wirth had come to see the city as the prototype of mass society, and because he examined the city from the distant vantage point of the folk society—from the wrong end of the telescope, so to speak—his view of urban life is not surprising. In addition, Wirth found support for his theory in the empirical work of his Chicago colleagues. As Greer and Kube [9] and Wilensky [10] have pointed out, the Chicago sociologists conducted their most intensive studies in the inner city.[11] At that time, these were slums recently invaded by new waves of European immigrants and rooming house and skid row districts, as well as the habitat of Bohemians and well-to-do Gold Coast apartment dwellers. Wirth himself studied the Maxwell Street Ghetto, an inner-city Jewish neighborhood then being dispersed by the acculturation and mobility of its inhabitants.[12] Some of the characteristics of urbanism which Wirth stressed in his essay abound in these areas.

Wirth's diagnosis of the city as *Gesellschaft* must be questioned on three counts. First, the conclusions derived from a study of the inner city cannot be generalized to the entire urban area. Second, there is as yet not enough evidence to prove—nor, admittedly, to deny—that number, density, and heterogeneity result in the social consequences which Wirth proposed. Finally, even if the causal relationship could be verified, it can be shown that a significant proportion of the city's inhabitants were, and are, isolated from these consequences by social structures and cultural patterns which they either brought to the city, or developed by living in it. Wirth conceived the urban population as consisting of heterogeneous individuals, torn from past social systems, unable to develop new ones, and therefore prey to social anarchy in the city. While it is true that a not insignificant proportion of the inner city population was, and still is, made up of unattached individuals,[13] Wirth's formulation ignores the fact that this population consists mainly of relatively homogeneous groups, with social and cultural moorings that shield it fairly effectively from the suggested consequences of number, density, and heterogeneity. This applies even more to the residents of the outer city, who constitute a majority of the total city population.

The social and cultural moorings of the inner city population are best described by a brief analysis of the five types of inner city residents. These are:

1) the "cosmopolites";
2) the unmarried or childless;
3) the "ethnic villagers";
4) the "deprived"; and
5) the "trapped" and downward mobile.

The "cosmopolites" include students, artists, writers, musicians, and entertainers, as well as other intellectuals and professionals. They live in the city in order to be near the special "cultural" facilities that can only be located near the center of the city. Many cosmopolites are unmarried or childless. Others rear children in the city, especially if they have the income to afford the aid of servants and governesses. The less affluent ones may move to the suburbs to raise their children, continuing to live as cosmopolites under considerable handicaps, especially in the lower-middle-class suburbs. Many of the very rich and powerful are also cosmopolites, although they are likely to have at least two residences, one of which is suburban or exurban.

The unmarried or childless must be divided into two subtypes, depending on the permanence or transience of their status. The temporarily unmarried or childless live in the inner city for only a limited time. Young adults may team up to rent an apartment away from their parents and close to job or entertainment opportunities. When they marry, they may move first to an apartment in a transient neighborhood, but if they can afford to do so, they leave for the outer city or the suburbs with the arrival of the first or second child. The permanently unmarried may stay in the inner city for the remainder of their lives, their housing depending on their income.

The "ethnic villagers" are ethnic groups which are found in such inner city neighborhoods as New York's Lower East Side, living in some ways as they did when they were peasants in European or Puerto Rican villages.[14] Although they reside in the city, they isolate themselves from significant contact with most city facilities, aside from workplaces. Their way of life differs sharply from Wirth's urbanism in its emphasis on kinship and the primary group, the lack of anonymity and secondary-group contacts, the weakness of formal organizations, and the suspicion of anything and anyone outside their neighborhood.

The first two types live in the inner city by choice; the third is there partly because of necessity, partly because of tradition. The final two types are in the inner city because they have no other choice. One is the "deprived" population: the very poor; the emotionally disturbed or otherwise handicapped; broken families; and, most important, the non-white population. These urban dwellers must take the dilapidated housing and blighted neighborhoods to which the housing market relegates them, although among them are some for whom the slum is a hiding place, or a temporary stop-over to save money for a house in the outer city or the suburbs.[15]

The "trapped" are the people who stay behind when a neighborhood is invaded by non-residential land uses or lower-status immigrants, because they cannot afford to move, or are otherwise bound to their present location.[16] The "downward mobiles" are a related type; they may have started life in a higher class position, but have been forced down in the socio-economic hierarchy and in the quality of their accommodations. Many of them are old people, living out their existence on small pensions.

These five types all live in dense and heterogeneous surroundings, yet they have such diverse ways of life that it is hard to see how density and heterogeneity could exert a common influence. Moreover, all but the last two types are isolated or detached from their neighborhood and thus from the social consequences which Wirth described.

When people who live together have social ties based on criteria other than mere common occupancy, they can set up social barriers regardless of the physical closeness of the heterogeneity of their neighbors. The ethnic villagers are the best illustration. While a number of ethnic groups are usually found living together in the same neighborhood, they are able to *isolate* themselves from each other through a variety of social devices. Wirth himself recognized this when he wrote that "two groups can occupy a given area without losing their separate identity because each side is permitted to live its own inner life and each somehow fears or idealizes the other." [17] Although it is true that the children in these areas were often oblivious to the social barriers set up by their parents, at least until adolescence, it is doubtful whether their acculturation can be traced to the melting pot effect as much as to the pervasive influence of the American culture that flowed into these areas from the outside.[18]

The cosmopolites, the unmarried, and the childless are *detached* from neighborhood life. The cosmopolites possess a distinct subculture which causes them to be disinterested in all but the most superficial contacts with their neighbors, somewhat like the ethnic villagers. The unmarried and childless are detached from their neighborhood because of their life-cycle stage, which frees them from the routine family responsibilities that entail some relationship to the local area. In their choice of residence, the two types are therefore not concerned about their neighbors, or the availability and quality of local community facilities. Even the well-to-do can choose expensive apartments in or near poor neighborhoods, because if they have children, these are sent to special schools and summer camps which effectively isolate them from neighbors. In addition, both types, but especially the childless and unmarried, are transient. Therefore, they tend to live in areas marked by high population turnover, where their own mobility and that of their neighbors creates a universal detachment from the neighborhood.[19]

The deprived and the trapped do seem to be affected by some of the consequences of number, density, and heterogeneity. The deprived population suffers considerably from overcrowding, but this is a consequence of low income, racial discrimination, and other handicaps, and cannot be considered an inevitable result of the ecological make-up of the city.[20] Because the deprived have no residential choice, they are also forced to live amid neighbors not of their own choosing, with ways of life different and even contradictory to their own. If familial defenses against the neighborhood climate are weak, as is the case among broken families and downward mobile people, parents may lose their children to the culture of "the street." The trapped are the unhappy people who remain behind when their more advantaged neighbors move on; they must endure the heterogeneity which results from neighborhood change.

Wirth's description of the urban way of life fits best the transient areas of the inner city. Such areas are typically heterogeneous in population, partly because they are inhabited by transient types who do not require homogeneous neighbors or by deprived people who have no choice, or may themselves be quite mobile. Under conditions of transience and heterogeneity, people interact only in terms of the segmental roles necessary for obtaining local services. Their social relationships thus display anonymity, impersonality, and superficiality.[21]

The social features of Wirth's concept of urbanism seem therefore to be a result of residential instability, rather than of number, density, or heterogeneity. In fact, heterogeneity is itself an effect of residential instability, resulting when the influx of transients causes landlords and realtors to stop acting as gatekeepers—that is, wardens of neighborhood homogeneity.[22] Residential instability is found in all types of settlements, and, presumably, its social consequences are everywhere similar. These consequences cannot therefore be identified with the ways of life of the city.

THE OUTER CITY AND THE SUBURBS

The second effect which Wirth ascribed to number, density, and heterogeneity was the segregation of homogeneous people into distinct neighborhoods,[23] on the basis of "place and nature of work, income, racial and ethnic characteristics, social status, custom, habit, taste, preference and prejudice." [24] This description fits the residential districts of the *outer city*.[25] Although these districts contain the majority of the city's inhabitants, Wirth went into little detail about them. He made it clear, however, that the socio-psychological aspects of urbanism were prevalent there as well.[26]

Because existing neighborhood studies deal primarily with the exotic sections of the inner city, very little is known about the more typical residential neighborhoods of the outer city. However, it is evident that the way of life in these areas bears little resemblance to Wirth's urbanism. Both the studies which question Wirth's formulation and my own observations suggest that the common element in the ways of life of these neighborhoods is best described as *quasi-primary*. I use this term to characterize relationships between neighbors. Whatever the intensity or frequency of these relationships, the interaction is more intimate than a secondary contact, but more guarded than a primary one.[27]

There are actually few secondary relationships, because of the isolation of residential neighborhoods from economic institutions and workplaces. Even shopkeepers, store managers, and other local functionaries who live in the area are treated as acquaintances or friends, unless they are of a vastly different social status or are forced by their corporate employers to treat their customers as economic units.[28] Voluntary associations attract only a minority of the population. Moreover, much of the organizational activity is of a sociable nature, and it is often difficult to accomplish the association's "business" because of the members' preference for sociability. Thus, it would appear that interactions in organizations, or between neighbors generally, do not fit the secondary-relationship model of urban life. As anyone who has lived in these neighborhoods knows, there is little anonymity, impersonality or privacy.[29] In fact, American cities have sometimes been described as collections of small towns.[30] There is some truth to this description, especially if the city is compared to the actual small town, rather than to the romantic construct of anti-urban critics.[31]

Postwar suburbia represents the most contemporary version of the quasi-primary way of life. Owing to increases in real income and the encouragement of home ownership provided by the FHA, families in the lower-middle class and upper working class can now live in modern single-family homes in low-density subdivisions, an opportunity previously available only to the upper and upper-middle classes.[32]

The popular literature describes the new suburbs as communities in which conformity, homogeneity, and other-direction are unusually rampant.[33] The implication is that the move from city to suburb initiates a new way of life which causes considerable behavior and personality change in previous urbanites. A preliminary analysis of data which I am now collecting in Levittown, New Jersey, suggests, however, that the move from the city to this predominantly lower-middle-class suburb does not result in any major behavioral changes for most people. Moreover, the changes which do occur reflect the move from the social isolation of a transient city or suburban apartment building to the quasi-primary life of a neighborhood of single-family homes. Also, many of the people whose life has changed reported that the changes were intended. They existed as aspirations before the move, or as reasons for it. In other words, the suburb itself creates few changes in ways of life. Similar conclusions have been reported by Berger in his excellent study of a working-class population newly moved to a suburban subdivision.[34]

A COMPARISON OF CITY AND SUBURB

If urban and suburban areas are similar in that the way of life in both is quasi-primary, and if urban residents who move out to the suburbs do not undergo any significant changes in behavior, it would be fair to argue that the differences in ways of life between the two types of settlements have been overestimated. Yet the fact remains that a variety of physical and demographic differences exist between the city and the suburb. However, upon closer examination, many of these differences turn out to be either spurious or of little significance for the way of life of the inhabitants.[35]

The differences between the residential areas of cities and suburbs which have been cited most frequently are:

1) Suburbs are more likely to be dormitories.

2) They are further away from the work and play facilities of the central business districts.

3) They are newer and more modern than city residential areas and are designed for the automobile rather than for pedestrian and mass-transit forms of movement.

4) They are built up with single-family rather than multi-family structures and are therefore less dense.

5) Their populations are more homogeneous.

6) Their populations differ demographically: they are younger; more of them are married; they have higher incomes; and they hold proportionately more white collar jobs.[36]

Most urban neighborhoods are as much dormitories as the suburbs. Only in a few older inner city areas are factories and offices still located in the middle of residential blocks, and even here many of the employees do not live in the neighborhood.

The fact that the suburbs are farther from the central business district is often true only in terms of distance, not travel time. Moreover, most people make relatively little use of downtown facilities, other than workplaces.[37] The downtown stores seem to hold their greatest attraction for the upper-middle class; [38] the same is probably true of typically urban entertainment facilities. Teen-agers and young adults may take their dates to first-run movie theaters, but the museums, concert halls, and lecture rooms attract mainly upper-middle-class ticket-buyers, many of them suburban.[39]

The suburban reliance on the train and the automobile has given rise to an imaginative folklore about the consequences of commuting on alcohol consumption, sex life, and parental duties. Many of these conclusions are, however, drawn from selected high-income suburbs and exurbs, and reflect job tensions in such hectic occupations as advertising and show business more than the effects of residence.[40] It is true that the upper-middle-class housewife must become a chauffeur in order to expose her children to the proper educational facilities, but such differences as walking to the corner drug store and driving to its suburban equivalent seem to me of little emotional, social, or cultural import.[41] In addition, the continuing shrinkage in the number of mass-transit users suggests that even in the city many younger people are now living a wholly auto-based way of life.

The fact that suburbs are smaller is primarily a function of political boundaries drawn long before the communities were suburban. This affects the kinds of political issues which develop and provides somewhat greater opportunity for citizen participation. Even so, in the suburbs as in the city, the minority who participate are the professional politicians, the economically concerned businessmen, lawyers and salesmen, and the ideologically motivated middle- and upper-middle-class people with better than average education.

The social consequences of differences in density and house type also seem overrated. Single-family houses on quiet streets facilitate the supervision of children; this is one reason why middle-class women who want to keep an eye on their children move to the suburbs. House type also has some effects on relationships between neighbors, insofar as there are more opportunities for visual contact between adjacent homeowners than between people on different floors of an apartment house. However, if occupants' characteristics are also held constant, the differences in actual social contact are less marked. Homogeneity of residents turns out to be more important as a determinant of sociability than proximity. If the population is heterogeneous, there is little social contact between neighbors, either on apartment-house floors or in single-family-house blocks; if people are homogeneous, there is likely to be considerable social contact in both house types. One need only contrast the apartment house located in a transient, heterogeneous neighborhood and exactly the same structure in a neighborhood occupied by a single ethnic group. The former is a lonely, anonymous building; the latter, a bustling micro-society. I have observed similar patterns in suburban areas: on blocks where people are homogeneous, they socialize; where they are heterogeneous, they do little more than exchange polite greetings.[42]

Suburbs are usually described as being more homogeneous in house type than the city, but if they are compared to the outer city, the differences are small. Most inhabitants of the outer city, other than well-to-do homeowners, live on blocks of uniform structures as well—for example, the endless streets of rowhouses in Philadelphia and Baltimore or of two-story duplexes and six-flat apartment houses in Chicago. They differ from the new suburbs only in that they were erected through more primitive methods of mass production. Suburbs are of course more predominantly areas of owner-occupied single homes, though in the outer districts of most American cities homeownership is also extremely high.

Demographically, suburbs as a whole are clearly more homogeneous than cities as a whole, though probably not more so than outer cities. However, people do not live in cities or suburbs as a whole, but in specific neighborhoods. An analysis of ways of life would require a determination of the degree of population homogeneity within the boundaries of areas defined as neighborhoods by residents' social contacts. Such an analysis would no doubt indicate that many neighborhoods in the city as well as the suburbs are homogeneous. Neighborhood homogeneity is actually a result of factors having little or nothing to do with the house type, density, or location of the area relative to the city limits. Brand new neighborhoods are more homogeneous than older ones, because they have not yet experienced resident turnover, which frequently results in population heterogeneity. Neighborhoods of low- and medium-priced housing are usually less homogeneous than those with expensive dwellings because they attract families who have reached the peak of occupational and residential mobility, as well as young families who are just starting their climb and will eventually move to neighborhoods of higher status. The latter, being accessible only to high-income people, are therefore more homogeneous with respect to other resident characteristics as well. Moreover, such areas have the economic and political power to slow down or prevent invasion. Finally, neighborhoods located in the path of ethnic or religious group movement are likely to be extremely homogeneous.

The demographic differences between cities and suburbs cannot be questioned, especially since the suburbs have attracted a large number of middle-class child-rearing families. The differences are, however, much reduced if suburbs are compared only to the outer city. In addition, a detailed comparison of suburban and outer city residential areas would show that neighborhoods with the same kinds of people can be found in the city as well as the suburbs. Once again, the age of the area and the cost of housing are more important determinants of demographic characteristics than the location of the area with respect to the city limits. . . .

CONCLUSION

Many of the descriptive statements made here are as time-bound as Wirth's.[43] Twenty years ago, Wirth concluded that some form of urbanism would eventually predominate in all settlement types. He was, however, writing during a time of immigrant acculturation and at the end of a serious depression, an era of minimal choice. Today, it is apparent that high-density, heterogeneous surroundings are for most people a temporary place of residence; other than for the Park Avenue or Greenwich Village cosmopolites, they are a result of necessity rather than choice. As soon as they can afford to do so, most Americans head for the single-family house and the quasi-primary way of life of the low-density neighborhood, in the outer city or the suburbs.[44]

Changes in the national economy and in government housing policy can affect many of the variables that make up housing supply and demand. For example, urban sprawl may eventually outdistance the ability of present and proposed transportation systems to move workers into the city; further industrial decentralization can forestall it and alter the entire relationship between work and residence. The expansion of present urban renewal activities can perhaps lure a significant number of cosmopolites back from the suburbs, while a drastic change in renewal policy might begin to ameliorate the housing conditions of the deprived population. A serious depression could once again make America a nation of doubled-up tenants.

These events will affect housing supply and residential choice; they will frustrate but not suppress demands for the quasi-primary way of life. However, changes in the national economy, society, and culture can affect people's characteristics—family size, educational level, and various other concomitants of life-cycle stage and class. These in turn will stimulate changes in demands and choices. The rising number of college graduates, for example, is likely to increase the cosmopolite ranks. This might in turn create a new set of city dwellers, although it will probably do no more than encourage the development of cosmopolite facilities in some suburban areas.

The current revival of interest in urban sociology and in community studies, as well as the sociologist's increasing curiosity about city planning, suggest that data may soon be available to formulate a more adequate theory of the relationship between settlements and the ways of life within them. The speculations presented in this paper are intended to raise questions; they can only be answered by more systematic data collection and theorizing.

GEORG SIMMEL

THE METROPOLIS AND MENTAL LIFE

THE deepest problems of modern life derive from the claim of the individual to preserve the autonomy and individuality of his existence in the face of overwhelming social forces, of historical heritage, of external culture, and of the technique of life. The fight with nature which primitive man has to wage for his *bodily* existence attains in this modern form its latest transformation. The eighteenth century called upon man to free himself of all the historical bonds in the state and in religion, in morals and in economics. Man's nature, originally good and common to all, should develop unhampered. In addition to more liberty, the nineteenth century demanded the functional specialization of man and his work; this specialization makes one individual incomparable to another, and each of them indispensable to the highest possible extent. However, this specialization makes each man the more directly dependent upon the supplementary activities of all others. Nietzsche sees the full development of the individual conditioned by the most ruthless struggle of individuals; socialism believes in the suppression of all competition for the same reason. Be that as it may, in all these positions the same basic motive is at work: the person resists to being leveled down and worn out by a social-technological mechanism. An inquiry into the inner meaning of specifically modern life and its products, into the soul of the cultural body, so to speak, must seek to solve the equation which structures like the metropolis set up between the individual and the super-individual contents of life. Such an inquiry must answer the question of how the personality accommodates itself in the adjustments to external forces. This will be my task today.

The psychological basis of the metropolitan type of individuality consists in the *intensification of nervous stimulation* which results from the swift and uninterrupted change of outer and inner stimuli. Man is a differentiating creature. His mind is stimulated by the difference between a momentary impression and the one which preceded it. Lasting impressions, impressions which differ only slightly from one another, impressions which take a regular and habitual course and show regular and habitual contrasts—all these use up, so to speak, less consciousness than does the rapid crowding of changing images, the sharp discontinuity in the grasp of a single glance, and the unexpectedness of onrushing impressions. These are the psychological conditions

Reprinted from *The Sociology of Georg Simmel,* ed. Kurt Wolff (Glencoe, Ill.: Free Press, 1950), pp. 409–21, by permission of The Macmillan Company.

which the metropolis creates. With each crossing of the street, with the tempo and multiplicity of economic, occupational and social life, the city sets up a deep contrast with small town and rural life with reference to the sensory foundations of psychic life. The metropolis exacts from man as a discriminating creature a different amount of consciousness than does rural life. Here the rhythm of life and sensory mental imagery flows more slowly, more habitually, and more evenly. Precisely in this connection the sophisticated character of metropolitan psychic life becomes understandable—as over against small town life which rests more upon deeply felt and emotional relationships. These latter are rooted in the more unconscious layers of the psyche and grow most readily in the steady rhythm of uninterrupted habituations. The intellect, however, has its locus in the transparent, conscious, higher layers of the psyche; it is the most adaptable of our inner forces. In order to accommodate to change and to the contrast of phenomena, the intellect does not require any shocks and inner upheavals; it is only through such upheavals that the more conservative mind could accommodate to the metropolitan rhythm of events. Thus the metropolitan type of man—which, of course, exists in a thousand individual variants—develops an organ protecting him against the threatening currents and discrepancies of his external environment which would uproot him. He reacts with his head instead of his heart. In this an increased awareness assumes the psychic prerogative. Metropolitan life, thus, underlies a heightened awareness and a predominance of intelligence in metropolitan man. The reaction to metropolitan phenomena is shifted to that organ which is least sensitive and quite remote from the depth of the personality. Intellectuality is thus seen to preserve subjective life against the overwhelming power of metropolitan life, and intellectuality branches out in many directions and is integrated with numerous discrete phenomena.

The metropolis has always been the seat of the money economy. Here the multiplicity and concentration of economic exchange gives an importance to the means of exchange which the scantiness of rural commerce would not have allowed. Money economy and the dominance of the intellect are intrinsically connected. They share a matter-of-fact attitude in dealing with men and with things; and, in this attitude, a formal justice is often coupled with an inconsiderate hardness. The intellectually sophisticated person is indifferent to all genuine individuality, because relationships and reactions result from it which cannot be exhausted with logical operations. In the same manner, the individuality of phenomena is not commensurate with the pecuniary principle. Money is concerned only with what is common to all: it asks for the exchange value, it reduces all quality and individuality to the question: How much? All intimate emotional relations between persons are founded in their individuality, whereas in rational relations man is reckoned with like a number, like an element which is in itself indifferent. Only the objective measurable achievement is of interest. Thus metropolitan man reckons with his merchants and customers, his domestic servants and often even with persons with whom he is obliged to have social intercourse. These features of intellectuality contrast with the nature of the small circle in which the inevitable knowledge of individuality as inevitably produces a warmer tone of behavior, a behavior which is beyond a mere objective balancing of service and return. In the sphere of the economic psychology of the small group it is of importance that under primitive conditions production serves the customer who orders the good, so that the producer and the consumer are acquainted. The modern metropolis, however, is supplied almost entirely by production for the market, that is, for entirely unknown purchasers who never personally enter the producer's actual field of vision. Through this anonymity the interests of each party acquire an unmerciful matter-of-factness; and the intellectually calculating economic egoisms of both parties need not fear any deflection because of the imponderables of personal relationships. The money economy dominates the metropolis; it has displaced the last survivals of domestic production and the direct barter of goods; it minimizes, from day to day, the amount of work ordered by customers. The matter-of-fact attitude is obviously so intimately interrelated with the money economy, which is dominant in the metropolis, that nobody can say whether the intellectualistic mentality first promoted the money economy or whether the latter determined the former. The metropolitan way of life is certainly the most fertile soil for this reciprocity, a point which I shall document merely by citing the dictum of the most eminent English constitutional historian: throughout the whole course of English history, London has never acted as England's heart but often as England's intellect and always as her moneybag!

In certain seemingly insignificant traits, which lie upon the surface of life, the same psychic currents characteristically unite. Modern mind has become more and more calculating. The calculative exactness of practical life which the money economy has brought about corresponds to the ideal of natural science: to transform the world into an arithmetic problem, to fix every part of the world by mathematical formulas. Only money economy has filled the days of so many people with weighing, calculating, with numerical determinations, with a reduction of qualitative values to quantitative ones. Through the calculative nature of money a new precision, a certainty in the definition of identities and differences, an unambiguousness in agreements and arrangements has been brought about in the relations of life-elements—just as externally this precision has been effected by the universal diffusion of pocket watches. However, the conditions of metropolitan life are at once cause and effect of this trait. The relationships and affairs of the typical metropolitan usually are so varied and complex that without the strictest punctuality in promises and services the whole structure would break down into an inextricable chaos. Above all, this necessity is brought about by the aggregation of so many people with such differentiated interests, who must integrate their relations and activities into a highly complex organism. If all clocks and watches in Berlin would suddenly go wrong in different ways, even if only by one hour, all economic life and communication of the city would be disrupted for a long time. In addition an apparently mere external factor: long distances, would make all waiting and broken appointments result in an ill-afforded waste of time. Thus, the technique of metropolitan life is unimaginable without the most punctual integration of all activities and mutual relations into a stable and impersonal time schedule. Here again the general conclusions of this entire task of reflection become obvious, namely, that from each point on the surface of existence—however closely attached to the surface alone—one may drop a sounding into the depth of the psyche so that all the most banal externalities of life finally are connected with the ultimate decisions concerning the meaning and style of life. Punctuality, calculability, exactness are forced upon life by the complexity and extension of metropolitan existence and are not only the most intimately connected with its money economy and intellectualistic character. These traits must also color the contents of life and favor the exclusion of those irrational, instinctive, sovereign traits and impulses which aim at determining the mode of life from within, instead of receiving the general and precisely schematized form of life from without. Even though sovereign types of personality, characterized by irrational impulses, are by no means impossible in the city, they are, nevertheless, opposed to typical city life. The passionate hatred of men like Ruskin and Nietzsche for the metropolis is understandable in these terms. Their natures discovered the value of life alone in the unschematized existence which cannot be defined with precision for all alike. From the same source of this hatred of the metropolis surged their hatred of money economy and of the intellectualism of modern existence.

The same factors which have thus coalesced into the exactness and minute precision of the form of life have coalesced into a structure of the highest impersonality; on the other hand, they have promoted a highly personal subjectivity. There is perhaps no psychic phenomenon which has been so unconditionally reserved to the metropolis as has the blasé attitude. The blasé attitude results first from the rapidly changing and closely compressed contrasting stimulations of the nerves. From this, the enhancement of metropolitan intellectuality, also, seems originally to stem. Therefore, stupid people who are not intellectually alive in the first place usually are not exactly blasé. A life in boundless pursuit of pleasure makes one blasé because it agitates the nerves to their strongest reactivity for such a long time that they finally cease to react at all. In the same way, through the rapidity and contradictoriness of their changes, more harmless impressions force such violent responses, tearing the nerves so brutally hither and thither that their last reserves of strength are spent; and if one remains in the same milieu they have no time to gather new strength. An incapacity thus emerges to react to new sensations with the appropriate energy. This constitutes that blasé attitude which, in fact, every metropolitan child shows when compared with children of quieter and less changeable milieus.

This physiological source of the metropolitan blasé attitude is joined by another source which flows from the money economy. The essence of the blasé attitude consists in the blunting of discrimination. This does not mean that the objects are not perceived, as is the case with the half-wit, but rather that the meaning and differing values of things, and thereby the things themselves, are experienced as insubstantial. They appear to the blasé person in an evenly flat and gray tone; no one object deserves preference over any other. This mood is the faithful subjective reflection of the completely internalized money economy. By being the equivalent to all the manifold things in one and the same way, money becomes the most frightful leveler. For money expresses all qualitative differences of things in terms of "how much?" Money, with all its colorlessness and indifference, becomes the common denominator of all values; irreparably it hollows out the core of things, their individuality, their specific value, and their incomparability. All things float with equal specific gravity in the constantly moving stream of money. All things lie on the same level and differ from one another only in the size of the area which they cover. In the individual case this coloration, or rather discoloration, of things through their money equivalence may be unnoticeably minute. However, through the relations of the rich to the objects to be had for money, perhaps even through the total character which the mentality of the contemporary public everywhere imparts to these objects, the exclusively pecuniary evaluation of objects has become quite considerable. The large cities, the main seats of the money exchange, bring the purchasability of things to the fore much more impressively than do smaller localities. That is why cities are also the genuine locale of the blasé attitude. In the blasé attitude the concentration of men and things stimulates the nervous system of the individual to its highest achievement so that it attains its peak. Through the mere quantitative intensification of the same conditioning factors this achievement is transformed into its opposite and appears in the peculiar adjustment of the blasé attitude. In this phenomenon the nerves find in the refusal to react to their stimulation the last possibility of accommodating to the contents and forms of metropolitan life. The self-preservation of certain personalities is brought at the price of devaluating the whole objective world, a devaluation which in the end unavoidably drags one's own personality down into a feeling of the same worthlessness.

Whereas the subject of this form of existence has to come to terms with it entirely for himself, his self-preservation in the face of the large city demands from him a no less negative behavior of a social nature. This mental attitude of metropolitans toward one another we may designate, from a formal point of view, as reserve. If so many inner reactions were responses to the continuous external contacts with innumerable people as are those in the small town, where one knows almost everybody one meets and where one has a positive relation to almost everyone, one would be completely atomized internally and come to an unimaginable psychic state. Partly this psychological fact, partly the right to distrust which men have in the face of the touch-and-go elements of metropolitan life, necessitates our reserve. As a result of this reserve we frequently do not even know by sight those who have been our neighbors for years. And it is this reserve which in the eyes of the small-town people makes us appear to be cold and heartless. Indeed, if I do not deceive myself, the inner aspect of this outer reserve is not only indifference but, more often than we are aware, it is a slight aversion, a mutual strangeness and repulsion, which will break into hatred and fight at the moment of a closer contact, however caused. The whole inner organization of such an extensive communicative life rests upon an extremely varied hierarchy of sympathies, indifferences, and aversions of the briefest as well as of the most permanent nature. The sphere of indifference in this hierarchy is not as large as might appear on the surface. Our psychic activity still responds to almost every impression of somebody else with a somewhat distinct feeling. The unconscious, fluid and changing character of this impression seems to result in a state of indifference. Actually this indifference would be just as unnatural as the diffusion of indiscriminate mutual suggestion would be unbearable. From both these typical dangers of the metropolis, indifference and indiscriminate suggestibility, antipathy protects us. A latent antipathy and the preparatory stage of practical antagonism effect the distances and aversions without which this mode of life could not at all be led. The extent and the mixture of this style of life, the rhythm of its emergence and disappearance, the forms in which it is satisfied—all these, with the unifying motives in the narrower sense, form the inseparable whole of the metropolitan style of life. What appears in the metropolitan style of life directly as dissociation is in reality only one of its elemental forms of socialization.

This reserve with its overtone of hidden aversion appears in turn as the form of the cloak of a more general mental phenomenon of the metropolis: it grants to the individual a kind and an amount of personal freedom which has no analogy whatsoever under other conditions. The metropolis goes back to one of the large developmental tendencies of social life as such, to one of the few tendencies for which an approximately universal formula can be discovered. The earliest phase of social formations found in historical as well as in contemporary social structures is this: a relatively small circle firmly closed against neighboring, strange, or in some way antagonistic circles. However, this circle is closely coherent and allows its individual members only a narrow field for the development of unique qualities and free, self-responsible movements. Political and kinship groups, parties and religious associations begin in this way. The self-preservation of very young associations requires the establishment of strict boundaries and a centripetal unity. Therefore they cannot allow the individual freedom and unique inner and outer development. From this stage social development proceeds at once in two different, yet corresponding, directions. To the extent to which the group grows—numerically, spatially, in significance and in content of life—to the same degree the group's direct, inner unity loosens, and the rigidity of the original demarcation against others is softened through mutual relations and connections. At the same time, the individual gains freedom of movement, far beyond the first jealous delimitation. The individual also gains a specific individuality to which the division of labor in the enlarged group gives both occasion and necessity. The state and Christianity, guilds and political parties, and innumerable other groups have developed according to this formula, however much, of course, the special conditions and forces of the respective groups have modified the general scheme. This scheme seems to me distinctly recognizable also in the evolution of individuality within urban life. The small-town life in Antiquity and in the Middle Ages set barriers against movement and relations of the individual toward the outside, and it set up barriers against individual independence and differentiation within the individual self. These barriers were such that under them modern man could not have breathed. Even today a metropolitan man who is placed in a small town feels a restriction similar, at least, in kind. The smaller the circle which forms our milieu is, and the more restricted those relations to others are which dissolve the boundaries of the individual, the more anxiously the circle guards the achievements, the conduct of life, and the outlook of the individual, and the more readily a quantitative specialization would break up the framework of the whole little circle.

The ancient *polis* in this respect seems to have had the very character of a small town. The constant threat to its existence at the hands of enemies from near and afar effected strict coherence in political and military respects, a supervision of the citizen by the citizen, a jealousy of the whole against the individual whose particular life was suppressed to such a degree that he could compensate only by acting as a despot in his own household. The tremendous agitation and excitement, the unique colorfulness of Athenian life, can perhaps be understood in terms of the fact that a people of incomparably individualized personalities struggled against the constant inner and outer pressure of a de-individualizing small town. This produced a tense atmosphere in which the weaker individuals were suppressed and those of stronger natures were incited to prove themselves in the most passionate manner. This is precisely why it was that there blossomed in Athens what must be called, without defining it exactly, "the general human character" in the intellectual development of our species. For we maintain factual as well as historical validity for the following connection: the most extensive and the most general contents and forms of life are most intimately connected with the most individual ones. They have a preparatory stage in common, that is, they find their enemy in narrow formations and groupings the maintenance of which places both of them into a state of defense against expanse and generality lying without and the freely moving individuality within. Just as in the feudal age, the "free" man was the one who stood under the law of the land, that is, under the law of the largest social orbit, and the unfree man was the one who derived his right merely from the narrow circle of a feudal association and was excluded from the larger social orbit—so today metropolitan man is "free" in a spiritualized and refined sense, in contrast to the pettiness and prejudices which hem in the small-town man. For the reciprocal reserve and indifference and the intellectual life conditions of large circles are never felt more strongly by the individual in their impact upon independence than in the thickest crowd of the big city. This is because the bodily proximity and narrowness of space makes the mental distance only the more visible. It is obviously only the obverse of this freedom if, under certain circumstances, one nowhere feels as lonely and lost as in the metropolitan crowd. For here as elsewhere it is by no means necessary that the freedom of man be reflected in his emotional life as comfort.

It is not only the immediate size of the area and the number of persons which, because of the universal historical correlation between the enlargement of the circle and the personal inner and outer freedom, has made the metropolis the locale of freedom. It is rather in transcending this visible expanse that any given city becomes the seat of cosmopolitanism. The horizon of the city expands in a manner comparable to the way in which wealth develops; a certain amount of property increases in a quasi-automatical way in ever more rapid progression. As soon as a certain limit has been passed, the economic, personal, and intellectual relations of the citizenry, the sphere of intellectual predominance of the city over its hinterland, grow as in geometrical progression. Every gain in dynamic extension becomes a step, not for an equal, but for a new and larger extension. From every thread spinning out of the city, ever new threads grow as if by themselves, just as within the city the unearned increment of ground rent, through the mere increase in communication, brings the owner automatically increasing profits. At this point, the quantitative aspect of life is transformed directly into qualitative traits of character. The sphere of life of the small town is, in the main, self-contained and autarchic. For it is the decisive nature of the metropolis that its inner life overflows by waves into a far-flung national or international area. Weimar is not an example to the contrary, since its significance was hinged upon individual personalities and died with them; whereas the metropolis is indeed characterized by its essential independence even from the most eminent individual personalities. This is the counterpart to the independence, and it is the price the individual pays for the independence, which he enjoys in the metropolis. The most significant characteristic of the metropolis is this functional extension beyond its physical boundaries. And this efficacy reacts in turn and gives weight, importance, and responsibility to metropolitan life. Man does not end with the limits of his body or the area comprising his immediate activity. Rather is the range of the person constituted by the sum of effects emanating from him temporally and spatially. In the same way, a city consists of its total effects which extend beyond its immediate confines. Only this range is the city's actual extent in which its existence is expressed. This fact makes it obvious that individual freedom, the logical and historical complement of such extension, is not to be understood only in the negative sense of mere freedom of mobility and elimination of prejudices and petty philistinism. The essential point is that the particularity and incomparability, which ultimately every human being possesses, be somehow expressed in the working-out of a way of life. That we follow the laws of our own nature—and this after all is freedom—becomes obvious and convincing to ourselves and to others only if the expressions of this nature differ from the expressions of others. Only our unmistakability proves that our way of life has not been superimposed by others.

Cities are, first of all, seats of the highest economic division of labor. They produce thereby such extreme phenomena as in Paris the remunerative occupation of the *quatorzième*. They are persons who identify themselves by signs on their residences and who are ready at the dinner hour in correct attire, so that they can be quickly called upon if a dinner party should consist of thirteen persons. In the measure of its expansion, the city offers more and more the decisive conditions of the division of labor. It offers a circle which through its size can absorb a highly diverse variety of services. At the same time, the concentration of individuals and their struggle for customers compel the individual to specialize in a function from which he cannot be readily displaced by another. It is decisive that city life has transformed the struggle with nature for livelihood into an inter-human struggle for gain, which here is not granted by nature but by other men. For specialization does not flow only from the competition for gain but also from the underlying fact that the seller must always seek to call forth new and differentiated needs of the lured customer. In order to find a source of income which is not yet exhausted, and to find a function which cannot readily be displaced, it is necessary to specialize in one's services. This process promotes differentiation, refinement, and the enrichment of the public's needs, which obviously must lead to growing personal differences within this public.

All this forms the transition to the individualization of mental and psychic traits which the city occasions in proportion to its size. There is a whole series of obvious causes underlying this process. First, one must meet the difficulty of asserting his own personality within the dimensions of metropolitan life. Where the quantitative increase in importance and the expense of energy reach their limits, one seizes upon qualitative differentiation in order somehow to attract the attention of the social circle by playing upon its sensitivity for differences. Finally, man is tempted to adopt the most tendentious peculiarities, that is, the specifically metropolitan extravagances of mannerism, caprice, and preciousness. Now, the meaning of these extravagances does not at all lie in the contents of such behavior, but rather in its form of "being different," of standing out in a striking manner and thereby attracting attention. For many character types, ultimately the only means of saving for themselves some modicum of self-esteem and the sense of filling a position is indirect, through the awareness of others. In the same sense a seemingly insignificant factor is operating, the cumulative effects of which are, however, still noticeable. I refer to the brevity and scarcity of the inter-human contacts granted to the metropolitan man, as compared with social intercourse in the small town. The temptation to appear "to the point," to appear concentrated and strikingly characteristic, lies much closer to the individual in brief metropolitan contacts than in an atmosphere in which frequent and prolonged association assures the personality of an unambiguous image of himself in the eyes of the other.

The most profound reason, however, why the metropolis conduces to the urge for the most individual personal existence—no matter whether justified and successful—appears to me to be the following: the development of modern culture is characterized by the preponderance of what one may call the "objective spirit" over the "subjective spirit." This is to say, in language as well as in law, in the technique of production as well as in art, in science as well as in the objects of domestic environment, there is embodied a sum of spirit. . . .

LESTER SINGER

ETHNOGENESIS AND NEGRO-AMERICANS TODAY

THE view of Negro-white relations in American society generally accepted by sociologists is that they are caste relations.[1] A competing view—which has not achieved wide acceptance, although the terminology persists—is that the phenomena are best understood as race relations.[2] The two approaches, regardless of the differences between them, deal primarily with the structure of Negro-white relations and with the factors serving to maintain that relational structure.[3]

My thesis is that these structural models fail to illuminate the character of the entities that occupy the various places in the relational structure. I had asked the question: In sociological terms, *what* are the Negroes in American society? And, at first, the answer appeared to be: They are a caste, or a race—and the whites must be one or the other also. Upon further consideration,

Reprinted from *Social Research* 29 (Winter 1962): 422–32 by permission of the author and the publisher.

however, it became clear that caste—as defined by Warner, Davis and Myrdal—and race—as defined by Cox—are not answers to the substantive question. The writers answer the question: *Where?* That is to say, they tell us the position of the Negroes in the structure of Negro-white relations. They do not indicate *what* the Negroes are, *what* they constitute as a social entity.

SOCIAL CATEGORY, SOCIAL ENTITY, AND NEGRO-AMERICANS

Let me make clear the notions that underlie the use of the term "social entity" as contrasted with the term "social category." "Social category" has been defined by Bennett and Tumin as referring to "numbers of people who constitute an aggregate because they have a common characteristic(s) *about which* [italics mine] society expresses some views and which therefore influences their life chances." [4] The "members" of a social category are not necessarily involved in any relationship among themselves. Thus the terms "men," "women," "immigrants," and "divorcees" stand for social categories.[5] The term "social entity," on the other hand, refers to a number of people manifesting such qualities as patterned relationships, shared values, and self-recognition. Thus a team, a gang, a community, an ethnic group, and a society all constitute recognizable social entities.

For this writer, the nub of the contrast between the two terms is the presence or absence of internal structure and the accompanying cultural, or ideological, elements. This is somewhat like the difference between a bin full of spare parts and an engine which has been assembled from such spare parts. As with the bin full of spare parts, the social cateogry contains elements that have no necessary relations to one another. The social entity, however, like the engine, can only be understood through an understanding of the elements in patterned relations.

In the work of both Myrdal and Cox are to be found the empirical generalizations that express the distinctive social attributes of Negro-Americans. These attributes, when viewed as elements in a pattern rather than separately, make up a picture of Negroes as a social entity on the order of an ethnic group.[6] Among these qualities are briefly, the following. 1) The existence of a separate Negro prestige continuum, that is "a social-class system." 2) the existence of a distinctive Negro culture pattern. [While it is true that the pattern is derived from that of the larger society—Myrdal calls it a "pathological" form and Cox a "truncated" form of the larger American culture pattern—such references to origin indicate, if we disregard their evaluative content, the distinctiveness of this pattern. This is not to say that it is completely different from the larger pattern but, rather, that it is not quite the same.] 3) The existence of various aspects of Negro solidarity vis-à-vis the "whites": its, heretofore, primarily defensive character; its tentativeness; the predominantly compromising nature of Negro leadership; the development of a self-image; the Negroes' conviction of their rights in the larger society; and the direction of collective Negro aspiration toward the realization of these rights within the larger society. 4) The existence of the uncountable relational networks and organizations (that is the internal structure) which, by virtue of discrimination and the defensive response of Negroes, are manned largely by Negroes.

If this pattern of qualities is significant it means, as I indicated at the outset, that Negroes and, therefore, Negro-white relations, cannot be fully understood if only category concepts are used. The employment of such concepts results in interpretations that are not as inclusive as the data allow. Another important limiting effect of the use of these concepts is an emphasis on the static. Consequently, while a model based on category concepts may prepare the ground, it certainly does not facilitate an examination of process, development, and change.[7] And yet it is in precisely this latter direction that we must search if social entities, which come into existence and which disappear, are to be adequately understood. Thus, if Negroes in the United States are to be understood as an entity, that is as an ethnic group, it is necessary to attempt to answer such questions as: How did this entity commence to form? What are the factors tending to maintain or change the formative process? What are the circumstances under which such an entity will cease to exist? The remainder of this essay attempts to answer these questions.

ETHNOGENESIS

During the seventeenth and eighteenth centuries the Africans who were brought to North America, as well as to other parts of the New World, were representatives of a variety of societies, cultural backgrounds, language groups, and so forth. Consequently, *as a totality* they can only be viewed as a social category; that is as "Africans" or "slaves." Removed from their various social contexts and thrown together as enslaved strangers they had, particularly in North America, no internal organization. In fact, the evidence indicates that virtually all traces of African social organization disappeared under the impact of American slavery.[8] One method of accomplishing this was the intentional separation of members of the same society. This was done, among other reasons, to diminish the possibility of revolt. As R. E. Park says, "It was found easier to deal with the slaves, if they were separated from their kinsmen." [9]

It should be realized that during the period of slavery the newly arrived Africans came into contact with whites and acculturated slaves. In this way, cut off from their own background, they came to take on the culture of American society in whatever form it was available to them. With Emancipation the former Africans and their descendants became, as the federal government put it, "freedmen." This legal term, however, stands only for a social category; the freedman lacked determinate social group characteristics. But indications of what was to come had been evidenced earlier by the numerous slave revolts,[10] and the participation of runaway slaves and free Negroes in the Abolitionist movement and the Underground Railroad.

Following Emancipation, the group-forming process moved with much greater speed and intensity than before. I propose that this formative process be referred to as "ethnogenesis," meaning by this term the process whereby a people, that is an ethnic group, comes into existence.[11] The process [12] appears to have the following form. 1) A portion of a population becomes distinguished, on some basis or bases, in the context of a power relationship. [The particulars are not important for the general outline of the process. The bases may be ideological differences, imputed intrinsic differences, particular functions in the division of labor, and so forth.] 2) The members of this distinguished population segment are "assigned" to a particular social role and fate; [13] that is, the division of labor becomes reorganized.[14] 3) As these people react to the situation in which they find themselves, they become involved with one another, if the situation permits. In other words, social structures develop among them; it is at this point that entity characteristics first become apparent. 4) Then these people become aware of their commonality of fate. The growth of such corporate self-awareness reinforces the structuring tendencies.[15] 5) The further development of the emerging ethnic group will then depend, in part, on the nature of the structures that develop, the content of the group's "self-image," and the shared conception of its destiny. This, of course, emphasizes internal development, which is our present concern. The other big area of causal factors—with which we are not here concerned except to indicate a context of power relations—is the specific character of the relationship with the other segment(s) of the population. Necessarily, internal group development and external (inter-group) relationships influence one another.

It has already been pointed out that the enslaved Africans were not a social group, although from the first they were a distinguishable portion, that is category, of the population. They were not merely physically distinguishable, as has been stressed in the literature, but also socially distinguishable by virtue of their depressed economic situation, with all of the occupational, educational, and associational consequences. With the passage of time it was the latter point that became the most important. Slavery, however, muted the overt, collective responses of the enslaved to the situation. Emancipation altered this picture.

Let us now briefly scan the phases of the processes of ethnogenesis as it has operated in the case of the Negroes.

Reconstruction. During the Reconstruction period the freedmen achieved physical mobility and, consequently, wide-ranging contacts with one another. Also, as a consequence of political participation and the struggle for land, some Negro political leaders emerged. A significant factor that influenced all of the subsequent developments was the failure of the freedman to obtain land and their consequent involvement in cotton farming on, for the most part, the lowest levels of tenancy of the plantation system. It is also significant that during the Reconstruction period the Negroes had to fight in a variety of ways, including organized militia, against the physical onslaughts of Southern whites.[16]

The National Compromise and After. The Reconstruction governments were overthrown by the end of the third quarter of the nineteenth century, and the last quarter of that century witnessed both the actual restoration and the political assertion of "white supremacy." During this period discriminatory practices increased and there developed a tendency to treat all Negroes alike regardless of social attributes. This marked a change from the previous period (Myrdal, note 1, pp. 578–82). *Pari passu,* behind the growing barriers there were developing distinctive structural and ideological attributes among the Negroes; for example, a "lower class" Negro family pattern, Negro businesses catering to Negroes, the expansion of the Negro Church (with the Negro ministers representing the status quo as supernaturally sanctioned), and the emergence of Booker T. Washington as a national Negro leader.

Early Twentieth Century. The peak, or depth, of discriminatory tendencies was reached in the early twentieth century. By this time the Reconstruction state constitutions had been changed, legislation requiring "separate but equal" facilities had been declared constitutional, and in the preceding three decades approximately 2,500 Negroes had been lynched in the South. This last is an index of the community-wide methods of violence and intimidation used by the whites to maintain the situation.

In the first decade of this century, however, there appeared other entity, or ethnic group, characteristics among the Negroes; for example, organized protest in the form of the Niagara Movement and the National Association for the Advancement of Colored People and the attempts to create economic opportunities for Negroes in the form of the National Urban League. These, in turn, give evidence of the Negroes' conception of their destiny—full and unhampered participation in the larger society. This was first expressed organizationally in the statement of the Niagara Movement in 1905.

World War I and After. At the time of World War I there commenced the "Great Migration" to the Northern cities.[17] The advent of the Negroes was accompanied by a number of severe "race riots." (A "race riot" is a situation in which Negroes fight back against extra-legal mob violence.) As a result of the disappointments and frustrations which followed the move North to "freedom" and the "war to save democracy," the 1920s saw the rise of the Garvey [Back-to-Africa] Movement. Although unsuccessful, it is significant because a fundamental part of the ideology of the movement was anti-white and separatist in orientation. (It is also significant that a movement with this orientation failed.) Further, no whites were involved in the leadership of the movement, as was not the case with the previously mentioned Negro lay organizations nor the Negro Church. This fact demonstrated, for the first time, that Negroes could organize Negroes as such. By this time, as Frazier points out (note 8, p. 531):

> The impact of urban living . . . [and] . . . conflicts in the North tended not only to intensify the consciousness of being a Negro, but . . . also gave new

meaning to being a Negro . . . [it] meant being a member of a group with a cause, if not a history. As in the case of the nationalistic struggles in Europe, the emergence of a . . . literature helped in the development of a consciousness. As we have seen, there appeared a Negro Renaissance following World War I. Much of the literature and art of the Negro Renaissance was not only militant but tended to give the Negro a conception of the mission and destiny of the Negro. The Negro newspapers, which began to influence the masses, tended to create a new . . . consciousness . . . [which] did not have separation from American life.[18]

The recent past and the present continue to yield evidence of both Negro ethnicity and the persistence of ethnogenesis. Such evidence is found in the assaults on the "white primary," segregated education at all levels, the increased efforts of Negroes to get Negroes out to vote, the increasing attempts of Southern Negroes to vote, the "sit-ins" in the South, and the "Freedom Riders." [19]

SUMMARY AND CONCLUSION

To sum up, then, Negro-Americans are an instance of a people:

1) whose ancestors, as recently as four generations ago, showed little in the way of ethnic group characteristics and who in this *ante bellum* period could only be conceptualized as a social category;

2) who now form a distinct social and subcultural entity within the American society and are in the process of becoming a full-fledged ethnic group;

3) whose character as an emergent ethnic group is the consequence of factors outside themselves as well as their response to these factors.

The earlier ways of conceptualizing Negroes in Negro-white relations in the United States were called into question because they are based on static category concepts and, as such, appear not to do justice to the phenomenon. The available data seem to require an entity concept that will allow the developmental factors to be taken into account. If the ethnogenesis concept, which has been offered to replace these other conceptual tools, is to be properly evaluated, two questions must be answered: Does this new approach encompass the data and relate them adequately to one another? The answer to this is, ultimately, the task of future investigators. How fruitful is the new approach as a source of hypotheses?

This paper will close with some suggestions in answer to the second question.

1. In the light of the above, it may be suggested that the people whom we call the Negroes in this society are not comparable to the so-called Negroes of Brazil, Haiti, or the British West Indies. Negro-Americans are different from Brazilians, Cubans, Haitians, or Jamaicans who may happen to possess some negroid physical traits. In these latter instances, the use of the term Negro lays stress on biological similarities and blurs the sociological, cultural, and psychological differences.

2. Although there have been and are many instances of ethnogenesis,[20]

this particular instance involving the Negroes has a contemporary uniqueness beyond the particularity which inheres in any single case. As stated above, the special aspect in this case resides in the forcible acculturation of the individual African progenitors in a completely strange setting and the loss of their African cultures in the process. An important consequence of this is that when ethnogenesis moved into high gear less than a century ago the freedmen were "social persons"; that is, they were socialized individuals since they had been members of plantation households. Taken as a whole, however, they comprised a collection of unrelated individuals. Further, this collection of unrelated individuals was without the community of tradition, sentiment, and so forth, that has marked other populations and given rise to ethnic groups such as the Italian immigrants.[21] Thus we have here a case of ethnogenesis starting *ab initio,* unlike all other current instances of ethnogenesis in which members of some ethnic groups become transformed into another ethnic group.[22]

3. To say that Negroes are involved in the process of ethnogenesis is not the same as saying that they are a full-fledged ethnic group. They are not. Full-fledged ethnicity would appear to be characterized by at least two qualities: long tradition and a marked, if not a general, tendency toward self-perpetuation. Concerning tradition, Negro-Americans have neither a legendary nor a long historic past. E. K. Francis writes of what would here be called a full-fledged ethnic group, "Since an ethnic group is based on an elementary feeling of solidarity, we must suppose that mutual adjustment has been achieved over a considerable length of time and that the memory of having possibly belonged to another system of social relationships must have been obliterated. Certainly, this is not the case with Negroes" (note 11, p. 396). As for self-perpetuation, this is usually achieved by endogamy. Now, despite the use of this term with reference to Negro-white relations in the United States, it is quite clear that pressures outside the Negro group are primarily effective in preventing marriages between Negroes and the members of other ethnic groups rather than self-imposed restrictions.[23] The current tendency among Negroes to frown on intermarriage is a defensive *re*action. It is suggested by various writers that this attitude is neither deep nor abiding (Myrdal, note 1, pp. 56–7 and 62–4 and Cox, note 2, pp. 447–50.).

On the basis of the two points of tradition and self-perpetuation, while it is proper to regard the Negroes as an ethnic group, it is also proper to say that they are still in the *process of becoming* an ethnic group, that is, their ethnicity is still *developing*. Paradoxical as the formulation may appear, it is no stranger than applying the term "tree" to a young tree, to a mature tree, and to an old tree. Let us now turn to some predictive notions based on the view of the Negroes as a developing ethnic group.

4. As the Negroes become more of an ethnic group—more focused and organized—it may be expected that rather than *re*acting to the actions of whites, they will increasingly *act* along paths of action chosen to achieve their goal of full, individual participation in the larger society. It can be added that

any successes can be expected to pave the way for increased activity. It may be further hypothesized that, as the barriers to full participation yield and slowly crumble, frustration and impatience over the differences between actuality and aspirations may prompt segments of the Negro group to manifest radical and separatist (anti-white) sentiments, such as the "black Muslim" movement. It is doubtful that any of these organizations will be large. Size, however, should not be confused with importance. By defining one end of the spectrum of Negro responses, such groups will affect the thinking of all Negroes. Further, because of the impact upon the whites, they may contribute to the general struggle for Negro aspirations despite their separatist orientations.

5. A related hypothesis concerns the character of Negro leadership. Negro leaders in the latter part of the nineteenth century typically played the role of justifying the condition of Negroes.[24] Many Negro ministers fell into this category. It is important to add that such leaders were, either directly or as a result of the "veto power," chosen by dominant whites. (This type of leadership is called "accommodating leadership" by Myrdal while Cox refers to it as "the spirit of Uncle Tom.") By the turn of the century, a new kind of leadership had emerged typified by Booker T. Washington. (This type of leadership is called "conciliatory leadership" by Frazier and "compromising leadership" by Myrdal.) It is Myrdal's thesis, as well as Cox's, that whites were, and are, also influential in the selection of this type of leader. If my thesis is correct, then we may expect that in the coming period the character of Negro leadership will have more of the qualities exemplified earlier by W. E. B. DuBois and today by Martin Luther King, Thurgood Marshall, and James Farmer. We may also expect that the "Uncle Toms" and the conciliators will become fewer and fewer.

6. As a final point, it may be suggested that there are implications for psychological research in the social entity approach to the American Negro. The effects of identifying oneself with an emergent group that is no longer on the defensive but is coming more and more to act for itself as well as with a group that has strong leaders and hero figures should make a significant difference in the personalities of Negro-Americans. Indeed, I believe that as the self-image of the Negroes is internalized by individual Negroes, a redirection and transformation of Negro resentment and hostility and a redefinition of individual Negro selves will surely take place.

Chapter **7**

THE STRATIFIED COMMUNITY

THE experience that sociologists study under the heading of stratification has as its most important expression social class in contemporary society. Sociologists have differed greatly in their understanding of class. It is not really possible to reconcile all extant sociological views on the subject; it is all the more important, then, to understand how each viewpoint conceptualizes the problem and which aspects of the phenomenon it focuses on. During the classical period of sociology, for example, there was a fundamental conflict between the Marxist view of class and various viewpoints that were developed as alternatives to Marxism. The Marxists approach class in terms of ownership or lack of ownership of the means of production. Consequently, any Marxist approach to stratification focuses heavily on its economic aspects. Max Weber, by contrast, tried to do equal justice to the noneconomic aspects of stratification (which are usually interpreted by Marxists as dependent variables of the economic forces). Specifically, Weber elaborated a threefold approach to stratification in terms of (1) class—a fundamentally economic category dealing with what he called the "life-chances" of a person or group, (2) power—which Weber defined as the ability to enforce one's wishes even against opposition, and (3) status—roughly equivalent to what is usually meant by prestige.

However stratification may be conceptualized, different strata have very different styles of social life, and much sociological effort has gone into the study of these differences. In American society stratification has been greatly complicated by the racial factor. The divisions between the major racial groups, particularly between whites and blacks, have been of such deep character as to explode the category of class (be it in a Marxist or non-Marxist sense). Some American sociologists have dealt with race under the category of caste, by which is meant a stratum from which the individual cannot escape, whatever his "life chances." There has been some suggestion recently that, perhaps as a result of affluence and growing leisure, styles of life are now developing that cut right across class lines.

Some of the most important studies of class in America were undertaken by W. Lloyd Warner and a group of social scientists working with him. The passage selected from *Social Class in America* deals with differences in stratification in different regions of the United States. The next two selections deal with two distinctive worlds of social class in America: an article by S. M. Miller and Frank Riessman, "The Working Class Subculture," and a passage from *White Collar: The American Middle Classes* by C. Wright Mills. An important feature of this experience of stratification is the encounter with other classes. Members of different classes perceive each other through a web of stereotypes. The last selection, William Saroyan's "Being Refined," gives a lively picture of how someone from an ethnic community (in this instance Armenian) perceives and experiences the world of upper-class WASPs.

W. LLOYD WARNER, MARCHIA MEECKER, AND KENNETH EELLS

SOCIAL CLASS IN AMERICA

CLASS AMONG THE NEW ENGLAND YANKEES

Studies of communities in New England clearly demonstrate the presence of a well-defined social-class system.[1] At the top is an aristocracy of birth and wealth. This is the so-called "old family" class. The people of Yankee City say the families who belong to it have been in the community for a long time—for at least three generations and preferably many generations more than three. "Old family" means not only old to the community but old to the class. Present members of the class were born into it; the families into which they were born can trace their lineage through many generations participating in a way of life characteristic of the upper class back to a generation marking the lowly beginnings out of which their family came. Although the men of this level are occupied gainfully, usually as large merchants, financiers, or in the higher professions, the wealth of the family, inherited from the husband's or the wife's side, and often from both, has been in the family for a long time. Ideally, it should stem from the sea trade when Yankee City's merchants and sea captains made large fortunes, built great Georgian houses on elm-lined Hill Steet, and filled their houses and gardens with the proper symbols of their high position. They became the 400, the Brahmins, the Hill Streeters to whom others looked up; and they, well-mannered or not, looked down on the rest. They counted themselves, and were so counted, equals of similar levels in Salem, Boston, Providence, and other New England cities. Their sons and daughters married into the old families from these towns and at times, when family fortune was low or love was great, they married wealthy sons and daughters from the newly rich who occupied the class level below them. This was a happy event for the fathers and mothers of such fortunate young people in the lower half of the upper class, an event well publicized and sometimes not too discreetly bragged about by the parents of the lower-upper-class children, an occasion to be explained by the mothers from the old families in terms of the spiritual demands of romantic love and by their friends as "a good deal and a fair exchange all the way around for everyone concerned."

The new families, the lower level of the upper class, came up through the

Reprinted from W. Lloyd Warner, Marchia Meecker, and Kenneth Eells, *Social Class in America* (New York: Harper & Row, 1960), pp. 11–21.

new industries—shoes, textiles, silverware—and finance. Their fathers were some of the men who established New England's trading and financial dominance throughout America. When New York's Wall Street rose to power, many of them transferred their activities to this new center of dominance. Except that they aspire to old-family status, if not for themselves then for their children, these men and their families have a design for living similar to the old-family group. But they are consciously aware that their money is too new and too recently earned to have the sacrosanct quality of wealth inherited from a long line of ancestors. They know, as do those about them, that, while a certain amount of wealth is necessary, birth and old family are what really matter. Each of them can cite critical cases to prove that particular individuals have no money at all, yet belong to the top class because they have the right lineage and right name. While they recognize the worth and importance of birth, they feel that somehow their family's achievements should be better rewarded than by a mere second place in relation to those who need do little more than be born and stay alive.

The presence of an old-family class in a community forces the newly rich to wait their turn if they aspire to "higher things." Meanwhile, they must learn how to act, fill their lives with good deeds, spend their money on approved philanthropy, and reduce their arrogance to manageable proportions.

The families of the upper and lower strata of the upper classes are organized into social cliques and exclusive clubs. The men gather fortnightly in dining clubs where they discuss matters that concern them. The women belong to small clubs or to the Garden Club and give their interest to subjects which symbolize their high status and evoke those sentiments necessary in each individual if the class is to maintain itself. Both sexes join philanthropic organizations whose good deeds are an asset to the community and an expression of the dominance and importance of the top class to those socially beneath them. They are the members of the Episcopalian and Unitarian and, occasionally, the Congregational and Presbyterian churches.

Below them are the members of the solid, highly respectable upper-middle class, the people who get things done and provide the active front in civic affairs for the classes above them. They aspire to the classes above and hope their good deeds, civic activities, and high moral principles will somehow be recognized far beyond the usual pat on the back and that they will be invited by those above them into the intimacies of upper-class cliques and exclusive clubs. Such recognition might increase their status and would be likely to make them members of the lower-upper group. The fact that this rarely happens seldom stops members of this level, once activated, from continuing to try. The men tend to be owners of stores and belong to the large proprietor and professional levels. Their incomes average less than those of the lower-upper class, this latter group having a larger income than any other group, including the old-family level.

These three strata, the two upper classes and the upper-middle, constitute the levels above the Common Man. There is a considerable distance socially between them and the mass of the people immediately below them. They comprise three of the six classes present in the community. Although in number of levels they constitute half the community, in population they have no more than a sixth, and sometimes less, of the Common Man's population. The three levels combined include approximately 13 per cent of the total population.

The lower-middle class, the top of the Common Man level, is composed of clerks and other white-collar workers, small tradesmen, and a fraction of skilled workers. Their small houses fill "the side streets" down from Hill Street, where the upper classes and some of the upper-middle live, and are noticeably absent from the better suburbs where the upper-middle concentrate. "Side Streeter" is a term often used by those above them to imply an inferior way of life and an inconsequential status. They have accumulated little property but are frequently home owners. Some of the more successful members of ethnic groups, such as the Italians, Irish, French-Canadians, have reached this level. Only a few members of these cultural minorities have gone beyond it; none of them has reached the old-family level.

The old-family class (upper-upper) is smaller in size than the new-family class (lower-upper) below them. It has 1.4 per cent, while the lower-upper class has 1.6 per cent, of the total population. Ten per cent of the population belongs to the upper-middle class, and 28 per cent to the lower-middle level. The upper-lower is the most populous class, with 34 per cent, and the lower-lower has 25 per cent of all the people in the town.

The prospects of the upper-middle-class children for higher education are not as good as those of the classes above. One hundred per cent of the children of the two upper classes take courses in the local high school that prepare them for college, and 88 per cent of the upper-middle do; but only 44 per cent of the lower-middle take these courses, 28 per cent of the upper-lower, and 26 per cent of the lower-lower. These percentages provide a good index of the position of the lower-middle class, ranking it well below the three upper classes, but placing it well above the upper-lower and the lower-lower.[2]

The upper-lower class, least differentiated from the adjacent levels and hardest to distinguish in the hierarchy, but clearly present, is composed of the "poor but honest workers" who more often than not are only semi-skilled or unskilled. Their relative place in the hierarchy of class is well portrayed by comparing them with the classes superior to them and with the lower-lower class beneath them in the category of how they spend their money.

A glance at the ranking of the proportion of the incomes of each class spent on ten items (including such things as rent and shelter, food, clothing, and education, among others) shows, for example, that this class ranks second for the percentage of the money spent on food, the lower-lower class being first and the rank order of the other classes following lower-middle according to their place in the social hierarchy. The money spent on rent and shelter by upper-lower class is also second to the lower-lower's first, the other classes' rank order and position in the hierarchy being in exact correspondence. To give a bird's-eye view of the way this class spends its money, the rank of the upper-lower, for the percentage of its budget spent on a number of common and important items, has been placed in parentheses after every item in the list which follows: food (2), rent (2), clothing (4), automobiles (5), education (4), and amusements (4–5). For the major items of expenditure the amount of money spent by this class out of its budget corresponds fairly closely with its place in the class hierarchy, second to the first of the lower-lower class for the major necessities of food and shelter, and ordinarily, but not always, fourth or fifth to the classes above for the items that give an opportunity for cutting down the amounts spent on them. Their feelings about doing the right thing, of being respectable and rearing their children to do better than they have, coupled with the limitations of their income, are well reflected in how they select and reject what can be purchased on the American market.[3]

The lower-lower class, referred to as "Riverbrookers" or the "low-down Yankees who live in the clam flats," have a "bad reputation" among those who are socially above them. This evaluation includes beliefs that they are lazy, shiftless, and won't work, all opposites of the good middle-class virtues belonging to the essence of the Protestant ethic. They are thought to be improvident and unwilling or unable to save their money for a rainy day and, therefore, often dependent on the philanthropy of the private or public agency and on poor relief. They are sometimes said to "live like animals" because it is believed that their sexual mores are not too exacting and that premarital intercourse, post-marital infidelity, and high rates of illegitimacy, sometimes too publicly mixed with incest, characterize their personal and family lives. It is certain that they deserve only part of this reputation. Research shows many of them guilty of no more than being poor and lacking in the desire to get ahead, this latter trait being common among those above them. For these reasons and others, this class is ranked in Yankee City below the level of the Common Man (lower-middle and upper-lower). For most of the indexes of status it ranks sixth and last.

CLASS IN THE DEMOCRATIC MIDDLE WEST AND FAR WEST

Cities large and small in the states west of the Alleghenies sometimes have class systems which do not possess an old-family (upper-upper) class. The period of settlement has not always been sufficient for an old-family level, based on the security of birth and inherited wealth, to entrench itself. Ordinarily, it takes several generations for an old-family class to gain and hold the prestige and power necessary to impress the rest of the community sufficiently with the marks of its "breeding" to be able to confer top status on those born into it. The family, its name, and its lineage must have had time to become identified in the public mind as being above ordinary mortals.

While such identification is necessary for the emergence of an old-family (upper-upper) class and for its establishment, it is also necessary for the community to be large enough for the principles of exclusion to operate. For example, those in the old-family group must be sufficiently numerous for all the varieties of social participation to be possible without the use of new-family members; the family names must be old enough to be easily identified; and above all there should always be present young people of marriageable age to become mates of others of their own class and a sufficient number of children to allow mothers to select playmates and companions of their own class for their children.

When a community in the more recently settled regions of the United States is sufficiently large, when it has grown slowly and at an average rate, the chances are higher that it has an old-family class. If it lacks any one of these factors, including size, social and economic complexity, and steady and normal growth, the old-family class is not likely to develop.

One of the best tests of the presence of an old-family level is to determine whether members of the new-family category admit, perhaps grudgingly and enviously and with hostile derogatory remarks, that the old-family level looks down on them and that it is considered a mark of advancement and prestige by those in the new-family group to move into it and be invited to the homes and social affairs of the old families. When a member of the new-family class says, "We've only been here two generations, but we still aren't old-family," and when he or she goes on to say that "they (old family) consider themselves better than people like us and the poor dopes around here let them get away with it," such evidence indicates that an old-family group is present and able to enforce recognition of its superior position upon its most aggressive and hostile competitors, the members of the lower-upper, or new-family, class.

When the old-family group is present and its position is not recognized as superordinate to the new families, the two tend to be co-ordinate and view each other as equals. The old-family people adroitly let it be known that their riches are not material possessions alone but are old-family lineage; the new families display their wealth, accent their power, and prepare their children for the development of a future lineage by giving them the proper training at home and later sending them to the "right" schools and marrying them into the "right" families.

Such communities usually have a five-class pyramid, including an upper class, two middle, and two lower classes.[4]

Jonesville, located in the Middle West, approximately a hundred years old, is an example of a typical five-class community. The farmers around Jonesville use it as their market, and it is the seat of government for Abraham County. Its population of over 6,000 people is supported by servicing the needs of the farmers and by one large and a few small factories.

At the top of the status structure is an upper class commonly referred to as "the 400." It is composed of old-family and new-family segments. Neither can successfully claim superiority to the other. Below this level is an upper-middle class which functions like the same level in Yankee City and is composed of the same kind of people, the only difference being the recognition that the distance to the top is shorter for them and the time necessary to get there much less. The Common Man level, composed of lower-middle- and upper-lower-class people, and the lower-lower level are replicas of the same classes in Yankee City. The only difference is that the Jonesville ethnics in these classes are Norwegian Lutherans and Catholic Poles, the Catholic Irish and Germans having been absorbed for the most part in the larger population; whereas in Yankee City the ethnic population is far more heterogeneous, and the Catholic Irish are less assimilated largely because of more opposition to them, and because the church has more control over their private lives.

The present description of Jonesville's class order can be brief and no more than introductory because all the materials used to demonstrate how to measure social class are taken from Jonesville. The interested reader will obtain a clear picture in the chapters which follow [in *Social Class in America*] of what the classes are, who is in them, the social and economic characteristics of each class, and how the people of the town think about their status order.

The communities of the mountain states and Pacific Coast are new, and many of them have changed their economic form from mining to other enterprises; consequently, their class orders are similar to those found in the Middle West. The older and larger far western communities which have had a continuing, solid growth of population which has not destroyed the original group are likely to have the old-family level at the top with the other classes present; the newer and smaller communities and those disturbed by the destruction of their original status structure by large population gains are less likely to have an old-family class reigning above all others. San Francisco is a clear example of the old-family type; Los Angeles, of the more amorphous, less well-organized class structure.

CLASS IN THE DEEP SOUTH

Studies in the Deep South demonstrate that, in the older regions where social changes until recently have been less rapid and less disturbing to the status order, most of the towns above a few thousand population have a six-class system in which an old-family elite is socially dominant.

For example, in a study of a Mississippi community, a market town for a cotton-growing region around it, Davis and the Gardners found a six-class system. . . .[5]

The people of the two upper classes make a clear distinction between an old aristocracy and an aristocracy which is not old. There is no doubt that the first is above the other; the upper-middle class views the two upper ones much as the upper classes do themselves but groups them in one level with two divisions, the older level above the other; the lower-middle class separates them but considers them co-ordinate; the bottom two classes, at a greater social distance than the others, group all the levels above the Common Man as "society" and one class. An examination of the terms used by the several classes for the other classes shows that similar principles are operating.

The status system of most communities in the South is further complicated by a color-caste system which orders and systematically controls the relations of those categorized as Negroes and whites.

Although color-caste in America is a separate problem and the present volume does not deal with this American status system, it is necessary that we describe it briefly to be sure a clear distinction is made between it and social class. Color-caste is a system of values and behavior which places all people who are thought to be white in a superior position and those who are thought of as black in an inferior status.

Characteristics of American Negroes vary from very dark hair and skin and Negroid features to blond hair, fair skin, and Caucasian features, yet all of them are placed in the "racial" category of Negro. The skin and physical features of American Caucasians vary from Nordic blond types to the dark, swarthy skin and Negroid features of some eastern Mediterranean stocks, yet all are classed as socially white, despite the fact that a sizable proportion of Negroes are "whiter" in appearance than a goodly proportion of whites. The members of the two groups are severely punished by the formal and informal rules of our society if they intermarry, and when they break this rule of "caste endogamy," their children suffer the penalties of our caste-like system by being placed in the lower color caste. Furthermore, unlike class, the rules of this system forbid the members of the lower caste from climbing out of it. Their status and that of their children are fixed forever. This is true no matter how much money they have, how great the prestige and power they may accumulate, or how well they have acquired correct manners and proper behavior. There can be no social mobility out of the lower caste into the higher one. (There may, of course, be class mobility within the Negro or white caste.) The rigor of caste rules varies from region to region in the United States.[6]

The Mexicans, Spanish Americans, and Orientals occupy a somewhat different status from that of the Negro, but many of the characteristics of their social place in America are similar.[7]

The social-class and color-caste hypotheses, inductively established as working principles for understanding American society, were developed in the researches which were reported in the "Yankee City" volumes, *Deep South,* and *Caste and Class in a Southern Town*. Gunnar Myrdal borrowed them, particularly color-caste, and made them known to a large, non-professional American audience.[8]

THE GENERALITIES OF AMERICAN CLASS

It is now time to ask what are the basic characteristics of social status common to the communities of all regions in the United States and, once we have answered this question, to inquire what the variations are among the several systems. Economic factors are significant and important in determining the class position of any family or person, influencing the kind of behavior we find in any class, and contributing their share to the present form of our status system. But, while significant and necessary, the economic factors are not sufficient to predict where a particular family or individual will be or to explain completely the phenomena of social class. Something more than a large income is necessary for high social position. Money must be translated into socially approved behavior and possessions, and they in turn must be translated into intimate participation with, and acceptance by, members of a superior class.

This is well illustrated by what is supposed to be a true story of what happened to a Mr. John Smith, a newly rich man in a far western community. He wanted to get into a particular social club of some distinction and significance in the city. By indirection he let it be known, and was told by his friends in the club they had submitted his name to the membership committee.

Mr. Abner Grey, one of the leading members of the club and active on its membership committee, was a warm supporter of an important philanthropy in this city. It was brought to his attention that Mr. Smith, rather than contributing the large donation that had been expected of him, had given only a nominal sum to the charity.

When Mr. Smith heard nothing more about his application, he again approached one of the board members. After much evasion, he was told that Mr. Grey was the most influential man on the board and he would be wise to see that gentleman. After trying several times to make an appointment with Mr. Grey, he finally burst into Grey's offices unannounced.

"Why the hell, Abner, am I being kept out of the X club?"

Mr. Grey politely evaded the question. He asked Mr. Smith to be seated. He inquired after Mr. Smith's health, about the health of his wife, and inquired about other matters of simple convention.

Finally, Mr. Smith said, "Ab, why the hell am I being kept out of your club?"

"But, John, you're not. Everyone in the X club thinks you're a fine fellow."

"Well, what's wrong?"

"Well, John, we don't think you've got the *kind* of money necessary for being a good member of the X club. We don't think you'd be happy in the X club."

"Like hell I haven't. I could buy and sell a half dozen of some of your board members."

"I know that, John, but that isn't what I said. I did not say the amount of money. I said the kind of money."

"What do you mean?"

"Well, John, my co-workers on the charity drive tell me you only gave a few dollars to our campaign, and we had you down for a few thousand."

For a moment Mr. Smith was silent. Then he grinned. So did Mr. Grey. Smith took out his fountain pen and checkbook. "How much?"

At the next meeting of the X club Mr. Smith was unanimously elected to its membership.

Mr. Smith translated his money into philanthropy acceptable to the dominant group, he received their sponsorship, and finally became a participant in the club. The "right" kind of house, the "right" neighborhood, the "right" furniture, the proper behavior—all are symbols that can ultimately be translated into social acceptance by those who have sufficient money to aspire to higher levels than they presently enjoy.

To belong to a particular level in the social-class system of America means that a family or individual has gained acceptance as an equal by those who belong in the class. The behavior in this class and the participation of those in it must be rated by the rest of the community as being at a particular place in the social scale.

Although our democratic heritage makes us disapprove, our class order helps control a number of important functions. It unequally divides the highly and lowly valued things of our society among the several classes according to their rank. Our marriage rules conform to the rules of class, for the majority of marriages are between people of the same class. No class system, however, is so rigid that it completely prohibits marriages above and below one's own class. Furthermore, an open class system such as ours permits a person during his lifetime to move up or down from the level into which he was born. Vertical social mobility for individuals or families is characteristic of all class systems. The principal forms of mobility in this country are through the use of money, education, occupation, talent, skill, philanthropy, sex, and marriage. Although economic mobility is still important, it seems likely now that more people move to higher positions by education than by any other route. We have indicated before this that the mere possession of money is insufficient for gaining and keeping a higher social position. This is equally true of all other forms of mobility. In every case there must be social acceptance.

Class varies from community to community. The new city is less likely than an old one to have a well-organized class order; this is also true for cities whose growth has been rapid as compared with those which have not been disturbed by huge increases in population from other regions or countries or by the rapid displacement of old industries by new ones. The mill town's status hierarchy is more likely to follow the occupational hierarchy of the mill than the levels of valuated participation found in market towns or those with diversified industries. Suburbs of large metropolises tend to respond to selective factors which reduce the number of classes to one or a very few. They do not represent or express all the cultural factors which make up the social pattern of an ordinary city.

Yet systematic studies from coast to coast, in cities large and small and of many economic types, indicate that, despite the variations and diversity, class levels do exist and that they conform to a particular pattern of organization.

S. M. MILLER AND FRANK RIESSMAN

THE WORKING CLASS SUBCULTURE: A NEW VIEW

INTRODUCTION

A decade and a half ago the working class was depicted by Allison Davis and Robert J. Havighurst [1] as permissive and indulgent toward their children and free of the emotional strain of impulse-inhibition which characterized the middle class in the United States. Indeed, it was felt by many that the middle class had much to envy and imitate in the working class.[2] This romantic view of the working class has faded. It is now asserted that the working class (usually termed the "lower class") is incapable of deferring gratification [3] and consequently unable to make major strides in improving their conditions. Frequently accompanying this view is the belief that this lower class is "immoral," "uncivilized," "promiscuous," "lazy," "obscene," "dirty," and "loud." With the rising plane and standard of living of workers has come the argument that workers are middle class in their outlook and desires; the difficulties in attaining full middle-class status lead to juvenile delinquency on the part of those youth who fall back into the working and lower classes and to authoritarianism on the part of those who rise into the middle class. Recently, a further vigorous blow has felled any notions of desirable characteristics of workers: their economic liberalism is not paralleled by political liberalism for workers are said to be more authoritarian in outlook than are members of the middle class. The free, spontaneous worker is now seen as an aggressive, authoritarian, yet fettered person. . . .

In this paper, we can only present a few elements of what we believe is a more realistic picture of workers. This analysis is severely compressed and truncated in this presentation and it might be helpful therefore to indicate at the outset an important element of our general orientation. Our stress is much more on cognitive and structural factors than on the more commonly cited affectual and motivational ones. The nature of the conditions of work-

Reprinted from *Social Problems* 9 (Summer 1961): 86–97 by permission of the authors and the publisher.

ing-class lives (jobs, opportunities, family structure) affects behavior more than has been frequently realized; similarly, modes of understanding the environment can be more important than deep-seated personality factors in behavioral patterns. (For example, workers' low estimates of opportunities and high expectations of risk and loss may be more crucial in the unwillingness to undertake certain long-term actions than personality inadequacies involved in a presumed inability to defer gratification.) This is not to argue that motivational-psychological-affectual variables are unimportant but that they have been overstressed while cognitive and structural variables have been underemphasized. The recognition of the importance of the internal life of man has sometimes overshadowed the significance of the more manifest aspects of his existence.

Our definition of working class is simple: regular members of the non-agricultural labor force in manual occupations. Thus, we exclude the "lower class," irregular working people, although the analysis has some relevance to the lower class as will be mentioned below. One of the greatest sources of difficulties in understanding non-upper and non-middle class behavior is that social scientists have frequently used the omnibus category of "lower class" to encompass the stable, and frequently mobile, fairly high income skilled workers, the semi-skilled factory worker, the worker in varied service trades, the unskilled worker and the irregular worker. This collection is probably more a congeries of fairly disparate groups than a category with similar life chances and circumstances. It is especially important to distinguish the segment which has irregular employment (and "voluntary" withdrawals from the labor force), unskilled jobs in service occupations (and is largely Negro and Puerto Rican now) from the other groupings, which are larger and have more of a commonness to them.

This latter group of regular workmen we call "working class" despite the reluctance of many social scientists today to use this historic term; the opprobrious term "lower class" might be applied to the irregular segment although it would probably be better all around if a less invidious term (perhaps "the unskilled") were employed.

The reluctance to make the distinction between "working class" and "lower class," despite useful discussions by Kahl [4] and others, not only is a topic worthy of independent study, but leads to error. For example, Hollingshead and Redlich in their important study have been interpreted as finding that the lower the class, the higher the rate of mental illness. Close examination of their data reveals, however, that the working class, Class IV, is closer to the upper and middle classes, Classes I, II and III, than to the lower class, Class V. Classes I through IV are similar, while Class V is quite dissimilar from all the other classes, including the working class.[5]

Within the working class, we are primarily interested in the *stable* working-class subculture. We believe there is considerable variation within the working class, but the differences probably are variations upon the theme of the stable working-class pattern. While we think in terms of working class subcultures, and, to some extent, lower-class subcultures, a key to understanding them, we believe, is likely to be the *stable* working-class subculture.

Our analysis is aimed at developing *themes* in working-class life. Thus, we are interpreting the *meaning* of findings rather than reporting new findings. We have utilized the published materials commonly employed plus our own interviews and observations of working-class people. . . .

BASIC THEMES

Before discussing a few of the themes which we think are basic in working-class life, we present a brief overall picture of what we believe are the essential characteristics of the stable American worker today.

He is traditional, "old fashioned," somewhat religious, and patriarchal. The worker likes discipline, structure, order, organization and directive, definite (strong) leadership, although he does not see such strong leadership in opposition to human, warm, informal, personal qualities. Despite the inadequacy of his education, he is able to build abstractions, but he does so in a slow, physical fashion.[6] He reads ineffectively, is poorly informed in many areas, and is often quite suggestible, although interestingly enough he is frequently suspicious of "talk" and "new fangled ideas."

He is family centered; most of his relationships take place around the large extended, fairly cooperative family. Cooperation and mutual aid are among his most important characteristics. While desiring a good standard of living, he is not attracted to the middle-class style of life with its accompanying concern for status and prestige. He is not class conscious although aware of class differences. While he is somewhat radical on certain economic issues, he is quite illiberal on numerous matters, particularly civil liberties and foreign policy. The outstanding weakness of the worker is lack of education. Strongly desiring education for his children, he shows considerable concern about their school work, although he feels estranged and alienated from the teacher and the school, as he similarly feels alienated from many institutions in our society.[7] This alienation is expressed in a ready willingness to believe in the corruptness of leaders and a general negative feeling toward "big shots."

He is stubborn in his ways, concerned with strength and ruggedness, interested in mechanics, materialistic, superstitious, holds an "eye for an eye" psychology, and is largely uninterested in politics.

Stability and Security. We suspect that one of the central determinants in working-class life is the striving for stability and security. External and internal factors promote instability and insecurity. Chief among the external factors is unemployment and layoff. Prosperity has of course barred the anguish of the prolonged depression of the 1930's, but the danger of occasional layoffs of some duration are not remote during the usually shaky prosperity conditions which are interlarded with episodes of recession, plant relocation, industry decline and strikes.[8]

Chief among the internal factors promoting instability are family discord, including divorce and desertion, intergenerational conflict, and the desire for excitement.

Coping with the instability threats becomes a dominant activity within the working-class family. Many practices, such as mutual aid and cooperation, extended family perspectives, are important as adjustive mechanisms. "Getting by" rather than "getting ahead" in the middle-class self-realization and advancement sense is likely to be dominant. For example, the limited desire to become foremen is partly a result of the economic insecurity resulting from the loss of job seniority in case of a layoff.[9]

Part of the ambivalence toward obtaining a college education reflects the same emphasis on security. Even a highly talented working-class youth is not sure what he can do with a college diploma, and he may fear the disruption of his familial, community and peer group security.[10]

The poll data indicating the unwillingness of workers to take economic risks and their greater concern for jobs with security, is part of the same pattern of a striving for stability.

Traditionalism. The American working class is primarily a migrant group; not only have people come from European farms and rural settlements to American factories but they also have migrated from America's rural life to the industrial scene.[11] Traditional practices, once thought to be infrequent in urbanized, industrialized, nuclear-oriented families, are very strong in working-class families. The pattern is patriarchal, extended (with many relevant cousins, grandparents, and aunts and uncles) and delineated by sharply separated sex roles. The family is not child-centered (or child-dominant or dominating), but parent-centered and controlled. Traditional values of automatic obedience by children are expected to be the norm even if not always observed in practice.

One probable consequence of this is that workers seem to be more authoritarian than they probably are. For while on the F-scale type of test, they tend to be "conventional," a characteristic of the authoritarian according to Adorno et al., it is doubtful, . . . that this conventionalism means the same in both the middle and working class.

The worker also has a traditional attitude toward discipline which again may be confused with authoritarianism. All the child-rearing data indicate that workers utilize physical punishment as a basic discipline technique. In the eyes of the worker punishment discourages people from wrong-doing whether the punishment is inflicted upon them or upon others who serve as "examples." There is also a "rightness" about punishment for a misdeed, for punishment is the other side of responsibility for one's actions. Thus, for example, acceptance of the death penalty may not be the result of a sado-masochistic character structure but the product of a belief in the efficacy of punishment in deterring others from misdeeds and in the value of attaching responsibility to people's actions. Workers consequently do not easily accept the notion that an individual is not responsible for his crimes because of his emotional state at the time of their occurrence.

Intensity. We believe that one of the most neglected themes in working-class life and one of the most difficult to understand and interpret is that of intensity. This intensity is expressed in a number of different ways. It is found in the areas in which workers have belief and emotional involvement. While there are numerous areas about which workers are confused, and lacking in opinion (e.g., the high percentage of "no answer" and "don't know" on public opinion polls), there are important spheres in which they have definite convictions, and indeed, are highly stubborn. Their beliefs about religion, morality, superstition, diet, punishment, custom, traditional education, the role of women, intellectuals, are illustrative here. Many of these attitudes are related to their traditional orientation and they are held unquestioningly in the usual traditional manner. They are not readily open to reason and they are not flexible opinions.

Other possible sources of this intensity may be their physical (less symbolic) relation to life,[12] their person centeredness (to be discussed below), and their lack of education.

Person-Centered. Threaded through much of working-class life is a person-centered theme. On one level this theme has an informal, human quality, of easy, comfortable relationship to people where the affectionate bite of humor is appreciated. The factory "horse-play," the ritualistic kidding, is part of this although by no means all of it. It is an expressive component of life.

At another level, it is the importance of personal qualities. One learns more from people than from books, it is said. At a political level, the candidate as a decent, human person is more important than the platform.

In the bureaucratic situation, the worker still tends to think of himself as relating to people, not to roles and invisible organizational structure. This orientation is an aspect of particularism, the reaction to persons and situations in terms of their personal qualities and relations to oneself rather than in terms of some universal characteristics of their social position. The neighbor or workmate who gets ahead is expected "not to put on airs"; he should like the "old gang" and accept them despite his new position. An individual is expected to transcend his office. A foreman is a s.o.b. not because he has stresses and demands on the job which force him to act forcibly and harshly, but because of his personal qualities. Contrariwise, one of the top executives is frequently regarded as one who would help the rank-and-file workers if he had the chance, because *he* is a "nice guy"; putting him in the stresses of a new position would not force him to act as others in that position have acted.[13] It is the man not the job that makes for behavior; this attitude is not a class-conscious one, far from it. Another example of particularism is the juvenile delinquent who reacts positively to the social worker or therapist who seems to be interested in him beyond the call of professional duty.

Pragmatism and Anti-Intellectualism. With workers, it is the end-result of action rather than the planning of action or the preoccupation with means that counts. An action that goes astray is not liked for itself; it has to achieve the goal intended to be satisfactory.[14] It is results that pay off. While this orientation has an anti-intellectual dimension, it does somewhat reduce the reliance on personality (person-centered theme) by its emphasis on results. Workers like the specific action, the clear action, the understood result. What can be seen and felt is more likely to be real and true in the workers' perspectives, which are therefore likely to be limited. The pragmatic orientation of workers does not encourage them to see abstract ideas as useful. Education, for what it does for one in terms of opportunities, may be desirable but abstract intellectual speculation, ideas which are not rooted in the realities of the present, are not useful, indeed may be harmful.

On the other hand, workers often have an exaggerated respect for the ability of the learned. A person with intellectual competence in one field is frequently thought to be a "brain" with ability in all fields; partly this is due to the general abstract nature of ideas regardless of field. If a real obstacle comes up, they may expect "the brain" to have a ready solution for it, even if they may not be willing to adopt it.

At first glance, the anti-words orientation may appear to be incompatible with the possible appeal of the charismatic. But it is not. For the charismatic are charismatic because they can be emotional and expressive, qualities not usually associated with abstract ideas. Also, the charismatic leader may promise "pie in the sky" but it is a very concrete, specific set of ingredients with a clear distribution of the pie.

Excitement. Another component in workers' lives is the appreciation of excitement, of moving out of the humdrum. News, gossip, new gadgets, sports, are consequently very attractive to workers. To some extent, the consumership of workers—the desire to have new goods, whether television sets or cars—is part of this excitement dimension. The excitement theme is often in contradiction with the traditional orientation.

It is worth noting that different subgroups within the working class may favor one theme rather than another. Thus younger groups, and especially juvenile delinquents, are probably much more attracted to the excitement theme, are more alienated and less traditional. On the other hand, workers with a more middle-class orientation are probably less alienated, more traditional and pragmatic.

Parsimony and Variation. In the preceding remarks we have touched only very fleetingly on a few themes of working-class life and ignored other important themes, like cooperation and a physical orientation, almost completely. While we can sum up our analysis in a relatively few descriptive adjectives, such as person centered, traditional, pragmatic, etc., we have been unable to develop a parsimonious conceptualization, such as a non-deferred gratification pattern which attempts to explain by this single formulation or theme a vast array of behavior. Perhaps the simplest shorthand, if one wishes to use it, would be Parsons'; employing his criteria, we could say that workers are particularistic rather than universalistic, affective rather than neutral, ascriptive rather than achievement-minded, diffuse in definition of role rather than specific. But this summary may obscure more than it reveals.

Indeed, our analysis contains a number of themes which may, in part, be in opposition to each other. For example, traditionalism and alienation have certain conflicting features, as do pragmatism and person centeredness, and the resulting strains and adjustive mechanisms are important to analyze.

Let us make just two points to indicate the general value of the orientation that we have only sketchily presented here: (1) It may be possible to understand other working-class and lower-class styles by looking for sources of variation from the stable working-class pattern. (2) The development of the stable working-class style among lower-class and working-class youth might be the goal of educational and other socializing and remedial forces rather than instilling the middle-class value structure.

VARIATIONS OF WORKING-CLASS CULTURE

By stating that we are describing the *stable* worker we imply that there are other worker subcultures. We feel that the stable worker has been relatively ignored in the emphasis on the "underprivileged," "lower class," unskilled, irregular worker and the middle-class oriented worker. By understanding the stable worker, important leads are provided for understanding other subcultural variations.

The unskilled, irregular (read "lower class") worker lacks the disciplined, structured and traditional approach of the stable worker and stresses the excitement theme. He does less to cope with insecurity and instability. In the large industrial and commercial centers today the lower-class style of life (as distinct from the stable working-class style) is found particularly among peoples relatively new to industrial and urban life: Negroes, Puerto Ricans, transplanted Southern whites. They have not been able so far to make the kind of adjustment that stable workers have. Frequently, they have special problems not only of discrimination but of fairly menial (service) jobs at low pay, extremely poor housing and considerable overcrowding. Some children of stable workers do not develop the stable pattern and assume the lower-class style. A few children of middle-class parents become lower class: they have unskilled jobs and adopt the lower-class style of life. But the bulk of individuals with the lower-class style come from those who are children of unskilled workers and of farmers, thus including many of the ethnic people of whom we spoke earlier.[15]

Another deviant group from the main working-class pattern are those workers who are very much concerned with achievement of success for children and for the symbols of success in consumership. In many cases the families are secure and stable and have been able to make a workable accommodation to the stresses of their lives. But this is not enough for the middle-class orientation; in many cases there is a vague opportunity and motivational factor present.

Those of working-class origins who do move into the middle class and into the middle-class style of life are likely to have a middle-class cross-pressure in that they more frequently than other working-class children have relatives who were or are middle class. Their grandparents may have been middle class; their parents though in working-class occupations are more likely to have more education than is typical in the working class and to have other attributes of middle-class life. If we may give a literary example, in *Sons and Lovers,* the hero, brought up in a mining community, had a working-class father but his mother was a teacher and came from a middle-class community. Undoubtedly, the hero, whose life follows that of D. H. Lawrence, received motivation from her to move into literary activities and probably also some early direct help in reading and school. The motivational factor is important but it is likely linked to the background and experiential factor of grandparental and paternal activities.

We have discussed these two styles in different ways. The lower-class style is considered to be the inability to develop an adequate measure of coping with the environment so that some degree of security and stability ensues. The origin of the middle-class style would seem to emerge from the stable pattern. A working-class family would likely first go through a stable period of accommodation before it or the children developed middle-class orientations. *It is not intrinsic in the stable pattern that a middle-class orientation emerge but the stable stage would seem to be a necessary step in most cases for the development of a middle-class orientation.*

Other variations in the subculture of workers exist. Religious, ethnic, educational, and regional factors are important in producing deviations from the pattern we have described.

THE STABLE STYLE AS GOAL

Explicitly as well as implicitly, many agents of educational and other institutions that deal with working-class and lower-class youth attempt to "middle-classize" them. When any effort is extended toward the juvenile delinquent, it is usually with this orientation. Such endeavors are largely a failure because the middle-class outlook is alien to the experiences, prospects and values of these youth. Possibly there is a better chance of emphasizing working-class values; for example cooperation—as happens in group therapy—rather than vocational success in middle-class terms. We recognize that it is not easy to develop some of the working-class values but they are probably much easier to develop than the middle-class ones. In addition, emphasis on the former may develop a more favorable attitude on the part of the youth to both the institution and its agents than does the insistence on the middle-class values.

A basic value question is involved here: Do we attempt to make the middle-class style a model for all to follow? Or do we adopt a rigid cultural relativity position that the lower class has a right to its way of life regardless of the social effects? Or do we attempt to develop what appear to be the most positive elements, from the point of view of society and the individuals involved, of the styles of life closest to them? While we have some doubts about the answer, the possibility of the stable working-class style as the goal adds a new dimension to a deep problem that deserves more forthright scrutiny than it has received.

Our attempts at interpreting working-class life will undoubtedly prove inadequate. But we are certain that without an attempt at analyzing the contexts and the genotypes of working-class behavior and attitude, the *description* (and there is faulty description) and interpretation of working-class life will remain a reflex of social scientists' changing attitudes toward the middle class.

C. WRIGHT MILLS

WHITE COLLAR: THE AMERICAN MIDDLE CLASSES

IN the early nineteenth century, although there are no exact figures, probably four-fifths of the occupied population were self-employed enterprisers; by 1870, only about one-third, and in 1940, only about one-fifth, were still in this old middle class. Many of the remaining four-fifths of the people who now earn a living do so by working for the 2 or 3 percent of the population who now own 40 or 50 percent of the private property in the United States. Among these workers are the members of the new middle class, white-collar people on salary. For them, as for wage-workers, America has become a nation of employees for whom independent property is out of range. Labor markets, not control of property, determine their chances to receive income, exercise power, enjoy prestige, learn and use skills.

OCCUPATIONAL CHANGE

Of the three broad strata composing modern society, only the new middle class has steadily grown in proportion to the whole. Eighty years ago, there

Reprinted from *White Collar: The American Middle Classes* by C. Wright Mills (New York: Oxford University Press, 1951), pp. 63–76. Reprinted by permission.

were three-quarters of a million middle-class employees; by 1940, there were over twelve and a half million. In that period the old middle class increased 135 percent; wage-workers, 255 percent; new middle class, 1600 percent.

The Labor Force

	1870	1940
Old Middle Class	33%	20%
New Middle Class	6	25
Wage-Workers	61	55
Total	100%	100%

The employees composing the new middle class do not make up one single compact stratum. They have not emerged on a single horizontal level, but have been shuffled out simultaneously on the several levels of modern society; they now form, as it were, a new pyramid within the old pyramid of society at large, rather than a horizontal layer. The great bulk of the new middle class are of the lower middle-income brackets, but regardless of how social stature is measured, types of white-collar men and women range from almost the top to almost the bottom of modern society.

The managerial stratum, subject to minor variations during these decades, has dropped slightly from 14 to 10 percent; the salaried professionals, displaying the same minor ups and down, have dropped from 30 to 25 percent of the new middle class. The major shifts in over-all composition have been in the relative decline of the sales group, occurring most sharply around 1900, from 44 to 25 percent of the total new middle class; and the steady rise of the office workers, from 12 to 40 percent. Today the three largest occupational groups in the white-collar stratum are schoolteachers, salespeople in and out of stores, and assorted office workers. These three form the white-collar mass.

New Middle Class

	1870	1940
Managers	14%	10%
Salaried Professionals	30	25
Salespeople	44	25
Office Workers	12	40
Total	100%	100%

White-collar occupations now engage well over half the members of the American middle class as a whole. Between 1870 and 1940, white-collar workers rose from 15 to 56 percent of the middle brackets, while the old middle class declined from 85 to 44 percent.

The Middle Classes

	1870	1940
Old Middle Class	*85%*	*44%*
Farmers	62	23
Businessmen	21	19
Free Professionals	2	2
New Middle Class	*15%*	*56%*
Managers	2	6
Salaried Professionals	4	14
Salespeople	7	14
Office Workers	2	22
Total Middle Classes	100%	100%

Negatively, the transformation of the middle class is a shift from property to no-property; positively, it is a shift from property to a new axis of stratification, occupation. The nature and well-being of the old middle class can best be sought in the condition of entrepreneurial property; of the new middle class, in the economics and sociology of occupations. The numerical decline of the older, independent sectors of the middle class is an incident in the centralization of property; the numerical rise of the newer salaried employees is due to the industrial mechanics by which the occupations composing the new middle class have arisen.

INDUSTRIAL MECHANICS

In modern society, occupations are specific functions within a social division of labor, as well as skills sold for income on a labor market. Contemporary divisions of labor involve a hitherto unknown specialization of skill: from arranging abstract symbols, at $1000 an hour, to working a shovel, for $1000 a year. The major shifts in occupations since the Civil War have assumed this industrial trend: as a proportion of the labor force, fewer individuals manipulate *things,* more handle *people* and *symbols.*

This shift in needed skills is another way of describing the rise of the white-collar workers, for their characteristic skills involve the handling of paper and money and people. They are expert at dealing with people transiently and impersonally; they are masters of the commercial, professional, and technical relationship. The one thing they do not do is live by making things; rather, they live off the social machineries that organize and co-ordinate the people who do make things. White-collar people help turn what someone else has made into profit for still another; some of them are closer to the means of production, supervising the work of actual manufacture and recording what is done. They are the people who keep track; they man the paper routines involved in distributing what is produced. They provide technical and personal services, and they teach others the skills which they themselves practice, as well as other skills transmitted by teaching.

As the proportion of workers needed for the extraction and production of things declines, the proportion needed for servicing, distributing, and co-ordinating rises. In 1870, over three-fourths, and in 1940, slightly less than one-half of the total employed were engaged in producing things.

	1870	1940
Producing	77%	46%
Servicing	13	20
Distributing	7	23
Co-ordinating	3	11
Total employed	100%	100%

By 1940, the proportion of white-collar workers of those employed in industries primarily involved in the production of things was 11 percent; in service industries, 32 percent; in distribution, 44 percent; and in coordination, 60 percent. The white-collar industries themselves have grown, and within each industry the white-collar occupations have grown. Three trends lie back of the fact that the white-collar ranks have thus been the most rapidly growing of modern occupations: the increasing productivity of machinery used in manufacturing; the magnification of distribution; and the increasing scale of co-ordination.

The immense productivity of mass-production technique and the increased application of technologic rationality are the first open secrets of modern occupational change: fewer men turn out more things in less time. In the middle of the nineteenth century, as J. F. Dewhurst and his associates have calculated, some 17.6 billion horsepower hours were expended in American industry, only 6 percent by mechanical energy; by the middle of the twentieth century, 410.4 billion horsepower hours will be expended, 94 percent by mechanical energy. This industrial revolution seems to be permanent, seems to go on through war and boom and slump; thus 'a decline in production results in a more than proportional decline in employment; and an increase in production results in a less than proportional increase in employment.'

Technology has thus narrowed the stratum of workers needed for given volumes of output; it has also altered the types and proportions of skill needed in the production process. Know-how, once an attribute of the mass of workers, is now in the machine and the engineering elite who design it. Machines displace unskilled workmen, make craft skills unnecessary, push up front the automatic motions of the machine-operative. Workers composing the new lower class are predominantly semi-skilled: their proportion in the urban wage-worker stratum has risen from 31 percent in 1910 to 41 percent in 1940.

The manpower economies brought about by machinery and the large-scale rationalization of labor forces, so apparent in production and extraction, have not, as yet, been applied so extensively in distribution—transportation, communication, finance, and trade. Yet without an elaboration of these means of distribution, the wide-flung operations of multi-plant producers could not be integrated nor their products distributed. Therefore, the proportion of people engaged in distribution has enormously increased so that today about one-fourth of the labor force is so engaged. Distribution has expanded more than production because of the lag in technological application in this field, and because of the persistence of individual and small-scale entrepreneurial units at the same time that the market has been enlarged and the need to market has been deepened.

Behind this expansion of the distributive occupations lies the central problem of modern capitalism: to whom can the available goods be sold? As volume swells, the intensified search for markets draws more workers into the distributive occupations of trade, promotion, advertising. As far-flung and intricate markets come into being, and as the need to find and create even more markets becomes urgent, 'middle men' who move, store, finance, promote, and sell goods are knit into a vast network of enterprises and occupations.

The physical aspect of distribution involves wide and fast transportation networks; the co-ordination of marketing involves communication; the search for markets and the selling of goods involves trade, including wholesale and retail outlets as well as financial agencies for commodity and capital markets. Each of these activities engage more people, but the manual jobs among them do not increase so fast as the white-collar tasks.

Transportation, growing rapidly after the Civil War, began to decline in point of the numbers of people involved before 1930; but this decline took place among wage-workers; the proportion of white-collar workers employed in transportation continued to rise. By 1940, some 23 percent of the people in transportation were white-collar employees. As a new industrial segment of the U.S. economy, the communication industry has never been run by large numbers of free enterprisers; at the outset it needed large numbers of technical and other white-collar workers. By 1940, some 77 percent of its people were in new middle-class occupations.

Trade is now the third largest segment of the occupational structure, exceeded only by farming and manufacturing. A few years after the Civil War less than 5 out of every 100 workers were engaged in trade; by 1940 almost 12 out of every 100 workers were so employed. But, while 70 percent of those in wholesaling and retailing were free enterprisers in 1870, and less than 3 percent were white collar, by 1940, of the people engaged in retail trade 27 percent were free enterprisers; 41 percent white-collar employees.

Newer methods of merchandising, such as credit financing, have resulted in an even greater percentage increase in the 'financial' than in the 'commercial' agents of distribution. Branch banking has lowered the status of many banking employees to the clerical level, and reduced the number of executive positions. By 1940, of all employees in finance and real estate 70 percent were white-collar workers of the new middle class.

The organizational reason for the expansion of the white-collar occupations is the rise of big business and big government, and the consequent trend of modern social structure, the steady growth of bureaucracy. In every branch of the economy, as firms merge and corporations become dominant, free entrepreneurs become employees, and the calculations of accountant, statistician, bookkeeper, and clerk in these corporations replace the free 'movement of prices' as the co-ordinating agent of the economic system. The rise of thousands of big and little bureaucracies and the elaborate specialization of the system as a whole create the need for many men and women to plan, co-ordinate, and administer new routines for others. In moving from smaller to larger and more elaborate units of economic activity, increased proportions of employees are drawn into co-ordinating and managing. Managerial and professional employees and office workers of varied sorts—floorwalkers, foremen, office managers—are needed; people to whom subordinates report, and who in turn report to superiors, are links in chains of power and obedience, co-ordinating and supervising other occupational experiences, functions, and skills. And all over the economy, the proportion of clerks of all sorts has increased: from 1 to 2 percent in 1870 to 10 or 11 percent of all gainful workers in 1940.

As the worlds of business undergo these changes, the increased tasks of government on all fronts draw still more people into occupations that regulate and service property and men. In response to the largeness and predatory complications of business, the crises of slump, the nationalization of the rural economy and small-town markets, the flood of immigrants, the urgencies of war and the march of technology disrupting social life, government increases its co-ordinating and regulating tasks. Public regulations, social services, and business taxes require more people to make mass records and to integrate people, firms, and goods, both within government and in the various segments of business and private life. All branches of government have grown, although the most startling increases are found in the executive branch of the Federal Government, where the needs for co-ordinating the economy have been most prevalent.

As marketable activities, occupations change (1) with shifts in the skills required, as technology and rationalization are unevenly applied across the economy; (2) with the enlargement and intensification of marketing operations in both the commodity and capital markets; and (3) with shifts in the organization of the division of work, as expanded organizations require co-ordination, management, and recording. The mechanics involved within and between these three trends have led to the numerical expansion of white-collar employees.

There are other less obvious ways in which the occupational structure is shaped: high agricultural tariffs, for example, delay the decline of farming as an occupation; were Argentine beef allowed to enter duty-free, the number of meat producers here might diminish. City ordinances and zoning laws abolish peddlers and affect the types of construction workers that prevail. Most states have bureaus of standards which limit entrance into professions and semi-professions; at the same time members of these occupations form associations in the attempt to control entrance into 'their' market. More successful than most trade unions, such professional associations as the American Medical Association have managed for several decades to level off the proportion of physicians and surgeons. Every phase of the slump-war-boom cycle influences the numerical importance of various occupations; for instance, the movement back and forth between 'construction worker' and small 'contractor' is geared to slumps and booms in building.

The pressures from these loosely organized parts of the occupational world draw conscious managerial agencies into the picture. The effects of attempts to manage occupational change, directly and indirectly, are not yet great, except of course during wars, when government freezes men in their jobs or offers incentives and compulsions to remain in old occupations or shift to new ones. Yet increasingly the class levels and occupational composition of the nation are managed; the occupational structure of the United States is being slowly reshaped as a gigantic corporate group. It is subject not only to the pulling of autonomous markets and the pushing of technology but to an 'allocation of personnel' from central points of control. Occupational change thus becomes more conscious, at least to those who are coming to be in charge of it.

WHITE-COLLAR PYRAMIDS

Occupations, in terms of which we circumscribe the new middle class, involve several ways of ranking people. As specific activities, they entail various types and levels of *skill,* and their exercise fulfils certain *functions* within an industrial division of labor. These are the skills and functions we have been examining statistically. As sources of income, occupations are connected with *class* position; and since they normally carry an expected quota of prestige, on and off the job, they are relevant to *status* position. They also involve certain degrees of *power* over other people, directly in terms of the job, and indirectly in other social areas. Occupations are thus tied to class, status, and power as well as to skill and function; to understand the occupations composing the new middle class, we must consider them in terms of each of these dimensions.

'Class situation' in its simplest objective sense has to do with the amount and source of income. Today, occupation rather than property is the source of income for most of those who receive any direct income: the possibilities of selling their services in the labor market, rather than of profitably buying and selling their property and its yields, now determine the life-chances of most of the middle class. All things money can buy and many that men dream about are theirs by virtue of occupational income. In new middle-class occupations men work for someone else on someone else's property. This is the clue to many differences between the old and new middle classes, as well as to the contrast between the older world of the small propertied entrepreneur and the occupational structure of the new society. If the old middle class once fought big property structures in the name of small, free properties, the new middle class, like the wage-workers in latter-day capitalism, has been, from the beginning, dependent upon large properties for job security.

Wage-workers in the factory and on the farm are on the propertyless bottom of the occupational structure, depending upon the equipment owned by others, earning wages for the time they spend at work. In terms of property, the white-collar people are *not* 'in between Capital and Labor'; they are in exactly the same property-class position as the wage-workers. They have no direct financial tie to the means of production, no prime claim upon the proceeds from property. Like factory workers—and day laborers, for that matter—they work for those who do own such means of livelihood.

Yet if bookkeepers and coal miners, insurance agents and farm laborers, doctors in a clinic and crane operators in an open pit have this condition in common, certainly their class situations are not the same. To understand their class positions, we must go beyond the common fact of source of income and consider as well the amount of income.

In 1890, the average income of white-collar occupational groups was about double that of wage-workers. Before World War I, salaries were not so adversely affected by slumps as wages were but, on the contrary, they rather steadily advanced. Since World War I, however, salaries have been reacting to turns in the economic cycles more and more like wages, although still to a lesser extent. If wars help wages more because of the greater flexibility of wages, slumps help salaries because of their greater inflexibility. Yet after each war era, salaries have never regained their previous advantage over wages. Each phase of the cycle, as well as the progressive rise of all income groups, has resulted in a narrowing of the income gap between wage-workers and white-collar employees.

In the middle 'thirties the three urban strata, entrepreneurs, white-collar, and wage-workers, formed a distinct scale with respect to median family income: the white-collar employees had a median income of $1,896; the entrepreneurs, $1,464; the urban wage-workers, $1,175. Although the median income of white-collar workers was higher than that of the entrepreneurs, larger proportions of the entrepreneurs received both high-level and low-level incomes. The distribution of their income was spread more than that of the white collar.

The wartime boom in incomes, in fact, spread the incomes of all occupational groups, but not evenly. The spread occurred mainly among urban entrepreneurs. As an income level, the old middle class in the city is becoming less an evenly graded income group, and more a collection of different strata, with a large proportion of lumpen-bourgeoisie who receive very low incomes, and a small, prosperous bourgeoisie with very high incomes.

In the late 'forties (1948, median family income) the income of all white-collar workers was $4000, that of all urban wage-workers, $3300. These averages, however, should not obscure the overlap of specific groups within each stratum: the lower white-collar people—sales-employees and office workers—earned almost the same as skilled workers and foremen,[1] but more than semi-skilled urban wage-workers.

In terms of property, white-collar people are in the same position as wage-workers; in terms of occupational income, they are 'somewhere in the middle.' Once they were considerably above the wage-workers; they have become less so; in the middle of the century they still have an edge but the over-all rise in incomes is making the new middle class a more homogeneous income group.

As with income, so with prestige: white-collar groups are differentiated socially, perhaps more decisively than wage-workers and entrepreneurs. Wage earners certainly do form an income pyramid and a prestige gradation, as do entrepreneurs and rentiers; but the new middle class, in terms of income and prestige, is a superimposed pyramid, reaching from almost the bottom of the first to almost the top of the second.

People in white-collar occupations claim higher prestige than wage-workers, and, as a general rule, can cash in their claims with wage-workers as well as with the anonymous public. This fact has been seized upon, with much justification, as the defining characteristic of the white-collar strata, and although there are definite indications in the United States of a decline in their prestige, still, on a nation-wide basis, the majority of even the lower white-collar employees—office workers and salespeople—enjoy a middling prestige.

The historic bases of the white-collar employees' prestige, apart from superior income, have included the similarity of their place and type of work to those of the old middle-classes' which has permitted them to borrow prestige. As their relations with entrepreneur and with esteemed customer have become more impersonal, they have borrowed prestige from the firm itself. The stylization of their appearance, in particular the fact that most white-collar jobs have permitted the wearing of street clothes on the job, has also figured in their prestige claims, as have the skills required in most white-collar jobs, and in many of them the variety of operations performed and the degree of autonomy exercised in deciding work procedures. Furthermore, the time taken to learn these skills and the way in which they have been acquired by formal education and by close contact with the higher-ups in charge has been important. White-collar employees have monopolized high school education—even in 1940 they had completed 12 grades to the 8 grades for wage-workers and entrepreneurs. They have also enjoyed status by descent: in terms of race, Negro white-collar employees exist only in isolated instances—and, more importantly, in terms of nativity, in 1930 only about 9 percent of white-collar workers, but 16 percent of free enterprisers and 21 percent of wage-workers, were foreign born. Finally, as an underlying fact, the limited size of the white-collar group, compared to wage-workers, has led to successful claims to greater prestige.

The power position of groups and of individuals typically depends upon factors of class, status, and occupation, often in intricate interrelation. Given occupations involve specific powers over other people in the actual course of work; but also outside the job area, by virtue of their relations to institutions of property as well as the typical income they afford, occupations lend power. Some white-collar occupations require the direct exercise of supervision over other white-collar and wage-workers, and many more are closely attached to this managerial cadre. White-collar employees are the assistants of authority; the power they exercise is a derived power, but they do exercise it.

Moreover, within the white-collar pyramids there is a characteristic pattern of authority involving age and sex. The white-collar ranks contain a good many women: some 41 percent of all white-collar employees, as compared with 10 percent of free enterprisers, and 21 percent of wage-workers, are women.[2] As with sex, so with age: free enterprisers average (median) about 45 years of age, white-collar and wage-workers, about 34; but among free enterprisers and wage-workers, men are about 2 or 3 years older than women; among white-collar workers, there is a 6- or 7-year difference. In the white-collar pyramids, authority is roughly graded by age and sex: younger women tend to be subordinated to older men.

The occupational groups forming the white-collar pyramids, different as they may be from one another, have certain common characteristics, which are central to the character of the new middle class as a general pyramid overlapping the entrepreneurs and wage-workers. White-collar people cannot be adequately defined along any one possible dimension of stratification—skill, function, class, status, or power. They are generally in the middle ranges on each of these dimensions and on every descriptive attribute. Their position is more definable in terms of their relative differences from other strata than in any absolute terms.

On all points of definition, it must be remembered that white-collar people are not one compact horizontal stratum. They do not fulfil one central, positive *function* that can define them, although in general their functions are similar to those of the old middle class. They deal with symbols and with other people, co-ordinating, recording, and distributing; but they fulfil these functions as dependent employees, and the skills they thus employ are sometimes similar in form and required mentality to those of many wage-workers.

In terms of property, they are equal to wage-workers and different from the old middle class. Originating as propertyless dependents, they have no serious expectations of propertied independence. In terms of income, their class position is, on the average, somewhat higher than that of wage-workers. The overlap is large and the trend has been definitely toward less difference, but even today the differences are significant.

Perhaps of more psychological importance is the fact that white-collar groups have successfully claimed more prestige than wage-workers and still generally continue to do so. The bases of their prestige may not be solid today, and certainly they show no signs of being permanent; but, however vague and fragile, they continue to mark off white-collar people from wage-workers.

Members of white-collar occupations exercise a derived authority in the course of their work; moreover, compared to older hierarchies, the white-collar pyramids are youthful and feminine bureaucracies, within which youth, education, and American birth are emphasized at the wide base, where millions of office workers most clearly typify these differences between the new middle class and other occupational groups. White-collar masses, in turn, are managed by people who are more like the old middle class, having many of the social characteristics, if not the independence, of free enterprisers.

WILLIAM SAROYAN

BEING REFINED

BEING refined is a very nice thing, and I have had some happy times noticing refinement in the members of my family, most of whom, especially those who were born in the old country, in Bitlis, finally learned that vocal modulation, for instance, constituted one of the many signs of being refined. Shouting was all right in the family, but out among Americans and people like that it was always a good idea to modulate the voice, at least until you found out that the Americans themselves weren't very refined, which my Uncle Shag seemed to be finding out all the time.

Another good sign of being refined was to look at a painting and not have your mouth hanging open in wonder because the fruit on the plate seemed so real you wanted to reach out and take some, which was pretty much the way paintings were appreciated by the immigrants who had only recently arrived in America.

Still another good idea was not to ask priests difficult questions about God, or biology, or about a stick becoming a snake, or a body of water dividing itself so that there would be a dry road running through it, or a dead man coming to life. Asking such questions really didn't demonstrate that you were an intelligent man, or that you had safely emerged from the Dark Ages, or that you knew how to think for yourself; all it seemed to do was make refined people look at you sideways, cock-eyed-like, by which they meant that you must be some sort of unrefined person, all of your success as a lawyer, for instance, and all of your wealth notwithstanding. In the presence of music you hated, something classical by Ethelbert Nevin, being played on the piano by somebody's wife, accompanied by somebody else's daughter on the cello, it was not a sign of refinement to blurt out, "Can't you play something lively, like *Dari Lolo?*" Or if somebody you had just met looked ill, worn-out from worry of some sort, sunk in spirit, it was not courteous to say, "What's the matter with you? Why don't you stand up straight?"

Shag, or as he had it in full on his card and on the door of his office, Arshag Bashmanian, by the time he was 55 and all the rest of us were in our early 20s, had picked up a wide variety of pointers, as he put it, on how to be refined; and whenever it was in order to demonstrate his refinement, he hardly ever failed to do so.

The year his first daughter, named by her mother Genevieve because the name was refined, became engaged to an American boy named Edmund Armbruster who was a premed student in San Francisco, Arshag was obliged to drive there from Fresno, so that Mr. and Mrs. Armbruster, the boy's father and mother, could meet Arshag and his wife Shushanik, who had a wide circle of friends disciplined to calling her Susan because Shushanik just wouldn't do. And of course the Armbrusters were dying to have a look at the girl their boy had fallen in love with.

Taking his wife and daughter in the Cadillac to San Francisco didn't appeal to Shag, so he asked me to sit up front with him, while they sat in the back, where they belonged, and somehow I wasn't able to get out of it.

"Be in front of your house at five minutes to six," he said. "I'll pick you up, and we'll go right on."

"Isn't that a little early? It's only a five-hour drive, with one stop for gasoline, comfort, and maybe a cup of coffee."

"The earlier we start the better," Shag said. "I've always believed that."

"Are they expecting you at ten in the morning?"

"Well, don't argue," Shag said. "Don't argue about *everything*. Just be in front of your house in your best suit at five minutes to six, and I'll pick you up."

"How long will we be gone?"

"Well, we don't know yet. These people want to see if I pass the inspection. If I do, *that* will be *that,* and we'll come right back. If I don't, we'll come back the next day. If you ask me, I think they're going to have a very pleasant little surprise for themselves. I suppose they think we're country people. I don't suppose they expect to see somebody like me, in the kind of clothes I wear, driving a Cadillac."

"OK, I'll be standing there."

"A white shirt, a tie, and shine your shoes. And when we get up there and go into their house, don't all of a sudden say, 'I'm so hungry I could eat a horse,' or hint around that you want them to give us lunch. I think they'll give us lunch anyway."

"OK."

"And if any of their women—besides the boy, I think they've got two daughters—are beautiful, just compliment them in a nice way, and if they like it and start flirting, flirt back, but *politely.*"

"OK."

"I don't like the idea of driving two hundred miles to have people I don't know inspect me, but what are you going to do when you've got a daughter who's in love and a wife who wants her daughter to marry into the best possible family? You've *got* to go, that's all. What kind of a father would I be if I wouldn't do my daughter and my wife a little favor like that? I've met the boy, Bobby, and he's got class, there's no question about that, and if his people are anything like him, *they've* got class, too."

"Isn't his name Edmund?"

"Is that what it is? Well, anyway, he's a nice boy, a slow boy, slow in the head, but nice. Every time he said Bashmanian, I almost didn't know what he was saying. He took too long. It's not a complicated name, all you've got to do is say Bashmanian, not Bash Man Ian. But Jenny thinks she loves him, so maybe she does, so maybe you better go get a haircut, too, and be sure to shave real neat. Here's fifty cents for the haircut, give the barber a dime tip, keep the rest."

"OK."

I didn't get a haircut, but I did all the other things he said, and I was in front of my house at five minutes to six. Less than two minutes later the big sky-blue Cadillac drew up and I got in and sat beside him. His wife and daughter looked very nice in their new dresses and coats, and Shag himself looked all right, too. He was wearing everything. Diamond stickpin in his tie. Silk handkerchief in his vest, gold watch attached to the chain. Red rose in the lapel buttonhole. Haircut, shampoo, manicure, shoeshine. Sen-Sen in his mouth—the damned smell nearly knocked me over.

We stopped for gas, comfort and coffee in Modesto, and we were in San Francisco at a quarter to eleven. At four minutes to eleven the Cadillac drew up in front of the house, which was in a neighborhood called Seacliff, where only rich people could afford to live, or as Shag put it, "They've got money all right, but let me tell you something. I can buy and sell them any day in the week, and don't ever forget it."

"Poppa," his daughter said. "Please don't talk that way. Just, please, forget that they've got money and that we've got money."

"All right, honey, for your sake I'll let it go this time, but I don't want these people putting on a lot of airs with *me*. I'm Arshag Bashmanian, who the hell do they think they are? Three-car garage. Why three? Why not make the whole house a garage?"

"Poppa, please."

"All right, all right, don't worry about your father."

The door was opened by a rather handsome woman in her late 40s, and from the expression of surprise on Shag's face I was sure he imagined that this was the mother of the boy his daughter had fallen in love with. He had never before visited anybody who had had a servant.

"Yes?" she said.

"Are you sure this is the right address?" Shag said to his daughter, who instead of answering him said to the woman, "I'm Genevieve Bashmanian, and this is my father, and my mother, and my cousin."

"Oh, yes, of course," the woman said. "Won't you please come in and sit down."

Well, the place was really swank. It was certainly the swankest place I had ever walked into, but it gave Shag an awful pain, because by comparison his mansion on Van Ness Avenue in Fresno was a remodeled barn full of Grand Rapids furniture and an original oil painting for which he had paid $1000 by somebody named Gaston Voillard, 1874—a meaningless landscape in dull colors. Three years before when Shag had asked me over to the house to see how a successful man was entitled to live, so that it would be a lesson to me, he showed me the picture, told me how much he had paid for it, and then said, "This Gaston Voillard, 1874, he's one of the *greatest* painters, isn't he?"

"Yes, he is," I said.

Encouraged by my lie, Shag then said, "For God's sake, look at that picture, will you? Look at those leaves on those trees. The man's a genius. I wouldn't take *five* thousand dollars for that picture, if you want to know the truth."

But of course I didn't want to know the truth, so the conversation collapsed and we went to the little bar just off the kitchen, where he poured each of us a drink of raki.

Well, on the walls of the room in which we were now sitting there was an original Cézanne, an original Matisse and an original Picasso. Shag looked from one to the other, and then at me. He leaned his head over slowly to the right, and at the same time lifted his eyebrows, by which he meant, "What kind of cockeyed paintings do you call those?"

Soon the father, the mother, the son and a daughter of 11 came into the room. They were nice people, very gracious, very warm, and yet somehow in spite of everything, even in spite of the fine paintings they owned, they seemed to lack something. I really didn't know what it was, but it was a rather large thing. I suppose it might have been wit of some kind, or maybe health of some kind, or maybe humor. At any rate, it was impossible to be really at ease with them.

The boy's mother said lunch would be at one, and she would be terribly let down if we had made other plans. As a matter of fact, she would insist that we change our plans. In the meantime, perhaps we'd like to see the rest of the house, and then the garden, and after that we might enjoy taking a short drive up to the Legion of Honor Palace to see the new show.

"What kind of a show is it?" Shag said, as if it just might unaccountably be burlesque or something, in which case he would let these people know he didn't take his women to places like that.

"Well, it's the Second Winter Invitational, and I think even better than the first, which was an enormous success."

Shag looked at me, so I said, "*California* painters?"

"Well, actually, *Northern* California painters."

"Oh, paintings," Shag said. "Sure, let's go see 'em."

And so first we saw the house, and then the garden, and every bit of each burned hell out of Shag, and then we were all asked to get in the chauffeur-driven Rolls, but Shag said, "No, let's not all of us try to get into one car. I'll drive up with my nephew."

"Just follow us, then," the father of the boy said, "unless you know the way."

Following the Rolls, Shag said, "What do you think?"

"They're nice people all right."

"No, I don't mean *them*. What do you think of the impression I've made so far?"

"So far it's pretty good, I must say."

"Voice modulated, smiles, politeness?"

"Yes, you showed them all those things all right."

"I'll show them plenty more, too."

At the museum the Second Winter Invitational wasn't bad, although not much good, either, mainly a lot of stuff without any style of any kind, most of it experimental and messy. Not one picture like the one painted by Gaston Voillard, 1874.

"Shall we look at the permanent collection as well?" the mother asked. Shag said, "Why not?"

Well, it was the older stuff, an El Greco, a Rembrandt, a Rubens, but none of it especially exciting, certainly not to Shag; but then in the first of the five small rooms just off the main hall, to the left, there was a painting that really impressed him. Years later I made a point of going back and getting the name of the thing, and of the painter. His name was Jean Marc Nattier, French, 1685–1766, and the name of the picture was *The Duchess of Chateauroux as Thalia, Muse of Comedy*. It was a rather big picture of this pretty girl whose right breast was delightfully exposed. It was as big as life, very white, with a nipple the size and color of a pink rosebud. The girl's face had a twinkle to it, as of mischief. All around her were foldings of dark velvet, and in the background was a small stage with actors upon it.

We all stopped in front of the picture, and after a moment, Shag said, "My goodness, that girl's chest is so real you could reach out and touch it."

Lunch was soup, fish, meat, raspberries with ice cream and coffee.

Somewhere near the end of lunch there was a moment of silence, whereupon Shag said, "I don't think I've ever seen anything more real than that girl's chest."

Less than an hour after lunch we got back into the Cadillac, and Shag began to drive back to Fresno. He had been refined every minute he had been with the elegant people. He had worked very hard at it, saying, for instance, my goodness, and chest, for instance, but what is a man to do about a daughter? A daughter is always a lot of trouble, and now all of a sudden she was crying.

The upshot of the whole thing was that the engagement was slowly broken, or possibly it was simply permitted by time and silence to fade away, and a year later Genevieve married a poor but ambitious boy, by whom she now has four sons and three daughters.

As for Shag, one day he said, "I never did like those Armstrongs."

"Isn't the name Armbruster?"

"That's *exactly* what I mean. There's such a thing as a name like Armstrong, but whoever heard of a name like Armbruster? Those people were phonies. They weren't *really* refined. They were performing, like those little trained dogs at vaudeville shows, and one thing I can't stand is a lousy performance."

Chapter **8**

THE STRATIFIED SOCIETY

THE experience of stratification in a local community leads the individual to an awareness of the way in which larger society is stratified. For most people below the upper strata of society this awareness is closely linked to the question of social mobility—that is, to the question of their chances to maintain or improve their own position and that of their children in the class system. This is not at all an abstract or merely intellectual matter. Rather it touches upon the most intimate questions of who a person is and what he would like to become.

Sociologists have made the distinction between ascribed and achieved status. The former term refers to a situation in which the individual's position in a stratification system is socially assigned to him and comes to him more or less regardless of his own efforts; the latter term refers to a situation in which the individual's own efforts have an important effect upon his position. The American class system has long been justified in terms of an achievement ethos, according to which the society is supposed to provide equal opportunity to all and the resulting inequalities are supposed to reflect, at least approximately, inequalities of effort. Needless to say, the realities of American society (even if one momentarily puts aside the gross handicap imposed by race) do not reflect the assumptions of this ethos.

A large investment of sociological thought and research has gone into the study of this discrepancy. A key question in all this has been that of social mobility: how much social mobility is there in American society? How does this compare with social mobility in other societies? Are the mobility patterns changing here or elsewhere? Again there is widespread disagreement among sociologists regarding these questions. Sociologists tending toward a left position politically have produced studies that seem to show the delusionary character of most mobility aspirations and the intractability of the class system. Other sociologists have pointed to the apparent success of many people in improving their class position and have accordingly stressed the flexibilities within the class system. Again, it is not possible to reconcile all existing views, but it is necesssary that students understand what concepts are being used and how these concepts influence the findings.

The first selection in this chapter, an article by Thomas Luckmann and Peter L. Berger, "Social Mobility and Personal Identity," deals precisely with this problem area. The next selection is from Norman Podhoretz's candid autobiography, *Making It,* in which he describes his own experiences in moving from a lower-middle-class Jewish background in Brooklyn to a larger American world of intellectual achievement and status. Christopher Jencks and David Riesman in my selection from their *Academic Revolution* discuss the role of the educational system in social mobility. In recent years, largely because of a general criticism of contemporary American society by liberal and radical social scientists, there has been much debate (not only within the social sciences but also in the political arena) of the facts concerning inequalities in the society. The selection from Herman P. Miller's *Rich Man, Poor*

Man, deals with some of these facts. The final selection is an essay by Seymour Martin Lipset, "Social Mobility and Equal Opportunity" in which the author compares social mobility in America with that in other contemporary societies.

THOMAS LUCKMANN AND PETER L. BERGER

SOCIAL MOBILITY AND PERSONAL IDENTITY

IT is now generally recognized that social stratification in contemporary societies on comparable levels of industrial development has strikingly similar over-all features.[1] It appears that what is in the making is a stratification system of international scope which characterizes societies with different traditions, different political systems and different official or semi-official ideologies of stratification. Its salient features are these: Class becomes the dominant form of stratification,[2] displacing earlier forms such as estates or castes. It established itself in societies in which the popular conception of class still contains remnants of the estate tradition, as in Western Europe, as well as in societies in which its reality is ideologically denied, relativized or embedded in an ethos of "equal opportunity," as in the "Marxist" world and the United States, respectively. The occupational structure is progressively transformed, with shifts occurring from rural to urban occupations, from primary to secondary and tertiary industries, and, more recently, from manual to white-collar occupations. Relatively high rates of mobility, social as well as geographic, make their appearance. Within a relatively small number of generations the middle strata have expanded significantly. The underlying causes of this transformation not only affected the stratification system; they changed the social order and left their mark on practically all social institutions.

It has been noted that this transformation produced tensions in the relationship between individual existence and the social order. We argue that, apart from the tensions which might be perhaps considered the transitory birth pangs of modern society, a fundamental change in the relationship between the individual and society resulted. We shall discuss a limited aspect of this change, concentrating on the effect of class structure, and especially of social mobility upon personal identity.

Reprinted by permission from the *European Journal of Sociology,* V (1964), 331–44.

One of the features most important for the problem we have taken up is a certain looseness in the structure of class stratification.[3] The segmentation of institutional spheres which characterizes modern society is accompanied by the emergence of relatively independent ranking systems within these institutions. The criteria for class position are multi-dimensional. There is no single criterion, be it legal or conventional, that clearly assigns an individual to a class and to this class only. This is especially true of the broad middle strata.[4] There are no clearcut lines dividing the classes and their style of life. Instead, there appears a sort of continuum of only mildly divergent class-subcultures, with the exception of the top and the bottom of the stratification system. . . .

What is emerging from this convergence is an interesting combination of three factors—weak to almost non-existent class-consciousness; increasing and, at the same time, increasingly ambivalent status consciousness; and increasing mobility orientation even where mobility-aspiration has not yet become a central cultural axiom. This combination constitutes a general problem for the sociology of modern societies. Our problem at hand, however, is to explore the meaning of this combination for the individual.

We are not the first, by any means, to look into the socio-psychological aspect of class. Ever since Durkheim remarked on the relation between mobility and anomic suicide,[5] there have been investigations which correlated class differences, differences in mobility and differences in status consistency with suicide, psycho-physiological stress, crime, mental illness, etc. Merton's topology of adaptations to the discrepancy between what he calls cultural goals and institutional means inspired much recent research in this area.[6] The notion of status inconsistency, especially as developed by Lenski and Landecker, was also productive, especially in the exploration of the relation between class and political attitudes, social participation, etc.[7] We have no quarrel with these approaches. They have defined an important problem and yielded interesting results. But most of them tend either to bracket the problem of individual existence, as is the case with the structural-functional approach, or to take the so-called psychological factors for granted rather than to account for them in the perspective of sociological theory. Our question, however, is how the modern class structure affects personal identity. We shall approach this question as one involving a "total social phenomenon" (Marcel Mauss), to be dealt with in terms of sociological analysis.[8]

The looseness of the class structure contributes to a relative *uncertainty* of status, at least in the middle strata broadly conceived, and increases the frequency of cases of *weak status consistency*.[9] The institutionally based multidimensionality of ranking criteria makes for situations in which a middle-level executive and an architect, for example, will be obligated to bargain with one another before defining the situation with respect to status. It also makes for situations in which individuals will be ranked higher with respect to some criteria as against other criteria. Thus, a skilled industrial worker may rank higher on income than a white-collar worker but lower on source of income.

Or, a college professor will rank higher on education but lower on income than a used car dealer.[10] It may be that not all situations require mutual and congruent definitions with respect to status. Performances in hierarchical institutional bureaucratics, for example, minimize status uncertainty *within that sphere*. Yet we may assert with confidence that in modern life individuals are not unequivocally perceived in a status hierarchy by others and, in consequence, by themselves. In view of the correlative importance of status consciousness in modern society, the class structure contains a built-in phenomenon of status anxiety. But beyond this it can be the root of a continuing crisis of identity.

If one takes seriously the sociological perspective on identity stemming from the work of Charles Cooley and George Herbert Mead, the nature of this crisis would be clear.[11] Individual identity is a configuration of self-conceptions which originate in social processes. The identity of an individual is a social construct as much as an individual creation; in fact it emerges in the dialectics between the two, or, as Cooley puts it, the self is shaped by a mirror-effect. Therefore the consistency and stability of the self depends on the internal "fit" of the various reflected images. It should be obvious, then, that the degree of status consistency and status certainty will be an important factor in the shaping of identity. If status is relatively uncertain and relatively inconsistent, conditions are created that are unfavorable for the consistency and stability of the self. There is, therefore, a causal connection between the looseness of the class structure, status uncertainty and low status consistency, on the one hand, and tenuous identity, on the other hand.[12] A wide range of observations and research findings make good sense if placed in this theoretical context. The general incidence of psychopathology in modern society, while undoubtedly the result of a complex combination of causes, can be also directly related to our theoretical considerations. More specifically, the correlations found between mobility, status consistency, career patterns, etc., and psychopathy, psycho-physiological stress, suicide, etc., not to mention the clinical recognition of a status-anxiety syndrome, fall into place. While the individual findings may not be definitive by themselves, since they offer rather complex patterns in detail, they are cumulatively convincing.[13] One may also refer here to the reflection of this phenomenon in contemporary literature, popular writing and in the mass media. In America one can point, for example, to novels and motion pictures such as *The Man in the Grey Flannel Suit, Marjorie Morningstar, Executive Suite,* or to the success of books such as Vance Packard's *The Status Seekers*.

The consequences of the class structure for the individual should be placed in a broader theoretical context. It would obviously be an oversimplification of the "total social phenomenon" with which we are concerned here to argue that class is the only factor shaping identity in modern society. Rather, we must view our problem in a perspective which only a theory of modern society can provide. It would be presumptuous to claim that such a theory exists, yet for our particular problem we may take as points of departure the classical theories of the division of labor by Durkheim, Weber's theory of bureaucracy, and more recently some perceptive analyses by David Riesman and Arnold Gehlen.[14] One of the most important characteristics of modern society is the sharp segmentation of institutional domains, especially of economics and politics, within the social structure. The norms within the domains become increasingly "rationalized," i.e., determined by the sheer functional requirements of the institution as such. Since the institutional domains became at the same time increasingly autonomous, the institutional norms, autonomous only within their own domain and "rational" only in relation to it, become of necessity restricted in their supra-institutional relevance in general and in their ability to provide meaning for individual existence in particular. At the same time there occurs a certain constriction and functional impoverishment of the institutions shaping the area of "private life".[15] The consequences of this for the individual are—paradoxical. As far as his involvement in what we may call primary public institutions is concerned, he is compelled to play narrowly defined functionary roles. As an actor on the social scene he moves through a series of situations defined bureaucratically. Subjectively, he is forced to define himself as an anonymous performer, as a "cog in the machine." In consequence of the functional rationality of institutional norms, the institutional domains need insist only on performance control, while taking minimal interest in the whole person or his presumptive "inner life." The primary public institutions tend to seem meaningless to the individual; their functional rationality cannot be converted into individual sense. Yet in his private life, a set of what we may call secondary institutions appears to him as highly flexible, permissive and actually geared to cater to his personal choices. The paradox of total conformity in one sector of individual existence and seemingly absolute autonomy in the other, therefore, has its structural roots in total performance control by the primary institutional domains combined with their indifference to the person.[16]

Compared with other historical situations this leads to an under-definition of identity. In one area of the social system the individual is functionalized as a performer, while in the other he is left to his own devices to discover a presumed "essential identity." Neither area is capable of shaping the highly profiled identities so characteristic of other historical periods (say, the peasant, the aristocrat, or even the proletarian). To satisfy the need for "essential identities" an identity market appears, supplied by secondary institutions. The individual becomes a consumer of identities offered on this market, some of them of reasonable durability, others so subject to fashion that one can speak of planned obsolescence. The secondary institutions, the suppliers on this market, are a variety of identity-marketing agencies, some of them in competition with each other—the mass media, the religious organizations, and the different combines of experts on marriage, child-rearing and other private activities.

We must return now from the general problem of identity in modern society to the narrower focus of the class system and its social-psychological dynamics. To do so, one must first look at the psychological consequences of mobility in itself. For the individual, mobility leads to changes in milieu, with whatever frequency. Nearly inevitably this entails a weakening, occasionally even a disruption of his relations with the primary groups that originally socialized him. Since the family is the most important of these groups in the life of the individual, the problem of generations becomes especially acute here.[17] The norms and values on which the individual has been brought up are no longer reaffirmed in the presently relevant social relationships. They are no longer backed by the authority of the old primary groups. Thus they become less and less "real" to the individual—as does his past identity itself. This causes an interesting reversal of the original socialization process. The norms that were originally internalized are now externalized once more, that is, they are located outside the self as belonging to the past or to others from whom one has become alienated. There appears a cleavage between past and present identity, with the former now being reinterpreted in terms of the latter.[18] Inevitably, there now exists an increasing dependence on the reaffirmation of one's identity by those with whom one shares one's present social situation—a social-psychological configuration aptly described by Riesman as "other-directed" (a term otherwise debatable).

Following Gehlen, we would argue that these processes are endemic in modern industrial society. However, we shall concentrate on American society, not only because the largest quantity of data is available on the latter, but also because it can probably be taken as paradigmatic. In America the processes of identity generation and transformation just indicated have gone on for a considerable time, and have had a readily visible effect on family structure and socialization patterns. Most important for our problem is the phenomenon of anticipatory socialization.[19] The family, reduced both in size and in functionality, loses its authority as the fundamental reference group as compared with the mass media that have become major socializing agencies. Thus, almost from the beginning, socialization anticipates mobility, that is, anticipates migration between reference groups. This means that the individual is socialized into milieux of which he has no direct experience but which he perceives in anticipation through the relatively empty stereotypes supplied by the mass media and reinforced by the peers with whom this anticipation is shared. But when the new milieux are in fact reached, the anticipatory stereotypes codetermine their "reality." Thus, the girl from a working-class background who has entered a white-collar occupation and is anticipating marriage into the upper middle class is bombarded through the mass media with stereotyped images of the upper-middle-class housewife. These images are internalized, so to speak, in the act of waiting. If and when her ambition is realized, she quite naturally slips into the roles pre-defined in this imagery, not only in her actions but in her tastes, opinions, attitudes and even emotions—in sum, she acquires a pre-fabricated identity, advertised, marketed and guaranteed by the identity-producing agencies.[20] The identity thus acquired, however, has a peculiar second-hand quality, not only because of its mass-communicated-stereotypes, but because it remains anticipatory even when mobility has taken place, a point to which we shall return. It can be said, of course, that identity is always a project into the future.[21] The peculiar character of this contemporary constellation, however, is not that it is oriented towards the future, but that this future is conceived in derivatory stereotypes that are not personally experienced and yet enter into each stage of socialization. . . .

In terms of our problem, one must next look at the pervasiveness and quasi-religious character of the mobility ethos, again most clearly and paradigmatically apparent in America.[22] One is confronted here with a secularized version of a key feature of the Protestant ethic.[23] Mobility becomes a sort of secular salvation. Some wish to better one's condition is probably fairly universal. But here it is legitimated in a generally pervasive ethos, which combines democratic and humanitarian idealism with the "materialism" of a consumer culture. It would be quite erroneous to think of this "materialism" in terms of an anti-spiritualistic *Weltanschauung,* as the massive fact of American religiosity clearly shows. Rather, material goods become visible symbols of an inner worth. In other words, one is dealing here with a sacramentalism of consumption. This is necessitated structurally by high geographical mobility, as a result of which conspicuous patterns of consumption take the place of continuous interpersonal contacts within an individual's biography. That is, material objects rather than human beings must be called upon to testify to the individual's worth. Simmel's observations on the social functions of the anonymity of money are pertinent here.

This has a significant psychological consequence, in that an important part of identity reaffirmation is directly played by material objects rather than by human beings.[24] That is, the processes of identity support are externalized, perishable and therefore of tenuous subjective plausibility. The predictable obsolescence of consumer goods may serve as an analogy for the obsolescence of identities already remarked upon. In an almost mystical fashion the inner world of the self reflects here the dynamics of industrial production. . . .

NORMAN PODHORETZ

THE BRUTAL BARGAIN

ONE of the longest journeys in the world is the journey from Brooklyn to Manhattan—or at least from certain neighborhoods in Brooklyn to certain parts of Manhattan. I have made that journey, but it is not from the experience of having made it that I know how very great the distance is, for I started on the road many years before I realized what I was doing, and by the time I did realize it I was for all practical purposes already there. At so imperceptible a pace did I travel, and with so little awareness, that I never felt footsore or out of breath or weary at the thought of how far I still had to go. Yet whenever anyone who has remained back there where I started—remained not physically but socially and culturally, for the neighborhood is now a Negro ghetto and the Jews who have "remained" in it mostly reside in the less affluent areas of Long Island—whenever anyone like that happens into the world in which I now live with such perfect ease, I can see that in his eyes I have become a fully acculturated citizen of a country as foreign to him as China and infinitely more frightening.

That country is sometimes called the upper middle class; and indeed I am a member of that class, less by virtue of my income than by virtue of the way my speech is accented, the way I dress, the way I furnish my home, the way I entertain and am entertained, the way I educate my children— the way, quite simply, I look and I live. It appalls me to think what an immense transformation I had to work on myself in order to become what I have become: if I had known what I was doing I would surely not have been able to do it, I would surely not have wanted to. No wonder the choice had to be blind; there was a kind of treason in it: treason toward my family, treason toward my friends. In choosing the road I chose, I was pronouncing a judgment upon them, and the fact that they themselves concurred in the judgment makes the whole thing sadder but no less cruel.

When I say that the choice was blind, I mean that I was never aware—obviously not as a small child, certainly not as an adolescent, and not even as a young man already writing for publication and working on the staff of an important intellectual magazine in New York—how inextricably my "noblest" ambitions were tied to the vulgar desire to rise above the class into which I was born; nor did I understand to what an astonishing extent these ambitions were shaped and defined by the standards and values and tastes of the class into which I did not know I wanted to move. It is not that I was or am a social climber as that term is commonly used. High society interests me, if at all, only as a curiosity; I do not wish to be a member of it; and in any case, it is not, as I have learned from a small experience of contact with the very rich and fashionable, my "scene." Yet precisely because social climbing is not one of my vices (unless what might be called celebrity climbing, which very definitely *is* one of my vices, can be considered the contemporary variant of social climbing), I think there may be more than a merely personal significance in the fact that class has played so large a part both in my life and in my career.

But whether or not the significance is there, I feel certain that my long-time blindness to the part class was playing in my life was not altogether idiosyncratic. "Privilege," Robert L. Heilbroner has shrewdly observed in *The Limits of American Capitalism,* "is not an attribute we are accustomed to stress when we consider the construction of *our* social order." For a variety of reasons, says Heilbroner, "privilege under capitalism is much less 'visible,' especially to the favored groups, than privilege under other systems" like feudalism. This "invisibility" extends in America to class as well.

No one, of course, is so naïve as to believe that America is a classless society or that the force of egalitarianism, powerful as it has been in some respects, has ever been powerful enough to wipe out class distinctions altogether. There was a moment during the 1950's, to be sure, when social thought hovered on the brink of saying that the country had to all intents and purposes become a wholly middle-class society. But the emergence of the civil-rights movement in the 1960's and the concomitant discovery of the poor —to whom, in helping to discover them, Michael Harrington interestingly enough applied, in *The Other America,* the very word ("invisible") that Heilbroner later used with reference to the rich—has put at least a temporary end to that kind of talk. And yet if class has become visible again, it is only in its grossest outlines—mainly, that is, in terms of income levels—and to the degree that manners and style of life are perceived as relevant at all, it is generally in the crudest of terms. There is something in us, it would seem, which resists the idea of class. Even our novelists, working in a genre for which class has traditionally been a supreme reality, are largely indifferent to it— which is to say, blind to its importance as a factor in the life of the individual.

In my own case, the blindness to class always expressed itself in an outright and very often belligerent refusal to believe that it had anything to do with me at all. I no longer remember when or in what form I first discovered that there was such a thing as class, but whenever it was and whatever form the discovery took, it could only have coincided with the recognition that criteria existed by which I and everyone I knew were stamped as inferior: we were in the *lower* class. This was not a proposition I was willing to accept, and my way of not accepting it was to dismiss the whole idea of class as a prissy triviality.

Given the fact that I had literary ambitions even as a small boy, it was inevitable that the issue of class would sooner or later arise for me with a sharpness it would never acquire for most of my friends. But given the fact also that I was on the whole very happy to be growing up where I was, that I was fiercely patriotic about Brownsville (the spawning-ground of so many famous athletes and gangsters), and that I felt genuinely patronizing toward other neighborhoods, especially the "better" ones like Crown Heights and East Flatbush which seemed by comparison colorless and unexciting—given the fact, in other words, that I was not, for all that I wrote poetry and read books, an "alienated" boy dreaming of escape—my confrontation with the issue of class would probably have come later rather than sooner if not for an English teacher in high school who decided that I was a gem in the rough and who took it upon herself to polish me to as high a sheen as she could manage and I would permit.

I resisted—far less effectively, I can see now, than I then thought, though even then I knew that she was wearing me down far more than I would ever give her the satisfaction of admitting. Famous throughout the school for her altogether outspoken snobbery, which stopped short by only a hair, and sometimes did not stop short at all, of an old-fashioned kind of patrician anti-Semitism, Mrs. K. was also famous for being an extremely good teacher; indeed, I am sure that she saw no distinction between the hopeless task of teaching the proper use of English to the young Jewish barbarians whom fate had so unkindly desposited into her charge and the equally hopeless task of teaching them the proper "manners." (There were as many young Negro barbarians in her charge as Jewish ones, but I doubt that she could ever bring herself to pay very much attention to them. As she never hesitated to make clear, it was punishment enough for a woman of her background—her family was old-Brooklyn and, she would have us understand, extremely distinguished—to have fallen among the sons of East European immigrant Jews.)

For three years, from the age of thirteen to the age of sixteen, I was her special pet, though that word is scarcely adequate to suggest the intensity of the relationship which developed between us. It was a relationship right out of *The Corn Is Green,* which may, for all I know, have served as her model; at any rate, her objective was much the same as the Welsh teacher's in that play: she was determined that I should win a scholarship to Harvard. But whereas (an irony much to the point here) the problem the teacher had in *The Corn Is Green* with her coal-miner pupil in the traditional class society of Edwardian England was strictly academic, Mrs. K.'s problem with me in the putatively egalitarian society of New Deal America was strictly social. My grades were very high and would obviously remain so, but what would they avail me if I continued to go about looking and sounding like a "filthy little slum child" (the epithet she would invariably hurl at me whenever we had an argument about "manners")?

Childless herself, she worked on me like a dementedly ambitious mother with a somewhat recalcitrant son; married to a solemn and elderly man (she was then in her early forties or thereabouts), she treated me like a callous, ungrateful adolescent lover on whom she had humiliatingly bestowed her favors. She flirted with me and flattered me, she scolded me and insulted me. Slum child, filthy little slum child, so beautiful a mind and so vulgar a personality, so exquisite in sensibility and so coarse in manner. What would she do with me, what would become of me if I persisted out of stubbornness and perversity in the disgusting ways they had taught me at home and on the streets?

To her the most offensive of these ways was the style in which I dressed: a tee shirt, tightly pegged pants, and a red satin jacket with the legend "Cherokees, S.A.C." (social-athletic club) stitched in large white letters across the back. This was bad enough, but when on certain days I would appear in school wearing, as a particular ceremonial occasion required, a suit and tie, the sight of those immense padded shoulders and my white-on-white shirt would drive her to even greater heights of contempt and even lower depths of loving despair than usual. *Slum child, filthy little slum child.* I was beyond saving; I deserved no better than to wind up with all the other horrible little Jewboys in the gutter (by which she meant Brooklyn College). If only I would listen to her, the whole world could be mine: I could win a scholarship to Harvard, I could get to know the best people, I could grow up into a life of elegance and refinement and taste. Why was I so stupid as not to understand? . . .

CHRISTOPHER JENCKS AND DAVID RIESMAN

THE ACADEMIC REVOLUTION

LIKE all stratified societies, America must engage in a constant struggle to prevent its elite from decaying into an hereditary aristocracy. In principle even the elite itself supports this goal. There is no vocal opposition to keeping American society open, to encouraging the poor-but-able to get ahead, or to equality of opportunity. Nevertheless, well-to-do parents also have an understandable impulse to make sure that their children will enjoy the same privileges that they do. While such parents hope their children will earn their

Reprinted from *The Academic Revolution* by Christopher Jencks and David Riesman (New York: Doubleday & Co., 1969), pp. 97–107, by permission of the publisher and the authors.

[Jencks and Riesman's notes deleted.—Ed.]

privileges in fair competition, the parental commitment to fairness seldom takes precedence over family loyalty. If a middle-class child cannot win out on merit alone, his parents will try to obtain some sort of special treatment for him. Many of these exceptions to meritocratic principles have been institutionalized and have acquired legitimacy even in the eyes of poorer people who do not benefit from them. The fact that a man can pass on his business to his children, for example, is taken for granted by almost everyone, both rich and poor—although inheritance taxes place some limitations on this process. Nor do many people rail against the fact that a child whose father can pay tuition has a better chance of getting to an elite college than does a child who needs a scholarship. Yet when all such exceptions to meritocratic principles are taken together, they provide upper- and upper-middle class children with enormous advantages over other people's children. The elite, in other words, gets what Everett Hughes calls "equality plus." Nor are the reasons for this hard to discover. There seems to be a tacit consensus among all classes that downward social mobility is more painful than the frustration of upward aspirations. Devices for preventing downward mobility, or at least preventing too great or conspicuous a fall, are therefore widely accepted.

Yet in a society where the elite grows no faster than the rest of the population, and where fertility differentials are relatively moderate, devices for preventing or limiting downward mobility also necessarily limit upward mobility. If a business firm is both willing and able to hire all the incompetent friends and relatives of its managers and owners, it will have few jobs left for the talented children of its clerks and janitors. If a profession allows the inept sons of its practitioners to earn licenses and take over their fathers' lucrative practices, outsiders without connections will be left with a very small potential clientele and little incentive to spend years getting licensed. It is in this context that education becomes critically important. In principle it is the great leveler, treating everyone alike, judging everyone by universalistic meritocratic standards, and providing those who meet these standards with passports to success.

When we first began working on the relationship of higher education to social class, we assumed (like almost everyone to whom we talked) that education was playing an ever-larger role in determining social position. The empirical research is, however, ambiguous.[4] The correlation between educational attainment and occupational status may be increasing, but if so the increase is quite slow. Whatever the real trend, however, there has clearly been a trend toward more popular *awareness* of the connection between education and adult success. Both rich and poor parents today assume that the social escalator begins in the first-grade classroom and progresses through other classrooms for many years before emerging into the "real" world. Almost all parents today want their children to get an extensive education. Indeed, modern parents show the same preoccupation with their children's higher education that an earlier generation showed with dowries and trust funds. They view diplomas as a peculiarly valuable sort of property and describe education as "insurance" against unemployment or as something on which a man can "trade" in the job market. Even parents who have themselves managed to get on without higher education generally feel that times are changing and that their children need credentials. Since teachers say the same thing, children have little ground for doubting it.

Few ambitious youngsters, rich or poor, today quit school to advance their careers. A boy may personally feel that he is learning nothing in the classroom and that he would learn more from "practical" experience. This feeling may be re-enforced by knowing that his father, or at least some of his father's friends and relatives, started work young, learned on the job, and rose through the ranks. But his family, his teachers, the mass media, and every other opinion-forming agency all tell him that this is no longer possible. Even dropouts usually accept this judgment. Whether they are lower-class boys who leave high school because "it just isn't worth it," lower-middle class girls who leave college to get married, or upper-middle class men who leave graduate school to work in the civil rights movement, those who quit the classroom assume their decision will hurt their career prospects. They justify it, proudly or sadly, on other grounds.

Census reports suggest that the dropouts may be more defensive than they need to be. A significant fraction of those who fail to attend college seem to do quite well both occupationally and economically. This is particularly true of those who enter business. The odds against success have, it is true, always been greater for dropouts than for those with extensive education, but the advantage of the educated over the uneducated appears to be increasing very slowly. What is changing, however, is the relative importance of different *kinds* of academic credentials. There was a time when a high school diploma sufficed to give a man the inside track on most non-manual jobs. Today this is no longer so. So too, there is some evidence that the economic value of a college degree may be rising, while that of a high school diploma may be relatively smaller than it once was.

The foregoing comments suggest that one of the central functions of higher education—along with providing jobs for scholars—is to control access to the upper-middle social strata. Since demand for upper-middle class jobs and living standards far exceeds the supply, colleges must (in Erving Goffman's terminology) cool out large numbers of youngsters whose ambitions exceed their ability. Not only that—these individuals must be eliminated in such a way as to preserve at least the appearance of fairness to all social strata. Since college diplomas are a key to future power and affluence, they cannot simply be sold at auction to the highest bidders or automatically conferred on the sons of previous alumni. Instead, the distribution system must be in keeping with traditional American mythology, which portrays America as a land of opportunity with unlimited room at the top. The mythology also dictates that failure never be ascribable to what is called "the system" but must rather seem to derive from individual skill, character, or luck.

Higher education seems very well suited to this role, for it is in principle almost infinitely expansive. There does not seem to be anything inherent in the system that limits the number who can earn degrees—if they "have what it takes." At the same time, most young people find academic work both difficult and disagreeable. Many will revise their career goals downward to escape it. They do not even have to be flunked out of college and told never to return. If they are regularly told that they are doing poor work, and if others continually excel in a competition on which colleges place enormous emphasis, an appreciable fraction of the losers will quit the game voluntarily. (Some, of course, merely slip into the "soft" academic options, like business and education. But even these require a certain amount of perseverance and academic drudgery to complete.) Both the dropout and his college can then portray his departure as a matter of personal choice. Even if he feels frustrated later and is resentful about the doors that remain closed to him, his anger is likely to be directed against himself rather than against "the system."

Reliance on colleges to preselect the upper-middle class obviously eliminates most youngsters born into lower-strata families, since they have "the wrong attitudes" for academic success. But it should be remembered that colleges also eliminate a substantial fraction of youngsters born into the upper-middle class, not primarily because they have the wrong attitudes but rather because they lack academic competence and dislike feeling like failures year after year. The Class of 1961, for example, represented the best educated sixth of its generation. If we look at the education of this elite's fathers, we find that 39 per cent came from the best-educated sixth of their parent's generation, 18 per cent came from the second sixth, 15 per cent from the third sixth, and 27 per cent from the bottom half.

These figures, however, understate the actual advantage of being born into a well-educated family. The best-educated parents have somewhat fewer children than average, so that if all other things were equal they would produce less than a sixth of the children who end up in the educational elite of the next generation. We do not have precise estimates of fertility differentials among fathers who produced children just before World War II, so we cannot compare our figures on the Class of 1961 to the "expected" distribution if parental education had no effect on college chances but only on fertility. We do, however, have data on men born between 1927 and 1936, most of whom attended college during the 1950s, which allow us to deal with this problem. We find that 39.4 per cent of the men who ended up in the best-educated sixth of the generation were among the most advantaged sixth at birth, as judged by their father's education. Thirty years earlier the amount of educational inheritance was microscopically greater. Among men born between 1897 and 1906, the best-educated sixth includes not only college graduates but college dropouts, as well as some men who had merely finished high school. We find that 37.8 per cent of this elite was born into the most advantaged sixth of its generation; the rest rose into it. Or to put it the other way, the rate of turnover in the educational elite (defined as the best-educated sixth) fell from 62.2 per cent for men born between 1897 and 1906 to 60.6 per cent for those born between 1927 and 1936.

In weighing the significance of these figures it is important to bear in mind that even if educational opportunity were completely equal and attainment depended entirely on innate ability, we would expect *some* correlation between the educational attainment of fathers and sons. This is because well-educated fathers tend to be somewhat more intelligent than poorly educated fathers. Part of this difference is genetic, and a man normally passes part of his genetic advantage along to his children. Assuming equality of opportunity, children would translate their inherited genetic advantage into high educational attainment, even if their parents could not provide them any special environmental advantages. A corresponding cycle of genetic deprivation presumably exists among poorly educated families. This being so, we must expect the best-educated sixth of any generation to include a disproportionately large number of individuals whose fathers were also in the best-educated sixth, and disproportionately small number whose fathers were in the worst-educated sixth. We have no basis for estimating the likely magnitude of these genetic effects, though we assume they are probably small.

Given the present state of genetic and sociological knowledge we cannot say with certainty how much intergenerational turnover would be expected in a "pure" meritocratic system. We can, however, say with considerable confidence that there would be more turnover than at present. Table 1 shows the influence of both academic aptitude and social background on a high school graduate's chances of getting to college. The table makes it clear that aptitude plays a larger role than class in determining who goes to college, but it is not *much* larger.

Having said all this about the degree of cultural mobility in America, we must still ask to what extent cultural mobility implies social mobility. Society does not, after all, always automatically accept academicians' verdicts even on intellectual questions; when the verdict is being passed on one's children its acceptance is even less likely. Upper-middle class parents can and often do resist educators' efforts to relegate their sons and daughters to mediocre jobs and modest incomes. Those with connections in the business world, for example, can often find a sinecure for their dropout son in which he can maintain both social respectability and reasonable income; indeed, he can sometimes do very well for himself, for sometimes he has valuable talents that school and college repressed or at least ignored.

Conversely, those who maintain or even improve on their parents' educational status may not do the same in occupational or economic terms. Some children from elite families "overreact" to their education and move in directions that, while often involving more education than their parents had, provide less income and occupational power. The college that manages to get the son of a corporation executive excited about medieval history, for example, may induce him to acquire considerably more education than his father; but

TABLE 1

Percent of High School Graduates Going to College the Following Year, by Academic Aptitude, Socioeconomic Background, and Sex, 1960

	SOCIOECONOMIC STATUS					
ACADEMIC APTITUDE	LOW	LOWER-MIDDLE	MIDDLE	UPPER-MIDDLE	HIGH	ALL
Males						
Low	10	13	15	25	40	14
Lower-Middle	14	23	30	35	57	27
Middle	30	35	46	54	67	46
Upper-Middle	44	51	59	69	83	63
Upper	69	73	81	86	91	85
All	24	40	53	65	81	49
Females						
Low	9	9	10	16	41	11
Lower-Middle	9	10	16	24	54	18
Middle	12	18	25	40	63	30
Upper-Middle	24	35	41	58	78	49
Upper	52	61	66	80	90	76
All	15	24	32	51	75	35

SOURCE: Project Talent 1960 High School Senior Sample, as reported in Folger. Data have been adjusted for non-respondent bias, but not for students who delay college entry. These latter are mostly lower status men, so that the long-term class bias of the system is slightly less than the table indicates. The socioeconomic measures were more diverse than those we have discussed, including many cultural indices, and were cumulated in a single scale. The quintiles for both socioeconomic status and ability are of equal size in the overall age grade but not among high school graduates. The cell sizes are quite unequal, because aptitude is not evenly distributed between social classes.

once he gets his Ph.D. he will have less power, less money, and hence will be socially downwardly mobile. . . .

Nonetheless, the striking fact about America, at least to us, is not the extent of these dissonances between social and cultural class but the extent to which the verdict of academicians on the young is accepted by men who have little apparent sympathy for academic values. Corporations, for example, evidently need a way to avoid hiring the office manager's dimwitted nephew without causing offence. They also need a preselection device that will help them pick out the more diligent, adaptable, and competent applicants from among the majority who have no inside connections. College degrees serve both these purposes, for college is a kind of protracted aptitude test for measuring certain aspects of both intelligence and character. The measurements are, it is true, very selective and very crude, and employers who put great faith in them are almost always disappointed. They continue in use only because other devices for predicting job performance are in most cases even less reliable and more cumbersome. On-the-job performance seems best predicted by previous on-the-job performance, when that can be observed. But companies do not want to hire large numbers of prospective managers, try them out on a mass scale, and then sack the majority, for incompetents get in the way and cause problems even if they are only around for a few months, and sacking them also causes morale problems.

When one turns to the question of why the rich and powerful have not monopolized college places, several factors must be taken into account. First there is the fact that some of the man who have run colleges have had sternly moral, self-denying temperaments. They have really believed that if their children or their friends' children could not measure up they should not get special treatment. Their capacity for self-deception about what constituted special treatment has, as we shall see, been considerable, but it has not been unlimited. One of the most important limits, especially in recent years, has derived from faculty pressure for strictly meritocratic selection. Faculty opposition has often been a decisive factor in reducing favoritism toward alumni (though not faculty) sons and daughters, and has sometimes pushed the administration into raising more scholarships for poor but talented applicants and doing more to solicit such applicants. Together these influences have kept a number of elite colleges fairly open to students from all classes and conditions. . . .

HERMAN P. MILLER

THE DISTRIBUTION OF PERSONAL INCOME IN THE UNITED STATES

FAMILIES BY INCOME LEVELS

Below are figures showing the spread of income in the United States. They come from a study conducted by the U.S. Bureau of the Census in March 1969. You may be interested in finding out where you fit in the income picture. Since only seven different income groups are shown, these figures give an unrealistic view of the actual spread of income. It is really much greater than most people imagine. The noted economist Paul Samuelson has de-

[Miller's notes deleted.—Ed.]

TABLE 2

Families by Income Level: 1968

INCOME LEVEL	NUMBER OF FAMILIES	PERCENTAGE	
		FAMILIES	INCOME
All families	50.5 million	100%	100%
Under $3,000	5.2 million	10	2
Between $3,000 and $7,000	13.4 million	27	14
Between $7,000 and $10,000	11.8 million	23	20
Between $10,000 and $15,000	12.6 million	25	31
Between $15,000 and $25,000	6.1 million	12	23
Between $25,000 and $50,000	1.2 million	2	7
$50,000 and over	150,000	0.3	2
Median income	$8,600		
Average (mean) income	$9,700		

Note: Sums of tabulated figures in this chapter may not equal totals because of rounding.

Derived from U.S. Bureau of the Census, *Current Population Reports,* Series P-60, No. 66.

scribed income distribution in the following terms: "If we made an income pyramid out of a child's blocks, with each layer portraying $1,000 of income, the peak would be far higher than the Eiffel Tower, but almost all of us would be within a yard of the ground." This statement gives you some idea of the diversity that is compressed within these seven groupings.

About 5 million families received less than $3,000 in 1968. They represented about one-tenth of all families and received one-fiftieth (2 percent) of the income. Some lived on farms where their cash incomes were supplemented by food and lodging that they did not have to purchase. But even if this income were added to the total, the figures would not change much.

At the top income level were about 150,000 families with incomes of $50,000 and over. They represented $^{3}/_{10}$ of 1 percent of all families and received 2 percent of all the income.

Another way to view these figures is to examine the share of income received by each fifth of the families, ranked from lowest to highest by income. Table 3 shows that in 1968, the poorest fifth of the families had incomes under $4,600; they received 6 percent of the total. In that same year, the highest fifth of the families had incomes over $13,500; they received 41 percent of the total.

Who sits at the top of the heap? The figures show that until you get to the very top the incomes are not so high. The top 5 percent of the families had incomes over $23,000. They received 14 percent of all the income. Families with incomes over $42,500 were in the top 1 percent and they received 5 percent of the total.

The figures in Table 3 show the distribution of income before taxes. Since

TABLE 3

Share of Income Received by Each Fifth of Families and by Top 5 Percent and Top 1 Percent: 1968

FAMILIES RANKED FROM LOWEST TO HIGHEST	INCOME RANGE	PERCENTAGE OF INCOME RECEIVED
Lowest fifth	Under $4,600	6%
Second fifth	Between $4,600 and $7,400	12
Middle fifth	Between $7,400 and $10,000	18
Fourth fifth	Between $10,000 and $13,500	24
Highest fifth	$13,500 and over	41
Top 5%	$23,000 and over	14
Top 1%	$42,500 and over	5

Derived from U.S. Bureau of the Census, *Current Population Reports,* Series P-60, No. 66.

families in the higher income groups pay a larger share of the taxes, their share of income should be smaller on an after-tax basis. It is, but not by as much as you might think. Table 4 shows the figures both ways for 1966, using data collected by the Survey Research Center of the University of Michigan. This table shows that the share of income received by the top fifth of the families and individuals is reduced by only two percentage points when taxes are taken into account. The reason taxes have such little impact is that our tax structure is not very progressive. In fact, in 1965, families at each income level between $2,000 and $15,000 paid the same proportion of their income in taxes (see Table 5). There is some progressivity in the federal tax structure. Families in the $2,000 to $4,000 income class pay 16 percent of their income in federal taxes whereas those in the $15,000 and over class turn over 32 percent of their income to the federal treasury. State and local taxes, however, are regressive from beginning to end. Families with incomes under $2,000 pay one-fourth of their income in state and local taxes (mostly

TABLE 4

Share of Income Received by Each Fifth of Families and Individuals, Before and After Taxes: 1966

FAMILIES AND INDIVIDUALS, RANKED FROM LOWEST TO HIGHEST	PERCENTAGE OF AGGREGATE INCOME RECEIVED	
	BEFORE TAXES	AFTER TAXES
Lowest fifth	5%	5%
Second fifth	11	11
Middle fifth	18	17
Fourth fifth	23	25
Highest fifth	43	41

Derived from University of Michigan, Survey Research Center, *Survey of Consumer Finances: 1967.*

TABLE 5

Taxes and Transfers as a Percentage of Income: 1965

	TAXES				
INCOME CLASS	FEDERAL	STATE AND LOCAL	TOTAL	TRANSFER PAYMENTS	TAXES LESS TRANSFERS
Under $2,000	19%	25%	44%	126%	−83%*
$2,000–$4,000	16	11	27	11	16
$4,000–$6,000	17	10	27	5	21
$6,000–$8,000	17	9	26	3	23
$8,000–$10,000	18	9	27	2	25
$10,000–$15,000	19	9	27	2	25
$15,000 and over	32	7	38	1	37
Total	22	9	31	14	24

* The minus sign indicates that families and individuals in this class received more from federal, state, and local governments than they, as a group, paid to these governments in taxes.

Joseph A. Pechman, "The Rich, the Poor and the Taxes They Pay," *The Public Interest,* November 1969. The data are from the *Economic Report of the President,* 1969, p. 161.

sales taxes at the bottom income class), whereas families in the top income class pay only 7 percent of their income in taxes to state and local governments.

The government not only takes money away from people, it also gives it back to some in the form of transfer payments like social security, unemployment compensation, public assistance, etc. When transfer payments are taken into account, a large measure of progressivity is added to the tax structure. Families at the very lowest income levels receive more in transfer payments than they pay in taxes to the federal, state, and local governments. The share of income paid in taxes, less transfer payments, does rise with income level.

TABLE 6

Percentage Change in Income, by Income Class, After Tax Payments and the Benefits of Government Expenditures: 1960

INCOME CLASS	PERCENTAGE CHANGE IN INCOME
Under $2,000	55%
$2,000 to $2,999	44
$3,000 to $3,999	19
$4,000 to $4,999	−1
$5,000 to $7,499	−3
$7,500 to $9,999	2
$10,000 and over	−13

Article by W. Irwin Gillespie, "Effect of Public Expenditures on the Distribution of Income" in Richard A. Musgrave, *Essays in Fiscal Federalism.* Washington, D.C.: The Brookings Institution, 1965, p. 162.

Table 6 shows how families at each income level are affected when the joint net impact of the burden of tax payments and the benefits of government expenditures are taken into account. The incomes of families at the lowest level (under $2,000) are increased by 55 percent and those at the $2,000–$3,000 level are increased by 44 percent. The incomes of families at the $10,000 and over level are reduced by 13 percent. Within the $4,000 to $10,000 there is very little change. It does appear, therefore, when all factors are taken into account, that the poor do benefit appreciably as a result of the government's efforts to redistribute income.

ARE U.S. INCOMES TOO UNEQUALLY DISTRIBUTED?

There is no objective answer to this question. It all depends on how equally you think incomes should be distributed.

Around the turn of the century, the French poet and philosopher Charles Péguy wrote: "When all men are provided with the necessities, the real necessities, with bread and books, what do we care about the distribution of luxury?" This point of view went out of style with spats and high-button shoes. There is an intense interest in the distribution of luxury in the modern world.

Since we all cannot have as many material things as we should like, many people are of the opinion that those who are more productive should get more both as a reward for past performance and as an incentive to greater output in the future. This seems like a reasonable view, consistent with the realities of the world. Lincoln said: "That some should be rich shows that others may become rich and hence is just encouragement to industry and enterprise." The fact is that all modern industrial societies, whatever their political or social philosophies, have had to resort to some forms of incentives to get the most work out of their people.

Despite its reasonableness, this view has its critics. Some have argued that a man endowed with a good mind, drive, imagination, and creativity, and blessed with a wholesome environment in which these attributes could be nurtured, has already been amply rewarded. To give him material advantages over his less fortunate fellows would only aggravate the situation. The British historian R. H. Tawney wrote in his book *Equality:* ". . . some men are inferior to others in respect to their intellectual endowments. . . . It does not, however, follow from this fact that such individuals or classes should receive less consideration than others or should be treated as inferior in respect to such matters as legal status or health, or economic arrangements, which are within the control of the community."

Since there is no objective answer to the question as it has been formulated, it may be fruitful to set it aside and turn to the comparison of income in the United States and other major countries for which such data are available.

Anyone who doubts that real incomes—purchasing power—are higher in the United States than in all other major countries just hasn't been around.

But how much higher? That is hard to say. How do you compare dollars, pounds, rubles, and francs? Official exchange rates are often very poor guides. Differences in prices, quality of goods, and living standards add to the complexity. In view of these problems, international comparisons are sometimes made in terms of the purchasing power of wages. But even this measure has serious limitations. What constitutes a representative market basket in different countries, and how does one compare the market basket in one country with another? For example, Italians may like fish, which is relatively cheap, whereas Americans may prefer beef, which is quite expensive. How then would one compare the cost of a "typical" meal for families in the two countries? Because of this kind of problem, and many others, international comparisons of levels of living must be made with great caution. One study that casts some light on the subject was published in 1959 by the National Industrial Conference Board. It shows the amount of work it would take to buy the following meal for a family of four in several different countries. The items were selected from an annual survey of retail prices conducted by the International Labor Office:

Beef, sirloin	150 grams
Potatoes	150 grams
Cabbage	200 grams
Bread, white	50 grams
Butter	10 grams
Milk	0.25 liter
Apples	150 grams

The results are shown in Table 7. The industrial worker in the United States had to work one hour to buy the meal above. The Canadian worker, whose level of living is not far behind that of his American cousin, had to work nine minutes more to buy the same meal. In Europe, the Danes came closest to the American standard, but even in Denmark the average worker

TABLE 7

Work Time Needed to Buy a Meal: 1958

COUNTRY	MINUTES OF WORK
United States	60
Canada	69
Denmark	88
West Germany	131
United Kingdom	138
Belgium	200
Austria	244
France	277
Italy	298

Zoe Campbell, "Food Costs in Work Time Here and Abroad," *Conference Board Business Record,* December 1959.

had to toil one-half hour longer to feed his family. In West Germany and Great Britain it took more than two hours of work to buy the same meal and in Italy it took five hours. These and many other figures of a similar nature show that American workers are paid more in real terms than the workers of any other major country.

COMPARISON WITH OTHER COUNTRIES

Do the rich get a larger share of income in the United States than they do in other countries? According to the available evidence this is not the case. The United States has about the same income distribution as Denmark, Sweden, and Great Britain and a much more equal distribution than most of the other countries for which data are shown.

The figures in Table 8 classify the top 5 percent as "rich." This is a rather low point on the income scale. In the United States it would include all families receiving more than $23,000 a year. A more interesting comparison would be the share of income received by the top 1 percent ($42,500 or more per year) or perhaps an even higher income group. Such information, however, is not available for most other countries.

A comprehensive study of international comparisons of income was made in 1960 by Professor Irving Kravis of the University of Pennsylvania. He

TABLE 8

Percentage of Income Received by Top 5 Percent of Families in Selected Countries

COUNTRY		INCOME PERCENTAGE
United States	(1950)	20%*
Sweden	(1948)	20
Denmark	(1952)	20
Great Britain	(1951–52)	21
Barbados	(1951–52)	22
Puerto Rico	(1953)	23
India	(1955–56)	24
West Germany	(1950)	24
Italy	(1948)	24
Netherlands	(1950)	25
Ceylon	(1952–53)	31
Guatemala	(1947–48)	35
El Salvador	(1946)	36
Mexico	(1957)	37
Colombia	(1953)	42
Northern Rhodesia	(1946)	45
Kenya	(1949)	51
Southern Rhodesia	(1946)	65

* The numbers represent total income before taxes received by families or spending units.

Simon Kuznets, "Quantitative Aspects of the Economic Growth of Nations," *Economic Development and Cultural Change,* Vol. XI, No. 2 (January 1963), Table 3.

summarized the income distribution among the countries for which data are available in the following way:

More nearly equal distribution than U.S.
- Denmark
- Netherlands
- Israel (Jewish population only)

About the same distribution as U.S.
- Great Britain
- Japan
- Canada

More unequal distribution than U.S.
- Italy
- Puerto Rico
- Ceylon
- El Salvador

It is of particular interest to compare the earnings of workers in the United States with those of workers in socialist countries where, in theory at least, the ravages of the marketplace have been eliminated and the goal is to pay workers in accordance with their need rather than their ability to produce. Such comparisons must be restricted to wage earners for whom comparable data are available in both places. The results may therefore differ significantly from figures which cover the entire population, particularly in the United States, where many people receive nonwage income from a business in which they are self-employed or from interest, dividends, and other sources. Nevertheless, it is useful to make the comparison. There is a difference between the United States and several socialist countries, but it is not

TABLE 9

Dispersion of Wages in Various Countries

COUNTRY	PERCENTAGE BY WHICH WAGES OF TOP 5% OF WORKERS EXCEEDS AVERAGE
West Germany	54%
Belgium	58
Sweden	60
Netherlands	62
United Kingdom	69
United States	78
Canada	85
France	100
Czechoslovakia	70%
Hungary	73
Poland	104
Yugoslavia	107

Harold Lydall, *The Structure of Earnings*. Oxford, Clarendon Press, 1968, p. 142. Most of the figures shown cover the wage or salary income or full-period, manual wage or salary workers in nonfarm industries. In some cases the figures represent all workers of this type rather than just full-period workers.

very great. In the United States the highest-paid (top 5 percent) manual workers in nonfarm industries earn about 78 percent more than the average worker in these industries. In socialist Czechoslovakia and Hungary the differential is somewhat less (70 and 73 percent), and in Poland and Yugoslavia it is somewhat greater (104 and 107 percent). In Canada and France the differential between the high-paid and average worker appears to be somewhat greater than in the United States, but in most western European countries it is somewhat less. Wage incomes appear to be appreciably more equally distributed in West Germany, Belgium, Sweden, the Netherlands, and the United Kingdom than in the United States. It must be remembered in all of these comparisons that the average level is much higher in the United States than in the other countries. It is only the spread of wages that is under consideration here.

SEYMOUR MARTIN LIPSET

SOCIAL MOBILITY AND EQUAL OPPORTUNITY

THE degree of social mobility in America has long been a point of dispute not only among social scientists but also among political advocates. The once popular acceptance of America as the golden land of opportunity, for example, rested on the belief that it offered unparalleled chances for those in the lower ranks of society to improve their position. This theory of social mobility was also adopted, though for different reasons, by socialists attempting to explain the relatively low level of class consciousness among workers in America. They argued that because American workers could realistically hope to improve their circumstances and move out of their class, they were less likely than their more deprived European counterparts to support revolutionary movements. Socialists who made these assumptions, from Marx on, anticipated the emergence of radicalism among the working classes at a later period, when changes in the economic system would have sharply reduced upward mobility. And the notion is by now widespread that the chances for upward mobility would diminish—and have in fact diminished—as a result of the ongoing processes of urbanization, industrialization, and bureaucratization.

Reprinted from *Public Interest,* No. 29 (Fall 1972): 90–98 by permission of the publisher.

In the past, most discussion of social mobility has focused upon the extent to which the lowly have a realistic hope of moving up the social ladder. The fact that it was indeed possible for some to go from rags to riches was generally taken as evidence that opportunity to rise was a reality in the United States. But today the traditional American notion of equality of opportunity has to a large extent given way to a newer, more radical version. According to this new understanding, real equality of opportunity requires not merely that the children of the lower classes have a realistic chance of achieving a high degree of success, but that they do in fact achieve such success in equal proportions with the children of the middle and upper classes. This view of equality of opportunity would demand, for example, that the children of the poorest tenth of the population be represented in institutions of higher learning or in privileged occupations in equal numbers with the children of the richest ten per cent.

Especially in the light of the stringent demands made upon the mechanisms of social mobility by this new ideal of social justice, it becomes a matter of importance to know something about the actual rates of social mobility in America. Has the rate of social mobility in America declined? What is the influence of racial and ethnic factors in this context? How does America compare with other industrialized countries in regard to social mobility? Have the Communist nations been more successful than their capitalist rivals in promoting equality of opportunity? What are the factors which seem to limit social mobility, and to what extent can they be mitigated or eliminated altogether? These are the questions which this essay attempts to answer.

HAS EQUALITY OF OPPORTUNITY IN AMERICA DECLINED?

It has often been argued that an advanced stage of industrialization, such as that achieved by contemporary America, is likely to produce an increasing "rigidification" of social classes. But a number of analyses of the pattern of opportunity in American society clearly indicate that there has been no decline in social mobility; in fact, in some respects American society today is *less* rigid in terms of social advancement than it was in the past. A *Scientific American* survey of the backgrounds of big business executives (presidents, chairmen, and principal vice-presidents of the 600 largest U.S. non-financial corporations) found that the business elite has been opened to entry from below in a way that had never been true before in American history. Since this study has never been widely disseminated, it may be worthwhile to reproduce some of its findings here:

> Only 10.5 per cent of the current generation of big business executives . . . [are] sons of wealthy families; as recently as 1950 the corresponding figure was 36.1 per cent, and at the turn of the century, 45.6 per cent . . . two thirds of the 1900 generation had fathers who were heads of the same corporation or who were independent businessmen; less than half of the current generation had fathers so placed in American society. On the other hand, less than 10 per cent of the 1900 generation

had fathers who were employees; by 1964 this percentage had increased to nearly 30 per cent.[1]

Surprisingly both to scholars in the field and to those radicals convinced that a mature capitalism would become increasingly immobile, particularly with respect to sharp jumps into the elite, the evidence indicates that the post-World War II period brought the greatest increase in the proportion of those from economically "poor" backgrounds (from 12.1 per cent in 1950 to 23.3 per cent in 1964) who entered the top echelons of American business; and there was a correspondingly great decline in the percentage from wealthy families (from 36.1 per cent in 1950 to 10.5 per cent in 1964). A number of underlying structural trends appear to be responsible for this development: the replacement of the family-owned enterprise by the public corporation; the bureaucratization of American corporate life; the recruitment of management personnel from the ranks of college graduates; and the awarding of higher posts on the basis of a competitive promotion process similar to that which operates in government bureaucracy. Because of the spread of higher education to the children of the working classes (almost one third of whom now attend college), the ladder of bureaucratic success is increasingly open to those from poorer circumstances. Privileged family and class backgrounds continue to be enormous advantages in the quest for corporate success, but training and talent can make up for them in an increasing number of cases.

Other, more broadly-focused studies provide further evidence that there has been no hardening of class lines in American society. According to the findings of Stephan Thernstrom (who has played the leading role among historians both in doing research and in stimulating work on the part of others), there has been a continuation of a high rate of social mobility over an 80-year period. In Boston, which he studied himself, there was "impressive consistency in . . . career patterns . . . between 1880 and 1968. About a quarter of all the men who first entered the labor market as manual workers ended their careers in a middle-class calling; approximately one in six of those who first worked in a white-collar job later skidded to a blue-collar post." Almost identical patterns to Boston's have been reported in a "dozen samples [from various cities] for the period from 1850 to World War I;" about 30 to 35 per cent from manual families moved into middle-class positions in various surveys. Rates of downward mobility also do not vary a great deal; the large majority of those from middle-class backgrounds (between 70 and 80 per cent) maintained middle class status.[2]

Thernstrom notes that these findings challenge the often-voiced belief that changes in American capitalism have created a permanent and growing class of the poor. In fact, all the available evidence points in the opposite direction. Statistical data from Poughkeepsie in the 1840's, Boston in five different samples from the 1880's to recent years, and Indianapolis in 1910, as well as various post-World War II surveys, local and national, indicate that most of the sons of unskilled workers either moved up into the ranks of the skilled or found middle-class jobs of various kinds.

These conclusions are reinforced by the most comprehensive and methodologically sophisticated national sample survey of the American population, that of Blau and Duncan in 1963. They analyzed the mobility patterns of American families over several generations by relating family occupational background to first job (thus permitting a comparison of the very young still on their first job with the experience of the very old when they were young), and found that "the influence of social origins has remained constant since World War I. There is absolutely no evidence of 'rigidification.' "[3]

Recent historical research has not only challenged the conventional wisdom about mobility rates, which assumed that the growth of large corporations would mean movement from greater to lesser equality of opportunity; it has also upset long-cherished notions about the direction of change in the distribution of income from the early 19th century on. The tentative conclusion which may be reached from a number of studies is that Jacksonian America —described by Tocqueville and others as an egalitarian social system (which, compared to Europe, it undoubtedly was)—was characterized by much more severe forms of social and economic inequality than the society of the 1970's. As Historian Edward Pessen points out:

> The explanation, popular since Karl Marx's time, that it was industrialization that pauperized the masses, in the process transforming a relatively egalitarian social order, appears wanting. Vast disparities between urban rich and poor antedated industrialization [in America]. . . . Even Michael Harrington and Gabriel Kolko whose estimates reveal the greatest amounts of [present-day] inequality, attribute percentages of income to the upper brackets that are far smaller than the upper one per cent of New York City controlled in income in 1863 or in wealth in 1845.[4]

Most other pre-Civil War American cities resembled New York in these respects, and even in rural areas the pattern of property distribution was extremely unequal.

Without entering into the issue of trends in the distribution of income since 1900, it is obvious that economic growth has brought with it an almost constant increase in the national income. Average per capita income has increased almost sixfold during the course of this century, and this dramatic growth has brought about a wide distribution of various social and economic benefits, usually wider than in any other country. Thus, a much larger proportion of the population graduates from high school (over 80 per cent) or enters college (close to 50 per cent) in America than in any other nation. The greater wealth of the United States also means that consumer goods such as automobiles, telephones, and the like are more evenly distributed here than elsewhere. A recent evaluation by the (London) *Economist,* using 12 social indicators to assess the relative advantages of different countries as places to live, placed the United States far in the lead over eight other non-Communist industrialized states. The wider distribution of consumer goods that inevitably accompanies greater wealth means that the gap in standards of living between American social classes, while still great absolutely, is small by comparative world standards.

DO MOBILITY PATTERNS DIFFER FOR MINORITY GROUPS?

The simple analysis of varying rates of opportunity or general consumption levels is far from the whole story with respect to the underlying pattern of opportunity over time. It is important to recognize that in America class position has been differentially distributed among ethnic and racial groups. For much of its history, the United States has been divided between "majority" and "minority" ethnic groups. The latter have, in effect, repeatedly provided new sets of recruits for the lowly paid, low-status positions, thus enabling others of less recent settlement to rise. An analysis of Census data by E. P. Hutchinson reported that in 1870 and in 1880, "the foreign-born were most typically employed in the factories, in heavy industry, as manual laborers and domestic servants. Clerical, managerial, and official positions remained largely inaccessible to them." [5] The Census of 1890 gathered information for the first time on the occupations of the native-born children of immigrants, thus permitting a comparison of the two generations. As Hutchinson shows, they varied considerably:

> Unlike the immigrant males who were in highest proportion among domestic and personal service workers, the second generation males were most numerous relatively among workers in trade and in transportation and manufacturing. It is also notable that those in the second generation were more successful in entering the professions, even though not as successful as members of the native stock (the native born of native parents). . . . Altogether, the second generation conformed more closely to the occupational distribution of the entire white labor force than did the foreign born.

This pattern, in which the second generation was, as a group, in a much better position than the immigrant generation, continued for the duration of mass immigration. Thus Hutchinson reports that in 1900 "the foreign born were no more widely distributed by occupation . . . than in 1890, but that the second generation became more widely distributed and moved closer to the occupational distribution of the entire labor force in 1900." The Census was not as comprehensive in gathering comparable occupational data from 1910 on, but the evidence clearly indicates comparable patterns to those summarized above for the remaining period of mass immigration through 1924.

More recently, particularly since the economy began a prolonged period of relatively full employment in the 1940's, migrants from various parts of North America—Blacks, Puerto Ricans, Mexicans, and, to a small extent, French Canadians—have furnished the bulk of the less skilled labor force. As Reinhard Bendix and I wrote in our analysis of mobility processes in the late 1950's:

Now, as before, there is a close relationship between low income and membership in segregated groups. A large proportion of seasonal farm laborers and sharecroppers in the South and Southwest come from them. In the cities, Negroes, Mexicans, and Puerto Ricans predominate in the unskilled, dirty, and badly paid occupations. These twenty million people earn a disproportionately low share of the national income; they have little political power and no social prestige; they live in ethnic ghettoes, in rural and urban areas alike, and they have little social contact with white Americans. Indeed, today there are two working classes in America, a white one and a Negro, Mexican, and Puerto Rican one. A real social and economic cleavage is created by widespread discrimination against these minority groups, and this diminishes the chances for the development of solidarity along class lines. In effect, the overwhelming majority of whites, both in the working class and in the middle and upper classes, benefit economically and socially from the existence of these "lower classes" within their midst. This continued splintering of the working class is a major element in the preservation of the stability of the class structure.[6]

The assumptions made in that analysis about the difference between the situations of initially underprivileged whites and blacks were given a more elaborate statistical confirmation in the largest study of American social mobility, that of Blau and Duncan. These authors found in 1963 that lowly social origin *had little negative effect on the chances of whites—including the children of white immigrants—to advance economically*. The mobility picture for whites is such that Blau and Duncan reject the notion that a "vicious cycle" perpetuates inequality "for the population at large."

But if whites, including working-class whites, experienced a fluid occupational class system, in which the able and ambitious could rise, the reverse was true for blacks. Blau and Duncan's data indicated "that Negroes are handicapped at every step in their attempts to achieve economic success, and these cumulative disadvantages are what produces the great inequalities of opportunities under which the Negro American suffers. . . . The multiple handicaps associated with being an American Negro are cumulative in their deleterious consequences for a man's career."

Education, which we have seen opens all sorts of doors for whites, even many of quite low social origin, did not work in the same way for blacks:

The difference in occupational status between Negroes and whites is twice as great for men who have graduated from high school or gone to college as for those who have completed no more than eight years of schooling. In short, the careers of well-educated Negroes lag even further behind those of comparable whites than do the careers of poorly educated Negroes. . . . Negroes, as an underprivileged group, must make greater sacrifices to remain in school, but they have less incentive than whites to make these sacrifices, which may well be a major reason why Negroes often exhibit little motivation to continue in school and advance their education.

The Blau-Duncan findings on mobility data have been reinforced by a detailed analysis of racial differences in income by Albert Wohlstetter and Sinclair Coleman, using data running through the end of the 1960's. Although they report that nonwhite personal income has increased twice as fast as white since 1948, they conclude that even at the end of the last decade "nonwhite family and personal income are much inferior to white incomes along the entire distribution of each." The differences between the two groups are greatest in the upper echelons: "In fact, compared with white, there has been little or no change in nonwhite income at the top." The evidence also indicates that nonwhites are more likely to be affected by economic dips than whites. While increases in educational attainment improve income possibilities for both groups, Wohlstetter and Coleman conclude that even if education were equalized, the economic return for nonwhites would still remain far behind that for whites.[7]

Still, we cannot lose sight of the significant social and economic progress made by black Americans during the past decade. They have made their most impressive gains in the area of education. If median school years are used as an indicator of formal educational attainment, then by 1970 blacks had come very close to parity with whites. More important, 1970 Census data suggest changes in the traditional pattern of discrimination noted earlier—the pattern whereby better educated blacks were worse off economically relative to comparable whites than were blacks with less education. During the 1960's, better educated blacks improved their market position.[8] It is still true, however, that at each educational level black men earn less than comparably educated whites; as in the 1950's, for example, young black males with high school diplomas still have a lower median income than young whites who have completed only grammar school. The earnings of black men relative to those of whites still decline as their educational levels increase. During the 1960's, this "earnings from education" gap began to narrow, particularly among college-educated men. The most striking transformation, however, occurred for black females. As of 1969, black females at both the high school and college level actually had a higher median income than their white counterparts. Yet, despite the considerable progress of certain segments of the black community, whites are still enormously advantaged by the presence of a racial minority which (together with other minority groups) handles a heavily disproportionate share of the less rewarded jobs and status positions. . . .

Chapter 9

WHAT IS SOCIAL CONTROL? THE CASE OF EDUCATION

A pervasive and continuing aspect of the individual's experience of society is social control. Put simply, this means that society models the individual into a particular shape and prevents him from doing many of the things that he would want to do if left to his own devices. This experience obtains from the earliest stages of socialization. But within the context of the family it is often mitigated by the emotional elements of affection and personal consideration. For most members of contemporary society the first massive experience of an impersonal control system occurs in school.

The perception of all social institutions differs in terms of the different groups experiencing them. Thus, the educators who manage the school will probably understand it in terms of its alleged educational function—such as the promulgation of general learning. The students who undergo the experience of school are more likely to be aware of its control aspects and of its sometimes quite crass relations to their hopes for social mobility. The sociologist, if he is not to become the agent of any particular group, must take all these differing experiences into account. In the case of education this means that the sociologists must try to understand its place within a variety of social forces.

In a significant way education relates to social control—that is, to the mechanisms by which society restrains deviance and pushes the individual into approved channels of conduct. A number of sociological studies of American education have shown how the school rewards specific traits and imposes sanctions upon others, and how these controls relate to the "management" of social mobility. Beyond all this, education has become an institution embodying the peculiarly modern myth of a collective and an individual good. This myth is one of the few truly universal ones in the contemporary world, cutting across a variety of ideological divisions.

My first selection, Charles E. Bidwell's essay "Youth in Modern Society," describes the centrality of schooling today. The next, a passage from Jonathan Kozol's *Death at an Early Age,* describes the control mechanism in a black American city school, and the third, from Edgar Z. Friedenberg's *Coming of Age in America,* deals with similar processes for white suburban children. These three selections are generally characterized by an "anti-school" viewpoint. In providing these selections here I am not necessarily suggesting that such a viewpoint is imperative for a sociological perspective on this matter, but this kind of critical or even polemical treatment is particularly useful in bringing out the control aspect of education. F. Musgrove, a British sociologist, in the passage selected from *Youth and the Social Order* deals with the psychological consequences of education as a control mechanism in contemporary societies. I have made my final selection, Sanford M. Dornbusch's "The Military Academy as an Assimilating Institution," to show that similar processes of molding and constraint can take place in educational processes beyond adolescence. This selection deals with some of the peculiarities of a Coast Guard Academy, but comparable processes of learning and "unlearning" are found in other cases of professional education.

CHARLES E. BIDWELL

THE CENTRALITY OF SCHOOLING

IN modern societies universal compulsory education is a necessity, and children leave the family not to enter the adult world, but to enter school. Although in these societies the family may serve as a prime agency for forming the values and beliefs of new generations, societal complexity and rationalization demand an education in the more technical aspects of social life that few families can provide. This is most immediately apparent in training for work, but can be seen as well in such school subjects as citizenship and consumer education, in which the student learns how to participate in large-scale political and economic systems.

The pressure toward universal education is reinforced by the variable occupational destinations of young persons. Since these destinations cannot be forecast specifically from parental social position, a system of schools that can provide training of diverse kinds is needed to mediate between families and the occupational destinations of their offspring.

From the viewpoint of the society, schooling is essential for two additional reasons. Because of the indeterminacy of adult destinations, which is accompanied by high levels of inter-generational mobility, all children and youth must receive a minimal, uniform education. Universal schooling cuts across disruptive variations in the ability of parents to educate their offspring. In addition, families tend to be parochial, reflecting the beliefs and ways of life characteristic of region, class, and religion and ethnicity. They tend to form children's loyalties in a similarly parochial fashion. Universal education, however, can strengthen loyalties to the society itself and foster commitment to its central values and sentiments.

Hence children in modern societies move out of the family into school, where, as they grow older, their lives increasingly center. Moreover, the complexity of training for occupations, and for other social roles as well, increases the volume of material to be taught, so that the school-leaving age tends to advance. Further higher education, as a necessary prerequisite to high-status occupations, holds large portions of young people (e.g., in 1961, 38 per cent, in 1939, 14 per cent of the eighteen-to-twenty-one age group in the United States).[1] These trends are reinforced by rising expectations among populations increasingly able to stand the costs of extended education and by the action of occupational groups—especially the professions—to control entry

Reprinted from *American Sociology*, ed. Talcott Parsons (New York: Basic Books, 1968), pp. 249–52, by permission of the publisher.

into their ranks by increasing the volume, complexity, and cost of the training that they require. In modern societies, in point of fact, it is the phenomenon of prolonged schooling that defines youth as a social category and lies at the root of the continued dependency of young people well past physical maturity.

Three aspects of prolonged education are especially important in the present context. First, schools and teachers have great power vis-à-vis their students, and their ability to dominate is not seriously lessened, for all but lower-class students, as education advances. In the secondary schools and colleges this power, in contrast to the diffuse moral authority of lower-grade teachers, is based on the ability of the teaching staff to control the occupational destinies of their students.

Teachers' grades and other assessments of students' capacities are central to their progress in the educational system and, later, in the world of work. But teacher authority, at the higher levels of education, is exercised intermittently. As students prepare more explicitly for autonomous adult performance and develop some competence, teachers supervise them less closely and exercise their authority by periodic examination.

Consequently youth in school are in an ambivalent position. Although they are in many ways independent, they are *permitted* this independence as part of their education. It is always hemmed about by teachers' expectations and judgments—from which there is no formal appeal. Moreover, the freedom of action that results as the youth move more clearly outside the family must be used in ways that will not hurt school work too seriously. Thus teachers and schools form for youth perhaps the most tangible barriers to adulthood, and young people will question their legitimacy to the degree that work in school seems unrelated to the adult future.

Second, despite the centrality of schooling to demanding adult roles, students even at the college level are not often exposed to excessive academic pressure. As David Matza notes, education in modern societies undoubtedly could proceed faster, with more rapid movement of youth into adulthood.[2] Yet the belief persists in modern societies that childhood and youth are carefree and happy times. This belief colors the timing of education, with its long vacations, and tempers the demands of teachers. As a result, youth is a time of unusual leisure, and young people have substantial reserves of time and energy to draw on as they wish. Students escape the consuming and determining impact of much adult work; youthful leisure persists into adulthood primarily among the lower occupations.

Third, schooling in modern societies contains contradictory tendencies toward the "impractical" and instrumental. On the one hand, resulting in part from the persistence of the aristocratic conception of liberal education and in part from the utility of personal flexibility in complex societies, many courses taken in secondary schools and colleges are offered as ends in themselves—as liberalizing—and have no direct part in occupational training.

On the other hand, schools present adult reality in an abstract and vicarious mode. The delay of adulthood means that adult roles must be experienced through talking and reading about the adult world or through simulated practice of adult activities. The complexity of modern societies forces abstractness, since children and youth must be prepared to master many diverse experiences by categorizing them and by learning quite generalized norms governing whole classes of situations. Florian Znaniecki argued that the abstract and vicarious quality of schooling destroys the immediacy of learning, and with it direct interest in the subject matter.[3] Instead, learning must always be for the sake of the future and appear artificial in the present. It becomes important and endurable only for the sake of more or less distant goals.

Consequently youth are likely to find school satisfying, to persist beyond the leaving age, and to do well to the extent that these distant goals have meaning and importance. These are the young people for whom parental expectations and teacher demands and sanctions seem sensible and legitimate; the vicarious quality of schooling, for these students, reinforces school authority. Others, for whom the link between present and future is more tenuous, find little that is sensible or legitimate in the expectations of parents or teachers, and the authority of these adults declines.

JONATHAN KOZOL

DEATH AT AN EARLY AGE

ALL books used in a school system, merely by the law of averages, are not going to be blatantly and consistently bad. A larger number of the books we had in Boston were either quietly and subtly bad, or else just devastatingly bad only in one part. One such book, not used in my school but at the discipline school, was entitled *Our World Today*. It seems useful to speak about it here because it exemplifies to perfection the book which might be remarkably accurate or even inspired in its good intentions in one section and then brutally clumsy, wrong and stupid in another. Right and wrong, good and bad, alternate in this book from sentence to sentence and from page to page:

"The people of the British Isles are, like our own, a mixed people. Their

Reprinted from *Death at an Early Age* by Jonathan Kozol (Boston: Houghton Mifflin, 1967), pp. 73–84.

ancestors were the sturdy races of northern Europe, such as Celts, Angles, Saxons, Danes, and Normans, whose energy and abilities still appear in their descendants. With such a splendid inheritance what could be more natural than that the British should explore and settle many parts of the world, and, in time, build up the world's greatest colonial empire?"

"The people of South Africa have one of the most democratic governments now in existence in any country."

"Africa needs more capitalists . . . White managers are needed . . . to show the Negroes how to work and to manage the plantations . . ."

"In our study of the nations of the world, we should try to understand the people and their problems from their point of view. We ought to have a sympathetic attitude toward them, rather than condemn them through ignorance because they do not happen always to have our ways."

"The Negro is very quick to imitate and follow the white man's way of living . . ."

". . . The white man may remain for short periods and direct the work, but he cannot . . . do the work himself. He must depend upon the natives to do the work."

"The white men who have entered Africa are teaching the natives how to live."

Something similar to this, though it was not in a printed textbook, was a mimeographed test about American history that the Fourth Grade teachers at my school had been using for several years. The test listed a number of attributes and qualities that were supposed to have been associated with George Washington: "courageous, rich, intelligent, wise, handsome, kind, good in sports, patient, believed in God, sense of humor, dressed in style, rode a horse well." From these the class were asked to underline the things that made George Washington "a great leader." The answers that would get points were only the noble virtues. "Rich," "handsome," "dressed in style," "rode a horse well" and "good in sports" were wrong. It was, I felt, not really a lesson on George Washington but a force-feeding of a particular kind of morality:

> These are good qualities.
> George Washington got someplace.
> These must be the things that made him great.

What had happened, very clearly, was that the right answers had never been derived from a real study of George Washington but rather they were taken from somebody's cupboard of good qualities ("moral builders") and then *applied* to George Washington exactly like plaster or paint. All the things listed were assumed to be true of him, but only the moral uplifters could be considered to be the things that helped him to be great. I thought this was wrong for several reasons. One reason simply was a matter of accuracy: George Washington was not really a very handsome man so it seemed unwise and dumbly chauvinistic to say he was. Another mistake, though it is a small one, was that he did not really have much of a sense of humor and people have even said that he was rather short-tempered. On the subject of his religion, it seemed presumptuous to me, and rather risky, to make any statements at all in regard to a belief in God about which, if it was really so, only George Washington himself could have known. On the opposite side, we do know well that good looks and lots of money have helped many men and do not even necessarily diminish them but have formed romantic parts of their greatness. This was true, for example, of President Kennedy, and it does him no dishonor to say so. What does do a man dishonor is to paint him up in false colors which we either do not know about or which we do know about and know that he did not have. I spoke to the Reading Teacher about this and I pointed out to her that it seemed to me Washington's wealth would not be at all a bad answer. The matter of his belief in God seemed questionable. The Reading Teacher did not often get openly angry with me, but she did on this occasion.

"That's out of the question! We are not going to start teaching cynicism here in the Fourth Grade."

I found myself equally angry. I said that I did not think that it would be teaching cynicism at all, but quite the reverse. I said I thought children should learn now, and the sooner the better, that money frequently, and more often than not, counted for more than religious intensity in the political world. I also said that I thought it was a far greater kind of cynicism to dish out to them at this age a fatuous and lyrical idealism which was going to get smashed down to the ground the first time that any one of them went out into the City of Boston and just tried to get himself a decent job, let alone try to follow George Washington on the strength of such qualities as "patience," "sense of humor," "belief in God," and hope to become President of the United States with the kind of education they were getting here. I said I thought that the highest cynicism of all was not to let the people in a running-race have any knowledge of the odds. I said that the only way in which one of these children ever *could* be President was if he understood absolutely and as soon as possible how many fewer advantages he had than had been the fortune of George Washington. With that knowledge first, not cynical in the least but having a true connection with the world, then those admired qualities might perhaps be the attributes of greatness, but not without a prior sense of the real odds.

When I began to talk about some of these things more openly, as I did now more frequently with the Reading Teacher, she would tell me sometimes that, temperamentally, she agreed with my ideas and did not believe that in themselves they were wrong but that to teach such things to the children "at this level and at this stage" was something that she could not allow because this was "not the proper age at which to start to break things down." The Fourth Grade, she told me in some real awkwardness and verbal confusion (which made me feel guilty for having brought the subject up), was the age "in which a teacher ought to be building up things and ideas." Later on, she seemed to be saying, there would be time to knock the same things down. I thought this a little like a theory of urban renewal, but it seemed a kind of renewal program that was going to cost somebody dearly. It was to erect first the old rotten building (pollyanna voyages, a nation without Negroes, suburban fairy tales, pictures in pastel shades), let it all stand a year or two until it began to sway and totter, then tear it down and, if the wreckage could be blasted, put up a new, more honest building in its spot. If a city planner ever came up with a theory like that, I should think he would be laughed out of business. Yet this was not very different from the Reading Teacher's view of children and of education. There was a lot in the Boston curriculum, furthermore, to support her.

I've said something about the social studies books and teacher attitudes already. One thing I've scarcely mentioned is the curriculum material in literature and reading. Probably this would be the best place to make some reference to it, for nothing could better typify the image of the crumbling school structure than the dry and deadly basic reading textbooks that were in use within my school. Most of these books were published by Scott Foresman. The volume aimed at Third Graders, used for slow Fourth Graders at my school, was called *New Streets and Roads*. No title could have been farther from the mark. Every cliché of bad American children's literature seemed to have been contained within this book. The names of the characters describe the flavor of the stories: Betty Jane Burns and Sarah Best and Miss Molly and Fluffy Tail and Miss Valentine of Maple Grove School . . . The children in my class had been hearing already for several years about Birmingham and Selma and tear-gas and cattle-prods and night-courts and slum-lords —and jazz. To expect these children to care about books which even very comfortable suburban children would probably have found irrelevant and boring seemed to be futile. Yet there were no other books around. These were the only ones we had. I wondered if it was thought that the proper way to teach reading to slow Fourth Graders was by foisting upon them a pablum out of nursery land. Possibly it was a benevolent school-lady's most dearly held belief that she could shut out the actual world in this manner and could make the world of these growing Negro children as neat and aseptic as her own. It may have been another belief that a few dozen similarly inclined white ladies in a few dozen Negro classrooms with a few hundred pure white texts would be able to overcome taste and appetite, sight and sound. I didn't think so. The books seemed so overwhelmingly boring. Wouldn't the children find them boring too? The look of boredom seemed always so apparent in the faces of those children. The books did not refer to them. What did refer to them was obvious.

Once I asked a class to think of a sentence using both the word "glass" and the word "house." The first answer I received told me that "there is a lot of glass out back of my house."

To a boy in the reading group: "Do you know an antonym for dry?"

"Mr. Kozol—isn't there something called a dry martini?"

The Reading Teacher's reactions to these kinds of student responses, when I related them to her, were generally pretty much the same:

"We cannot use books that are sordid."

"Children do not like gloomy stories."

"I haven't seen any evidence that children like to read especially about things that are real."

I suppose it is true that the children she took out for reading generally were attentive, at least on a temporary basis, to the texts that she was using with them, but I think there were two special reasons for this. One reason was that she could "sell" almost anything to anyone if she wanted, being such a very experienced and such an intensely persuading teacher. The other was that the children, in reading with docility and in writing without their own imagination, were always more than willing to confirm a white teacher's idea of them and to put forward in their writings and conversations not what they really felt or dreamed but what they had good reason by now to know that she wanted to believe about them. I was not asking for children's books to be sordid, either, but the Reading Teacher was pretty much suggesting an identity between *sordid* and *real.* I had the idea that to build upon some of the things the children already knew would be more fruitful than to deny them. I also had a suspicious, ungenerous feeling about the reluctance of the white teachers to make use of more realistic books. Their argument, stated one way or another, was that such books might be bad for the children but I thought that that was not what they really believed. I thought the denial came not from a fear that such things might be bad for the child but rather from a certainty that they would be bad for the teacher. I thought that the Reading Teacher and the Deputy Superintendent and many others like them would have been confused to be told that the world of those Negro children was in a great many cases a good deal more interesting and more vital than their own. It seemed to me that what they were trying ineffectively to do was to replace a very substantial and by no means barren lower-class culture with a concoction of pretty shopworn middle-class ideas. The ideas they introduced, moreover, did not even have the joy of being exuberant, for they were mainly the values of a parched and parochial and rather grim and beaten lower middle-class and were, I felt, inferior by many times to that which the children and their parents already had. More succinctly, what I mean is that the real trouble with perpetrating such colorless materials upon very colorful children was not only that the weak culture they purveyed was out of kilter with the one the children already had, but that it also was mediocre by comparison.

EDGAR Z. FRIEDENBERG

COMING OF AGE IN AMERICA

THE school, as schools continually stress, acts *in loco parentis;* and children may not leave home because their parents are unsatisfactory. What I have pointed out is no more than a special consequence of the fact that students are minors, and minors do not, indeed, share all the rights and privileges—and responsibilities—of citizenship. Very well. However one puts it, we are still discussing the same issue. The high school, then, is where you really learn what it means to be a minor.

For a high school is not a parent. Parents may love their children, hate them, or, like most parents, do both in a complex mixture. But they must, nevertheless, permit a certain intimacy and respond to their children as persons. Homes are not run by regulations, though the parents may think they are, but by a process of continuous and almost entirely unconscious emotional homeostasis, in which each member affects and accommodates to the needs, feelings, fantasy life, and character structure of the others. This may be, and often is, a terribly destructive process; I intend no defense of the family as a social institution. Salmon, actually, are much nicer than people: more dedicated, more energetic, less easily daunted by the long upstream struggle and less prudish and reticent about their reproductive functions, though inclined to be rather cold-blooded. But children grow up in homes or the remnants of homes, are in physical fact dependent on parents, and are too intimately related to them to permit their area of freedom to be precisely defined. This is not because they have no rights or are entitled to less respect than adults, but because intimacy conditions freedom and growth in ways too subtle and continuous to be defined as overt acts.

Free societies depend on their members to learn early and thoroughly that public authority is *not* like that of the family; that it cannot be expected—or trusted—to respond with sensitivity and intimate perception to the needs of individuals but must rely basically, though as humanely as possible, on the impartial application of general formulae. This means that it must be kept functional, specialized, and limited to matters of public policy; the meshes of the law are too coarse to be worn close to the skin. Especially in an open society, where people of very different backgrounds and value systems must function together, it would seem obvious that each must understand that he

Reprinted from *Coming of Age in America* by Edgar Z. Friedenberg (New York: Random House, 1965), pp. 43–49.

may not push others further than their common undertaking demands or impose upon them a manner of life that they feel to be alien.

After the family, the school is the first social institution an individual must deal with—the place in which he learns to handle himself with strangers. The school establishes the pattern of his subsequent assumptions as to which relations between the individual and society are appropriate and which constitute invasions of privacy and constraints on his spirit—what the British, with exquisite precision, call "taking a liberty." But the American public school evolved as a melting pot, under the assumption that it had not merely the right but the duty to impose a common standard of genteel decency on a polyglot body of immigrants' children and thus insure their assimilation into the better life of the American dream. It accepted, also, the tacit assumption that genteel decency was as far as it could go. If America has generally been governed by the practical man's impatience with other individuals' rights, it has also accepted the practical man's respect for property and determination to protect it from the assaults of public servants. With its contempt for personal privacy and individual autonomy, the school combines a considerable measure of Galbraith's "public squalor." The plant may be expensive—for this is capital goods; but nothing is provided graciously, liberally, simply as an amenity, either to teachers or students, though administrative offices have begun to assume an executive look. In the schools I know, the teachers' lounges are invariably filled with shabby furniture and vending machines. Teachers do not have offices with assigned clerical assistance and business equipment that would be considered satisfactory for, say, a small-town, small-time insurance agency. They have desks in staffrooms, without telephones.

To justify this shabbiness as essential economy and established custom begs the question; the level of support and working conditions customarily provided simply defines the status of the occupation and the value the community in fact places on it. An important consequence, I believe, is to help keep teachers timid and passive by reminding them, against the contrasting patterns of commercial affluence, of their relative ineffectiveness; and to divert against students their hostilities and their demands for status. Both teachers and students, each at their respective levels, learn to regard the ordinary amenities and freedoms of middle-class life as privileges. But the teacher has a few more of them. He hasn't a telephone, but he may make calls from a phone in the general office, while, in some schools, the public pay phone in the hallway has a lock on it and the student must get a key from the office before he can dial his call. Where a hotel or motel, for example, provides in its budget for normal wear and tear and a reasonable level of theft of linens and equipment and quietly covers itself with liability insurance, the school—though it may actually do the same thing—pompously indoctrinates its students with "respect for public property," "good health habits," and so forth before it lets them near the swimming pool. In a large city, the pool may have been struck out of the architect's plans before construction began, on the grounds that it would be unfair to provide students in a newer school with a costly facility that students in older schools do not have.

If the first thing the student learns, then, is that he, as a minor, is subject to peculiar restraints, the second is that these restraints are general, and are not limited to the manifest and specific functions of education. High school administrators are not professional educators in the sense that a physician, an attorney, or a tax accountant are professionals. They are not practitioners of a specialized *instructional* craft, who derive their authority from its requirements. They are specialists in keeping an essentially political enterprise from being strangled by conflicting community attitudes and pressures. They are problem-oriented, and the feelings and needs for growth of their captive and disfranchized clientele are the least of their problems; for the status of the "teenager" in the community is so low that even if he rebels the school is not blamed for the conditions against which he is rebelling. He is simply a truant or juvenile delinquent; at worst the school has "failed to reach him." What high school personnel become specialists in, ultimately, is the *control* of the large groups of students even at catastrophic expense to their opportunity to learn. These controls are not exercised primarily to facilitate instruction, and, particularly, they are in no way limited to matters bearing on instruction. At several schools in our sample boys had, for example, been ordered by the assistant principal—sometimes on the complaint of teachers—to shave off beards. One of these boys, who had played football for the school all season, was told that, while the school had no legal authority to require this, he would be barred from the banquet honoring the team unless he complied. Dress regulations are another case in point.

Of course these are petty restrictions, enforced by petty penalties. American high schools are not concentration camps; and I am not complaining about their severity but about what they teach their students concerning the proper relationship of the individual to society. The fact that the restrictions and penalties are petty and unimportant in themselves in one way makes matters worse. Gross invasions are more easily recognized for what they are; petty restrictions are only resisted by "troublemakers." What matters in the end, however, is that the school does not take its own business of education seriously enough to mind it.

The effects on the students of the school's diffuse willingness to mind everybody's business but its own are manifold. The concepts of dignity and privacy, notably deficient in American adult folkways, are not permitted to develop here. The high school, certainly, is not the material cause of this deficiency, which is deeply rooted in our social institutions and values. But the high school does more than transmit these values—it exploits them to keep students in line and develop them into the kinds of people who fit the community that supports it.

A corollary of the school's assumption of custodial control of students is that power and authority become indistinguishable. If the school's authority is not limited to matters pertaining to education, it cannot be derived from educational responsibilities. It is a naked, empirical fact, to be accepted or controverted according to the possibilities of the moment. In this world power counts more than legitimacy; if you don't have power it is naïve to think you have rights that must be respected; wise up. High school students experience regulation only as control, not as protection; they know, for example, that the principal will generally uphold the teacher in any conflict with a student, regardless of the merits of the case. Translated into the high school idiom, *suaviter in modo, fortiter in re* becomes "If you get caught, it's just your ass."

Students, I find, do not resent this; that is the tragedy. All weakness tends to corrupt, and impotence corrupts absolutely. Identifying, as the weak must, with the more powerful and frustrating of the forces that impinge upon them, they accept the school as the way life is and close their minds against the anxiety of perceiving alternatives. Many students like high school; others loathe and fear it. But even these do not object to it on principle; the school effectively obstructs their learning of the principles on which objection might be based; though these are among the principles that, we boast, distinguish us from totalitarian societies.

Yet, finally, the consequence of submitting throughout adolescence to diffuse authority that is not derived from the task at hand—as a doctor's orders, or the training regulations of an athletic coach, for example, usually are—is more serious than political incompetence or weakness of character. There is a general arrest of development. An essential part of growing up is learning that, though differences of power among men lead to brutal consequences, all men are peers; none is omnipotent, none derives his potency from magic but only from his specific competence and function. The policeman represents the majesty of the State, but this does not mean that he can put you in jail; it means, precisely, that he cannot—at least not for long. Any person or agency responsible for handling throngs of young people—especially if it does not like them or is afraid of them—is tempted to claim diffuse authority and snare the youngster in the trailing remnants of childhood emotion, which always remain to trip him. Schools are permitted to infantilize adolescence and control pupils by reinvoking the sensations of childhood punishment, effective because it was designed, with great unconscious guile, to dramatize the child's weakness in the face of authority. In fact, they are strongly encouraged to do so by the hostility to "teen-agers" and the anxiety about their conduct that abound in our society.

In the process, the school affects society in two complementary ways. It alters individuals: their values, their sense of personal worth, their patterns of anxiety and sense of mastery and ease in the world on which so much of what we think of as our fate depends. But it also performs a Darwinian function. The school endorses and supports the values and patterns of behavior of certain segments of the population, providing their members with the credentials and shibboleths needed for the next stages of their journey, while instilling in others a sense of inferiority and warning the rest of society against them as troublesome and untrustworthy. In this way, the school contributes simultaneously to social mobility and social stratification. It helps to see to it that the kinds of people who get ahead are those who will support the social system it represents; while those who might, through intent or merely by their being, subvert it are left behind as a salutary moral lesson.

This leads immediately to two questions: what patterns of values are developed and ratified through the experience of compulsory school attendance, and what kinds of people and what social groups do succeed best in school and find most support there? The issue of bias in the schools is old and familiar; but it cannot be stated as a simple tendency of the schools to favor the "economically advantaged" over the "culturally deprived." The school's bias is both more and less diverse than this statement suggests. The school quite often—indeed, traditionally—supports the respectable and ambitious poor against the presumptions of the privileged; its animus is directed against youngsters who possess certain common features regardless of the widest possible variation in status and income.

F. MUSGROVE

YOUTH AND THE SOCIAL ORDER

WE have fashioned institutions of higher education which can achieve their ends most easily with a certain personality type: which can relatively easily be made to feel guilt and self-reproach and to drive itself without overmuch steering. This type of person is characterized by what Eysenck has described as the introverted and neurotic 'dimensions of personality'.[1] He is rewarded for, and confirmed in, these particular dimensions. Individuals who are stable and extravert seem to do less well in the activities which are valued by grammar schools and universities, even though they are of similar intelligence.[2] (It is not a matter of 'intelligence': there is no question of high introversion or neuroticism necessarily connoting high intelligence as measured by intelligence tests.) And particularly at the borderline of university admission extraverts probably tend more often than introverts to be denied a place.[3]

It has been suggested that extraversion is not in itself a handicap in aca-

Reprinted from *Youth and the Social Order* by F. Musgrove (Bloomington: Indiana University Press, 1965), pp. 4–9, by permission of the publisher.

demic pursuits below the level of university education, but there is evidence that our present system is already penalizing the (intelligent) extravert before this stage.[4] Any human society is highly selective of those aspects of human activity and achievement which its educational system takes seriously into account. Our own system is almost exclusively concerned with those activities and exercises in which 'neurotic introverts' tend to excel. Unfortunately their excellence in these matters is thought to qualify them for future roles for which they may be quite unsuited.[5]

Our emphasis in higher education is on a narrow range of a person's capacities and possibilities. These have the advantage of being fairly easily assessed. Other societies place their emphasis differently. The society which produced the *Kama Sutra* (like Aldous Huxley's fictional island of Pala [6]) would take into account in its educational provision aspects of the human personality and capacity for experience which we have chosen to ignore or regret.

Our formal educational system concerns itself with a highly selective range of human attributes which receive attention because of tradition and the pressures of contemporary social and economic circumstances. Of central concern are written-verbal skills and extended preparation for their assessment. (Oral-verbal skills have no such prominence. It is still possible to pass with distinction through many an English grammar school—and through some courses in some of our universities—virtually without speaking; indeed, overmuch speaking, however much to the point, would make a distinguished grammar-school career highly unlikely.)

In many other societies this is not so. Nor was it with us when the *viva voce* examination figured far more prominently in the examination of students. The Nyakyusa of Tanganyika place emphasis in their educational procedures on *oral*-verbal skills, the ability to communicate effectively with one's fellows through the spoken word. It is perhaps a corollary that they prize most highly the virtues of comradeship and the arts of social intercourse.[7]

There may be some doubt about the effectiveness of anxiety as a drive in accomplishing intellectual tasks—or more precisely about different kinds of anxiety [8]—but the broad picture regarding the value of introversion and neuroticism is reasonably clear. (It is not altogether clear, however, *why* a degree of neuroticism should be an asset in intellectual pursuits.) The introvert will gear himself for sustained application to tasks which involve words, ideas and abstractions; but the extravert will probably attain a more powerful 'drive' in solving concrete problems involving people. He will not be so interested in reducing humanity to a formula; but he may achieve more in dealing with specific and concrete human predicaments. (This book is unlikely to appeal to him.) Unfortunately for him our system of academic assessment does not take his interests and capacities into account.

The present-day university population obtains higher scores on introversion and neuroticism than equally intelligent people who have not been interested in attending a university or who have been denied admission.[9] Even in

America, where university entrance is more open to those without strong academic inclinations, university graduates tend to suffer more from serious social-personal maladjustment than those of similar intelligence who have remained outside the university. In Terman's most recent follow-up (1950–5) of his sample of highly gifted (intelligent) Americans, now in their forties, almost twice the proportion of college graduates were found to suffer from 'serious maladjustment' compared with those who had not enjoyed a college education.[10]

As a university degree becomes a necessary prerequisite for entry into an ever wider range of occupations, we are preparing for positions of high status a great army of young people who have not necessarily those attributes of personality needed to occupy them effectively—or even very happily. It is perhaps of little consequence that the universities should be staffed predominantly by neurotic introverts. (But if both staff and students score high in these dimensions of personality, the result is unlikely to be a particularly hilarious community. No doubt hilarity is an irrelevant criterion by which to judge the life of a university; but it is doubtful whether work of the highest intellectual originality will be produced where there is no dash of frivolity.) [11] In other institutions this personality type might survive less effectively and perform less useful work.

In the past sustained performance over a narrow range of intellectual skills has not been a general requirement for high office and positions of leadership in English society. In the learned professions this has, quite properly, been required of entrants; and since the later nineteenth century, perhaps equally properly, it has been a requirement of the higher civil service. But industry and commerce, social work and public life, and a wide variety of administrative services, have not commonly recruited from the same source. Appointments have been made through influence and patronage, or through proven competence at lower levels of the enterprise. The widespread employment of graduates over a greater range of managerial work is of relatively recent date.

And formerly the graduate was not necessarily, or even perhaps commonly, endowed with the type of personality which alone seems able to secure university entrance, and an adequate degree, today. Indeed, the university man who reached the topmost positions in the nation's life had often, in the eighteenth and nineteenth centuries, failed to take a degree at all.[12] The Colonial Service throughout the twentieth century has been comparatively indifferent to a candidate's performance in academic work and has taken other evidence of capacity into account.[13] Increasingly those who are permitted to pursue higher academic studies have no other capacities which *can* be taken into account.

Any return to methods of selecting the nation's élite which ignored merit as estimated by our institutions of higher education would be offensive to contemporary notions of social justice. If a university degree is to be a necessary qualification for entry into responsible positions in virtually all departments of the nation's life, then men with highly valuable personal qualities—

which don't happen to be those which ensure success in academic work—must be admitted to degree courses and helped to make a reasonable showing in the examinations. Much closer supervision of students might be necessary to achieve this. The intelligent individual who is unfortunate enough to be a stable extravert must not be denied the qualifications necessary for entering positions of high responsibility.

But other solutions are infinitely preferable.[14] More diversified routes into the national élite should be available, in which abilities in other than narrowly defined intellectual exercises are assessed: particularly the skills of human relationship—quiet unruffled confidence in human crises; an outgoing and reassuring approach to colleagues, subordinates and clients; tact and capacity to tolerate uncertain and ambiguous human situations; concern without hypersensitivity; an interest in and concern for actual, concrete individuals, instead of ideas about individuals and formulae which embrace an abstracted aspect of a faceless multitude. Some of our new institutions of higher education must recast their curricula in these terms.

Our society cannot afford to dispense with the services of the intelligent extravert at the highest levels of leadership. Our schools and institutions of higher learning must cater for him and cease to eliminate him along the educational route. If he is more often found in our grammar schools, and following a curriculum which exercises and rewards his special skills (even though these are difficult to assess), we may find a smaller proportion of the young who are accorded high status so resentful and in such apparent conflict as a result of their good fortune.

SANFORD M. DORNBUSCH

THE MILITARY ACADEMY AS AN ASSIMILATING INSTITUTION

THE function of a military academy is to make officers out of civilians or enlisted men. The objective is accomplished by a twofold process of transmit-

Reprinted from Sanford M. Dornbusch, "The Military Academy as an Assimilating Institution," *Social Forces* 33 (May 1955): 316–21 by permission of the University of North Carolina Press.

The writer is indebted to Harold McDowell, Frank Miyamoto, Charles Bowerman, and Howard S. Becker for their constructive criticism of this paper.

ting technical knowledge and of instilling in the candidates an outlook considered appropriate for members of the profession. This paper is concerned with the latter of these processes, the assimilating function of the military academy. Assimilation is viewed as "a process of interpenetration and fusion in which persons and groups acquire the memories, sentiments, and attitudes of other persons and groups, and, by sharing their experience and history, are incorporated with them in a common cultural life. . . . The unity thus achieved is not necessarily or even normally like-mindedness; it is rather a unity of experience and of orientation, out of which may develop a community of purpose and action." [1]

Data for this study consist almost entirely of retrospective material, based on ten months spent as a cadet at the United States Coast Guard Academy. The selective nature of memory obviously may introduce serious deficiencies in the present formulation. Unfortunately, it is unlikely that more objective evidence on life within the Academy will be forthcoming. Cadets cannot keep diaries, are formally forbidden to utter a word of criticism of the Academy to an outsider, and are informally limited in the matters which are to be discussed in letters or conversations. The lack of objective data is regrettable, but the process of assimilation is present here in an extreme form. Insight into this process can better be developed by the study of such an explicit, overt case of assimilation.

The Coast Guard Academy, like West Point and Annapolis, provides four years of training for a career as a regular officer. Unlike the other service academies, however, its cadet corps is small, seldom exceeding 350 cadets. This disparity in size probably produces comparable differences in the methods of informal social control. Therefore, all the findings reported here may not be applicable to the other academies. It is believed, however, that many of the mechanisms through which this military academy fulfills its assimilating function will be found in a wide variety of social institutions.

THE SUPPRESSION OF PRE-EXISTING STATUSES

The new cadet, or "swab," is the lowest of the low. The assignment of low status is useful in producing a correspondingly high evaluation of successfully completing the steps in an Academy career and requires that there be a loss of identity in terms of pre-existing statuses. This clean break with the past must be achieved in a relatively short period. For two months, therefore, the swab is not allowed to leave the base or to engage in social intercourse with non-cadets. This complete isolation helps to produce a unified group of swabs, rather than a heterogeneous collection of persons of high and low status. Uniforms are issued on the first day, and discussions of wealth and family background are taboo. Although the pay of the cadet is very low, he is not permitted to receive money from home. The role of the cadet must supersede other roles the individual has been accustomed to play. There are few clues left which will reveal social status in the outside world.[2]

It is clear that the existence of minority-group status on the part of some cadets would tend to break down this desired equality. The sole minority group present was the Jews, who, with a few exceptions, had been informally excluded before 1944. At that time 18 Jews were admitted in a class of 162. Their status as Jews made them objects of scrutiny by the upper classmen, so that their violations of rules were more often noted. Except for this "spotlight," however, the Jews reported no discrimination against them—they, too, were treated as swabs.

LEARNING NEW RULES AND ADJUSTMENT TO CONFLICTS BETWEEN RULES

There are two organized structures of rules which regulate the cadet's behavior. The first of these is the body of regulations of the Academy, considered by the public to be the primary source of control. These regulations are similar to the code of ethics of any profession. They serve in part as propaganda to influence outsiders. An additional function is to provide negative sanctions which are applied to violations of the second set of expectations, the informal rules. Offenses against the informal rules are merely labeled as breaches of the formal code, and the appropriate punishment according to the regulations is then imposed. This punitive system conceals the existence of the informal set of controls.

The informal traditions of the Academy are more functionally related to the existing set of circumstances than are the regulations, for although these traditions are fairly rigid, they are more easily forgotten. Unlike other informal codes, the Academy code of traditions is in part written, appearing in a manual for entering cadets.

In case of conflict between the regulations and tradition, the regulations are superseded. For example, it is against the regulations to have candy in one's room. A first classman orders a swab to bring him candy. Caught en route by an officer, the swab offers no excuse and is given 15 demerits. First classmen are then informally told by the classmate involved that they are to withhold demerits for this swab until he has been excused for offenses totaling 15 demerits. Experience at an Academy teaches future officers that regulations are not considered of paramount importance when they conflict with informal codes—a principle noted by other observers.[3]

Sometimes situations arise in which the application of either form of control is circumvented by the commanding officer. The following case is an example. Cadets cannot drink, cannot smoke in public, can never go above the first floor in a hotel. It would seem quite clear, therefore, that the possessor of a venereal disease would be summarily dismissed. Cadets at the Academy believed that two upper-class cadets had contracted a venereal disease, were cured, and given no punishment. One of the cadets was an outstanding athlete, brilliant student, and popular classmate. Cadets were told that a direct appeal by the commanding officer to the Commandant of the Coast Guard resulted in the decision to hush up the entire affair, with the second cadet getting the same treatment as his more popular colleague. The event indicated the possibility of individualization of treatment when rules are violated by officers.

THE DEVELOPMENT OF SOLIDARITY

The control system operated through the class hierarchy. The first class, consisting of cadets in their third or fourth year at the Academy, are only nominally under the control of the officers of the Academy. Only one or two officers attempt to check on the activities of the first classmen, who are able to break most of the minor regulations with impunity. The first class is given almost complete control over the rest of the cadet corps. Informally, certain leading cadets are even called in to advise the officers on important disciplinary matters. There are one or two classes between the first classmen and the swabs, depending on the existence of a three- or four-year course. These middle classes haze the swabs. Hazing is forbidden by the regulations, but the practice is a hallowed tradition of the Academy. The first class demands that this hazing take place, and, since they have the power to give demerits, all members of the middle classes are compelled to haze the new cadets.

As a consequence of undergoing this very unpleasant experience together, the swab class develops remarkable unity. For example, if a cadet cannot answer an oral question addressed to him by his teacher, no other member of his class will answer. All reply, "I can't say, sir," leaving the teacher without a clue to the state of knowledge of this student compared to the rest of the class. This group cohesion persists throughout the Academy period, with first classmen refusing to give demerits to their classmates unless an officer directly orders them to do so.

The honor system, demanding that offenses by classmates be reported, is not part of the Coast Guard Academy tradition. It seems probable that the honor system, if enforced, would tend to break down the social solidarity which the hazing develops within each class.

The basis for interclass solidarity, the development of group feeling on the part of the entire cadet corps, is not so obvious. It occurs through informal contacts between the upper classmen and swabs, a type of fraternization which occurs despite the fact it traditionally is discouraged. The men who haze the swab and order him hazed live in the same wing of the dormitory that he does. Coming from an outside world which disapproves of authoritarian punishment and aggressiveness, they are ashamed of their behavior. They are eager to convince the swab that they are good fellows. They visit his room to explain why they are being so harsh this week or to tell of a mistake he is making. Close friendships sometimes arise through such behavior. These friendships must be concealed. One first classman often ordered his room cleaned by the writer as a "punishment," then settled down for an uninterrupted chat. Such informal contacts serve to unite the classes and spread a "we-feeling" through the Academy.

In addition, the knowledge of common interests and a common destiny serves as a unifying force that binds together all Academy graduates. This is expressed in the identification of the interest of the individual with the interest of the Coast Guard. A large appropriation or an increase in the size of the Coast Guard will speed the rate of promotion for all, whether ensign or captain. A winning football team at the Academy may familiarize more civilians with the name of their common alma mater. Good publicity for the Coast Guard raises the status of the Coast Guard officer.

The Coast Guard regulars are united in their disdain for the reserves. There are few reserve officers during peacetime, but in wartime the reserve officers soon outnumber the regulars. The reserves do not achieve the higher ranks, but they are a threat to the cadets and recent graduates of the Academy. The reserves receive in a few months the rank that the regulars reach only after four grueling years. The Academy men therefore protectively stigmatize the reserves as incompetents. If a cadet falters on the parade ground, he is told, "You're marching like a reserve." Swabs are told to square their shoulders while on liberty, "or else how will people know you are not a reserve?" Myths spring up—stories of reserve commanders who must call on regular ensigns for advice. The net effect is reassurance that although the interlopers may have the same rank, they do not have equal status.

Another out-group is constituted by the enlisted men, who are considered to be of inferior ability and eager for leadership. Segregation of cadets and enlisted men enables this view to be propagated. Moreover, such segregation helps to keep associations within higher status social groups. There is only one leak in this insulating dike. The pharmacist mates at sick bay have direct contact with the cadets, and are the only enlisted personnel whom cadets meet on an equal basis. The pharmacist mates take pleasure in reviling the Academy, labeling it "the p---k factory." Some of the cadets without military experience are puzzled by such an attitude, which is inconsistent with their acquired respect for the Academy.

THE DEVELOPMENT OF A BUREAUCRATIC SPIRIT

The military services provide an excellent example of a bureaucratic structure. The emphasis is upon the office with its sets of rights and duties, rather than on the man. It is a system of rules with little regard for the individual case. The method of promotion within the Coast Guard perfectly illustrates this bureaucratic character. Unlike the Army or Navy, promotions in the Coast Guard up to the rank of lieutenant-commander do not even depend on the evaluation of superior officers. Promotion comes solely according to seniority, which is based on class standing at the Academy. The 50th man in the 1947 class will be lieutenant-commander before the 51st man, and the latter will be promoted before the 1st man in the 1948 class.

The hazing system contributes directly to acceptance of the bureaucratic structure of the Coast Guard, for the system is always viewed by its participants as not involving the personal character of the swab or upper classman. One is not being hazed because the upper classman is a sadist, but because one is at the time in a junior status. Those who haze do not pretend to be superior to those who are being hazed. Since some of those who haze you will also try to teach you how to stay out of trouble, it becomes impossible to attribute evil characteristics to those who injure you. The swab knows he will have his turn at hazing others. At most, individual idiosyncrasies will just affect the type of hazing done.[4]

This emphasis on the relativity of status is explicitly made on the traditional Gizmo Day, on which the swabs and their hazers reverse roles. The swabs-for-a-day take their licking without flinching and do not seek revenge later, for they are aware that they are under the surveillance of the first classmen. After the saturnalia, the swabs are increasingly conscious of their inability to blame particular persons for their troubles.

Upper classmen show the same resentment against the stringent restrictions upon their lives, and the manner in which they express themselves indicates a feeling of being ruled by impersonal forces. They say, "You can't buck the System." As one writer puts it, "The best attitude the new cadet can have is one of unquestioning acceptance of tradition and custom."

There is a complete absence of charismatic veneration of the Coast Guard heroes of the past and present. Stirring events are recalled, not as examples of the genius of a particular leader, but as part of the history of the great organization which they will serve. A captain is a cadet thirty years older and wiser. Such views prepare these men for their roles in the bureaucracy.

NEW SATISFACTIONS IN INTERACTION

A bureaucratic structure requires a stable set of mutual expectations among the occupants of offices. The Academy develops this ability to view the behavior of others in terms of a pre-ordained set of standards. In addition to preparing the cadet for later service as an officer, the predictability of the behavior of his fellows enables the cadet to achieve a high degree of internal stability. Although he engages in a continual bustle of activity, he always knows his place in the system and the degree to which he is fulfilling the expectations of his role.

Sharing common symbols and objects, the cadets interact with an ease of communication seldom found in everyday life. The cadet is told what is right and wrong, and, if he disagrees, there are few opportunities to translate mental reservations into action. The "generalized other" speaks with a unitary voice which is uncommon in modern societies. To illustrate, an upper classman ordered a swab to pick up some pieces of paper on the floor of a washroom. The latter refused and walked away. There were no repercussions. The swab knew that, if he refused, the upper classman would be startled by the choice of such an unconventional way of getting expelled from the Academy. Wondering what was happening, the upper classman would redefine his own behavior, seeing it as an attack on the high status of the cadet. Picking up litter in a washroom is "dirty work," fit only for enlisted men. The swab was sure that the upper classman shared this common universe of discourse and never considered the possibility that he would not agree on the definition of the situation.

Interaction with classmates can proceed on a level of confidence that only intimate friends achieve in the outside world. These men are in a union of sympathy, sharing the same troubles, never confiding secrets to upper classmen, never criticizing one another to outsiders. Each is close to only a few but is friendly with most of the men in his class.

When interacting with an upper classman in private, a different orientation is useful. The swab does not guess the reason why he is being addressed, but instead assumes a formal air of deference. If the upper classman says, "Aw cut it out," the swab relaxes. In this manner the role of the upper classman is explicitly denoted in each situation.

In addition to providing predictability of the behavior of others, the Academy provides a second set of satisfactions in the self-process. An increase in the cadet's self-esteem develops in conjunction with identification in his new role. Told that they are members of an elite group respected by the community, most cadets begin to feel at ease in a superordinate role. One may be a low-ranking cadet, but cadets as a group have high status. When cadets visit home for the first time, there is a conflict between the lofty role that they wish to play and the role to which their parents are accustomed. Upon return to the Academy, much conversation is concerned with the way things at home have changed.

This feeling of superiority helps to develop self-confidence in those cadets who previously had a low evaluation of themselves. It directly enters into relationships with girls, with whom many boys lack self-confidence. It soon becomes apparent that any cadet can get a date whenever he wishes, and he even begins to feel that he is a good "catch." The cadet's conception of himself is directly influenced by this new way of viewing the behavior of himself and others. As one cadet put it, "I used to be shy. Now I'm reserved."

SOCIAL MOBILITY

A desire for vertical social mobility on the part of many cadets serves as one means of legitimizing the traditional practices of the Academy. The cadets are told that they will be members of the social elite during the later stages of their career. The obstacles that they meet at the Academy are then viewed as the usual barriers to social mobility in the United States, a challenge to be surmounted.

Various practices at the Academy reinforce the cadets' feeling that they are learning how to enter the upper classes. There is a strong emphasis on etiquette, from calling cards to table manners. The Tactics Officer has been known to give long lectures on such topics as the manner of drinking soup from an almost empty bowl. The cadet must submit for approval the name of the girl he intends to take to the monthly formal dance. Girls attending the upper-class college in the vicinity are automatically acceptable, but some cadets claim that their dates have been rejected because they are in a low status occupation such as waitress.

Another Academy tradition actively, though informally, encourages contact with higher status girls. After the swabs have been completely isolated for two months, they are invited to a dance at which all the girls are relatives or friends of Coast Guard officers. A week later the girls at the nearby college have a dance for the swabs. The next week end finds the swab compelled to invite an acceptable girl to a formal reception. He must necessarily choose from the only girls in the area whom he knows, those that he met during the recent hours of social intercourse.

JUSTIFICATION OF INSTITUTIONAL PRACTICES

In addition to the social mobility theme which views the rigors of Academy life as obstacles to upward mobility, there is a more open method of justifying traditionally legitimated ways of doing things. The phrase, "separating the men from the boys" is used to meet objections to practices which seem inefficient or foolish. Traditional standards are thus redefined as further tests of ability to take punishment. Harsh practices are defended as methods by which the insincere, incompetent, or undisciplined cadets are weeded out. Cadets who rebel and resign are merely showing lack of character.[5]

Almost all cadets accept to some extent this traditional view of resignations as admissions of defeat. Of the 162 entering cadets in 1944, only 52 graduated in 1948. Most of the 110 resignations were entirely voluntary without pressure from the Academy authorities. Most of these resignations came at a time when the hazing was comparatively moderate. Cadets who wish to resign do not leave at a time when the hazing might be considered the cause of their departure. One cadet's history illustrates this desire to have the resignation appear completely voluntary. Asked to resign because of his lack of physical coordination, he spent an entire year building up his physique, returned to the Academy, finished his swab year, and then joyously quit. "It took me three years, but I showed them."

Every cadet who voluntarily resigns is a threat to the morale of the cadet corps, since he has rejected the values of the Academy. Although cadets have enlisted for seven years and could theoretically be forced to remain at the Academy, the usual procedure is to isolate them from the swabs and rush acceptance of their resignation. During the period before the acceptance is final, the cadets who have resigned are freed from the usual duties of their classmates, which action effectively isolates them from cadets who might be affected by their contagious disenchantment.

REALITY SHOCK

Everett C. Hughes has developed the concept of "reality shock," the sudden realization of the disparity between the way a job is envisaged before beginning work and the actual work situation.[6] In the course of its 75-year history the Coast Guard Academy has wittingly or unwittingly developed certain measures to lessen reality shock in the new ensign. The first classmen, soon to be officers, are aided in lessening the conflict between the internalized rules of the Academy world and the standards for officer conduct.

On a formal level the first classmen are often reminded that they are about to experience a relative decline in status. On their first ship they will be given the most disagreeable duties. The first classmen accept this and joke about how their attitudes will change under a harsh captain. On a more concrete level, first classmen are given week-end leaves during the last six months of their stay at the Academy. These leaves allow them to escape from the restrictive atmosphere of the nearby area. It is believed wise to let them engage in orgiastic behavior while still cadets, rather than suddenly release all controls upon graduation.

Rumors at the Academy also help to prepare the cadets for their jobs as officers. Several of the instructors at the Academy were supposed to have been transferred from sea duty because of their incompetence. Such tales protect the cadets from developing a romantic conception of the qualities of Coast Guard officers, as well as providing a graphic illustration of how securely the bureaucratic structure protects officers from their own derelictions. In addition, many stories were told about a junior officer whose career at the Academy had been singularly brilliant. He had completely failed in his handling of enlisted men because he had carried over the high standards of the Academy. The cadets were thus oriented to a different conception of discipline when dealing with enlisted personnel.

CONCLUSION

The United States Coast Guard Academy performs an assimilating function. It isolates cadets from the outside world, helps them to identify themselves with a new role, and thus changes their self-conception. The manner in which the institution inculcates a bureaucratic spirit and prevents reality shock is also considered in this analysis.

The present investigation is admittedly fragmentary. Much of the most relevant material is simply not available. It is also clear that one cannot assume that this analysis applies completely to any other military academy. However, as an extreme example of an assimilating institution, there is considerable material which can be related to other institutions in a comparative framework.

Chapter 10

BUREAUCRACY

ONE of the most prominent features of life in a modern society is the experience of bureaucracy. Weber saw bureaucracy as perhaps the most important institution related to what he called the "rationalization" of modern society—that is, the process by which rational patterns of thought and conduct come to dominate ever larger segments of social life. Bureaucracy is the most distinctive modern form of organization. Originating first in the political sphere, it has by now spread to virtually every area of contemporary social life. Bureaucracy not only has very distinctive traits in the way it patterns human activity (traits that Weber dealt with in considerable detail) but it also shapes human thought and even human personality. It does this most fully for those who are actually engaged in bureaucratic jobs. But the ethos of bureaucracy also affects its clients; and since these include almost everyone in contemporary society, almost everyone is affected. This fact has produced significant social tensions.

On a superficial level, there are always differences of interests between a bureaucracy and the population it seeks to administer. In the political sphere itself these differences are often related to conflict between classes and other broad social groups. There are also deeper conflicts, which are rooted in the aforementioned relation of bureaucracy to the "rationalization" of life. Probably because of impulses that are deeply rooted in the constitution of human beings, people have foreseen that the growing "rationalization" of modern life would call forth "irrational" reactions against it. Many current problems in American society make better sense if one understands them in terms of a profound antagonism between bureaucracy and various stubbornly nonbureaucratic qualities of human beings.

The spread of bureaucracy in the past century is the subject of the first selection, Reinhard Bendix's "Bureaucratization in Industry." Bureaucracy is not only a type of social institution but the producer of a very specific type of person, one who successfully operates in a bureaucratic setting. He is the subject of the second selection, by William H. Whyte, Jr. It is from *The Organization Man,* a book that aroused great attention when it was first published in the 1950s, and it provides a sharply drawn portrait of the bureaucratic type of person. Another book of the 1950s that attracted much attention proves that sometimes the inner workings of a social situation are illuminated with surprising clarity through satire. In *Parkinson's Law,* C. Northcote Parkinson, a British historian, presents in the guise of laughter some important sociological insight. The passage selected makes the consequential point that a bureaucratic organization very often achieves a dynamic of its own, which after a certain point has little to do with the purpose for which the organization was originally set up. In a more serious vein, Erving Goffman, in the final selection from *Asylums,* describes how human relations are bureaucratized in the setting of a mental hospital.

REINHARD BENDIX

BUREAUCRATIZATION IN INDUSTRY

THE bureaucratization of modern industry has increased over the last half century. At the same time the changes of industrial organization which have accompanied this development have contributed to industrial peace. The following essay is designed to explore some background factors which tend to support these two propositions. It seeks to establish that industrial entrepreneurs considered as a class have undergone major changes since the beginning of the 19th century. These changes have culminated in the development of an industrial bureaucracy. The consequences of this bureaucratization may be observed in the changing system of supervision as well as in the transformation of the prevailing ideology of industrial managers.

INTRODUCTION

Webster's Collegiate Dictionary defines "bureaucracy" as routine procedure in administration, as a system of carrying on the business of government by means of bureaus, each controlled by a chief. This definition reflects the fact that it has not been customary to speak of bureaucracy in industry. Traditionally, the term has been applied to the activities of government; it has been broadened to include large-scale organizations generally only in recent years.[1]

The polemic implications of the term bureaucracy obscure its use in a descriptive sense, and yet it is important to use it in that sense. "A system of carrying on the business of industry and government by means of bureaus" is a definition of "administration" as well as of bureaucracy. Yet the two terms are not synonymous. Bureaucracy suggests in addition that the number of bureaus has increased, that their functions have become specialized as well as routinized, and that increasing use is made of technical apparatus in the performance of these specialized functions, which is in turn related to the increasing use of expert, technical knowledge. The use of technical knowledge in the administration of industry implies the employment of specialists, whose work presupposes the completion of a course of professional training. The work of these specialists entails the subdivision and consequent elaboration of the managerial functions of planning, production organization, personnel selection, and supervision.

Reprinted from *Industrial Conflict*, ed. by Arthur Kornhauser, Robert Dubin, and Arthur M. Ross (New York: McGraw-Hill, 1954), pp. 164–75. Used with permission of the McGraw-Hill Book Company.

These developments have many ramifications. They depend, on the one side, on the growth of training facilities in many fields of applied science. They make possible a centralization of authority in industrial management, which can be made effective only by a simultaneous delegation of circumscribed authority to specialized bureaus or departments. This encourages the substitution of deliberately planned methods of procedure for rule-of-thumb "methods," and this in turn promotes the utilization of mechanical devices. But the adoption of rational procedures achieves greater operating efficiency than is possible in less elaborated organizational structures at the constant risk of more bureaucracy, in the negative sense.

These remarks give an idea of the complexity of the process called "bureaucratization." Little is gained, however, by adopting a concise definition of this term. It is rather intended as the common denominator of many related tendencies of administrative procedure which have characterized government and industry in recent decades. But while the term itself remains vague, its component elements do not.

The following aspects of bureaucratization will be considered in this essay. In the first section it is shown that industrial entrepreneurs as a group have been transformed since the inception of modern industry at the beginning of the 19th century. This has resulted in the elaboration of managerial functions in industry. In the second section an attempt is made to sketch the changes of managerial ideology which have accompanied bureaucratization. The first section characterizes the development of industrial bureaucracy. In the second section certain ideological consequences are analyzed which are especially relevant for an understanding of peace and conflict in industrial relations.

CHANGES IN MANAGERIAL FUNCTIONS

The growing bureaucratization of industry may be analyzed in a variety of ways. The role of the employer has changed fundamentally since the rise of modern industry. The manager or owner of old, who knew and directed every detail of his enterprise, has become the modern industrialist who is above all else a specialist in business administration. Evidence for this transformation of the entrepreneur may be reviewed briefly.

In a report for 1792 Robert Owen states that it took him 6 weeks of careful observation to become thoroughly familiar with every detail of an enterprise employing 500 men.[2] An enterprise employing 500 men could be comprehended and managed at one time by a man of talent and experience. It is improbable that the same could be done today. A manager of a plant with 500 employees cannot be in daily touch with the details of the manufacturing process as Owen was. He will have various subordinates to supervise this process for him; also, this manager has lost most or all of his personal contacts with the workers in the plant. And although labor may be as efficient as it had been before this depersonalization of the employment relationship, this same efficiency is now obtained "at an increased cost in supervisory staff, complicated accounting methods, precise wage systems, liberal welfare provisions, checks and balances, scheduling and routine." [3] The point to emphasize is that bureaucratization of industry is not simply synonymous with the increasing size of the enterprise but with the growing complexity of its operation.

The bureaucratization of industry is, therefore, not simply the outcome of a recent development. In his analysis of the Boulton and Watt factory in 1775–1805 Erich Roll has described an elaborate system of keeping records, which was used as a basis of wage-determination, of cost-calculation, and of planning new methods of production. It is probable that this system was introduced when the firm passed from the original founders, who were in close contact with every operation of their enterprise, into the hands of the younger generation, who were not in touch with every operation and who, therefore, needed such a system of control.[4] At that time few firms were organized as efficiently as Boulton and Watt, but the case illustrates the fact that the bureaucratization of industry is not synonymous with the recent growth in the size of the large enterprises.

Corroborative evidence on this point is also contained in a study of American business leaders in the railroads and in the steel and textile industries, in the decade 1901–1910. The careers of 185 prominent industrialists were classified in terms of whether they had made their way in business by their own efforts exclusively, whether they had made their way in a family-owned enterprise, or whether they had risen through the ranks of an industrial bureaucracy; the results of the classification are shown in Table 10. These data make it apparent that prominent industrialists have had a bureaucratic career pattern at a relatively early time.[5]

Occupational statistics reflect this decline of the independent enterpriser and the increase of the "industrial bureaucrat," especially in the heavy industries. In the period 1910–1940 the number of independent industrial enterprisers declined from about 425,000 in 1910 to 390,000 in 1930 and 257,-

TABLE 10

American Business Leaders, by Type of Career and Date of Birth *

TYPE OF CAREER	BEFORE 1841 %	1841–1850 %	1851–1860 %	AFTER 1860 %
Independent	26	19	11	8
Family	22	24	42	36
Bureaucratic	52	57	47	56
	—	—	—	—
Total cases (=100%)	23	59	55	25

* William Miller, "The Business Elite in Business Bureaucracies," in William Miller (ed.), *Men in Business,* Cambridge, Mass., Harvard University Press, 1952, p. 291. Reprinted by permission.

000 in 1940, in manufacturing, construction, and mining. In the same industries the number of managerial employees increased from 375,976 in 1910 to 769,749 in 1930 and 802,734 in 1940.[6]

The bureaucratization of industry has also profoundly altered the job environment of the lowest rung on the managerial ladder.[7] Until about a generation or two ago the foreman occupied a position of real importance in industry, especially with regard to the management of labor. In the majority of cases the foreman would recruit workers, he would train them on the job, he would supervise and discipline them, which included such handling of grievances as was permitted, and he would pay their wages on a time basis.[8]

Today the foreman performs the functions of the immediate supervisor of the workers, who is in effect the executive agent of various supervisory departments. And it is increasingly a matter of discretion for these departments whether or not they decide to consult the foreman. The following summary based on a study of 100 companies which were sampled for the purpose of analyzing the *best* practices in American industry, illustrates this point clearly:

> *Hiring.* In two-thirds of the companies replying, the personnel department interviews and selects new employees, while the foreman has final say; but in one-third the foreman has no voice in hiring.
>
> *Discharge.* Foremen have some say in discharge, but only in one-tenth of all cases can they discharge without any consultation.
>
> *Pay Increases and Promotion.* These must almost always be approved by other authorities.
>
> *Discipline.* In only one-tenth of all cases do foremen have complete charge of discipline.
>
> *Grievances.* Discussion with the foreman is generally the first step in the grievance procedure, but the extent to which he settles grievances is not clear. A small sample in the automotive-aircraft industries shows that this may range from 45 to 80%.
>
> *Policy-making.* Only 20% of the companies replying held policy meetings with foremen.[9]

These findings make it apparent that the "average" foreman's responsibilities have remained, while his authority has been parceled out among the various supervisory departments. It is not surprising that this bureaucratization of supervisory functions has entailed inescapable tensions between the various departments performing these functions as well as tensions between these departments and the foremen. The latter have had to surrender their authority to the supervisory departments, but their responsibility for the execution of decisions has remained.

The changes in managerial functions which have grown out of the increasing division of labor within the plant and which are evident in the changing activities and career patterns of business executives and foremen are reflected also in the rise of "administrative overhead." A recent study of the rise of administrative personnel in American manufacturing industries since 1899

TABLE 11

All Manufacturing Industries: Composition of Work Force in Administration and Production Categories, 1899–1947 (*In Thousands*) *

PERSONNEL	1899	1909	1923	1929	1937	1947
Administration	457	886	1,345	1,562	1,567	2,672
Production	4,605	6,392	8,261	8,427	8,602	12,010
Administration personnel, as per cent of production personnel	9.9%	13.9%	16.3%	18.5%	18.2%	22.2%

* Seymour Melman. "The Rise of Administrative Overhead in the Manufacturing Industries of the United States 1899–1947," *Oxford Economic Papers,* Vol. 3, No. 1 (February, 1951), p. 66. Reprinted by permission.

makes it clear that this rise has occurred throughout the economy (Table 11).

A detailed examination reveals that this over-all increase in the ratio of administrative and production personnel is *not* systematically related to any one factor except size. Melman finds, somewhat paradoxically, that administrative cost as a proportion of production cost is lower in large than in small firms, despite the general upward trend in administrative personnel. His explanation is that all firms have shown an absolute increase in administrative overhead but that at any one time large firms as a group have a proportionately lower administrative overhead than small firms. This relative advantage of the larger firms is attributed to skill in organization. In the long run, however, all firms must anticipate an increase in administrative cost.

It should be added that a comparative study of administrative personnel in the manufacturing industries of other countries reveals similar trends, though it is noteworthy that the ratio of administrative as compared with productive personnel has increased more in the United States than in France, Germany, or England.[10] Melman's summary figures for a nationwide sample of manufacturing industries do not reveal the striking differences between industries and it may therefore be helpful to cite a few sample figures from his data (Table 12). These figures make it apparent that the over-all upward trend of administrative overhead covers a great diversity of particular developments. Although it is true that the average proportion of administration to production personnel has increased, there are significant differences between industries, and important fluctuations of this ratio have also occurred within an industry over time.

It may be useful to enumerate, in addition, some of the factors which are relevant for the interpretation of these statistics. Economically, it makes a great difference whether administrative personnel in industry increases together with a rapidly or a slowly expanding work force. That is to say, increasing administrative expenditures can be easily sustained in a rapidly expanding industry. Also, the increases of personnel in administration are accompanied by capital investments. Today, a given number of clerks can do

TABLE 12

Sampled Manufacturing Industries: Administration and Production Personnel, 1899–1937 *

INDUSTRY	ADMINISTRATION PERSONNEL AS PER CENT OF PRODUCTION PERSONNEL				
	1899	1909	1923	1929	1937
Agricultural implements	22.1	18.6	19.8	18.3	20.6
Boots and shoes	6.2	8.1	10.8	11.0	8.5
Boxes, paper	6.3	9.1	13.6	14.4	16.1
Cash registers and business machines	15.9	23.8	23.2	36.9	37.7
Drugs and medicines	45.7	69.2	51.3	61.5	58.1
Electrical machinery	12.5	20.7	31.4	24.6	26.7
Explosives	17.3	12.3	33.6	38.2	44.4
Glass	4.5	5.3	8.6	9.7	11.5
Lighting equipment	12.8	20.1	22.9	20.6	19.2
Locomotives, not built in railroad shops	3.9	13.6	9.5	20.9	29.8
Meat packing, wholesale	15.3	20.1	24.8	22.8	29.2
Motor vehicles	13.1	13.3	11.5	14.8	16.9
Petroleum refining	10.0	19.3	24.1	35.6	45.3

* Seymour Melman. "The Rise of Administrative Overhead in the Manufacturing Industries of the United States 1899–1947," *Oxford Economic Papers,* Vol. 3, No. 1 (February, 1951), p. 66. Reprinted by permission.

a great deal more work than formerly, with the aid of various computing and multigraphing machines. As a group they also do a greater variety of work, owing to specialization and partial mechanization. However, neither the greater complexity of administrative work nor the various efforts at standardization and routinization can be measured by the number of clerks employed. The increase of administrative personnel is, therefore, only a proximate measure of bureaucratization.

MANAGERIAL IDEOLOGY AND BUREAUCRACY

The general trend is in the direction of an increase in the complexity of managerial tasks. To assess the problems created by this trend, it is not sufficient, for example, to describe how the functions of hiring and discharge, of administering an equitable wage structure, and of processing grievances and disciplining workers have become the special tasks of separate departments. In order to understand the modern problems of management, we must realize that this separation of functions has created for all ranks of management an ambiguity which is in many respects similar to, though it is not so intense as, that of the foreman. The over-all managerial problem has become more complex because each group of management specialists will tend to view the "interests of the enterprise" in terms which are compatible with the survival and the increase of its special function. That is, each group will have a trained capacity for its own function and a "trained incapacity" to see its relation to the whole.

The problem of industrial management is to subdivide, as well as to coordinate, the tasks of administration and production and then to maximize the efficiency of each operation. In so doing it employs specialists, and each group of specialists must exercise considerable discretion in order to get the work done. That is to say, with each step toward specialization the centrifugal tendencies and, hence, the coordinating tasks of central management increase. Bureaucratization has accompanied the whole development of industry, but it has increased more rapidly since the inception of scientific management in the 1890's. The major development of trade-unions has occurred during the same period. If we consider these parallel changes it becomes apparent that the greater complexity of the managerial task has consisted in the need for intramanagerial coordination at a time when managerial leadership was challenged by the organizing drives of trade-unions as well as by the ideological attacks of the muckrakers. Hence the ideology of business leaders, their justification of the authority they exercise and of the power they hold, has gradually assumed a double function: (1) to demonstrate that the authority and power of the industrial leader is legitimate and (2) to aid the specific job of managerial coordination.

Until recently the ideologies of the industrial leaders did not serve this double function. In the past their leadership was justified by a reiteration of time-tested shibboleths which would make clear what was already self-evident to all but the most die-hard radicals. Success is virtue, poverty is sin, and both result from the effort or indolence of the individual. Together with this belief went the idea that every use made of property was beneficial to the social welfare, as long as it resulted in an increase of wealth. These ideas, which justified the authority and power of the industrial leaders, established a goal in life for everyone. The tacit assumption was that in the prevailing economic order the chances of each "to get to the top" were the same. Hence the success of the industrial leader was itself the token of his proved superiority in a struggle between equals. To question this was to bar the way of those who would succeed after him.

These ideas have never really died; there is much contemporary evidence to show that the beliefs of industrial leaders have remained essentially the same. Successful industrialists as a group have always tended to express views which ranged from the belief that their virtue had been proved by their works and that their responsibilities were commensurate with their wealth to the assertion that their eminence was self-evident and that their privileges could not be questioned. They would speak with Andrew Carnegie of the "trusteeship of wealth" and point to their benevolent relations to their employees, their philanthropic activities, and their great contributions to the nation's wealth as evidence of their worth. Others would think of themselves as "Christian men to whom God in his infinite wisdom has given the control of the property interests of the country." Nor can we dismiss the possibility that some of these industrial giants would say the first and think the second.

Fifty years later the same opinions are expressed, albeit in modern dress. Alfred P. Sloan [11] writes:

> . . . those charged with great industrial responsibility must become industrial statesmen. . . . Industrial management must expand its horizon of responsibility. It must recognize that it can no longer confine its activities to the mere production of goods and services. It must consider the impact of its operations on the economy as a whole in relation to the socail and economic welfare of the entire community.

On the other hand, Tom Girdler [12] has written of his role in the company town which he had helped to develop:

> In fact I suppose I was a sort of political boss. Certainly I had considerable power in politics without responsibility to "the people." But who were the people in question? An overwhelming majority of them were the men for whom the company aspired to make Woodlawn the best steel town in the world . . . What did it matter if the taxes were soundly spent? What did it matter if Woodlawn had just about the best school system in Pennsylvania? What did it matter if there were no slums, no graft, no patronage, no gambling houses, no brothels? What did it matter if it was a clean town?

If all these wonderful things were done by the company for the people of Woodlawn, what did it matter that the company and its managers were not responsible to the people? As an Episcopalian vestry man, Girdler could also speak of Christian men who, by the grace of God, controlled the property interests of the community on behalf of the people.

Businessmen express themselves with the intention of demonstrating statesmanship and intransigeance, then as now.[13] Yet even the celebration of the industrial leader has had to accommodate, albeit tardily, the Puritan virtues of hard work, frugality, and unremitting effort to the qualities useful in a bureaucratic career. As the size and bureaucratization of business increased, this ideological accommodation could no longer be accomplished on the model of the Horatio Alger story. Of course, the idea of success as a reward of virtue is as much in evidence today as it was 100 years ago. But the celebration of the industrial leader can no longer suffice; it is accompanied today by a celebration of the organization and of the opportunities it has to offer. When A. P. Sloan writes that "the corporation [is] a pyramid of opportunities from the bottom to the top with thousands of chances for advancement" he refers to the promise of a bureaucratic career not to the earlier image of the individual enterpriser. And when he adds that "only capacity limited any worker's chance to improve his own position," [14] he simply ignores the fact that the methods of promotion themselves are bureaucratized, that they are regarded as a legitimate object of collective-bargaining strategies between union and management, and that under these circumstances minimum rather than maximum capacity is often a sufficient basis for promotion. At any rate, the idea of thousands of chances for promotion is different from the idea of individual success. Outstanding industrial leaders of today will reflect this difference in their attempts to define the image of success in an era of bureaucratization.

It is important to recognize that today managerial ideology performs a second function. While it is still designed to inspire confidence in the leaders of industry, it should also aid modern managers to achieve effective coordination within their enterprises, which is today a far more difficult task than it was formerly. There is a literature of advice to the ambitious young man which has accompanied the development of industry. In this literature the hero cult of the industrial leader has been abandoned gradually, and advice well suited to the industrial bureaucrat has taken its place.[15] Hero cult and advice to the industrial bureaucrat involve partly incompatible themes. The qualities of ruthlessness and competitive drive, while appropriate for the "tycoon," are ill suited for his managerial employees. This does not mean that these qualities are no longer useful but that they no longer provide a workable rationale for the majority of industrial managers.[16]

It may be useful to put formulations of these two themes side by side. The classic text of the individual enterpriser is *Self-help with Illustrations of Character, Conduct and Perseverance,* written by Samuel Smiles in 1859 and copied interminably ever since. Its purpose was,

> . . . to re-inculcate these old-fashioned but wholesome lessons . . . that youth must work in order to enjoy—that nothing creditable can be accomplished without application and diligence—that the student must not be daunted by difficulties, but conquer them by patience and perseverance—and that, above all, he must seek elevation of character, without which capacity is worthless and worldly success is naught.[17]

The classic text of the industrial bureaucrat is *Public Speaking and Influencing Men in Business,* written by Dale Carnegie in 1926 and used as the "official text" by such organizations as the New York Telephone Company, the American Institute of Banking, the YMCA schools, the National Institute of Credit, and others.[18] Though there is no single statement of purpose which can be cited, the following summary statement will suffice: "We have only four contacts with people. We are evaluated and classified by four things: by what we do, by how we look, by what we say, and how we say it." [19] In his foreword to this book Lowell Thomas [2] has written a testimonial to Dale Carnegie which gives the gist of this and many similar books with admirable clarity:

> Carnegie started at first to conduct merely a course in public speaking: but the students who came were businessmen. Many of them hadn't seen the inside of a class room in thirty years. Most of them were paying their tuition on the installment plan. They wanted results; and they wanted them quick—results that they could use the next day in business interviews and in speaking before groups.
>
> So he was forced to be swift and practical. Consequently, he has developed a system of training that is unique—a striking combination of Public Speaking, Salesmanship, Human Relationship, Personality Development and Applied Psychology. . . . Dale Carnegie . . . has created one of the most significant movements in adult education.

This new ideology of personality salesmanship appeared to put within reach of the average person the means by which to climb the ladder to success. No doubt this accounts for its popularity. But it should be added that its public acceptance implied a prior disillusion with the more old-fashioned methods of achieving success. The bureaucratization of modern industry has obviously increased the number of steps from the bottom to the top at the same time that it has made the Puritan virtues largely obsolete. It is probable, then, that the techniques of personality salesmanship became popular when the ideal of individual entrepreneurship ceased to be synonymous with success, while the image of a career of promotions from lower to higher positions became of greater significance. From the standpoint of the individual these techniques became a means of career advancement; from the standpoint of management they seemed to facilitate the coordination of a growing and increasingly specialized staff. In the context of American society this new ideology reflected the increasing importance of the service trades as well as the growing demand for skill in personnel relations.[21]

These considerations place the human-relations approach to the problems of labor management in a historical perspective. Attention to human relations has arisen out of the managerial problems incident to the bureaucratization of industry. It has also arisen out of the discrepancy between a people's continued desire for success and the increasing disutility of the Puritan virtues or of the tenets of Darwinian morality. But whatever their origins, the "personality cult" as well as the more sophisticated philosophies of personnel management have helped to make more ambiguous the position of the industrial manager. In giving orders to his subordinates in the past, the manager could claim to derive his authority from the rights of ownership conferred on him. For a long time the managerial employee had represented the "heroic entrepreneur," and he had justified his own actions by the right which success had bestowed upon him. But with the dispersion of ownership this justification became increasingly tenuous. Strictly speaking, the old ideology of success no longer applied to the managers since theirs was a bureaucratic, not an entrepreneurial, success. As the human-relations approach is extended downward from the office staff to the work force, managers come to attenuate their tough-minded conception of authority. But in so doing they are never single-minded. Their careers are often inspired by the older belief in the self-made man, though this belief is more and more at variance with their own experience in industry. In asserting their authority over subordinates as if they were the successful entrepreneurs of old, they come into conflict with the bureaucratic reality of their own careers. Yet if they adjust their beliefs to that reality, then they are faced with the dilemma of exercising authority while they deny the traditional claims which had hitherto justified this authority.[22]

It is at this point that managers are divided today in their attitudes toward their employees and toward their own exercise of authority. Many continue to believe in the heroic entrepreneur whose success is justification in itself, and they consequently resist the "tender-minded" approach to human relations in industry. They also resist recognition of the fact that the industrial environment has changed. Others have begun to reformulate the older statements of "business statesmanship" and "business responsibility" in keeping with the realities of industry in an era of bureaucratization. But in their attempts to do so they have had to demonstrate the self-evident truths once more that the economy provides ample opportunities, given drive and talent, and that those who succeed deserve to do so and provide a model to be followed. To develop an ideology along these lines by advertising the techniques of personality salesmanship and by celebrating the career opportunities of an industrial bureaucracy implies an interest in industrial peace, for these techniques and opportunities are beside the point under conditions of conflict. The new ideology is less combative than the old; but it is also insufficient because its appeals are more readily applicable to the salaried employee than to the industrial worker. The idea that all employees are members of "one big happy family" is a case in point, for the efforts to make this idea meaningful to the workers frequently take the form of personalizing an impersonal employment relationship. Perhaps this is appropriate for the managerial and ideological coordination of the salaried employees. It is, moreover, not surprising that the idea of the "family of employers and employees" often becomes the fighting creed of hard-pressed executives who seek to solidify their enterprises against the competing appeals of the trade-unions. But there is an element of cant in this approach which does not make it a promising foundation for a new ideology as long as democratic institutions prevail. Perhaps Horatio Alger is so reluctant to pass into limbo because his image implied an idealistic message. Perhaps it is the absence of such a message which makes the appeal to employees as members of a family so questionable. The ideological rationale of an economic order should have a positive meaning for everyone. The fact is that in this era of bureaucratization the industrialist does not have a fighting creed.

WILLIAM H. WHYTE, JR.

THE ORGANIZATION MAN

THIS [book] is about the organization man. If the term is vague, it is because I can think of no other way to describe the people I am talking about.

Reprinted from *The Organization Man* by William H. Whyte, Jr. (New York: Simon & Schuster, 1956), pp. 3–15.

They are not the workers, nor are they the white-collar people in the usual, clerk sense of the word. These people only work for The Organization. The ones I am talking about *belong* to it as well. They are the ones of our middle class who have left home, spiritually as well as physically, to take the vows of organization life, and it is they who are the mind and soul of our great self-perpetuating institutions. Only a few are top managers or ever will be. In a system that makes such hazy terminology as "junior executive" psychologically necessary, they are of the staff as much as the line, and most are destined to live poised in a middle area that still awaits a satisfactory euphemism. But they are the dominant members of our society nonetheless. They have not joined together into a recognizable elite—our country does not stand still long enough for that—but it is from their ranks that are coming most of the first and second echelons of our leadership, and it is their values which will set the American temper.

The corporation man is the most conspicuous example, but he is only one, for the collectivization so visible in the corporation has affected almost every field of work. Blood brother to the business trainee off to join Du Pont is the seminary student who will end up in the church hierarchy, the doctor headed for the corporate clinic, the physics Ph.D. in a government laboratory, the intellectual on the foundation-sponsored team project, the engineering graduate in the huge drafting room at Lockheed, the young apprentice in a Wall Street law factory.

They are all, as they so often put it, in the same boat. Listen to them talk to each other over the front lawns of their suburbia and you cannot help but be struck by how well they grasp the common denominators which bind them. Whatever the differences in their organization ties, it is the common problems of collective work that dominate their attentions, and when the Du Pont man talks to the research chemist or the chemist to the army man, it is these problems that are uppermost. The word *collective* most of them can't bring themselves to use—except to describe foreign countries or organizations they don't work for—but they are keenly aware of how much more deeply beholden they are to organization than were their elders. They are wry about it, to be sure; they talk of the "treadmill," the "rat race," of the inability to control one's direction. But they have no great sense of plight; between themselves and organization they believe they see an ultimate harmony and, more than most elders recognize, they are building an ideology that will vouchsafe this trust.

. . . America has paid much attention to the economic and political consequences of big organization—the concentration of power in large corporations, for example, the political power of the civil-service bureaucracies, the possible emergence of a managerial hierarchy that might dominate the rest of us. These are proper concerns, but no less important is the personal impact that organization life has had on the individuals within it. A collision has been taking place—indeed, hundreds of thousands of them, and in the aggregate they have been producing what I believe is a major shift in American ideology.

Officially, we are a people who hold to the Protestant Ethic. Because of the denominational implications of the term many would deny its relevance to them, but let them eulogize the American Dream, however, and they virtually define the Protestant Ethic. Whatever the embroidery, there is almost always the thought that pursuit of individual salvation through hard work, thrift, and competitive struggle is the heart of the American achievement.

But the harsh facts of organization life simply do not jibe with these precepts. This conflict is certainly not a peculiarly American development. In their own countries such Europeans as Max Weber and Durkheim many years ago foretold the change, and though Europeans now like to see their troubles as an American export, the problems they speak of stem from a bureaucratization of society that has affected every Western country.

It is in America, however, that the contrast between the old ethic and current reality has been most apparent—and most poignant. Of all peoples it is we who have led in the public worship of individualism. One hundred years ago de Tocqueville was noting that though our special genius—and failing—lay in cooperative action, we talked more than others of personal independence and freedom. We kept on, and as late as the twenties, when big organization was long since a fact, affirmed the old faith as if nothing had really changed at all.

Today many still try, and it is the members of the kind of organization most responsible for the change, the corporation, who try the hardest. It is the corporation man whose institutional ads protest so much that Americans speak up in town meeting, that Americans are the best inventors because Americans don't care that other people scoff, that Americans are the best soldiers because they have so much initiative and native ingenuity, that the boy selling papers on the street corner is the prototype of our business society. Collectivism? He abhors it, and when he makes his ritualistic attack on Welfare Statism, it is in terms of a Protestant Ethic undefiled by change—the sacredness of property, the enervating effect of security, the virtues of thrift, of hard work and independence. Thanks be, he says, that there are some people left—e.g., businessmen—to defend the American Dream.

He is not being hypocritical, only compulsive. He honestly wants to believe he follows the tenets he extols, and if he extols them so frequently it is, perhaps, to shut out a nagging suspicion that he, too, the last defender of the faith, is no longer pure. Only by using the language of individualism to describe the collective can he stave off the thought that he himself is in a collective as pervading as any ever dreamed of by the reformers, the intellectuals, and the utopian visionaries he so regularly warns against.

The older generation may still convince themselves; the younger generation does not. When a young man says that to make a living these days you must do what somebody else wants you to do, he states it not only as a fact of life that must be accepted but as an inherently good proposition. If the American Dream deprecates this for him, it is the American Dream that is going to have to give, whatever its more elderly guardians may think. People grow restive with a mythology that is too distant from the way things actually are, and as more and more lives have been encompassed by the organization way of life, the pressures for an accompanying ideological shift have been mounting. The pressures of the group, the frustrations of individual creativity, the anonymity of achievement: are these defects to struggle against—or are they virtues in disguise? The organization man seeks a redefinition of his place on earth—a faith that will satisfy him that what he must endure has a deeper meaning than appears on the surface. He needs, in short, something that will do for him what the Protestant Ethic did once. And slowly, almost imperceptibly, a body of thought has been coalescing that does that.

I am going to call it a Social Ethic. With reason it could be called an organization ethic, or a bureaucratic ethic; more than anything else it rationalizes the organization's demands for fealty and gives those who offer it wholeheartedly a sense of dedication in doing so—*in extremis,* you might say, it converts what would seem in other times a bill of no rights into a restatement of individualism.

But there is a real moral imperative behind it, and whether one inclines to its beliefs or not he must acknowledge that this moral basis, not mere expediency, is the source of its power. Nor is it simply an opiate for those who must work in big organizations. The search for a secular faith that it represents can be found throughout our society—and among those who swear they would never set foot in a corporation or a government bureau. Though it has its greatest applicability to the organization man, its ideological underpinnings have been provided not by the organization man but by intellectuals he knows little of and toward whom, indeed, he tends to be rather suspicious.

Any groove of abstraction, Whitehead once remarked, is bound to be an inadequate way of describing reality, and so with the concept of the Social Ethic. It is an attempt to illustrate an underlying consistency in what in actuality is by no means an orderly system of thought. No one says, "I believe in the social ethic," and though many would subscribe wholeheartedly to the separate ideas that make it up, these ideas have yet to be put together in the final, harmonious synthesis. But the unity is there.

In looking at what might seem dissimilar aspects of organization society, it is this unity I wish to underscore. The "professionalization" of the manager, for example, and the drive for a more practical education are parts of the same phenomenon; just as the student now feels technique more vital than content, so the trainee believes managing an end in itself, an *expertise* relatively independent of the content of what is being managed. And the reasons are the same. So too in other sectors of our society; for all the differences in particulars, dominant is a growing accommodation to the needs of society—and a growing urge to justify it.

Let me now define my terms. By social ethic I mean that contemporary body of thought which makes morally legitimate the pressures of society against the individual. Its major propositions are three: a belief in the group as the source of creativity; a belief in "belongingness" as the ultimate need of the individual; and a belief in the application of science to achieve the belongingness.

. . . The gist can be paraphrased thus: Man exists as a unit of society. Of himself, he is isolated, meaningless; only as he collaborates with others does he become worthwhile, for by sublimating himself in the group, he helps produce a whole that is greater than the sum of its parts. There should be, then, no conflict between man and society. What we think are conflicts are misunderstandings, breakdowns in communication. By applying the methods of science to human relations we can eliminate these obstacles to consensus and create an equilibrium in which society's needs and the needs of the individual are one and the same.

Essentially, it is a utopian faith. Superficially, it seems dedicated to the practical problems of organization life, and its proponents often use the word *hard* (versus *soft*) to describe their approach. But it is the long-range promise that animates its followers, for it relates techniques to the vision of a finite, achievable harmony. . . .

Like the utopian communities, it interprets society in a fairly narrow, immediate sense. One can believe man has a social obligation and that the individual must ultimately contribute to the community without believing that group harmony is the test of it. In the Social Ethic I am describing, however, man's obligation is in the here and now; his duty is not so much to the community in a broad sense but to the actual, physical one about him, and the idea that in isolation from it—or active rebellion against it—he might eventually discharge the greater service is little considered. In practice, those who most eagerly subscribe to the Social Ethic worry very little over the long-range problems of society. It is not that they don't care but rather that they tend to assume the ends of organization and morality coincide, and on such matters as social welfare they give their proxy to the organization.

It is possible that I am attaching too much weight to what, after all, is something of a mythology. Those more sanguine than I have argued that this faith is betrayed by reality in some key respects and that because it cannot long hide from organization man that life is still essentially competitive the faith must fall of its own weight. They also maintain that the Social Ethic is only one trend in a society which is a prolific breeder of counter-trends. The farther the pendulum swings, they believe, the more it must eventually swing back.

I am not persuaded. We are indeed a flexible people, but society is not a clock and to stake so much on counter-trends is to put a rather heavy burden on providence. . . .

. . . No one can say whether these trends will continue to outpace the counter-trends, but neither can we trust that an equilibrium-minded providence will see to it that excesses will cancel each other out. Counter-trends there are. There always have been, and in the sweep of ideas ineffectual many have proved to be.

It is also true that the Social Ethic is something of a mythology, and there is a great difference between mythology and practice. An individualism as stringent, as selfish as that often preached in the name of the Protestant Ethic would never have been tolerated, and in reality our predecessors co-operated with one another far more skillfully than nineteenth-century oratory would suggest. Something of the obverse is true of the Social Ethic; so complete a denial of individual will won't work either, and even the most willing believers in the group harbor some secret misgivings, some latent antagonism toward the pressures they seek to deify.

But the Social Ethic is no less powerful for that, and though it can never produce the peace of mind it seems to offer, it will help shape the nature of the quest in the years to come. The old dogma of individualism betrayed reality too, yet few would argue, I dare say, that it was not an immensely powerful influence in the time of its dominance. So I argue of the Social Ethic; call it mythology, if you will, but it is becoming the dominant one. . . .

This . . . is not a plea for nonconformity. Such pleas have an occasional therapeutic value, but as an abstraction, nonconformity is an empty goal, and rebellion against prevailing opinion merely because it is prevailing should no more be praised than acquiescence to it. Indeed, it is often a mask for cowardice, and few are more pathetic than those who flaunt outer differences to expiate their inner surrender.

I am not, accordingly, addressing myself to the surface uniformities of U.S. life. There will be no strictures . . . against "Mass Man"—a person the author has never met—nor will there be any strictures against ranch wagons, or television sets, or gray flannel suits. They are irrelevant to the main problem, and, furthermore, there's no harm in them. I would not wish to go to the other extreme and suggest that these uniformities per se are good, but the spectacle of people following current custom for lack of will or imagination to do anything else is hardly a new failing, and I am not convinced that there has been any significant change in this respect except in the nature of the things we conform to. Unless one believes poverty ennobling, it is difficult to see the three-button suit as more of a straitjacket than overalls, or the ranch-type house than old law tenements.

And how important, really, are these uniformities to the central issue of individualism? We must not let the outward forms deceive us. If individualism involves following one's destiny as one's own conscience directs, it must for most of us be a realizable destiny, and a sensible awareness of the rules of the game can be a condition of individualism as well as a constraint upon it. The man who drives a Buick Special and lives in a ranch-type house just like hundreds of other ranch-type houses can assert himself as effectively and courageously against his particular society as the bohemian against his particular society. He usually does not, it is true, but if he does, the surface uniformities can serve quite well as protective coloration. The organization people who are best able to control their environment rather than be controlled by it are well aware that they are not too easily distinguishable from the others in the outward obeisances paid to the good opinions of others. And that is one of the reasons they do control. They disarm society.

I do not equate the Social Ethic with conformity, nor do I believe those who urge it wish it to be, for most of them believe deeply that their work will help, rather than harm, the individual. I think their ideas are out of joint with the needs of the times they invoke, but it is their ideas, and not their good will, I wish to question. As for the lackeys of organization and the charlatans, they are not worth talking about.

Neither do I intend . . . a censure of the fact of organization society. We have quite enough problems today without muddying the issue with misplaced nostalgia, and in contrasting the old ideology with the new I mean no contrast of paradise with paradise lost, an idyllic eighteenth century with a dehumanized twentieth. Whether or not our own era is worse than former ones in the climate of freedom is a matter that can be left to later historians, but . . . I write with the optimistic promise that individualism is as possible in our times as in others.

I speak of individualism *within* organization life. This is not the only kind, and someday it may be that the mystics and philosophers more distant from it may prove the crucial figures. But they are affected too by the center of society, and they can be of no help unless they grasp the nature of the main stream. Intellectual scoldings based on an impossibly lofty ideal may be of some service in upbraiding organization man with his failures, but they can give him no guidance. The organization man may agree that industrialism has destroyed the moral fabric of society and that we need to return to the agrarian virtues, or that business needs to be broken up into a series of smaller organizations, or that it's government that needs to be broken up, and so on. But he will go his way with his own dilemmas left untouched.

I . . . argue that he should fight the organization. But not self-destructively. He may tell the boss to go to hell, but he is going to have another boss, and, unlike the heroes of popular fiction, he cannot find surcease by leaving the arena to be a husbandman. If he chafes at the pressures of his particular organization, either he must succumb, resist them, try to change them, or move to yet another organization.

Every decision he faces on the problem of the individual versus authority is something of a dilemma. It is not a case of whether he should fight against black tyranny or blaze a new trail against patent stupidity. That would be easy—intellectually, at least. The real issue is far more subtle. For it is not the evils of organization life that puzzle him, *but its very beneficence*. He is imprisoned in brotherhood. Because his area of maneuver seems so small and because the trapping so mundane, his fight lacks the heroic cast, but it is for all this as tough a fight as ever his predecessors had to fight.

Thus to my thesis, I believe the emphasis of the Social Ethic is wrong for him. People do have to work with others, yes; the well-functioning team is a whole greater than the sum of its parts, yes—all this is indeed true. But is it the truth that now needs belaboring? Precisely because it *is* an age of organization, it is the other side of the coin that needs emphasis. We do need to know how to co-operate with The Organization but, more than ever, so do we need to know how to resist it. Out of context this would be an irresponsible statement. Time and place are critical, and history has taught us that a philosophical individualism can venerate conflict too much and co-operation too little. But what is the context today? The tide has swung far enough the other way, I submit, that we need not worry that a counteremphasis will stimulate people to an excess of individualism.

The energies Americans have devoted to the co-operative, to the social, are not to be demeaned; we would not, after all, have such a problem to discuss unless we had learned to adapt ourselves to an increasingly collective society as well as we have. An ideal of individualism which denies the obligations of man to others is manifestly impossible in a society such as ours, and it is a credit to our wisdom that while we preached it, we never fully practiced it.

But in searching for that elusive middle of the road, we have gone very far afield, and in our attention to making organization work we have come close to deifying it. We are describing its defects as virtues and denying that there is—or should be—a conflict between the individual and organization. This denial is bad for the organization. It is worse for the individual. What it does, in soothing him, is to rob him of the intellectual armor he so badly needs. For the more power organization has over him, the more he needs to recognize the area where he must assert himself against it. And this, almost because we have made organization life so equable, has become excruciatingly difficult.

To say that we must recognize the dilemmas of organization society is not to be inconsistent with the hopeful premise that organization society can be as compatible for the individual as any previous society. We are not hapless beings caught in the grip of forces we can do little about, and wholesale damnations of our society only lend a further mystique to organization. Organization has been made by man; it can be changed by man. It has not been the immutable course of history that has produced such constrictions on the individual as personality tests. It is organization man who has brought them to pass and it is he who can stop them.

The fault is not in organization, in short; it is in our worship of it. It is in our vain quest for a utopian equilibrium, which would be horrible if it ever did come to pass; it is in the soft-minded denial that there is a conflict between the individual and society. There must always be, and it is the price of being an individual that he must face these conflicts. He cannot evade them, and in seeking an ethic that offers a spurious peace of mind, thus does he tyrannize himself.

There are only a few times in organization life when he can wrench his destiny into his own hands—and if he does not fight then, he will make a surrender that will later mock him. But when is that time? Will he know the time when he sees it? By what standards is he to judge? He does feel an obligation to the group; he does sense moral contraints on his free will. If he goes against the group, is he being courageous—or just stubborn? Helpful—or selfish? Is he, as he so often wonders, right after all? It is in the resolution of a multitude of such dilemmas, I submit, that the real issue of individualism lies today.

C. NORTHCOTE PARKINSON

PARKINSON'S LAW, OR THE RISING PYRAMID

WORK expands so as to fill the time available for its completion. General recognition of this fact is shown in the proverbial phrase "It is the busiest man who has time to spare." Thus, an elderly lady of leisure can spend the entire day in writing and dispatching a postcard to her niece at Bognor Regis. An hour will be spent in finding the postcard, another in hunting for spectacles, half an hour in a search for the address, an hour and a quarter in composition, and twenty minutes in deciding whether or not to take an umbrella when going to the mailbox in the next street. The total effort that would occupy a busy man for three minutes all told may in this fashion leave another person prostrate after a day of doubt, anxiety, and toil.

Granted that work (and especially paperwork) is thus elastic in its demands on time, it is manifest that there need be little or no relationship between the work to be done and the size of the staff to which it may be assigned. A lack of real activity does not, of necessity, result in leisure. A lack of occupation is not necessarily revealed by a manifest idleness. The thing to be done swells in importance and complexity in a direct ratio with the time to be spent. This fact is widely recognized, but less attention has been paid to its wider implications, more especially in the field of public administration. Politicians and taxpayers have assumed (with occasional phases of doubt) that a rising total in the number of civil servants must reflect a growing volume of work to be done. Cynics, in questioning this belief, have imagined that the multiplication of officials must have left some of them idle or all of them able to work for shorter hours. But this is a matter in which faith and doubt seem equally misplaced. The fact is that the number of the officials and the quantity of the work are not related to each other at all. The rise in the total of those employed is governed by Parkinson's Law and would be much the same whether the volume of the work were to increase, diminish, or even disappear. The importance of Parkinson's Law lies in the fact that it is a law of growth based upon an analysis of the factors by which that growth is controlled.

The validity of this recently discovered law must rest mainly on statistical proofs, which will follow. Of more interest to the general reader is the explanation of the factors underlying the general tendency to which this law gives definition. Omitting technicalities (which are numerous) we may distinguish at the outset two motive forces. They can be represented for the present purpose by two almost axiomatic statements, thus: (1) "An official wants to multiply subordinates, not rivals" and (2) "Officials make work for each other."

To comprehend Factor 1, we must picture a civil servant, called A, who finds himself overworked. Whether this overwork is real or imaginary is immaterial, but we should observe, in passing, that A's sensation (or illusion) might easily result from his own decreasing energy: a normal symptom of middle age. For this real or imagined overwork there are, broadly speaking, three possible remedies. He may resign; he may ask to halve the work with a colleague called B; he may demand the assistance of two subordinates, to be called C and D. There is probably no instance in history, however, of A choosing any but the third alternative. By resignation he would lose his pension rights. By having B appointed, on his own level in the hierarchy, he would merely bring in a rival for promotion to W's vacancy when W (at long last) retires. So A would rather have C and D, junior men, below him. They will add to his consequence and, by dividing the work into two categories, as between C and D, he will have the merit of being the only man who comprehends them both. It is essential to realize at this point that C and D are, as it were, inseparable. To appoint C alone would have been impossible. Why? Because C, if by himself, would divide the work with A and so assume almost the equal status that has been refused in the first instance to B; a status the more emphasized if C is A's only possible successor. Subordinates must thus number two or more, each being thus kept in order by fear of the other's promotion. When C complains in turn of being overworked (as he certainly will) A will, with the concurrence of C, advise the appointment of two assistants to help C. But he can then avert internal friction only by advising the appointment of two more assistants to help D, whose position is much the same. With this recruitment of E, F, G, and H the promotion of A is now practically certain.

Seven officials are now doing what one did before. This is where Factor 2 comes into operation. For these seven make so much work for each other that all are fully occupied and A is actually working harder than ever. An incoming document may well come before each of them in turn. Official E decides that it falls within the province of F, who places a draft reply before C, who amends it drastically before consulting D, who asks G to deal with it. But G goes on leave at this point, handing the file over to H, who drafts a minute that is signed by D and returned to C, who revises his draft accordingly and lays the new version before A.

What does A do? He would have every excuse for signing the thing unread, for he has many other matters on his mind. Knowing now that he is to succeed W next year, he has to decide whether C or D should succeed to his own office. He had to agree to G's going on leave even if not yet strictly entitled to it. He is worried whether H should not have gone instead, for reasons of health. He has looked pale recently—partly but not solely because of his domestic troubles. Then there is the business of F's special increment of salary for the period of the conference and E's application for transfer to the Ministry of Pensions. A has heard that D is in love with a married typist and that G and F are no longer on speaking terms—no one seems to know why. So A might be tempted to sign C's draft and have done with it. But A is a conscientious man. Beset as he is with problems created by his colleagues for themselves and for him—created by the mere fact of these officials' existence—he is not the man to shirk his duty. He reads through the draft with care, deletes the fussy paragraphs added by C and H, and restores the thing back to the form preferred in the first instance by the able (if quarrelsome) F. He corrects the English—none of these young men can write grammatically—and finally produces the same reply he would have written if officials C to H had never been born. Far more people have taken far longer to produce the same result. No one has been idle. All have done their best. And it is late in the evening before A finally quits his office and begins the return journey to Ealing. The last of the office lights are being turned off in the gathering dusk that marks the end of another day's administrative toil. Among the last to leave, A reflects with bowed shoulders and a wry smile that late hours, like gray hairs, are among the penalties of success.

From this description of the factors at work the student of political science will recognize that administrators are more or less bound to multiply. Nothing has yet been said, however, about the period of time likely to elapse between the date of A's appointment and the date from which we can calculate the pensionable service of H. Vast masses of statistical evidence have been collected and it is from a study of this data that Parkinson's Law has been deduced. Space will not allow of detailed analysis but the reader will be interested to know that research began in the British Navy Estimates. These were chosen because the Admiralty's responsibilities are more easily measurable than those of, say, the Board of Trade. The question is merely one of numbers and tonnage. Here are some typical figures. The strength of the Navy in 1914 could be shown as 146,000 officers and men, 3249 dockyard officials and clerks, and 57,000 dockyard workmen. By 1928 there were only 100,000 officers and men and only 62,439 workmen, but the dockyard officials and clerks by then numbered 4558. As for warships, the strength in 1928 was a mere fraction of what it had been in 1914—fewer than 20 capital ships in commission as compared with 62. Over the same period the Admiralty officials had increased in number from 2000 to 3569, providing (as was remarked) "a magnificent navy on land." These figures are more clearly set forth in tabular form.

Admiralty Statistics

YEAR	CAPITAL SHIPS IN COMMISSION	OFFICERS AND MEN IN R.N.	DOCKYARD WORKERS	DOCKYARD OFFICIALS AND CLERKS	ADMIRALTY OFFICIALS
1914	62	146,000	57,000	3249	2000
1928	20	100,000	62,439	4558	3569
Increase or Decrease	−67.74%	−31.5%	+9.54%	+40.28%	+78.45%

The criticism voiced at the time centered on the ratio between the numbers of those available for fighting and those available only for administration. But that comparison is not to the present purpose. What we have to note is that the 2000 officials of 1914 had become the 3569 of 1928; and that this growth was unrelated to any possible increase in their work. The Navy during that period had diminished, in point of fact, by a third in men and two-thirds in ships. Nor, from 1922 onward, was its strength even expected to increase; for its total of ships (unlike its total of officials) was limited by the Washington Naval Agreement of that year. Here we have then a 78 per cent increase over a period of fourteen years; an average of 5.6 per cent increase a year on the earlier total. In fact, as we shall see, the rate of increase was not as regular as that. All we have to consider, at this stage, is the percentage rise over a given period.

Can this rise in the total number of civil servants be accounted for except on the assumption that such a total must always rise by a law governing its growth? It might be urged at this point that the period under discussion was one of rapid development in naval technique. The use of the flying machine was no longer confined to the eccentric. Electrical devices were being multiplied and elaborated. Submarines were tolerated if not approved. Engineer officers were beginning to be regarded as almost human. In so revolutionary an age we might expect that storekeepers would have more elaborate inventories to compile. We might not wonder to see more draughtsmen on the payroll, more designers, more technicians and scientists. But these, the dockyard officials, increased only by 40 per cent in number when the men of Whitehall increased their total by nearly 80 per cent. For every new foreman or electrical engineer at Portsmouth there had to be two more clerks at Charing Cross. From this we might be tempted to conclude, provisionally, that the rate of increase in administrative staff is likely to be double that of the technical staff at a time when the actually useful strength (in this case, of seamen) is being reduced by 31.5 per cent. It has been proved statistically, however, that this last percentage is irrelevant. The officials would have multiplied at the same rate had there been no actual seamen at all.

It would be interesting to follow the further progress by which the 8118 Admiralty staff of 1935 came to number 33,788 by 1954. But the staff of the Colonial Office affords a better field of study during a period of imperial decline. Admiralty statistics are complicated by factors (like the Fleet Air Arm) that make comparison difficult as between one year and the next. The Colonial Office growth is more significant in that it is more purely administrative. Here the relevant statistics are as follows:

1935	1939	1948	1947	1954
372	450	817	1139	1661

Before showing what the rate of increase is, we must observe that the extent of this department's responsibilities was far from constant during these twenty years. The colonial territories were not much altered in area or population between 1935 and 1939. They were considerably diminished by 1943, certain areas being in enemy hands. They were increased again in 1947, but have since then shrunk steadily from year to year as successive colonies achieve self-government. It would be rational to suppose that these changes in the scope of Empire would be reflected in the size of its central administration. But a glance at the figures is enough to convince us that the staff totals represent nothing but so many stages in an inevitable increase. And this increase, although related to that observed in other departments, has nothing to do with the size—or even the existence—of the Empire. What are the percentages of increase? We must ignore, for this purpose, the rapid increase in staff which accompanied the diminution of responsibility during World War II. We should note rather, the peacetime rates of increase: over 5.24 per cent between 1935 and 1939, and 6.55 per cent between 1947 and 1954. This gives an average increase of 5.89 per cent each year, a percentage markedly similar to that already found in the Admiralty staff increase between 1914 and 1928.

Further and detailed statistical analysis of departmental staffs would be inappropriate in such a work as this. It is hoped, however, to reach a tentative conclusion regarding the time likely to elapse between a given official's first appointment and the later appointment of his two or more assistants.

Dealing with the problem of pure staff accumulation, all our researches so far completed point to an average increase of 5.75 per cent per year. This fact established, it now becomes possible to state Parkinson's Law in mathematical form: In any public administrative department not actually at war; the staff increase may be expected to follow this formula—

$$x = \frac{2k^m + 1}{n}$$

k is the number of staff seeking promotion through the appointment of subordinates; l represents the difference between the ages of appointment and retirement; m is the number of man-hours devoted to answering minutes within the department; and n is the number of effective units being administered. x will be the number of new staff required each year. Mathematicians will realize, of course, that to find the percentage increase they must multiply x by 100 and divide by the total of the previous year, thus:

$$\frac{100\,(2k^m + 1)}{yn}\,\%$$

where y represents the total original staff. This figure will invariably prove to be between 5.17 per cent and 6.56 per cent, irrespective of any variation in the amount of work (if any) to be done.

The discovery of this formula and of the general principles upon which it is based has, of course, no political value. No attempt has been made to inquire whether departments *ought* to grow in size. Those who hold that this growth is essential to gain full employment are fully entitled to their opinion. Those who doubt the stability of an economy based upon reading each other's minutes are equally entitled to theirs. It would probably be premature to attempt at this stage any inquiry into the quantitative ratio that should exist between the administrators and the administered. Granted, however, that a maximum ratio exists, it should soon be possible to ascertain by formula how many years will elapse before that ratio, in any given community, will be reached. The forecasting of such a result will again have no political value. Nor can it be sufficiently emphasized that Parkinson's Law is a purely scientific discovery, inapplicable except in theory to the politics of the day. It is not the business of the botanist to eradicate the weeds. Enough for him if he can tell us just how fast they grow.

ERVING GOFFMAN

ASYLUMS

THE special requirements of people-work establish the day's job for staff; the job itself is carried out in a special moral climate. The staff is charged with meeting the hostility and demands of the inmates, and what it has to meet the inmates with, in general, is the rational perspective espoused by the institution. We must therefore look at these perspectives.

The avowed goals of total institutions are not great in number: accomplishment of some economic goal, education and training; medical or psychiatric treatment; religious purification; protection of the wider community from pollution; and, as a student of prisons suggests, . . . *"incapacitation, retribution, deterrence and reformation."* . . .[1] It is widely appreciated that total institutions typically fall considerably short of their official aims. It is less well appreciated that each of these official goals or charters seems admirably suited to provide a key to meaning—a language of explanation that the staff, and sometimes the inmates, can bring to every crevice of action in the institution. Thus, a medical frame of reference is not merely a perspective through which a decision concerning dosage can be determined and made meaningful; it is a perspective ready to account for all manner of decisions, such as the hours when hospital meals are served or the manner in which hospital linen is folded. Each official goal lets loose a doctrine, with its own inquisitors and its own martyrs, and within institutions there seems to be no natural check on the licence of easy interpretation that results. Every institution must not only make some effort to realize its official aims but must also be protected, somehow, from the tyranny of a diffuse pursuit of them, lest the exercise of authority be turned into a witch hunt. The phantom of "security" in prisons and the staff actions justified in its name are instances of these dangers. Paradoxically, then, while total institutions seem the least intellectual of places, it is nevertheless here, at least recently, that concern about words and verbalized perspectives has come to play a central and often feverish role.

The interpretative scheme of the total institution automatically begins to operate as soon as the inmate enters, the staff having the notion that entrance is *prima facie* evidence that one must be the kind of person the institution was set up to handle. A man in a political prison must be traitorous; a man in a prison must be a lawbreaker; a man in a mental hospital must be sick. If not traitorous, criminal, or sick, why else would he be there?

Reprinted from *Asylums* by Erving Goffman (New York: Doubleday & Co., Anchor Books, 1961), pp. 83–92, by permission of the publisher.

This automatic identification of the inmate is not merely name-calling; it is at the center of a basic means of social control. An illustration is provided in an early community study of a mental hospital:

The chief aim of this attendant culture is to bring about the control of patients—a control which must be maintained irrespective of patient welfare. This aim is sharply illuminated with respect to expressed desires or requests of patients. All such desires and requests, no matter how reasonable, how calmly expressed, or how politely stated, are regarded as evidence of mental disorder. Normality is never recognized by the attendant in a milieu where abnormality is the normal expectancy. Even though most of these behavioral manifestations are reported to the doctors, they, in most cases, merely support the judgments of the attendants. In this way, the doctors themselves help to perpetuate the notion that the essential feature of dealing with mental patients is in their control.[2]

When inmates are allowed to have face-to-face contact with staff, the contact will often take the form of "gripes" or requests on the part of the inmate and justification for the prevailing restrictive treatment on the part of staff; such, for example, is the general structure of staff-patient interaction in mental hospitals. Having to control inmates and to defend the institution in the name of its avowed aims, the staff resort to the kind of all-embracing identification of the inmates that will make this possible. The staff problem here is to find a crime that will fit the punishment.

Further, the privileges and punishments the staff mete out are often phrased in a language that reflects the legitimated objectives of the institution, as when solitary confinement in prisons is called "constructive meditation." Inmates or low-level staff will have the special job of translating these ideological phrasings into the simple language of the privilege system, and vice versa. Belknap's discussion of what happens when a mental patient breaks a rule and is punished provides an illustration:

In the usual case of this kind, such things as impudence, insubordination, and excessive familiarity are translated into more or less professional terms, such as "disturbed" or "excited," and presented by the attendant to the physician as a medical status report. The doctor must then officially revoke or modify the patient's privileges on the ward or work out a transfer to another ward where the patient has to begin all over to work up from the lowest group. A "good" doctor in the attendants' culture is one who does not raise too many questions about these translated medical terms.[3]

The institutional perspective is also applied to actions not clearly or usually subject to discipline. Thus Orwell reports that in his boarding school bedwetting was seen as a sign of "dirtiness" and wickedness,[4] and that a similar perspective applied to disorders even more clearly physical.

I had defective bronchial tubes and a lesion in one lung which was not discovered till many years later. Hence I not only had a chronic cough, but running was a torment to me. In those days however, "wheeziness," or "chestiness," as it was called, was either diagnosed as imagination or was looked on as essentially a moral disorder, caused by overeating. "You wheeze like a concertina," Sim [the headmaster] would say disapprovingly as he stood behind my chair; "You're perpetually stuffing yourself with food, that's why." [5]

Chinese "thought reform" camps are claimed to have carried this interpretative process to the extreme, translating the innocuous daily events of the prisoner's past into symptoms of counterrevolutionary action.[6]

Although there is a psychiatric view of mental disorder and an environmental view of crime and counterrevolutionary activity, both freeing the offender from moral responsibility for his offense, total institutions can little afford this particular kind of determinism. Inmates must be caused to *self-direct* themselves in a manageable way, and, for this to be promoted, both desired and undesired conduct must be defined as springing from the personal will and character of the individual inmate himself, and defined as something he can himself do something about. In short, each institutional perspective contains a personal morality, and in each total institution we can see in miniature the development of something akin to a functionalist version of moral life.

The translation of inmate behavior into moralistic terms suited to the institution's avowed perspective will necessarily contain some broad presuppositions as to the character of human beings. Given the inmates of whom they have charge, and the processing that must be done to them, the staff tend to evolve what may be thought of as a theory of human nature. As an implicit part of institutional perspective, this theory rationalizes activity, provides a subtle means of maintaining social distance from inmates and a stereotyped view of them, and justifies the treatment accorded them.[7] Typically, the theory covers the "good" and "bad" possibilities of inmate conduct, the forms that messing up takes, the instructional value of privileges and punishments, and the "essential" difference between staff and inmates. In armies, officers will have a theory about the relation between discipline and the obedience of men under fire, the qualities proper to men, the "breaking point" of men, and the difference between mental sickness and malingering. And they will be trained into a particular conception of their own natures, as one ex-Guardsman suggests in listing the moral qualities expected of officers:

While much of the training was inevitably designed to promote physical fitness, there was nevertheless a strongly held belief that an Officer, whether fit or not, should always have so much in the way of pride (or "guts") that he would never admit to physical inadequacy until he dropped dead or unconscious. This belief, a very significant one, was mystical both in its nature and intensity. During a crippling exercise at the end of the course two or three Officers fell out complaining of blisters or other mild indispositions. The Chief Instructor, himself a civilized and self-indulgent man, denounced them in round terms. An Officer, he said, simply could not and did not fall out. Will-power, if nothing else, should keep him going for ever. It was all a matter of "guts." There was an unspoken implication that, since other ranks could and did fall out, even though they were often physically tougher, the Officer belonged to a superior caste. I found it an accepted belief among Officers later on that they could perform physical feats or endure physical discomforts without it being in the least necessary for them to train or prepare for such things in the manner required of the private soldier. Officers, for example, just did not do P.T.; they did not need it; they were Officers and would endure to the very end, had they stepped straight on to the field from a sanatorium or a brothel.[8]

In prisons, we find a current conflict between the psychiatric and the moral-weakness theories of crime. In convents, we find theories about the ways in which the spirit can be weak and strong and the ways in which its defects can be combated. Mental hospitals stand out here because the staff pointedly establish themselves as specialists in the knowledge of human nature, who diagnose and prescribe on the basis of this intelligence. Hence in the standard psychiatric textbooks there are chapters on "psychodynamics" and "psychopathology" which provide charmingly explicit formulations of the "nature" of human nature.[9]

An important part of the theory of human nature in many total institutions is the belief that if the new inmate can be made to show extreme deference to staff immediately upon arrival, he will thereafter be manageable—that in submitting to these initial demands, his "resistance" or "spirit" is somehow broken. (This is one reason for the will-breaking ceremonies and welcome practices discussed earlier.) Of course, if inmates adhere to the same theory of human nature, then staff views of character will be confirmed. Recent studies of the conduct of American army personnel taken prisoner in the Korean War provide an example. In America there is a current belief that once a man is brought to the "breaking point" he will thereafter be unable to show any resistance at all. Apparently this view of human nature, reinforced by training injunctions about the danger of any weakening at all, led some prisoners to give up all resistance once they had made a minor admission.[10]

A theory of human nature is of course only one aspect of the interpretative scheme offered by a total institution. A further area covered by institutional perspectives in work. Since on the outside work is ordinarily done for pay, profit, or prestige, the withdrawal of these motives means a withdrawal of certain interpretations of action and calls for new interpretations. In mental hospitals there are what are officially known as "industrial therapy" and "work therapy"; patients are put to tasks, typically mean ones, such as raking leaves, waiting on table, working in the laundry, and washing floors. Although the nature of these tasks derives from the working needs of the establishment, the claim presented to the patient is that these tasks will help him to relearn to live in society and that his capacity and willingness to handle them will be taken as diagnostic evidence of improvement.[11] The patient may himself perceive work in this light. A similar process of redefining the meaning of work is found in religious institutions, as the comments of a Poor Clare suggest:

This is another of the marvels of living in obedience. No one is ever doing anything more important than you are, if you are obeying. A broom, a pen, a needle are all the same to God. The obedience of the hand that plies them and the love in the heart of the nun who holds them are what make an eternal difference to God, to the nuns, and to all the world.[12]

People in the world are forced to obey manmade laws and workaday restrictions. Contemplative nuns freely elect to obey a monastic Rule inspired by God. The girl pounding her typewriter may be pounding for nothing but dollars' sake and wishing she could stop. The Poor Clare sweeping the monastery cloisters is doing it for God's sake and prefers sweeping, at that particular hour, to any other occupation in the world.[13]

Although heavily institutionalized motives such as profit or economy may be obsessively pursued in commercial establishments,[14] these motives, and the implied frames of reference, may nevertheless function to restrain other types of interpretation. When the usual rationales of the wider society cannot be invoked, however, the field becomes dangerously open to all kinds of interpretative flights and excesses and, in consequence, to new kinds of tyranny.

I would like to add a final point about institutional perspectives. The management of inmates is typically rationalized in terms of the ideal aims or functions of the establishment, which entail humane technical services. Professionals are usually hired to perform these services, if only to save management the necessity of sending the inmates out of the institution for servicing, it being unwise "for the monks to go abroad, for this is not at all healthful for their souls." [15] Professionals joining the establishment on this basis are likely to become dissatisfied, feeling that they cannot here properly practise their calling and are being used as "captives" to add professional sanction to the privilege system. This seems to be a classic cry.[16] In many mental hospitals there is a record of disgruntled psychiatrists asserting they are leaving so that they can do psychotherapy. Often a special psychiatric service, such as group psychotherapy, psychodrama, or art therapy, is introduced with great support from higher hospital management; then slowly interest is transferred elsewhere, and the professional in charge finds that gradually his job has been changed into a species of public relations work—his therapy given only token support except when visitors come to the institution and higher management is concerned to show how modern and complete the facilities are.

Professionals, of course, are not the only staff grouping in a somewhat difficult relation to the official goals of the establishment. Those members of staff who are in continuous contact with inmates may feel that they, too, are being set a contradictory task, having to coerce inmates into obedience while at the same time giving the impression that humane standards are being maintained and the rational goals of the institution realized.

Chapter 11

YOUTH

SOCIETIES differ in the manner in which the biography of the individual is divided into distinct phases. There are, of course, biological determinants of the stages of an individual's biography. Within these broad biological limits, however, different societies have very different ways of defining and organizing these stages. We have already seen this in earlier selections dealing with childhood. What is known as *youth* in modern society is another highly distinctive conceptualization and patterning of such a biographical phase. Partly because of the functional requirements of an industrial economy, and partly because of the development of new values in the spheres of the family and of education, youth was "invented" as an intermediary phase between the phases of childhood. Over the last century or so, the duration of this phase has been constantly expanding in all advanced industrial societies. As a result of this expansion, ever larger numbers of people have spent many years of their life in a sort of "waiting room," the organization of which has been largely in the hands of an astronomically expansive educational system. Both as a result of these facts and in conscious resistance to their frustrating aspects, the phase of youth has come to be characterized in quite distinctive social, cultural, and psychological ways. These distinctive characteristics make up what is now commonly called the youth culture—a phenomenon that may have been exaggerated or distorted by popular descriptions of it, particularly in the mass media, but about whose existence there can be little doubt.

Contemporary youth has been marked by considerable autonomy from the world of adults (a phenomenon that sociologists have called "peer socialization"). Contemporary youth, for reasons closely related to its segregated status within the larger society, has also been characterized by emotional intensity, by a propensity to experiment with different roles and possibilities of identity, and by a mood of opposition to adult institutions. The last of these characteristics is almost certainly related to resistance to bureaucracy.

The first selection from the seminal work *Centuries of Childhood* by the French historian Philippe Ariès shows how a highly distinctive reality of childhood came about in the emergence of modern society. The phenomenon of youth as we know it today must be understood sociologically as growing out of this modern "invention" of childhood and as receiving further shape by some basic processes of modern industrial society. F. Musgrove in the chapter "Demographic Influences" taken from his *Youth and the Social Order* discusses the fact of both demographic and technological changes in the emergence of youth. One of the first sociologists to collect a large body of data on the culture of adolescence was James Coleman, from whose *Adolescent Society* the third selection comes. An important feature of the emerging youth culture in recent years has been its collision with a variety of bureaucratic structures. This phenomenon, which in the 1960s came to be known as the "counterculture" is dealt with in the last selection, "Baltimore," from Elia Katz's *Armed Love*.

PHILIPPE ARIÈS

CENTURIES OF CHILDHOOD

In the Middle Ages, at the beginning of modern times, and for a long time after that in the lower classes, children were mixed with adults as soon as they were considered capable of doing without their mothers or nannies, not long after a tardy weaning (in other words, at about the age of seven). They immediately went straight into the great community of men, sharing in the work and play of their companions, old and young alike. The movement of collective life carried along in a single torrent all ages and classes, leaving nobody any time for solitude and privacy. In these crowded, collective existences there was no room for a private sector. The family fulfilled a function; it ensured the transmission of life, property and names; but it did not penetrate very far into human sensibility. Myths such as courtly and precious love denigrated marriage, while realities such as the apprenticeship of children loosened the emotional bond between parents and children. Medieval civilization had forgotten the *paideia* of the ancients and knew nothing as yet of modern education. That is the main point: it had no idea of education. Nowadays our society depends, and knows that it depends, on the success of its educational system. It has a system of education, a concept of education, an awareness of its importance. New sciences such as psycho-analysis, pediatrics and psychology devote themselves to the problems of childhood, and their findings are transmitted to parents by way of a mass of popular literature. Our world is obsessed by the physical, moral and sexual problems of childhood.

This preoccupation was unknown to medieval civilization, because there was no problem for the Middle Ages: as soon as he had been weaned, or soon after, the child became the natural companion of the adult. The age groups of Neolithic times, the Hellenistic paideia, presupposed a difference and a transition between the world of children and that of adults, a transition made by means of an initiation or an education. Medieval civilization failed to perceive this difference and therefore lacked this concept of transition.

The great event was therefore the revival, at the beginning of modern times, of an interest in education. This affected a certain number of churchmen, lawyers and scholars, few in number in the fifteenth century but increas-

Reprinted from *Centuries of Childhood* by Philippe Ariès, translated by Robert Baldick (New York: Alfred A. Knopf, 1962), pp. 411–15.

[Ariès's notes deleted.—Ed.]

ingly numerous and influential in the sixteenth and seventeenth centuries when they merged with the advocates of religious reform. For they were primarily moralists rather than humanists: the humanists remained attached to the idea of a general culture spread over the whole of life and showed scant interest in an education confined to children. These reformers, these moralists, whose influence on school and family we have observed in this study, fought passionately against the anarchy (or what henceforth struck them as the anarchy) of medieval society where the Church, despite its repugnance, had long ago resigned itself to it and urged the faithful to seek salvation far from this pagan world in some monastic retreat. A positive moralization of society was taking place: the moral aspect of religion was gradually triumphing in practice over the sacred or eschatological aspect. This was how these champions of a moral order were led to recognize the importance of education. We have noted their influence on the history of the school, and the transformation of the free school into the strictly disciplined college. Their writings extended from Gerson to Port-Royal, becoming increasingly frequent in the sixteenth and seventeenth centuries. The religious orders founded at that time, such as the Jesuits or the Oratorians became teaching orders, and their teaching was no longer addressed to adults like that of the preachers or mendicants of the Middle Ages, but was essentially meant for children and young people. This literature, this propaganda, taught parents that they were spiritual guardians, that they were responsible before God for the souls, and indeed the bodies too, of their children.

Henceforth it was recognized that the child was not ready for life and that he had to be subjected to a special treatment, a sort of quarantine before he was allowed to join the adults.

This new concern about education would gradually install itself in the heart of society and transform it from top to bottom. The family ceased to be simply an institution for the transmission of a name and an estate. It assumed a moral and spiritual function, it moulded bodies and souls. The care expended on children inspired new feelings, a new emotional attitude, to which the iconography of the seventeenth century gave brilliant and insistent expression: the modern concept of the family. Parents were no longer content with setting up only a few of their children and neglecting the others. The ethics of the time ordered them to give all their children, and not just the eldest—and in the late seventeenth century even the girls—a training for life. It was understood that this training would be provided by the school. Traditional apprenticeship was replaced by the school, an utterly transformed school, an instrument of strict discipline, protected by the law-courts and the police-courts. The extraordinary development of the school in the seventeenth century was a consequence of the new interest taken by parents in their children's education. The moralists taught them that it was their duty to send their children to school very early in life: 'Those parents', states a text of 1602, 'who take an interest in their children's education [*liberos erudiendos*] are more worthy of respect than those who just bring them into the world. They give them not only life but a good and holy life. That is why those parents are right to send their children at the tenderest age to the market of true wisdom [in other words to college] where they will become the architects of their own fortune, the ornaments of their native land, their family and their friends.'

Family and school together removed the child from adult society. The school shut up a childhood which had hitherto been free within an increasingly severe disciplinary system, which culminated in the eighteenth and nineteenth centuries in the total claustration of the boarding-school. The solicitude of family, Church, moralists and administrators deprived the child of the freedom he had hitherto enjoyed among adults. It inflicted on him the birch, the prison cell—in a word, the punishments usually reserved for convicts from the lowest strata of society. But this severity was the expression of a very different feeling from the old indifference: an obsessive love which was to dominate society from the eighteenth century on. It is easy to see why this invasion of the public's sensibility by childhood should have resulted in the now better-known phenomenon of Malthusianism or birth-control. The latter made its appearance in the eighteenth century just when the family had finished organizing itself around the child, and raised the wall of private life between the family and society.

The modern family satisfied a desire for privacy and also a craving for identity: the members of the family were united by feeling, habit and their way of life. They shrank from the promiscuity imposed by the old sociability. It is easy to understand why this moral ascendancy of the family was originally a middle-class phenomenon: the nobility and the lower class, at the two extremities of the social ladder, retained the old idea of etiquette much longer and remained more indifferent to outside pressures. The lower classes retained almost down to the present day the liking for crowds. There is therefore a connection between the concept of the family and the concept of class. Several times in the course of this study we have seen them intersect. For centuries the same games were common to the different classes; but at the beginning of modern times a choice was made among them: some were reserved for people of quality, the others were abandoned to the children and the lower classes. The seventeenth-century charity schools, founded for the poor, attracted the children of the well-to-do just as much; but after the eighteenth century the middle-class families ceased to accept this mixing and withdrew their children from what was to become a primary-school system, to place them in the *pensions* and the lower classes of the colleges, over which they established a monopoly. Games and schools, originally common to the whole of society, henceforth formed part of a class system. It was all as if a rigid, polymorphous social body had broken up and had been replaced by a host of little societies, the families, and by a few massive groups, the classes; families and classes brought together individuals related to one another by their moral resemblance and by the identity of their way of life, whereas the old unique social body embraced the greatest possible variety of ages and classes. For these classes were all the more clearly distinguished and graded for being close together in space. Moral distances took the place of physical distances. The strictness of external signs of respect and of differences in dress counterbalanced the familiarity of communal life. The valet never left his master, whose friend and accomplice he was, in accordance with an emotional code to which we have lost the key today, once we have left adolescence behind; the haughtiness of the master matched the insolence of the servant and restored, for better or for worse, a hierarchy which excessive familiarity was perpetually calling in question.

People lived in a state of contrast; high birth or great wealth rubbed shoulders with poverty, vice with virtue, scandal with devotion. Despite its shrill contrasts, this medley of colours caused no surprise. A man or woman of quality felt no embarrassment at visiting in rich clothes the poor wretches in the prisons, the hospitals or the streets, nearly naked beneath their rags. The juxtaposition of these extremes no more embarrassed the rich than it humiliated the poor. Something of this moral atmosphere still exists today in southern Italy. But there came a time when the middle class could no longer bear the pressure of the multitude or the contact of the lower class. It seceded: it withdrew from the vast polymorphous society to organize itself separately, in a homogeneous environment, among its families, in homes designed for privacy, in new districts kept free from all lower-class contamination. The juxtaposition of inequalities, hitherto something perfectly natural, became intolerable to it: the revulsion of the rich preceded the shame of the poor. The quest for privacy and the new desires for comfort which it aroused (for there is a close connection between comfort and privacy) emphasized even further the contrast between the material ways of life of the lower and middle classes. The old society concentrated the maximum number of ways of life into the minimum of space and accepted, if it did not impose, the bizarre juxtaposition of the most widely different classes. The new society, on the contrary, provided each way of life with a confined space in which it was understood that the dominant features should be respected, and that each person had to resemble a conventional model, an ideal type, and never depart from it under pain of excommunication.

The concept of the family, the concept of class, and perhaps elsewhere the concept of race, appear as manifestations of the same intolerance towards variety, the same insistence on uniformity.

F. MUSGROVE

POPULATION CHANGES AND THE STATUS OF THE YOUNG

IN the later eighteenth century and throughout the nineteenth, child mortality rates declined. While the significance of the eighteenth-century decline for population growth may be disputed,[1] no leading demographer would deny that it occurred. The middle- and upper-class could avail themselves more easily than the working classes of improved housing, sanitation and medical care; [2] the survival rate among their children up to 15 years of age was 83 per cent in 1871; in the population at large, while it was much improved on the rough estimate of 50 per cent a century earlier, it was still only 63 per cent.[3] It seems likely that the social-class differential had widened in the course of the nineteenth century. In 1830 79 per cent of the children of clergymen in the diocese of Canterbury survived their first 15 years, in 1871 85 per cent did so.[4]

There is no necessary correspondence between falling mortality rates among young people and the growth of suitable employment opportunities for them. There can be little doubt that by the 1870s middle-class children, by surviving in greater numbers, constituted a growing burden on their parents while they were growing up and an increasing problem to place in acceptable work when their education was completed. It is probable that a social-class differential in fertility existed much earlier in the century—Glass has computed negative correlation coefficients between fertility and status in twenty-eight London boroughs which were not notably smaller in 1851 than in 1911 or 1931; [5] nevertheless, it is from the seventies that the average size of the middle-class family began its steep decline. The birth-control movement was a symptom of the superabundance of the young in relation to family resources and to the needs of the economy. 'It may be possible to bring ten children into the world, if you only have to rear five, and, while one is "on the way", the last is in the grave, not in the nursery. But if the doctor preserves seven or eight of the ten, and other things remain equal, the burden may become intolerable.' [6] But other things did not even remain equal: it was unfortunate for the young that they were most abundant when the economy, whether at the level of professional or of manual employment, offered diminishing opportunities for youth and relative inexperience.

Reprinted from *Youth and the Social Order* by F. Musgrove (Bloomington: Indiana University Press, 1965), pp. 64–74, by permission of the publisher.

THE NEEDS OF THE ECONOMY

In the later eighteenth century and the first half of the nineteenth, parents valued children more highly as their chance of survival improved; employers valued them more highly as technological changes gave them a position of pivotal importance in new industries. As the traditional system of apprenticeship broke down because of its irrelevance in the eighteenth century, and the legal requirement to serve an apprenticeship to a trade was repealed in 1814, the young were liberated to find their true level of importance in the changing economy. Debased forms of 'apprenticeship'—particularly parish apprenticeship—within industry still often prevented the young worker from achieving his true economic wage and the social independence that went with it. But this form of exploitation, often by parents and relatives rather than by plant owners, became less common in the early decades of the nineteenth century. The new industries were heavily dependent on the skills and agility of the young.

Moreover, before the population aged in the later nineteenth and the twentieth centuries, there were greater opportunities for young people to secure top appointments: promotion was not blocked by a glut of older men. Jousselin has contrasted the opportunities which the young enjoyed in the later eighteenth and early nineteenth centuries with their frustrations today: 'Du fait du vieillissement de la population, la situation de jeunesse s'est profondément modifiée. Il leur est beaucoup plus difficile d'accéder aux postes de responsibilité et d'initiative. On ne connaît plus de generaux de 20 a 30 ans, ni de préfets ayant l'âge des gouverneurs de l'ancien régime.' [7]

In remote areas, away from large urban reservoirs of labour, young people were a particularly large proportion of the labour force in early-nineteenth-century factories, partly because the pauper apprentice was the most freely mobile economic unit. But even where labour was abundant and there was less need to employ parish apprentices, young people were often a high proportion of the employees. Forty-eight per cent of the 1,020 workpeople of M'Connel and Kennedy in Manchester in 1816 were under 18 years of age.[8]

The demand at this time was particularly for working-class youth; but middle-class youth—at least the males—were also needed as commerce expanded even more rapidly than industry and called for a great army of white-collar workers.[9] Only gradually, in the closing decades of the century, were 'accountants', for example, distinguished from 'book-keepers'—(and Upper Division civil servants from Lower Division[10])—and required to undertake prolonged education and training before receiving the economic rate for the work they did.[11]

In the new industries parents were often appendages to their children, heavily dependent on their earnings. When the farm labourer moved his family to the town, it was commonly for what his children could earn: his own employment might be as a porter, or in subsidiary work such as road-making, at a wage of 10 to 13 shillings a week; his child (and wife) could earn more on the power looms or in throstle-spinning.[12] As Mr Carey commented in Disraeli's *Sybil* (1845): 'Fathers and mothers goes for nothing. 'Tis the children gets the wages, and there it is.' The fathers of the poor families imported from Bedfordshire and Buckinghamshire were fit only for labourers. There must have been many a Devilsdust 'who had entered life so early that at 17 he combined the experience of manhood with the divine energy of youth'.

Apprenticeship and experience in traditional industries were a handicap in James Keir's chemical works at Tipton, founded in the 1780s; [13] Andrew Ure noted in 1836 that 'Mr Anthony Strutt, who conducts the chemical department of the great cotton factories at Belper and Milford . . . will employ no man who has learned his craft by regular apprenticeship.' [14] Samuel Oldknow's spinning mill at Mellor depended on youthful labour; subsidiary industries (line-kilns, coal-mining, farming) had to be provided for redundant fathers; [15] at Styal the Gregs had to develop an industrial colony to provide employment for the adult dependants of their juvenile and female workers. What shocked middle-class commentators on factory life in mid-Victorian England as much as the alleged immorality was the independence of the young. In London girls of 14 working in the silk or trimming departments earned 8 or 10 shillings a week: 'if they had cause to be dissatisfied with the conduct of their parents, they would leave them.' [16] Similar independence was to be found in the Birmingham metal trades: 'The going from home and earning money at such a tender age (of seven or thereabouts) has—as might be expected—the effect of making the child early independent of its parents . . .' [17] The Factory Commissioners reported in 1842 that by the age of fourteen young people 'frequently pay for their own lodgings, board and clothing. They usually make their own contracts, and are in the proper sense of the word free agents.' (Even the Poor Law provisions of the Speenhamland System had a similar effect in the early decades of the century. As Mr Assistant Commissioner Stuart stated in the *Report* from the Commissioners on the Poor Laws (1834): 'Boys of 14, when they become entitled to receive parish relief on their own account, no longer make a common fund of their income with their parents, but buy their own loaf and bacon and devour it alone. Disgraceful quarrels arise within the family circle from mutual accusations of theft.') [18]

While factory legislation was at least a potential threat to the earnings and power of the young, the Act of 1833 extended their independence by destroying the vestigial authority of parents in the textile—particularly the spinning—mills. Virtually autonomous family units, under the headship of the father, had infiltrated intact into some of the textile factories: Andrew Ure [19] reported in the 1830s that 'Nearly the whole of the children of 14 years of age, and under, who are employed in cotton mills, belong to the mule-spinning department, and are, in forty-nine cases out of fifty, the immediate dependants, often the offspring or near relations of the spinner, being hired and dismissed at his option.' The spinner paid his piecers and scavengers from his own wages. (In the mines outside Northumberland and Durham the young worker in 1840 was often even more completely under his father's authority; in South Wales 'the collier boy is, to all intents and purposes, the property of his father (as to wages) until he attains the age of 17 years, or marries.' Butties received 'apprentices' at the age of 9 for twelve years—a system likened by witnesses to the African slave trade.[20]) The early regulation of the employment of the young, and particularly their shortened and staggered hours of work, had the effect, along with technological change, of removing them from the control of the head of the family who developed a more specialized role which 'no longer implied co-operation with, training of, and authority over dependent family members.' [21]

The importance of the young to the economy is reflected in the high birth rate which was particularly buoyant in the later eighteenth and early-nineteenth centuries. This is not simply to say that children were begotten for the benefit of what they could earn whilst still children; in an expanding economy they were valued as a longer term proposition. Talbot Griffith rejected as 'scarcely tenable' the theory that the high birth rate of the period [22] was caused by the economic value of children,[23] but conceded that 'The feeling that the new industries would provide employment for the children at an early age and enable them possibly to help the family exchequer would tend, undoubtedly, to make parents contemplate a large family with equanimity and may have acted as a sort of encouragement to population without the more definite incentive implied in the theory that it was the value of children's work which led to the increase of the population.' [24] Marshall was more inclined to see significance for the birth rate in children's earnings: 'By 1831 the birth rate, measured in proportion to women aged 20–40, got back for the first time to the level of 1781 (this is a guess); by 1841 it had slumped far below it (this is a fact). Now it is only fair to older theories to point out how this fall by stages, slow at first and then rapid, reflects the history of child labour and the Poor Law.' [25] Glass, on the other hand, found little or no connection between the employment of children aged 10 to 15 and fertility between 1851 and 1911 in the forty-three registered counties.[26] The value of children for their earnings *whilst still children* is doubtless an inadequate explanation of the changing birth rate; but the value of children more broadly conceived, as likely, in a buoyant economy, to constitute an insurance against misfortune in later life and old age, is an explanation not inconsistent with Arthur Young's argument that population is proportional to employment.[27]

Habakkuk has attributed the growth in population during the Industrial Revolution primarily to 'specifically economic changes', and in particular to 'an increase in the demand for labour', but has pointed out that the way in which this demand operated (whether directly on the birth rate or indirectly through the lower age of marriage) remains open to question.[28] The experience of Ireland, as Talbot Griffith realized, provides the key to this problem.

It was an embarrassment to Griffith's argument that in relatively insanitary Ireland population increased between 1780 and 1840 at almost twice the rate experienced in England; inadequate statistics made it impossible for him to prove or disprove his contention that a declining rate of child mortality was the primary reason. On the other hand, the comparative lack of industrial development in Ireland seemed to nullify the argument that the primary reason was increased demand for labour.[29] This latter difficulty is overcome if we regard Ireland-with-England—or at least Ireland-with-Lancashire—as a single field of employment, as the Irish themselves clearly did. The relatively unskilled jobs available in the textile industry were particularly suited to Irish immigrants; and before 1819 movement into Lancashire was easy—easier than moving in from elsewhere in England—since the Irish were regarded as having no place of 'settlement' and so could not be removed by the Poor Law authorities if they became a charge upon the rates. But these were not young children seeking employment. Undue concentration on the earnings of *young* children has bedevilled the question of the value of offspring. At this time in Ireland—and in England too—'a large family was regarded less as a strain upon resources than as the promise of comfort and material well being in middle and old age.' [30]

Children were of value even and perhaps particularly as they grew up into adult life and work, as an insurance against misfortune, against sickness and old age. They were not an entirely reliable insurance, particularly with increased geographical mobility and the dispersal of the family; their unreliability, at least in London, was commented on by Mayhew in the middle of the nineteenth century [31] and by Booth at the end. Booth was under the impression that this unreliability had become more marked in recent years: 'The great loss of the last twenty years is the weakening of the family ties between parents and children. Children don't look after their old people according to their means. The fault lies in the fact that the tie is broken early. As soon as a boy earns 10 shillings a week he can obtain board and lodging in some family other than his own, and he goes away because he has in this great liberty.' [32]

The importance of children was undermined in the later nineteenth century by the growth of alternative forms of insurance. Indeed, 'insurance' in a broad sense—whether paid-up premiums, private means or a working wife—has demographic significance as a substitute for children. Children seen as insurance help to explain the apparent paradox, discussed by Stevenson, that in the nineteenth century high mortality appeared to promote large families rather than vice versa. Many were born when comparatively few survived: additional births were necessary to effect replacements. This was particularly the case when there was no other form of insurance against misfortune: amongst miners, whose wives were excluded from employment after the 1840s, child mortality rates were high, but so was fertility. High child mortality rates were also experienced in the families of the textile workers, but fertility was low also; it is arguable that replacements were not so necessary when wives were commonly at work. The low fertility of couples of independent means had perplexed nineteenth-century demographers. Stevenson considered that their low fertility was the most remarkable case of all: 'In their case, presumably, those anxieties and difficulties which militate against fertility are at a minimum, but fertility is also at a minimum.' [33] Once the rates of infant mortality had fallen, it was safe to assume that even a small family would survive to carry on the family name and estate.

Children were of diminishing value to couples who were covered by insurance. The birth rate at the end of the nineteenth century slumped not only among the professional and middle classes, but among artisans and skilled mechanics, many of them among the infertile textile families, who in large numbers joined Friendly Societies such as the Oddfellows, Manchester Unity (1810), the Foresters (1834), the Rechabites, Salford Unity (1835), the Hearts of Oak (1842), and the National Deposit Friendly Society (1868). By 1872 the Friendly Societies probably had some four million members, compared with one million trade unionists—and the latter had sickness, employment, and sometimes superannuation schemes too.[34] The decline in claims for lying-in benefit by members of the Hearts of Oak gives some indication of their declining fertility: between 1881 and 1904 the proportion of claims to membership declined by 52 per cent.[35] It is likely that they had less need of this benefit precisely because they were members of a provident society.

By the last quarter of the nineteenth century the status of the young was being undermined as a consequence of their earlier importance. Their value had resulted in their super-abundance. As Yule and Habakkuk have observed, the children who are produced in response to favourable economic circumstances may still be there when the circumstances have deteriorated: 'the present demand is met only by a delivery of the commodity some twenty years later; by that time the 'commodity' may not be required.' [36] Yule speaks of 'a very large and quite abnormal increase' in the labour force aged 20 to 55 years in the last twenty years of the century; as young workers the 'bulge' has entered the labour market in the late sixties and the seventies.[37] This increase was 'not produced by any present demands for labour, but in part by the "demand" of 1863–73 . . .': the birth rate in the sixties and seventies was as high as in the quarter-of-a-century after 1780.[38] Moreover, in the second half of the nineteenth century the survival rates among older children and adolescents improved much more rapidly than among children aged 0 to 4 years. While the annual mortality per thousand declined by 11·3 per cent among boys aged 0 to 4 years (from 71 to 63) between 1841–5 and 1891–1900, the decline among boys aged 5 to 9 declined by 53·2 per cent (from 9·2 to 4·3), and among boys aged 10 to 14 by 53·0 per cent (from 5·1 to 2·4).[39] Thus whilst adolescents were a better 'proposition', since they were more likely to live and so justify what was spent on their upbringing and education, the wastage among them was small at the very time that the economy had a diminishing need for their services.

This is the prelude to the introduction of compulsory education between

1870 and 1880. Not only was there a 'bulge' in young people, but advances in technology were in any case displacing the young worker. In some industries, too, extended factory legislation greatly diminished his value in the eyes of employers: the administrative complications raised by part-time schooling deterred mine-owners from employing boys under 12 after the Mines Act of 1860; [40] the Factory Acts Extension Act of 1867 and the Workshops Regulation Act of the same year had similar consequences in a wide range of industries including the metal trades, glass and tobacco manufacture, letterpress printing and bookbinding.[41] The textile industry, on the other hand, which had greatly reduced its child labour force after the Act of 1833, maintained a high proportion of juvenile workers after the Act of 1844 which simplified the administration of part-time work.[42] Agriculture also took the Gang Act of 1867 and the Agricultural Children Act of 1873 in its stride—chiefly by ignoring them.

Technical changes in many industries were in any case breaking the earlier dependence on juvenile labour: steam power in the lace[43] and pottery [44] industries was being substituted for children's energy and dexterity; the dramatic decline in the proportion of young people engaged in agriculture in the second half of the century has been similarly attributed in part to technical development: 'A new class connected with the application of science to agriculture has sprung into being . . .' [45] Young people were no longer central to the economy; they were moving ever more onto the periphery, into marginal and relatively trivial occupations: street-trading, fetching and carrying, and particularly indoor domestic service. . . . Compulsory education was a necessity by the 1870s not because children were at work, but because increasingly they were not.

JAMES COLEMAN

THE EMERGENCE OF AN ADOLESCENT SUBCULTURE IN INDUSTRIAL SOCIETY

PERHAPS it is self-evident that the institutional changes that have set apart the youth of our society in high schools should produce an "adolescent cul-

Reprinted from *The Adolescent Society* by James Coleman (Glencoe, Ill.: Free Press, 1961), pp. 5–9, by permission of The Macmillan Company.

ture," with values of its own. These changes have been discussed speculatively by numerous authors.[1] Whether or not there is a separate adolescent subculture is partly a matter of definition as to what constitutes a separate subculture. However, there are several items from the present study that give a sense of the degree to which these adolescents are oriented to parents and peers. In one set of questions, they were asked whether they would join a club in school (1) if their parents disapproved, (2) if their favorite teacher disapproved, and (3) if it would mean breaking with their closest friend . . . .[2] Then they were asked whose disapproval would be most difficult to accept—parents', teacher's, or friend's (see Table 13).[3]

The responses indicate a rather even split between friend and parent, while the teacher's disapproval counts most for only a tiny minority. The balance between parents and friends indicates the extent of the state of transition that adolescents experience—leaving one family, but not yet in another, they consequently look both forward to their peers and backward to their parents.

TABLE 13

Which One of These Things Would Be Hardest for You to Take—Your Parents' Disapproval, Your Teacher's Disapproval, or Breaking with Your Friend?

	BOYS	GIRLS
Parents' disapproval	53.8%	52.9%
Teacher's disapproval	3.5	2.7
Breaking with friend	42.7	43.4
Number of cases (excluding non-responses)	(3,621)	(3,894)

Thus, teen-agers are not oriented solely to one another; yet the pulls are extremely strong, as the responses in Table 13 show. It seems reasonable, however, that those adolescents who are more oriented to their parents might "set the standard" in school, while those more oriented to their peers would tend toward delinquency or at least enjoy less esteem than those who are parent-oriented. If this were so, then the adolescent cultures existing in the school would be oriented toward parents to a greater degree than the individual responses indicate, because the central persons in the schools were more oriented to parents. But this is not at all so.

This can be seen by looking at those students who are named most often (ten times or more) by their fellows in response to the following question: . . . "If a fellow came here to school and wanted to get in with the leading crowd, what fellows should he get to be friends with?" Quite reasonably, we can infer that the students named in response to this question include most of the "leaders" or the "elite" of the adolescent culture in the schools. For this group and for the students as a whole, the proportion who say parents' disapproval would be hardest for them to take is shown in Table 14.

TABLE 14

Proportion Who Say That "Parents' Disapproval Would Be Hardest to Take"

BOYS		GIRLS	
ALL BOYS	LEADING CROWD	ALL GIRLS	LEADING CROWD
53.8% (5,621)	50.2% (167)	52.9% (3,894)	48.9% (264)

The elites in the school are not closer to their parents than are the students as a whole, but are pulled slightly farther from parents, closer to fellow-adolescents as a source of approval and disapproval. Thus, those who "set the standard" are more oriented than their followers to the adolescent culture itself. The consequences of this fact are important, for it means that those students who are highly regarded by others are themselves committed to the adolescent group, thus intensifying whatever inward forces the group already has.

Turning back to the suggestion that the existence of an adolescent culture is more pronounced than it once was, there is little the present research can do to document this possibility. However, our data does afford some insight into changes that have taken place. The present study includes five small-town or rural schools and five city and suburban ones, and it is possible to study the stable small-town schools so as to look back into the past a short distance. These small-town schools represent a segment of American society that was once far more important than it is now: They are located in small-town mid-American market centers for the surrounding farmland, with a few industries, but not highly industrialized. At the other extreme, in keeping with the times, are two homogeneous suburbs of Chicago: one a new working-class suburb with a school only seven years old (to be called Newlawn throughout the book), the other an older, but similarly homogeneous, upper-middle-class suburb with a mushrooming population (to be called Executive Heights throughout the book).

Before examining the statistical data which can give some evidence of these changes, let us quote a passage of a letter from a particularly perceptive parent, a lawyer living in Executive Heights:

Mrs. ——— and I do not believe we could give you much useful information about the teen-age culture as it exists in Executive Heights. We see this culture in a very limited aspect, through our sophomore daughter, and our limited familiarity with her contemporaries. It is truly surprising how little one person knows about the inner thinking of another, particularly when the other is an adolescent. I suspect that the interests and values current in the high school group are more independent of those of the adult community than we generally believe.

Most children of this age, at least in Executive Heights, I believe do not have serious responsibilities, such as helping to support a family or to prepare to pay their way through college, and therefore are not acquainted with people engaged in making a living, and do not understand the problems and responsibilities which parents have. As a result, they do not obtain the experience which goes with such responsibility, . . . and, in effect, live in a world separate from the adult community.

This letter indicates what may be a general feeling among many parents today: that their teen-age children are in "a world apart." An index of such "apartness" might be the degree to which boys want to follow in their father's occupation. They were asked: . . . "What kind of work do you plan to go into when you finish your schooling?" The answers were coded (except in some cases where it was not possible) as to whether the occupation named was the same as their father's or different. The results are tabulated in Table 15, for the five small-town or rural schools and for the four public high schools in larger towns, cities, and suburbs.

TABLE 15

Percent of Boys Who Want to Go into Their Father's Occupation

	FIVE SMALL-TOWN SCHOOLS		FOUR CITY AND SUBURBAN SCHOOLS	
Same	23.0%		9.8%	
Different	77.0		90.2	
Total classifiable	100.0	(710)	100.0	(2,177)
(Unable to classify)		(66)		(173)
(No answer to own occupation)		(252)		(643)
Total		(1,028)		(2,993)

Twenty-three per cent of the boys in the small towns want to enter their father's occupation; only 9.8 percent of the boys in the city and suburban schools want to do so.[4] This difference is even more impressive when one realizes that it is the small-town occupations that will be less numerous in the next generation, as an industrial society more nearly replaces a rural one.

In examining the school of Executive Heights, we can see more directly the greater orientation to peers on the part of boys and girls who are fully a part of modern society. One question asked all students to rank four items according to these items' importance for him. The items were: "pleasing my parents"; "learning as much as possible in school"; "living up to my religious ideals"; and "being accepted and liked by other students." The ranking of "being accepted and liked by other students" is considerably higher in this school than in the average of all schools, as Table 16 shows.

This comparison and the comparison in Table 15 show differences suggesting a movement toward a separate adolescent culture. But it is likely that the attempt to "reach back into history" by examining these small towns or to

TABLE 16

Average Rank of "Being Accepted and Liked by Other Students" for All Schools and Executive Heights

	(RANK 1 IS HIGH; 4 IS LOW)			
	BOYS		GIRLS	
	ALL SCHOOLS	EXECUTIVE HEIGHTS	ALL SCHOOLS	EXECUTIVE HEIGHTS
Average rank	2.63	2.34	2.53	2.17
Number of cases	(4,020)	(932)	(4,134)	(898)

"reach forward" by examining an upper-middle-class suburb is impeded by the very shifts that have occurred in small towns themselves. The general prosperity, rapid transportation, and the mass communication media have radically altered the style of life in small towns, bringing them into the general culture. In particular, small-town adolescents have far greater access to one implement of our modern culture than do teen-agers in a large city or even a suburb: the automobile.

Despite the many ways in which small towns are representative of an older, non-industrial society, in this one way—the prevalence of automobiles among teen-agers—they are very much a part of modern culture. . . . It seems that only where there have been conscious efforts to maintain an older way of life—as with the Amish—or where there are physical or economic barriers—as in the center of a large city—or in economically depressed areas have the implements and culture of modern society failed to penetrate deeply.

In sum, then, the general point is this: our adolescents today are cut off, probably more than ever before, from the adult society. They are still oriented toward fulfilling their parents' desires, but they look very much to their peers for approval as well. Consequently, our society has within its midst a set of small teen-age societies, which focus teen-age interests and attitudes on things far removed from adult responsibilities, and which may develop standards that lead away from those goals established by the larger society.[5]

Given this general condition, there are several directions in which educational efforts could turn. One is toward a channeling of the adolescent societies so that the influence they exert on a child is in the directions adults desire. Rather than attempting to motivate children one by one, each parent (with teachers' assistance) exhorting his own child in one direction while the adolescent culture as a whole pulls in other directions, efforts can be made to redirect the whole society of adolescents itself, so that *it* comes to motivate the child in directions sought by the adult society. This is not a new device; "playing by ear," perceptive principals and teachers have long attempted to do this. Yet it has never been the focus of any general philosophy of education, nor have any general means for redirecting the adolescent society been set forth in schools of education, perhaps because we know too little about the ways in which subsocieties, such as those of our teen-agers, can be guided and directed. . . .

ELIA KATZ

BALTIMORE

December 1.

HERE it is freezing cold and we are surrounded by a huge lawn of snow, and a semicircular line of cars and vans are snowed in on the semicircular driveway in front of the house. Rateyes spends the mornings getting dressed, trying on all of his clothes, belts, heads, headbands, wrist binders, sashes, coats, every day and mixing them with articles he finds in rooms throughout the house. We are wasting our time, getting stoned and more stoned, walking every night to the blank highway for chocolate bars, peanuts, Cokes, Fritos, and sandwiches, eating ferociously disgusting quantities of food and falling asleep in front of the TV.

Where we are: This is a mansion that Doug rents in a suburb of Baltimore called Pikesville, where he is living with his brother Dennis, George, somebody named Bruce who gives the impression of being sly at odd scattered moments, Betty's new husband John and her sister Jane, and John's brother Stephen (who has short hair as he is just out of the Navy and feels the others resent his presence), as well as a rotating population of transient crashers, mostly runaways from other suburbs and older visitors, occult and peaceful, from a place called the Heathcote School for Life, which none of them feels qualified to talk about but which all of them say is an incredible place—a commune of mystics and farmers—in Pennsylvania. Of this crowd, never fewer than ten are always in the house, and we spend most of the time in the library, which is large and lined with bookshelves and has cushions spread over a soft carpet, watching the TV and the fireplace. There is always plenty of food, always beer, always grass and hash that Crazy goes out in the morning to get for everyone who comes throughout the day to buy it. Crazy, who has a long, calm face and a thin moustache, and wears a flat-brimmed Spanish hat with little puffballs hanging over his eyes, acts like a sidekick with a

Reprinted from "Baltimore," from *Armed Love* by Elia Katz (New York: Bantam Books, 1971), pp. 18–27.

full range of sidekick mannerisms. Most of the permanent residents here impress one as being subordinate and peripheral, incapable of independent existence, and, like Crazy, they all seem to take advice and instruction from Doug.

It is Doug whom Rateyes and I have come here to see. He is our old friend Doug, my old roommate Doug, who has grown rich dealing and gathered all these people around him in his magnificent mansion. Rateyes and I have decided to start our travels slowly, piercing the interior with caution, going first to visit our friends, and then maybe *their* friends, and so on, making this a natural, easy trip, in part because we cannot imagine ourselves thrusting, shoving newsmanly into strangers' collectives, getting those scoops, digging for those facts, and in part because we want them all to come with us, all our old friends, and theirs, in buses, cars, a caravan, having a good time, joining ourselves, our rambling group, to the millions of hippie insects crawling and darting through the fur of the Big Ugly Bear, all aimlessly traveling, the great Brownian migration on the coast-to-coast pelt, of the world's largest, newest leisure class.

Doug's room is on the third floor of the house. It is actually two adjacent rooms—an apartment—one of which is covered in purple velvet and has an altar in the middle of the floor. In this room he keeps the shrine to himself: all the photographs taken of him since his childhood, poems he has written, letters to him from various chicks, and the release forms from his two stays in mental institutions. The floor is covered with dogshit, because his dog, Zero, is too young to go outside in the cold. And in the center of the floor, on the altar before the shrine, incense is constantly burning in answer to the dogshit.

"Doug," we said, "we are like babies. Neither of us knows how to drive. Now we have to travel around the country seeing all these communes and we will either have to hitch or take planes." We have told him about the book and the money and have offered to cut him in on everything. "Come with us. Drive us. We'll buy the car. Don't you want your name on a book, man?"

"Fuck the book," said Rateyes, "this is a year for doing nothing except getting stoned and getting laid like."

"Fuck getting laid," said Doug, "I get laid right here. I get all the dope I need." But he has said he will think about it. He has been a friend of ours for years, though neither of us has seen him for a long time. He has changed. Looks excellent—handsome, arrogant, sleepy—with his hair very long and thick worn in the manner of Linda Darnell. Being a warlord has done him nothing but good. He observes that he no longer finds it necessary to write poetry because he has personal power over the people and occurrences of his daily life. In fact, having Rateyes and me around, though he says it is good to have someone he can talk to in a normal tone of voice without giving instruction or making complaint, is beginning to get on his nerves. However, there is a chance he will leave with us because he is becoming bored with this place and because the father of one of his little chicks, a colonel in the Air Force, makes daily drunken phone calls threatening his life.

This is a huge, partly furnished mansion they are renting, the shell of a mansion around us like a cave. The people here act toward the house with the bland malevolence of great gods. They are using it up, as though it were packaging. No thought to the extensive life of the house itself. It is incredible and powerful to watch this happen to a house. The windows break one by one from being slammed shut and from rocks; the banisters crack and are left hanging, poised in sections over the stairwell; and there are cigarette burns in the rugs and furniture. One day someone found a hatchet, so there are hatchet designs in the heavy front door and the walls dividing the rooms. Casual, peripheral destructive events in the moments of the tenants. They do not interrupt the flow or halt a moment to note or regret the event, or to worry about the piecemeal destruction of the shell they are in. When it is used up they are going to get another one. Treat all housing as raw shelter. Do not learn to depend on the shell and objects of survival. Realize their destructibility. The beauty of the tenants' attitude is in the purity of the lack of sentimental attachment to their home. Anarchic urban crash pad mentality. Doug feels that once there is developed, either through regulated chores or a sense of permanence, a sort of preservation (ecology) tropism toward one's home and possessions, one is ripe to be governed. He is always ready to leave within a few hours of the decision to leave. He has shown me how his possessions are practically all in a suitcase and ready to go.

You feel in this place the same pure energy of children who are openly ungrateful to their parents, and you feel these housebusters are in open defiance of even the laws of gratitude for comfort and shelter. The thrill of waste and waste and waste in search of the unwastable, the *worthwhile,* absolute *Home.*

This house is aging and coming down. It is ridiculed with supermarket banners and a permanent Grand Opening flag on the portico. It is painted hideously and cheaply. The grounds are gutted and littered with trash and papers. When I woke up on my first morning here I said I must have died and this must be heaven. Today the kitchen table exploded beneath a dance and we destroyed the long benches just out of formal necessity. Most of the furniture has been used in the fireplace. Now there is nowhere to eat in the kitchen, and meals are creeping outward over the three floors of the house. Dishes age and cake beside every bed and chair and along the walls. Cigarettes decomposing in glasses of milk. Whole organisms of reading material —newspapers, comics, books—growing as if from inner life in the living room. The smell of menacing bad food. This is a settlement of transience. A boom town. We are stripping all the value off its ribs.

December 5.

We have lain in wait for days now and tomorrow we will go someplace—maybe Heathcote, maybe Washington—to find and join a commune there. The news is about the My Lai Massacre; and they have killed another Black

Panther leader, this time in Chicago, and people come through all the time terrified. A guy named Albert said the Baltimore police stopped his car last night, stuck a rifle in the window and told him to get out. They searched him and his car, roughed him up a little, and left. He doesn't know where to go or what to do. He thinks he is going to die soon. He is small and nervous with yellow skin and long white hair. He talks about the pigs killing, jaywalkers, litterers, and vagrants.

In the Earl of Sandwich, like an emblem on our day, was a crazy girl. She was thin and blonde with gigantic ticking eyes and chopped hair in tangled shoots. No one in the place could look at her. She was filthy and deranged. Her clothes were torn up and she wore them askew, like a drunk. Soon we were all staring at her. On the table in front of her was a stack of newspapers; her hand and arm were inserted into the stack and she was watching her fingers as they came through on its far side. Grinning, querulous, her head tilted to watch her moving fingers, she presented herself to the five of us somehow as an important woman, a stately figure. I felt this way, and the other four did also. It is something I have noticed in myself and my friends that we respect insanity and, in a way, admire the insane. "What's the news today?" said Rateyes, to talk to her. "O the nose," she said, "yes I took a nose dive too. And so-o did you. That's why yours is broken and you have spoken." She lowered her head and looked at Rateyes from under the shelf of her brow, raising and lowering the brows pretending to be a judge on the bench. She was slightly contemptuous of us. "No," said Doug, "he said *news,*" though he knew immediately it was not the right thing to say, because it was dull and it suggested the girl had made some error. She was angry. Laughed harshly. "How *many* are we! How *many?*" she demanded. "Let's see. O my! O yes! We're the five dwarfs, ha ha." She was almost screaming, and everyone at the small tight tables in the Earl of Sandwich was watching her. We stared into our plates, not laughing, not looking. We got up to leave, because she had started to make shrill tittering noises that we felt showed her annoyance with us, and as we passed her Crazy said, "You're nuts." She screamed back at him across the restaurant, "That's the nut at the top of *your* penistree!" Then she followed us into the street, crying, her face the opposite mask of what it had been a moment before. "Please," she said, "be simple, honest, and straight. Please be simple, honest, and straight."

"We are," I said. I wanted her to stop crying.

"I'm honest, I *am* honest," she said, crying, harder than before, choking on her tears, as though we had accused her.

"We will be," said Doug, "don't worry about us."

We walked away from this crazy chick, gesturing with our hands as we went for her to calm down and be soothed, stay where she was. She was holding onto all these newspapers. We didn't know why. She followed us a little way, still crying, begging us to do things we didn't understand—not to tell something to somebody, not to go somewhere, not to do something again that we had never done. "She was on her own trip man," said Dennis afterwards with great respect, "her *own* fuckin' trip . . . *per*manently spaced." And that, the way she seemed incredibly and irrevocably stoned, was the thing we could not get over all afternoon.

Tonight we wanted to go find her again. Doug thought he should invite her to the house before they catch her. Rateyes said, "It would be amazing to ask that chick questions about this trip. Like what's going to happen? A head like that like . . . they can really tell you some amazing shit." We drove into the city, went up and down the streets downtown. The only car around. New snow coming down over the old shocked-out sludge. Couldn't find her.

Rateyes is getting a gun pretty soon. I try to stop him, but he is determined. "Look Elia, what do you do when some pig stops you in Alabama and there's nobody on the road and he pulls his piece out? All that pig has to say is you were resisting arrest. That's all he has to say and you're just dead. Just. Dead. You know it happens. Right? You have to be able to blow him away." Indeed, the stories Rateyes reports about the police and the citizens in the South and the West make me feel the need for protection almost as acutely as he does. It becomes more and more apparent that to attempt journalism in the United States right now is as dangerous as war reportage in Asia. Everyone is armed. Everyone is crazy. Everyone is afraid of dying and so many, so many can't wait to die.

Criminality is around here in Pikesville. Rateyes is adamant about his gun. When he was in San Francisco he was riding down the street in the Haight and he saw two Hell's Angels ripping off a hippie. They had the hippie at knife point and they wanted his jacket. It was the middle of the afternoon. Rateyes happened to have in the trunk of his car a shotgun which someone had given him to hold for a couple of days. He stopped the car and ran around back to get his gun, then he apppeared beside the open trunk, shotgun at hip level, a cigar in the corner of his mouth, and said, *"Move,* motherfuckers." According to Rateyes, the eyes of the thugs were open wide and white as baseballs and they ran down the street crouching and dodging in case this maniac should open fire. Rateyes chased them around the corner screaming, "Move! Move, fuckers!" He says there are so many people in this country armed with so many things the only way to assure peace among us is to arm everyone.

"Ask anyone, Elia. What the pigs are really scared of is long-haired freaks with guns. You got a piece and they got a piece and you get respect, you know?" Rateyes is the son of rich parents and there is in his family the instinct for controlling the situation, so Rateyes has probably considered this question at length. He asks me what I intend to do if six rednecks in a Buick stop me in a parking lot and try to squash me against a garage door. "Look, Elia, we're not playing around here. There are some pretty fucked up people roaming around, you know?"

December 6.

There is a growing struggle for dominance here between the dope dealers and the holyroller macrobiotic choir children. The latter group has slowly adopted nudity as one of its qualities in the never-ending search for freer freedom and you are in danger of walking into them anywhere, plopping around the hallways of the capacious shell we are in. They come into the bathroom when you're sitting on the toilet, sit themselves on the window sill and start a conversation. Sometimes they will throw open the shades and say, "We're *sun* people you know!" These people who are hung up on their diets are always a little draggy, I have noticed. There is a character named John, a nude Santa Claus who always wears a knit cap with a pompon on it and walks around the house farting, his fingers bunched and exploratory in the depths of a bowl of brown rice he carries with him wherever he goes. He has a benign smile that everyone hates. His ass is astoundingly shockingly immense. All his skin is orange. He has just moved in recently, after an unattended wedding in which he joined himself to a girl who has lived here for quite a while, Betty. Her last old man was a photographer and she's selling all his camera equipment to buy a loom. She does not look too good in the nude—so she has modified the canon and she wears a free-hanging shift in her wanderings. Doug says last night he was making love to some chick and in walks John with his cap and his bowl, taps Doug on the shoulder and asks if he can use the record player. Doug says sure. John spends a few minutes looking for a record—the *right* record, finally chooses a good one. While this is happening Doug and his girl are hiding beneath the covers watching him in amazement and dazzled awe. He leaves, the record plays and Doug resumes his balling. Fifteen minutes later the record stops and John comes back to turn it over. Benign smile by candlelight. . . .

Chapter 12

WORK AND LEISURE

ONE of the crucial characteristics of modern industrial society is the enormous complexity of the division of labor. Each of its vast array of highly distinctive types of work provides the individual with a distinctive form of social experience. Although a wide variety of work exists in pre-industrial society, sociological interest has been concentrated in the areas of industrial work. Industrial sociology in America began in earnest in the late 1920s when a number of important studies were made. One of the significant results of these studies were findings that showed how the formal organization of industry, as set up by management, was an escape of complicated interaction with the informal organization set up by the workers themselves. This insight was carried over from industry to other types of work, and incidentally served to modify the Weberian view of bureaucracy itself. Increasingly, sociologists looked at work not just from the point of view of management but also from that of management's "victims."

In an offshoot of the Chicago School of urban sociology, there developed a subdiscipline known as the sociology of occupations. At first its focus was primarily on marginal and less than respectable occupations. Subsequently, its emphasis on the way in which work influences the overall life and thought of people was also applied to more ordinary occupations. The problem of leisure in contemporary society has, of course, been brought to the fore by the shrinkage in the amount of time required on most jobs. This problem, too, has been a consequence of "rationalization," in this case of the increasing efficiency of various productive processes.

At the same time, the leisure part of the individual's life has become the major arena in which he seeks compensations for the frustrations of his work-life. As work continued to shrink both in the amount of time it consumed and in the meaning it could provide for the individual, there arose the increasing demand that leisure time should be filled with activities that would provide the deeper satisfactions and values required by the individual. Since this is an expectation not easily fulfilled, contemporary leisure has imposed problems and tensions of its own.

The first selection, W. F. Whyte's "Human Relations in Industry," discusses the implications of one of the first great experiments in industrial sociology, the so-called Hawthorne experiment in a Western Electric plant in the 1920s. The work experience of the merchant seaman and, by contrast, the advertising executive, are discussed in the selections from Mariam G. Sherar's *Shipping Out* and Ian Lewis's "In the Courts of Power: The Advertising Man."

In a modern industrial society the difficulty of most adults consists of a kind of "commuting" between the world of work and the world of private leisure. Sebastian de Grazia in "The Problems and Promise of Leisure" discusses the general problem of leisure time, and Orrin E. Klapp in a selection entitled "Fun," from his book *Collective Search for Identity,* discusses the way in which individuals seek their identities through leisure-time activities.

WILLIAM FOOTE WHYTE

HUMAN RELATIONS IN INDUSTRY

WHAT are the causes of conflict in industry? How may more harmonious relations be achieved? What are the satisfactions and dissatisfactions men find in industrial work? What factors lead men to increase their productive efforts? What factors lead them to hold back production?

These are some of the questions that students of human relations in industry have been investigating. The questions are not new. Men in industry and social philosophers outside of industry have discussed them for decades and even centuries. But the effort to examine these questions scientifically is new.

THE WESTERN ELECTRIC RESEARCH PROGRAM

Human Relations in Industry as a field of scientific inquiry is little more than twenty years old. The selection of any date for its beginning would be arbitrary, but we may call upon the twelfth experimental period in the Western Electric Research Program as a convenient starting point.

Harvard University's Graduate School of Business Administration under the leadership of Elton Mayo and F. J. R. Roethlisberger had been collaborating with executives of Western Electric's Hawthorne plant in Cicero, Illinois. They were investigating the factors that led workers to produce more or less. First attention was given to some of the physical conditions of work. The first experiment showed that there was no correlation between variations in lighting and productivity.

This led the researchers on to a new experiment designed to measure the effects of rest periods and refreshments upon productivity. Six girls of average previous performance were selected for the test room experiment. The purpose of the experiment was explained to them. They were asked to work at a normal pace and not to speed up or hold back. They were then segregated from the rest of the work force in a room especially provided for them. During the months they worked in the test room, they were under constant observation by a member of the research staff with whom they became quite friendly. Their task was to assemble telephone relays. Each completed relay was dropped into a box where it was automatically recorded so that the experimenters had a record of production day by day, hour by hour.

After providing for a period in which the girls were to get used to their new surroundings without the introduction of any changes in the way of rest

Reprinted from *The Delphian Quarterly* 39 (Spring 1956): 1–5, 40, by permission of the author.

periods, refreshments or working hours, the experimental changes were begun. Each change was discussed with the girls first before it was put into effect. Each change remained in effect for enough weeks so that its effects could presumably be measured.

In a summary article of this nature, there is no need to detail the exact sequence of these experimental changes. It is enough to report that rest periods were introduced, that refreshments were provided at certain periods of the day, and that the length of the working day was also shortened at one point. Various combinations of changes in these three areas were also tried out.

Through the first eleven periods of this experimental program, the productivity of the test room girls rose steadily. A student less cautious and scientifically minded than the Harvard researchers might have concluded from this experience simply that the introduction of rest periods and refreshments leads to higher productivity, even though the results gave no clue as to which combination of rest periods and refreshments seemed to be most effective. It was precisely at this point that Elton Mayo introduced the crucial test of the effect of rest periods and refreshments. The twelfth experimental period marked a return to working conditions prevailing at the beginning of the experiment—no rest periods, no refreshments and the customary eight hour working day. Productivity in this twelfth period surpassed all pre-existing records.

Some might have been inclined to toss aside the results of the twelfth period as an irrelevant fluke and conclude that after all in general these improvements in working conditions did increase productivity. But Mayo and his co-workers insisted that the results of the twelfth period demanded a revolution in thinking regarding the factors influencing productivity. If productivity rose to a new high upon withdrawal of these worker benefits (rest periods and refreshments), then it could no longer be seriously argued that that remarkable rise in productivity of the test room girls was due to these working conditions.

In abandoning the working conditions theory, the researchers were forced to cast about for new explanations. It was in that search that they took the decisive steps in founding this field of human relations in industry.

The experiment had been set up in the best scientific tradition so that certain working conditions would be varied while every other important factor was held constant. Now the twelfth period proved that the changes in rest periods and refreshments did not explain the changes in productivity. It therefore followed that there must be other factors in the experiment which had not been held constant. The researchers asked themselves what other significant changes had entered into the world of the test room girls.

As they came to look upon the test room as a social situation, some of the answers became evident.

Let us contrast the situation faced by the six test room girls with that faced by other girls on similar jobs in the regular factory departments. Obviously these six girls were singled out for special attention. They were not selected because they were outstanding workers, but nevertheless they were selected. They were not asked to produce an extraordinary amount, but they were informed that they were playing a part in an important experimental program. The evidence of the test room observer shows that they took considerable pride in their part in the experiment.

Furthermore, the girls were removed from the regular channels of factory supervision. In the factory departments, their work had been laid out for them, they had been told what to do and had been closely supervised. No one had consulted them about the conditions of their work. Now, while they were under the constant observation of one of the research team, they were almost completely cut off from the regular channels of supervision. They were consulted about each stage of the experiment. The consultation was more than an empty form as evidenced by the fact that at one point the girls vetoed a research proposal and did not have it imposed upon them. Finally, they were working in the company of an observer who took a friendly personal interest in them.

This picture suggested that while the experiment had focused upon changes in certain working conditions, actually the most important changes introduced were in the field of human relations. The test room girls experienced a marked change in their relations with management and with each other. They built up a tightly knit social group with very considerable pride in their membership in it.

This conclusion applies to the field of productivity but seems to have a much more general relevance. It can be stated in this way. In order to change the activities and attitudes of people, change the relations among them.

This aspect of the Western Electric program has been attacked by some scholars as simply being a discovery of the obvious. The researchers discovered that industry is a society in itself, that individuals are not solely concerned with money or wtih the physical conditions of work. They are also strongly influenced by the relations that grow up among them. In a sense, the criticism has weight but it fails to recognize the state of knowledge regarding human behavior in industry that existed at the time of the experimental program. At that time, the sociologists and social anthropologists who were busily studying primitive and modern industrial communities had not ventured inside the gates of the factory. A few psychologists were active in industry, but most of them were engaged in developing aptitude and other psychological tests regarding the relation between a man's abilities and the physical and mental work required in different jobs. While it might have occurred to the sociologist if you had asked him that a factory could be looked upon as a social system or community, he did nothing about such an idea until the way had been laid open for him by the Hawthorne experimental program.

Furthermore, at the time of the Western Electric program, popular thinking about human behavior in industry was dominated by certain ideas developed by economists and engineers. The prevailing notion was that man—at least in the factory setting—was an economically motivated individual. He responded as an individual to the financial rewards offered him or to the threats of the withdrawal of such rewards. In such thinking, groups did not exist. Each man responded rationally to his calculations of profit and loss.

The Hawthorne experiments exploded these ideas. The studies of the test room and other studies growing out of it demonstrated that it is futile to think and act in terms of these individualistic economic assumptions. Men live in a society when they are in the plant just as when they are in their communities. If we are to understand their behavior in industry, we must therefore study the relations among the men and women who work together in this industrial society. It was that conclusion that opened up for study the field of human relations in industry and led to a rapidly expanding popular interest in it.

THE HARWOOD EXPERIMENTS

If the Hawthorne experiment showed that changes in human relations were accompanied by changes in productivity, the next logical step would be to set up an experiment in which there were introduced changes in human relations with the deliberately planned objective of affecting productivity by this means. In the sense that the productivity changes were accidentally introduced in the Hawthorne program, it remained now to determine whether social scientists could affect productivity through the deliberate planning and execution of human relations changes. This next step was taken by researchers in the Harwood Manufacturing Company under the leadership of Kurt Lewin, a social psychologist.

The Western Electric study of the Bank Wiring Room had revealed and documented in detail a phenomenon well known to experienced factory management people: restriction of output. It was found that workers did not ordinarily produce up to the limits of their physical and mental capacities even when they were paid on incentive rates, which meant more money for more production. Instead of going all out individually to produce as much as they can, workers seem to get together in informal groups to decide how much constitutes a "fair day's work." They then produce up to that level and not beyond it. When asked why they do not go beyond the agreed-upon point even when they acknowledge that in many cases it would be readily possible, they explain the restriction in terms of fears that management would cut the rates, paying less per unit produced if they stepped up productivity or that they would work themselves out of a job.

This rate cutting argument is used even when no one can cite a specific example of a time when this particular management cut an incentive rate. Of course, there have been cases in which certain managements have cut rates and stories of what has happened in other plants may reach the workers in the plant in question.

Why do workers restrict output? The reasons would vary from case to case, but certain general answers can be given. Workers on the bottom of the chain of command are subject traditionally to the orders and to the changes introduced by supervisors, engineers, time and motion study men and other management officials. Traditional operating procedures of industry call for treating men as isolated individuals and make no provision for the groupings among them that inevitably arise. It is in reaction to this subordinate position, to this constant necessity of responding to changes introduced from above, that men band together informally to establish their own standards of approved and disapproved conduct. Like people in communities everywhere, they build up a culture or way of life. A good part of this culture arises either in opposition to management or else to interpose a cushioning effect upon management's activities. Where conflict between workers and management exists, these standards of what constitutes a "fair day's work" are likely to be most rigidly adhered to and informally enforced by the workers upon each other.

If this explanation of restriction of output is correct, it would then follow that a change in the relations between workers and management would result in changes in worker attitudes toward productivity and would also result in changes in the productive activity itself.

The experiments to be reported here were carried out by Alex Bavelas, John R. P. French, Jr. and Lester Coch with groups of sewing machines operators in the Harwood Manufacturing Company plant. These women operated individual machines and were paid on a piece-rate basis. That is, there was a guaranteed minimum hourly wage, but the employees were paid so much per piece and regularly made more per hour than the guaranteed minimum. However, in this case, as in others, there seemed to be a group standard that they would not produce beyond a certain amount, in spite of all management guarantees that piece rates would not be changed.

The first experiment carried out by Bavelas was designed to test the effect of group discussion methods upon productivity. Bavelas met with several groups of workers to discuss with them the problems they found in their jobs, their attitudes toward productivity, and the possibilities that they might want to produce more. The discussions were designed to provide a maximum opportunity for the women to express their feelings about the job and come to their own conclusions. Under Bavelas' discussion leadership, several of the groups did decide to raise their level of productivity, and determine upon new goals. Other groups decided against such an increase, and no effort was made to push them toward a decision to produce more. After the discussions, it was found that the groups that reached a decision to produce more did achieve substantially higher productivity whereas the groups where no such decision was reached remained at approximately the pre-existing levels. Thus the group process had very marked effects upon the activities of these people even though each one operated a machine separate from all the others.

Unfortunately for this experiment, we have only the barest outline of what was done and results achieved. We know very little of the process of discussion carried on under Bavelas' leadership, nor do we know why certain groups decided to increase productivity whereas others made the contrary decision. Despite these limitations, it is evident that changes in human relations led directly to these changes in work activity.

Another Harwood experiment involved the study of the effects of worker participation upon adjustment to changes in work methods. French and Coch had found that when changes in methods of work were introduced by management in the customary manner, one could expect a sharp drop in worker productivity and a concomitant increase in expressions of worker dissatisfaction. In some cases, the workers in time reached higher levels of productivity, but in other cases even when the new methods appeared to be more efficient from an engineering standpoint the workers did not even attain their previous levels of output. Since American industry is highly dynamic both in technology and in work processes, it is important to study the reaction of workers to such changes.

French and Coch began their study with three matched groups of employees who were introduced to similar changes in job method in three different ways. The first or "non-participating," was merely ordered to make the change and given instructions as to the new method. The second group participated through election of representatives who met with management officials to discuss the new methods and to help plan their introduction with management. The third group participated on a total basis with all members of the group meeting with the management people to discuss and plan the changes.

Results of these three different methods of introducing work changes were sharply contrasting. The group which had had changes imposed on it in the customary manner, dropped off sharply in its productivity at the time of the introduction of the change and its productivity remained at least fifteen percent below its pre-existing records for thirty days following the change. The group which had chosen representatives to participate in planning the change showed the sharpest drop of all at the time of the occurrence of the job change. Productivity dropped off about 35 percent on the first day but rose almost steadily each day thereafter. It reached pre-existing levels on approximately the twelfth day following the change and reached the level of approximately 10 percent higher than pre-existing averages by the end of the thirtieth day following the change. The group that had participated on a total basis experienced only a slight drop in productivity on the day of the job change and rose above the pre-existing average on the day immediately after the change. From then on, its productivity fluctuated considerably but showed a steady upward trend so that the workers in this group were more than 15 percent higher than previous averages well before the end of the experimental period.

The change in job methods had other marked effects. Expressions of dissatisfaction with management and with the job change were greatest in the non-participating group of workers, and non-existent in the group which was involved in total participation. The level of morale, then, seemed to be markedly affected by the social process whereby changes were introduced.

MARIAM G. SHERAR

THE SHIP

Every young boy who becomes a sailor does so because he wants to see the world; his first shore-leave will bring home to him that the ship has become his world, and that he'll never see foreign lands, strange peoples and ancient cultures without the commentary of the bosun or chief engineer. Java and Tahiti, New York and Reyjkavik—all they will ever be are backdrops to a bulky friend.[1]

All ship assignments are handled through the Union hiring halls. The Union hiring hall, therefore, has become the lifeline for the seaman.

Shipping Companies affiliated with a Union notify the Union of vacancies as they fall open among unlicensed personnel and vacancies are then posted on the Board, also accompanied by such information as name of ship, shipping company, type of vessel, and destination. Thus the seaman, at a glance, establishes what vacancies are available for which he may qualify.

Job preferences are based upon seniority and other requirements. For the N.M.U. there are four seniority ratings, known as Group 1, Group 2, Group 3 and Group 4.

For the S.I.U. seniority is based upon three ratings: Class A, Class B, and Class C. Thus men with highest seniority are given preference for jobs over men with lower seniorities. The seaman may reject the first job offer without losing his place on the rotary hiring list, but is expected to accept the second job offer.

It was 8:30 A.M. when Comanche reached the hiring hall, and the first job call was for 9:00 A.M. Thereafter, until 4:00 P.M. in the afternoon there would be a job call every hour, changing as men accepted jobs or as new openings were placed on the Board.

After years of seafaring a man becomes familiar with ships, shipping companies, even the old-timer captains who command certain vessels. Ships be-

Reprinted from *Shipping Out* by Mariam G. Sherar (Cambridge, Maryland: Cornell Maritime Press, 1973), pp. 7–13, by permission of the publisher and the author.

come known as "good feeders" (where the food served is good with good stewards, etc.) or "old rust-buckets" (ships in poor physical condition). Some men prefer deep-sea or long trips, others coastwise or intercoastal trips. In an issue of the *Seafarer's Log,* the newspaper of the S.I.U., seamen were asked: "Do you prefer long-distance, deep-sea trips, or short coastwise ones, and why?" [2] Of six seamen asked, five preferred deep-sea and one coastwise. Reasons given were:

1. Chance to understand and appreciate other cultures.
2. Time to do a great deal of thinking.
3. Can really get away from everything.
4. You make more money.
5. Better companionship in foreign ports.

The one man preferring coastwise trips, lists the following as his reasons:

1. No language problem.
2. You know what to expect in American ports.
3. Have a bunch of guys who are pretty much alike and most of them are stable fellows.
4. Can always get fresh milk.

Whatever the preference, sooner or later the particular job a seaman desires is bound to turn up on the board.

Comanche, who like many has seniority and is a member of the Black Gang (works in the engine department), prefers deep-sea trips, and was lucky that the type of job he preferred was listed on the board. By 10:00 A.M., job assignment in hand, he was ready to report for his new berth, a six week trip to Europe and back on a ship recently out of dry dock after her annual inspection and refitting.

Ashore employment is unlike that of the sea, for the ship, any ship, is unique. Ashore men work and live and recreate in separate establishments. They usually commute to a place of work, or travel to their favorite recreational area. But men afloat depend entirely on the ship as a place of business, temporary home, and for whatever recreational facilities there may be. Not only that, but the ship is afloat upon the threatening sea, and under the steel-plates of a ship's hull are many fathoms of water. The ship, therefore, is a place of business, a home and provider of recreation.

> For some reason a ship, like a tired jacket, becomes comfortable only when fondly referred to as 'old'.[3]

Old ships give the seaman a tremendous amount of security. As might be imagined, this sense of security is most needed at sea. An old ship means: one that has weathered many storms; that has passed through hurricanes; that has been everywhere and done everything. She is experienced if she has successfully passed through all these trials and she will do it again. Or at least, that is the sailor's belief. "Treat the old-girl good," many a generation of seamen has said, and so, over the years, a mysterious relationship has developed between seamen and their ships.

Joseph Conrad, in his book, *Mirror of the Sea,* commenting on ships, writes as follows: [4]

> There are good ships and bad ships, comfortable ships and ships, when, from the first day to last of voyage, there is no rest for the chief mate's body and soul. And ships are what men make them, this is a pronouncement of sailor wisdom, and no doubt, in the main it is true.

Yes, there ships which obey the man, and those that do not; some that constantly leak and feel damp and cold, many that do not. For the space of a roundtrip voyage, the seaman is dependent upon the ship for his comfort and his safety. Perhaps that is why seamen throughout the centuries have given ships a feminine nature. She is their "sea mother."

Again, in *Mirror of the Sea,* Conrad describes the relationship in the following way: [5]

> Yes, your ship wants to be humoured with knowledge. You must treat, with an understanding consideration, the mysteries of her feminine nature, and then she will stand by you faithfully in the unceasing struggle with forces wherein defeat is no shame. It is a serious relation, that in which a man stands to his ship. She has her rights as though she could breathe and speak; and indeed, there are ships that, for the right man, will do anything but speak, as the saying goes.

So the seaman, preparing for departure to sea, is always aware, consciously or not, that each trip on a ship has lurking uncertainty in it. There is much more security ashore. Yet, after each sojourn ashore the seaman must prepare himself once more for his sea voyage and the uncertainties of the venture.

A favorite expression of Comanche's is "I am on a ship," which, literally translated, means: my life is uncertain; I am on the ship, and what happens to me is at the whim of the shipping companies' decisions, the nature of weather conditions and the ship herself. Until I see you again, *I am on a ship.*

> Departures, except when the ship is running on a regular ferry service, are always depressing. Everyone is full of the worries or the joys of his shore-life; women weep, children are a nuisance, the lavatories are full of dirty water and some sailors are always missing, up to the last minute when they are carried on board blind drunk by the man-trader. The only ones who enjoy all this are the members of the crew who have no home and who have been eagerly waiting for the reunion.[6]

So said Jan de Hartog in *A Sailor's Life* and no one could have expressed it better than he.

Although Comanche had been aware that it was time to ship out, so, too, was he depressed at the thought of shipping out. Conrad once stated [7] that no sailor is ever really good-tempered during the first few days of a voyage. As he puts it, there are too many regrets, memories, an "instinctive longing for the idleness of life ashore" and the resentment of the work ahead. Perhaps, even deeper than this lies the realization that one is voluntarily withdrawing oneself from the major social world that exists primarily ashore. Radio, radar, improved communication helps to cut the isolation, but there is no getting away from the fact that once the ship leaves port, one must put aside all thoughts of life ashore and adjust to the reality of life on a ship. "The ship is like a prison," Comanche once said, "a prison without bars."

One cannot walk down to the corner bar for a drink, or walk in the woods, or shop at the local stores. Physical space is limited from stem to stern and from port to starboard beam. Social life is restricted to fellow shipmates, and these you must take or leave: one's cabin mate, one's watchpartner, the men congregating in the messroom, these are your companions. There is no wife to be with for weeks on end, no children to play with, no girl friend to visit. For weeks the ship will dominate the sea domain. Until the first landfall life becomes the routine of watches—4 hours on, 8 hours off.

"He who loves the sea, loves also the ship's routine," [8] said Conrad in *Mirror of the Sea,* and he was right. Not only must one's life become routinized to sea watches, but it must also adjust itself to the man in command and his ways and preferences for getting things done.

> In practice, every captain looks like an old fool and never is. His very presence determines the nature of the community that sails the ship.[9]

Alvin Moscow, in his book *Collision Course,* corroborates this by saying: [10]

> 'Every ship has its own splice' is a saying familiar in many forms to men of the sea. It means, in effect, that each ship is a reflection of the 'old man' or each ship is operated differently, according to the working philosophy and habits of the Captain.

The captain, in effect, must be likened to a leader. Captains, of course, are officers, and are apt to remain on a particular ship for years. Their duties and obligations are to the shipping company. As personalities, they may be liked or disliked. But, unlike the old days, a captain generally unliked may have difficulty getting a crew. Rumor has it that one ship was held up in a port several days finding it difficult to get a crew because of the particular captain involved. Men simply would not sign on, and there are no crimps or shanghaiing today.

Captains may be liked because of their proficiency or they may be liked because of their charismatic qualities. Some may be liked because they are young or humane, or considered "real gentlemen—well educated." For whatever reason, while afloat, his personality affects the ship's routine.

And what is ship's routine? Four hours on, eight hours off; four hours on, eight hours off. That's ship's routine.

The four hours on are easy. Some men prefer to work from 12–4, others from 4–8, and still others from 8–12. As each man signs on the ship, he designates his preference, and is so given if possible to do so. These, of course, are sea watches. In port, if a vessel remains for 24 hours, port watches are set, which are eight hours on, sixteen off.

Meals are served as follows:

Breakfast	7:30 A.M. to 8:30 A.M.
Dinner	11:30 A.M. to 12:30 P.M.
Supper	5:00 P.M. to 6:00 PM.

Men are also allowed breaks for coffee in the morning and afternoon and midnight snacks.

At sea, the eight hours off spaced by the four-hour watch, and then another eight hours off, constitutes a man's life. It is in the hours off that all else must be done—sleep, personal chores, recreation, relaxation, housekeeping etc.

We ashore who live a routine 9–5 existence know of the monotony of the 5-day work week. Yet we have the option of taking a day off, of calling in sick or taking personal leave, if need be, and escaping from the physical work week if we feel the psychological need to do so. We have the freedom of escaping from our physical boundaries of home or office and "going out to dinner" or "taking in a show." Or, to really cut loose we can relieve the monotony by taking a "night-on-the-town." We can, but not so a seaman on a ship.

Work life aboard a ship is probably one of the most perfect examples of Durkheim's "Theory of Organic Solidarity." Interdependence is essential for the running of any ship, and it is a seven-day job (with overtime, of course, during weekends and holidays). The ship continues to run, weekends, holidays, as well as daily. As long as engines are running, men to be fed, weather and navigation problems to be encountered, these must be men on hand. So seafaring is a seven-day job. Also, during those 8 hours off, there are limited places to go. The decks; the messroom; one's foc's'l. The messroom is the social club of the ship. Here there are television and radio, the ship's library (seamen are avid readers and all ships carry a library whose books are rotated every so often). Here the men congregate to chat, play cards, drink coffee, write letters. But messrooms are small, and space is naturally confining on any freighter.

"Ah, but there is the sea," you say. Passengers who travel by ship are ever involved with the mysteries of the sea. But passengers are not seamen. Conrad once wrote, "Does the passenger ever feel the life of the ship in which he is being carried?" [11]

I wondered about that question when I started writing this book. True I had been on ships before (passenger freighters as well as ocean liners), but I had always been on these ships as a passenger. In truth, I could not remember any sensation but that of the "passenger." So, having nothing more seaworthy than a Staten Island Ferry available, I nonetheless set off to sea across New York Bay to St. George, Staten Island. This time I tried to imagine myself, not as a passenger, but as a crew member might experience the trip.

Passengers are a different breed! They were sightseers on a ferry boat, true, but oblivious to the fact that even ferry boats have engines, radar, can get lost in fog and have to battle tides and rough seas or storms from time to time. As I stood there I tried to cut out the passenger feelings and picture life on a ferry as the crew members must. The throb of the engine becomes more than a machine. It is the beat of a ship's heart, and I am sure a ferry boat crew member must react constantly to that vibration. Rust must still be chipped away; sea traffic is heavy and rules of the road must be obeyed. In short, as the focus shifts, one enters a new dimension of time and space—as one shifts status from passenger to crew, so one must shift role, even though the physical body of the ship is the same.

So, too, afloat, the sea and recreating on it means two different things to seamen and passengers. Of this Hartog writes: [12]

> You may have been at sea for years without ever really looking at it. Many a landsman would be surprised to know that, after one cruise of a fortnight, he has a better picture in his mind of the sea than many a sailor, for he has been a sailor, and a sailor on the bridge, or on the lookout is on duty. When a sailor scans the sea, he is looking for something or at something, rarely at the sea itself.

Thus, to the seaman, days at sea become monotonous, and men look forward to landfalls with great anticipation. Landfall means a shift of the monotony. Routines will shift. Life will vary and all will change by a sojourn ashore.

Perhaps the most poignant description of landfall has been written in a novel by Wolfe Reese [13] who writes:

> There is something about sighting land after the long watches at sea that is like waking after a deep sleep. These days and nights of plodding across the water toward its ever receding rim are at an end, for there, at the edge of the sea is a segment of green thread. As the thread thrusts up into clearer view and takes new shape, there is a renewed sense of reality. In facing it, each one is affected according to his nature. A loquacious man will fall silent, withdrawn into himself to contemplate no nobler vision, perhaps, than of women and liquor, but a private vision nevertheless. Another, not given to levity will smile and joke and give unasked for advice about the ways of a port to anyone who will listen. A tight ship becomes a happy one, new alliances are formed for the sally ashore, old debts remembered, old worries revived. It is a sea change that happens with each new landfall over and over again; a small and perhaps insignificant miracle, but a never ending one.

Perhaps the most important significance of a landfall is that it breaks the routine at sea. It releases a man from the confining barriers of the ship, and provides him with the chance for change. Yet, too frequently, the change means nothing more than a trek to the nearest bar, usually less than a block away, where tensions are released amid the raucous voice of juke boxes and the boisterous shouts of fellow sailors and barmaids.

Change from routine, any kind of change, requires a difficult adjustment. Shipboard life also tends to become monotonous, and this monotony leads to the building up of tensions among seamen. Both the difficulty of adjusting to change, and the tensions resulting from shipboard monotony, require outlets, and nearing a port, any port, creates an excitement of anticipation and release.

In a way the ship might be likened to a *total environment,* that is, an environment which is complete in and of itself, such as prisons, boarding schools, mental institutions. The adjustment to living in such complete universes excludes being part of or relating to outside influences and because of this readjustment to life outside of one's total environment takes quite a while.

Regardless of whether the ship is small or a large liner, expectations build up as soon as the possibility of going ashore develops. In a letter written from the English Channel, Comanche once wrote as follows:

> Everyone has a thing called Channel Fever and are nervous, confused, can't stop talking and sit around the tables nervously twitching their feet, all talking of nothing exactly but all talking simultaneously nevertheless. I can't take any more of this—I am leaving the messroom now.

The next morning, with a landfall in sight, he wrote,

> I am up this morning—with Channel Fever.

Today, with quick turnabout due to automation, ability to load and unload ships quickly and high cost of dock facilities, a seaman may have only a day or two, if that, in a port. He must still work his port watch of eight hours, he still must sleep, so the time left over for shore visiting is brief indeed. Ships also tend to dock in port towns, not the cities themselves and are miles away, in many cases, from the center of tourist attractions. Thus ships dock in Southampton—rarely in London, or Cherbourg, and not Paris, Bremerhaven, Bayonne, Port Newark, or in other *boon docks* here and there, for the most part, hours away from the center of town. So, of the seaman's precious few hours of shore leave, an hour or so of it must be spent in commuting.

In an article in *Lookout,* the magazine of the Seamen's Church Institute of New York, recently, Reverend G. Basil Hollas wrote: [14]

> The merchant seaman today spends very little time ashore; by far the greater part of his working life must be spent at sea. There he lives, within the confines of his ship—often within the confines of his own particular department in the ship. Thrown together by chance with a few other men of differing ages, interests and outlook, cut off from the amenities of normal life ashore.

So, the corner bar is more convenient, and familiar to the seaman. Then, too, the very exotic quality of world travel tends to pall after awhile. Comanche has traveled thirteen times around the world and upon all oceans. A jaunt to Europe, after awhile, means no more than a trip to the local candy store to buy a pack of cigarettes.

All experiences tend to break down into the personal, it is the only way that man can cope. As Reese puts it: [15]

> Much closer to the truth, I decided, was the fact that (he) and others had seen too much, too fast, too often, so that over the years, drugged by the overwhelming magnitude of it all, they were made numb by recollection. A great exotic port, a country, a whole people were often remembered, if remembered at all, by a treacherous dock side floozie, an over-priced bottle of liquor, or a leaky valve chest that caused a missed shore leave.

Landfalls break the monotony of a ship's routine, but unless they are home ports, their effect is temporary.

A ship, in spite of it all, is a seaman's security, especially abroad, and although ports are looked forward to with great anticipation, so too are departures. Because of the nature of his calling, changeability and restlessness seem to be characteristics of a man long at sea, so that although landfalls are eagerly looked forward to, so too there is relief to be clear of the land and again back at sea.

The ship, after all, is a seaman's security. It is his temporary home, and because this is so, all else is foreign—especially abroad. Thus, when all money has been spent, when the last drop of liquor has been consumed, when the last puzzling exchange of currency has been made, the routine of shipboard becomes extremely desirable again. When all has been said and done, the ship remains, solid, faithful, and enduring.

> To be back on board my ship, to feel her deck underfoot, her rail under my hand, seemed to lift a load off me. Sickness, loneliness, my cold, even the rain seemed to vanish the moment I was back in the comforting, exhilarating freedom of my own world.[16]

That, to me, summarizes the relationship of a man and his ship. The ship is his world, his total environment, and despite its confining limitations of space it is his life, and therefore, the routinized existence is not binding at all, but *free*. And so man enters into a mystical relationship between man and matter, personalizing a steel structure, linking himself to it by a bond of loyalty and devotion, and in return receiving from it a mother's heart of steel.

IAN LEWIS

SKILL REQUIREMENTS FOR AGENCY PERSONNEL

ARTISTS, copywriters, designers, and other creative personnel to be successful in agency work need to be talented, trained in the exercise of their skill and, if possible, creative. Talent and training are attributes that can be judged (if not measured) by qualified experts, if these are available. Creativity in advertising can be judged only under pressure, if it can be judged at all.

A talented copywriter, without ideas, can go far if he is able to develop the necessary appearances of creativity. Thus when the "Be sociable, have a Pepsi" idea made Pepsi-Cola a major competitor to Coca-Cola, hundreds of brands in all product classes launched advertising that pictured the consumers of these brands as young, modern, carefree, sophisticated, fun-loving, sociable, prosperous, upper-middle-class suburbanites. This image became in a short time a cliché but a highly successful cliché—successful in raising the unspoken wishes of millions of Americans into a self-conscious "model for the millions." The cliché had such wide distribution and penetration that those copywriters and agencies who indiscriminately copied it were not able to establish any distinguishing characteristics for the brands they advertised.

To copy early in the copying cycle may be profitable; to copy late is suicidal. To copy a campaign that is so old that everybody has forgotten it may be a stroke of near genius. To create a genuine concept, a totally new idea that works, requires as much creativity in advertising as it does in other fields of art, letters, and sciences. Unfortunately, the collective character of the creative enterprise in advertising, and the speed at which both good and bad ideas are copied, make it almost impossible to recognize the original creator.

It is relatively easy to judge the skills of technicians in research, in film editing, in audio departments, in camera work, because there are established techniques in these fields. But even in technical fields the mechanical application of technique results in dull, plodding work. Insight, imagination, the ability to apply technique to the solution of a "business" or aesthetic problem require, even in advertising, the same kinds of artistry that are required of creative personnel. And in these aspects of the work, the criteria for judging a man's work become just as subjective as those used for judging the creative artist.

Reprinted from "In the Courts of Power: The Advertising Man" by Ian Lewis in *The Human Shape of Work*, ed. Peter L. Berger (New York: Macmillan, 1963), pp. 141–47, by permission of the publisher.

Media planners, time and space buyers, and buyers of television programs usually develop a great deal of detailed knowledge of their respective fields merely by working in them. No special training is required of the neophyte to enter the field. But once he is in such a field, special talents for detail work, administration, and ability to exercise "good business judgment" are qualities that bring a good man to the fore.

These nontechnical, business and administrative requirements resemble those appropriate to the account. The personnel account executive and supervisor are required to be "good businessmen," good administrators, and to have "good business judgment." In addition, the account administrators are required to be tactful, likable, charming, and to have, to a very high degree, all the qualities of successful salesmen. The similarity in the personal qualities required of the media buyer and the account man frequently results in the promotion of the time and space buyer to account executive, thence to account supervisor, and finally to agency president.

This career pattern is not fixed. Creative personnel, research men, media men, and a wide variety of other occupational types can become account executives; and all types can become agency presidents. But if a specialist is to become an account supervisor or an agency president, he must first exhibit the qualities of the general administrator, businessman, and man of judgment. He has the opportunity to develop these qualities as a department head and administrator, as a member of the board of directors, or in making an impression on important representatives of the client in agency-client meetings or at lunch and after-hours meetings.

Yet a description of the above qualities does not come near to setting forth all the qualities necessary to success in occupational mobility. As indicated above, talent and technical knowledge are desirable qualities for creative and technical personnel. But possession of such qualities results only in the acquisition of rank and position as a staff official. Administrative skill, judgment, and business sense are the terms used to describe the special talents of the nonspecialized businessmen of the agency, who include media and program buyers, account executives and supervisors, and general agency officials. Such lists of qualifications, however, represent only the *public* side of an occupational description. Other, more personal qualities are necessary to complete the description.

Nerve is a central quality that any agency man must possess in order to survive the pressures. Since much of agency work is done under the constant pressure of volume, deadlines, possible criticism, and the ever-present image of total failure, men who cannot "stand the grind" are quickly recognized. Those who can are selected as "comers" and are pushed along as long as they can keep the respect of those who do the pushing.

Nerve means more than the failure to crack under pressure. It means the capacity to exhibit, regardless of the pressures placed on one, calmness, tact, proper deference, good humor, and loyalty to the right people. But these latter qualities are independent of nerve. The ability to exhibit calmness, tact, deference, good humor, and loyalty to the right people is considered to be a basic personality requirement for any account officer, agency officer, or any creative or technical specialist who wants to move upward. For these qualities are client-pleasing qualities. They are also boss-pleasing qualities. Being in a business where one's very existence depends on the "favors" bestowed on one by the client gives advertising a court-like atmosphere. The chief officer of the client is the "king," and the agency personnel are only one set of his courtiers. But the courtiers on each level by virtue of their acceptance by the throne and by individuals who have access to the throne, receive deference in relationship to their imputed proximity to the throne. Ability to be successful as a courtier becomes objectified, is treated as a psychological trait in and of itself, and becomes the basis for further success in courtsmanship.

The quality of likability (other-orientedness with a purpose) is subject to limitations. In a business where costs are a factor, where pressures are greater than one can absorb by oneself, the ability to resist pressures which are destructive to the individual is a necessary trait for success and for survival. For instance, a client may make demands that are too expensive, too time-consuming, or are impossible of fulfillment. The account executive in this case must either talk the client into modifying his demands or convince him that his demands have been fulfilled when they have not. At times he may be able to convince the client that the demands are totally unreasonable. But whatever he does, the account executive must do so in a manner that gains the respect or the liking of the client. If the account executive gives in to such demands, he may embarrass the agency and its profit structure, overload the technical, administrative, and creative staffs who have to meet the demands, and find himself in trouble with his own office. Moreover, if the account executive indicates to the client that he can be pushed around, he invites the resentful or sadistic client to do just that. Finally, some clients expect top agency personnel to have convictions, policies, and beliefs of their own. They are paid to have these attributes. The account executive who is so likable that he accedes to whatever the client wishes is simply not doing his job. He is not providing the client with the counsel which is part of the service of the agency. If he does disagree, however, he must be tactful and, above all, he must know when to stop disagreeing. This *realistic toughness* is the third major quality necessary for success in an agency.

The three major personality traits so far discussed become the basis for types of agency personnel. The types are as follows:

Type 1. The Creative Genius. This is the official who creates or attempts to create the impression that his technical knowledge or creative ability is so great that he can ignore the other realities of the agency business. If he succeeds (by being almost as good as he claims to be), he can go far toward a top staff position. He is usually viewed as irresponsible in positions that require "judgment" or administrative ability.

Type 2. The Likable Chap. He works hard and is continually on the run.

He is pleasing, amiable, entertaining, and quite frequently provides more services than the client or account supervisor asks for. He is perfectly eager and willing to handle all details for his superiors, or to get someone else to do so. But he is at a total loss when two or more superiors disagree, or when he is asked to have an opinion before an official opinion exists. He lacks nerve. Such a person is not likely to reach a position higher than account executive because he finds it difficult to work without lines and because he is likely to give away his own, his subordinates' and, worse, the agency's shirt in the desire to please. He needs to work under Type 3.

Type 3. The Tough Realist. He may not have creative or technical ability but knows how to please when that quality is necessary. He knows when to get tough with his subordinates, with himself ("discipline"), and with the client. He knows in any given situation what his self-interest is, what the client's interests are, and what the agency's, account group's, or service department's interests are. No matter what the social situation is, he knows how to juggle these interests so that he, in the final analysis, will come out on top. Of course, such results do not always occur. When two or more tough realists come into open conflict, one may be forced to go (even if this means his occupational demise). One can make an error in judgment, such as cultivating and supporting an important official in the client firm, only to discover later that this "personal" client has been squeezed out. The enemies of the personal client may now be running the client firm, and the agency man is *persona non grata.* If this happens, the tough realist may be out of a job. A tough realist may spend a decade cultivating a particular individual in the client firm, nursing him up the ladder of success in his own firm. Shortly after the client "arrives," he dies. The tough realist then loses his "contact" and may have no other assets of value when he faces his own agency.

The three types can be conceived of as polar types. Most personnel in an agency, however, exhibit combinations of the traits of the three polar types. The creative genius can also be a likable chap or a tough realist (if he is willing to suppress his needs for personal recognition) but he cannot be both. The likable chap is likely to be a pure type. However, a likable account executive, after achieving that position, may suddenly begin to sound like a tough realist. If he can maintain that attitude under pressure or adversity, he may grow up to be a genuine tough realist. The tough realist may have had, in the earlier years of his career, creative or technical ability which, by and large, he is not able to exercise in the present because of the pressures of other work. He can always be likable when necessary. But the quality that accounts for his success is his realistic toughness.

SEBASTIAN DE GRAZIA

THE PROBLEMS AND PROMISE OF LEISURE

INTRODUCTION

Over the past fifty or one hundred years the "idleness problem" has been supplanted by the "leisure problem." Though people once excoriated idleness and today laud leisure, the problem is one and the same—the problem of free time. Moreover the problem is not the existence of too little free time but the threat of too much. Medically speaking, in Karl Marx's day a problem of too little free time did exist. Today no labor union would claim that its workers needed more free time for their health's sake. If anything, the thing they, but more particularly ministers, physicians, and psychiatrists fear, is the effect on their health of free time in too large a dose, the effect, that is, on people's mental health or on the social health of the community.

Most Americans when they use the word *leisure* have free time in mind. The confusion in usage is of recent origin. It dates from the rise of the commodity mentality. In selling products for use in free time, advertisers have wished to associate their goods with the ethical connotations of leisure. More recently others, like social workers, psychiatrists, priests, and ministers, worried about free time spent solely along commercial guidelines, have tried to add a self- or social-improvement twist to free time, again by calling it *leisure*. Now it has reached a new stage where, as the whole economy trembles at the slightest threat of unemployment, the word leisure is being groomed for impending, forced free time.

There is another question, the question of real leisure. Here the difficulty is not of too much but of too little. And the penalty for lack of leisure is not the unhealthy individual or a disturbed community, but an uncreative, unlovely country.

FREE TIME: MORE, LESS OR THE SAME

To begin, will there be more free time for Americans in the future? Theoretically, yes. Projections have been made of how much free time there will be in coming decades. Similarly, calculations can be made as to whom the increase will go and as to whether some classes of the population will be fa-

Reprinted from *Environment and Policy: The Next Fifty Years,* ed. William R. Ewald, Jr. (Bloomington: Indiana University Press, 1968), pp. 114–26, by permission of the publisher.

vored above others. The work week is to get shorter by about an hour and a half each decade. By the year 2000, the overall average of hours worked is to be 31 per week or 1600 hours per year. It may go down as far as 21 per week (1100 per year). In the year 2020, on-the-job hours are to average about 26 per week, or 1370 per year, or (as above) as low as 16 hours per week, or 870 hours per year. With such figures at hand, it is not surprising that many wonder what people are going to do with their time.

Regarding who will have this free time, it seems that workers in all industries, be they manufacturing, trade, mining, construction, will benefit equally. Since the population will be both aging and growing younger, those over 65 will not benefit (presumably they are today already out of the labor force), while those at work under 25 will reap the harvest of shorter work weeks. Women, though, are to have less free time, since many more of them are to take jobs in 2000 (43 percent as against a participation rate of 37 percent in 1964), unless one calculates by the usual labor statistics, in which case women have no free time at home but acquire it, paradoxically, when they are counted as working women.

Breakdowns on amounts of free time for the managerial and official classes have received less attention from these projectors for 2000 and 2020, but by the logic of extrapolation alone they will still be working the longest hours. In 1960 they put in on the job an average of ten or twelve hours more than other categories of workers and an on-and-off the job total of over sixty business hours per week.

Studies have also been made of the impact that future expanses of time will have on the physical environment, focusing on how much money and acreage should be set aside for parks, woods and wilds, for roads leading to them, and for saving them all from congestion. Generally, the conclusion a reader might easily draw from them is that the demand for recreational space is as inexhaustible as the money and acreage are not; and that all roads lead to congestion. . . .

TODAY'S TIME

One must bear in mind that readings into the future of free time, which have been tried ever since the industrial revolution, have been scandalously wrong. The problems of more time, how and by whom it will be enjoyed, cannot be separated from people's thoughts, hopes, fears, and goals. Recent forecasting has generally credited Americans with a simplistic psychology in which they appear as craving more free time, no matter what. A scrutiny of the available evidence reveals that they have not taken more free time, do not appear to wish more (though it be available to them), and do not seem likely to change in this respect over the next decade or two. The American full-time male worker puts in an average of nearly eight hours a day, six days a week. Contrary to what is widely believed, he has taken his gains in productivity not in the coin of free time but in coin of the realm. He has no more time free of work than men have had elsewhere when not caught in times of crisis or rough transition.

Technology and free time have sides to them that oppose each other. An age that breeds technology must be one that is beguiled by and desirous of material things and therefore must buy or somehow acquire them, to do which it must use up the time any one machine or device may save. Any primitive tribe enjoys more free time than a resident of the United States today. It is doubtful that any civilization ever had as little free time as we do. The commodity mentality, fascinated by the made and purchasable thing, holds the American worker in a vise of working overtime to buy time-saving devices and "leisure goods."

THE FUTURE: FORCED FREE TIME

Perhaps, independent of what people desire, there are forces coming into being that will propel them toward more time off the job. We have indicated already that increased time and distance between home and job have influenced the demand for more time off the job. In the future, automation, a larger population among which to spread available work, and a rising GNP with which to buy time may lead to increased free time. This may not be the case, however, so long as consumption continues to spiral upward.

An upward and onward spiral is just what the forecasters forecast. I shall not assemble another little table here but merely recall that forecasters see automation marching through business, government, and industry, population mounting lustily, and the graphic peak of the GNP at 2020 disappearing in a roseate cloud. If income doubles, let us say, and expenditures double too, it is hard to see how free time can increase. The worker will be working full-time and overtime as today to spend on what he wants to have. But statistics show that he has more free time today than ever before in history, do they not? No, they do not. They show that the American worker cannot be working any harder than he is now. He has in effect no time for free time. . . .

If consumption keeps pace with production, then, there should be no increase in free time or general unemployment. Yet some of the factors just mentioned may lead to unemployment. Apparently automation, many fear, cannot spread jobs as widely as industrial economy has hitherto been able to do, because the skills it requires demand an above-average intelligence. If this be true (and much of the rest of this paper is based on this possibility) unemployment may be in the offing. But will it necessarily? Instead of calling it unemployment, why not call it free time? Workers may not be asking for more free time, but let us cut down their work year without cutting down their pay, let us call this time off the job, "leisure," and, lo!, unemployment has disappeared.

Politically this is a triumph. Where before the economic system could be charged with failure, now it can be and is already heralded as the great provider of undreamed of leisure. "Undreamed of" is correct. The kinship of unemployment and free time, when both have elements of unwillingness in them, deserves close attention. Workers will take this "leisure," but today they do not ask for it. On the other hand they did not ask to be caught in the swamp of commodities either; yet today they find themselves there, waist-deep. It may all be a question of the proper psychological preparation.

PREPARING FOR THE FUTURE

From one view the future is being adequately anticipated. Business, government, and industry are selling the future as an age of riches and leisure for everyone. There are some, however, who sense that all is not right, that workers are being faced with more time off the job while having no desire to take it. If workers had a desire for more free time, would they not have some idea of what they want to do with it? This is where the stubborn thought reaches in; "the problem of leisure" is not the problem of not enough but of too much free time. On reviewing the arguments pro and con made by various groups—labor and business in their more thoughtful moments, the medical profession and clergy, welfare workers and educators—one can hardly come away without feeling that free time is a threat and a problem rather than the gateway to an exciting age of leisure.

The doubts these voices raise demand a reconsideration of the impact of more time on the individual. By and large the solution they talk most about is not to cut free time down—for that seems to them impossible—but to provide the worker with ways of filling his future empty time. The foreseers of riches and leisure promise the worker all sorts of intriguing new equipment and facilities for sports, travel, comfort and entertainment—at a flick of the finger.

Personally I am weary of reading such predictions of the future. Their fascination with devices and products is incredibly naive. Can they really believe the stuff they put out? Yes, they can. They are testimony to the strength of the technological religion, which in the midst of so much disillusion, conserves its grand illusion, that a great industrial-scientific day is still a-borning. In 2020, reads one projection, our standard of living will be three to seven times higher than that of 1965 and in tune with this increase will be the increase in leisure time. I ask, will we be able to stand it?

The more apprehensive writers do not look to the cornucopia of commodities to fill men's time. For this reason they place more of their faith in government and non-profit institutions. They recommend programs for churches and synagogues, universities, housing authorities, settlement houses, clubs and neighborhood circles. The lion's share goes to the government, but not merely to the central government. A state government should have a Department of Leisure with architects, planners, statisticians, engineers, demographers, and social scientists, while subordinate leisure agencies should include psychologists, sportsmen, naturalists, and hobbyists. Also, each city of, say, 75,000 or more inhabitants, should boast a municipal Department of Leisure. Preparation for leisure would begin no later than high school.

These writers are right to be concerned, but the institutions they propose lack underpinnings. The preparation they would give are in things like aesthetic appreciation, social ease, prowess in games, sightseeing, outdoor life, nature study, skill in hobbies. They may also include preparation in "intellectual growth," but here they reveal a most serious weakness. The only philosophy they have to bank on is that exalting the fullest development of the individual's potentialities. What kind of clue does this give them except to let the individual be an individual, to let him develop his potentialities by himself? If you are going to tell him how to fill his empty time, you will be pushing him along the old road of self-betterment dearly loved by the nineteenth century. On the whole, however, that century kept the government off to one side. The present proposals may end by recommending that state legislatures enforce local standards, that departments of leisure should never try to give people what they want, for most people do not know what they want and should be encouraged or compelled to learn to like things they dislike.

The difficulty is even greater than such writers believe. They think it involves complexity and the possibility of undemocratic compulsion. This would be true were it a simple task like sewage disposal or air pollution, both of which are problems amenable to a rational plan, an organized attack, and an efficient system of execution. Free time cannot easily be handled this way. The word "free" in free time refers to uncommitted or unobligated time. If there is any constraint at all, the freeness vanishes.

The specter is that of an economic system constraining workers not to work. To solve the problem of filling the time then emptied, the system applies the same method it applied successfully to work—rationality, organization, and efficiency. Except that this time the method will not work. Preparing for the future in this way will bring trouble.

For all the fun that is poked at the gospel of work today, it has been a crucial part of American belief. As such it served and still serves to make sense and give meaning to existence. But it is weakening. All the nice talk itself of free time and leisure is a detraction. Writers jeer at the conformity and methodicalness work requires. Their painting a future workless world as a utopia is almost the last straw.

If the economic system is producing unemployment that it must justify as free time, it must yet produce a substitute for the work ethic. What is there to substitute for it? It is useless simply to recommend that men get the same satisfaction from leisure that they get from work. A hard worker used to be an object of esteem. Is he today? Work used to be, and still is, good for you, a remedy for pain, loneliness, the death of a dear one, a disappointment in love, doubts about the purpose of life. Steady methodical work built a great and powerful and prosperous nation. Can leisure do the same?

Religiously, politically, economically, militarily, and mentally it is still thought better to work than to do what you please. The men who go to work in the morning and come home at night are still the pillars of the nation. If these pillars tumble, the country has lost an important part of its cohesion.

To sum up, free time today does not exist in any great quantity. Projections based on the conviction that an extraordinary amount is being enjoyed are in error. Yet free time can be had in the future and in large quantities. It has already been available for a long time but few wished to take it or felt they could afford to. To get them to take it or wish for it, the worker-consumer pattern has to be broken. The "consumer" part of the dualism is not deep. It is a recently acquired belief, post-Depression, to be exact. The "worker" part of the pattern is a longer, ingrained tradition. It may take decades yet to break. Certainly it will not go at the speed projected, but the process may have already begun.

What may shatter the work ethic like Humpty Dumpty will be the need to conceal latent unemployment. One way already in evidence is to call it free time. But since it is not wanted, it is not free but forced free time. The constraint is hidden.

DISAFFECTION

There is another place in which unemployment even today may be inaccurately assessed—in universal college education. Here the young are removed from the labor force for four or five years. When accompanied by a military draft prospect for those not making the grade at school, college becomes for many not a choice but an extension of compulsory education. This may be hidden unemployment too, and not all the young are taken in by it. Possibly for this reason disaffection with the work ethic is appearing most dramatically today among youth.

Among the young there is a vocal minority whose antipathy to work is more ideological than the bum's or the philosophic hobo's once was. Their doctrines include an antipathy to technology and science as well, and to what is sometimes called the Establishment, namely all those in positions of influence in the political, military, industrial, and scientific world. To talk to these persons of a higher standard of living, mounting GNP, the American way of life, larger incomes, more free time, and the cornucopia of commodities reaps nothing but scorn. They no longer worship these gods. What they want in exchange is not clear but what they reject is easy to see—the world prophesied by the projections we have been studying.

Disaffection will not remain restricted to youth. Deprived of a sense of purpose, with the old gods gone and without new ones, other people will fall into a hedonistic attitude toward life, wherein all they can justify is to get what they can out of pleasure. Historically this is at times accompanied but typically followed by some form of religious revival. Until then there is reason to fear, as some do, that more free time, forced free time, will bring on the restless tick of boredom, idleness, immorality, and increased personal violence.

As more and more free time is forced on adults, disaffection will spread. If the cause is identified as automation and its preference for higher intelligence, nonautomated jobs may increase, as it is thought, in the categories of trade and services but they will carry the stigma of stupidity. Men will prefer not to work rather than to accept them. Those who do accept will increasingly come to be a politically inferior class composed of women, immigrants, imported aliens, and humanoids.

There may then be three classes: (1) an elite that works on top policy, administrative or manufacturing tasks with automated accoutrements, or in the arts and communication—writing, filming, designing, etc.; (2) a free time citizenry accepting its dole status but disenchanted and even rebellious; and (3) those of inferior political status who will perform the unpopular trades and services.

There are other alternatives of course, but this one reveals that a whole future world is possible on which today's statistical projections have little bearing.

Should it continue, the process of disaffection will spread most rapidly in the cities. Projections for free time in the future indicate that regional fluctuation in the length of the work week or work year will be small. This may be so but its significance will have a regional difference nonetheless.

One may look on the capital area of the United States as a megalopolis, the urbanized region extending from Boston to Washington. Eventually it may arc over to Chicago. The rest of the country, including climatic resorts like Florida and California, can be regarded as the provinces. From these areas, for decades to come, there will appear in the cities of the capital persons as yet relatively untouched by the disillusion of forced free time. The flow of provincials may provide a breathing spell of stability for the rest of the country, a time in which solutions to the problem of hidden unemployment, forced free time, and disaffection will have to be found.

There should be lots of time free for thinking up a solution, but a problem of such magnitude requires a creative solution. Free time can bring more recreation and an appetite for spectacles, but genius or creativity—that is doubtful. Free time leads to recreation; it is leisure that bears directly on creation.

THE RHETORIC OF LEISURE

For those concerned with the sanity of the individual or the integration of the country, it is natural to worry about too much free time. As we have seen, the problem does not exist at present but well may in the future. Furthermore, it is more serious than anyone thought. Advertisers sell wares within the worker-and-his-job pattern. Social workers may feel that other leadership is necessary besides that of commercialism, but at least it is some guidance. The threat to come is that the worker model which up to now has supported the industrial system will bite the grime and grease, leaving no heirs. In the ideological vacuum, there will be those who will look back on advertising and think nostalgically of the good old days. The advertiser was bound by some standards, slight reeds though they may then have seemed.

Now leisure is a state of being, of being free of everyday necessity. Distinct from free time, it requires freedom from time and work—not hourly, daily, or monthly freedom, but freedom from the necessity to work, preferably over a lifetime. By contrast the present American free time is one-half of a pendulum—jobtime/free time. First you work, then you rest and recreate yourself.

Leisure has no particular activities. Men in a leisure condition may do anything; much of what they do may seem to an outsider suspiciously like work. It is modern usage to refer to such activity as work. Creative work ought not to be called work at all. Not having anything to do, these men do something. Often they may turn to religious ritual, music, wining and dining, friends and poetry, and notably to the play of ideas and theory; in short, to the theoretical life, to contemplation.

With the lack in America of a strong tradition of leisure, it is not surprising that we must ask, "What can leisure do for us?" The benefits are the benefits of cultivating the free mind. If persons have been brought up with a liberal education and have no need to work at anything except what they choose, they enjoy a freedom that lays the conditions for the greatest objectivity (for example in science), the greatest beauty (for example in art), and the greatest creativeness (for example in politics). Leisure is the mother of philosophy, said Hobbes. If such are its benefits, and we need them sorely, can we increase leisure?

To increase leisure is difficult. It is not contained, as is free time, by time (off work) and space (for recreation). To increase free time it is usually enough to send a man, any man, home early from work. For his recreation it is usually enough to give him some space to play in. How to provide leisure?

All steps that can be taken by the government through legislation and institutions, by business organizations, schools, and churches, steps that have a limited value even for free time have much less value for leisure. There are traces of the leisure ideal in some recent attempts by government and universities to provide in centers and institutes a creative setting, for scientists in particular. These efforts and others can help only inasmuch as they diffuse an appreciative climate, through teaching and example. Much more than this cannot be done directly.

There are two important limits to face. First, not everyone has the temperament for leisure. According to the Encyclical *Rerum Novarum,* all mankind would be capable of enjoying leisure were it not for the Fall. For most people, leisure lacks sufficient guidance and sense of purpose; the leisure life is too hard. Those who have the toughness or psychological security for it are not many. Second, since leisure will have nothing to do with work (except that freely chosen, which then by definition should not be called work), it involves having means of support. In modern terms this means that whoever is to lead a life of leisure should have some form of economic independence.

The objectives in creating more leisure should be these: to allow the greatest number of those who have the temperament for it to develop to their fullest extent; to allow them to secure the means of existence without work; and to create an atmosphere more friendly than hostile in which they may live their kind of life.

A number of the developments we have been discussing do affect these objectives. A liberal education is almost a sine qua non for the growth of the leisure temperament. Universal education today may soon see to it that all will have a college education. On the negative side, however, the education is not being freely chosen; military service is the alternative. Education, moreover, has declined in quality and will continue to because of the great numbers of students in compulsory attendance and the nursery school climate of the college as a place to put grown-up children while the adults go to work.

Also, forced free time will not have to expand much to reach a net separation of income from work. Recent proposals for a guaranteed annual wage or salary intimate the separation already. Should this happen, the wherewithal for a life of leisure will be there for all who think they have the temperament. Many will try; many will drop out; among the survivors the right few will be found.

The last-mentioned prerequisite, an atmosphere friendly to leisure, may be brought about by the increase in free time whether forced or not. If both parts of the worker-consumer model break down, if more free time is not only forced upon men but in time also sought and taken, the accompanying change in attitude may well be receptive to true leisure. A more relaxed pace to life may bring about a more favorable view of the whole ideal, as well as more reflection, more refinement, and less ambitious political, military, and economic projects. Play in man's free time is a taste of leisure. In turn the ideal and practice of leisure create standards for the enjoyment of free time. Indeed without leisure the outlook for the resolution of the problems of hidden unemployment and forced free time seems desperate: hedonism, disintegration of social and political ties, crises of law and order, a cynical and callous foreign policy.

There are other developments that may influence the future attitude toward leisure, such as population growth, urbanization, increased prosperity and commercialism, and military events. . . .

ORRIN E. KLAPP

FUN

THE NEW ROMANTICISM

It is not necessary to wait for the day of leisure to reach high noon to see that already people are playing harder than ever, longer than ever, and in new ways. Fun is breaking out all over. There is a search for sensations in ever new kinds of amusement: from skin diving and surfing to antique car racing and beer-and-banjo fun; from mountain climbing to parachute jumping; to camping with lounge chairs and colored tents. Business executives come home with broken legs from skiing, while we have full-dress English-style hunts in Indiana and group safaris for polar bears in Norway and elephants in Kenya. Tourists regard the world as an oyster which they will open with their cameras. Remote vacation spots like Majorca, Puerto Vallarta, and Tahiti are being overrun by tourists as magazines like *Holiday* and *Sunset* publicize one spot after another.

The enthronement of the playboy and of his queen, the playgirl, has had its effect even on those hard-working puritans, the businessmen. The bunny clubs were, I think, an answer to the wish of businessmen—"We have a right to fun too"—an effort to harmonize the style of the playboy with that of the hard worker, a place where one might go, possibly with a client, to feel like a playboy vicariously for awhile, buy drinks at a dollar and a half apiece, look at the girls ("but you mustn't touch")—all without risk to reputation or fear that the wife will be mad. I see the bunny clubs, therefore, as an offshoot of the playboy ideal, an effort to provide some of the style to "expense-account aristocrats," a compromise for those who are making enough money to want some of the fun but who are not free to take up the playboy life.

This is just one of the changes we may expect in a new era of leisure when people seek identities that cannot be based on work or which deny the importance of work. It is a good question at what point a social type that was originally a model of something not to be becomes a prophet, a serious exemplar of a new identity for modern man. The "swinging life" is already here; so its heroes will claim their place in the pantheon along with the self-made men, strong men, patriots, culture heroes, martyrs, and saints. Moreover, loss of prestige in the work role—loss perhaps of even the opportunity to work—means that the identity search must find new avenues in recreation; the number of play identities will increase, work (boring, anyway) will take more and more a back seat to hobbies. Then, will more words that originally signified a wasted or worthless life—philanderer, adventurer, idler, beatnik, beach rat, beachcomber, tennis bum, ski bum—come to have a heroic significance? An indignant newspaper editorial treats the large number of traveling "students" who have become "scroungers," adept at tricks for getting free meals and lodgings that used to be the expertise of hoboes. At least, there can be little doubt that as people play harder more of the time, there will be an increase in the number of "nonworking" models for identity.

Such a change in roles is not to be expected without a change in ethos, a change in the sense of what is right. It will be right to have kinds of fun formerly considered wrong. Movies such as *Tom Jones, Zorba the Greek, Never On Sunday* and *Goldfinger* (James Bond) clearly state the philosophy that "sex is fun." But there is more to this than appears on the surface, more than the idea that sex is right or that fun is right.[1] It is a deeper mystique behind the notion of a right to have kicks. It is the right to find oneself, to realize oneself, through fun as "peak experience." [2] Hedonism receives new validation as the search for and affirmation of identity: the *right to* identity. This I call the new romanticism. The older romanticism was a cult which asserted the primacy of emotion, intuition, and energy over reason—the heart over the head.[3] The new romanticism asserts the right to a self; fun is legitimized as a way of fulfilling oneself, not just because it is part of one's right to happiness. The kick has a new validity based on one's right to a self. This isn't saying that if you aren't having fun you are missing something in life; it is saying that if you aren't having fun you aren't really there. This is Rousseau improved by Camus. To say "I'm doing it for myself" sounds selfish, but to say "I'm doing it for my identity" makes it a basic right with which no one can argue. So the new romanticism extends even to the really obscene attitude exhibited by those such as Norman Mailer that any feeling is as valid as any other—even a murderous one, a depraved one—so long as one realizes oneself by it. Hence, the writer's artistic obligation to turn his insides out to the public. I hope I have managed to explain that this romanticism is not just the old one of Quixote or Rousseau. The new romanticism is not necessarily poetic or chivalrous or beautiful. Its only claim is that something shall be real for me. This makes it superior to everything else in a world of shams and clichés. A feeling which does something for my identity is right because I have a right to an identity. But when everyone in the mass begins to assert this right, according to his own ideas, it begins to look like the rebellion that so alarmed Ortega y Gasset. One can assert this right by going back to nature and discarding civilized forms; one can assert it on the beach, at a wild party, by perversion, crime, or rebellion. There is really no point at which the authority can look the rebel in the eye and say, "You have no right to develop your identity in this way."

So a society which becomes pleasure-seeking in this way is more than just hedonistic. It has a deeply serious intent, a mystique, by which to try to protect itself from the meaninglessness of *La Dolce Vita.*

When play acquires this mystique it becomes cultic. Such a mystique is seen in "nature camps" established at remote and beautiful sites like Cefalu (Sicily) and Corfu (Greece) in Tahitian style with bamboo huts, sarongs, and bikinis, devoted to *"la formule: la desintoxication mentale et physique. . . .* Joy revives essential truths. The individual rediscovers and remodels himself." [4]

CULTIC TENDENCY IN PLAY

Many sports today are beginning to develop mystiques and add a cultic burden which goes beyond what used to be called a character-developing value of sport. Devotees become centered in new ways and develop mystiques which even sportsmen in neighboring fields cannot understand. Their devotion goes beyond mere amusement, even beyond professionalism, to a kind of zealous piety. Sports then provide an "it" experience which is somewhat more than the joy of the game. This is perhaps especially true of daring and thrilling activities which set a person apart by a kind of charisma.

Examples of such mystique can be found among dare-devils, mountain climbers, parachute jumpers,[5] balloonists, high-wire performers, high steel workers, karate enthusiasts, and surfers. Statements by mountain climbers indicate such mystiques:

> Life is at its best when risked. . . . Mysterious impulses which cause men to peer into the unknown. . . . Like all profound experiences it is a paradox: both challenge and escape. There is certainly an element of escapism in most climbers. I climb partly to get away, for a time, from the life of the city and some of the values of contemporary living. But there is the reality even in the escape. You climb to discover things about yourself. To be on your own, to be with yourself, facing yourself in situations of stress and danger. . . . This is the . . . compulsion towards self discovery. Once you've tasted it you forever feel the urge to see where your limit lies.[6]

A sky-diver tries to explain his mystique:

> We're a closely knit group, we sky divers. . . . It's a mental thing, jumping out without a chute. You've got one thing on your mind—you've got to convince yourself it can be done and then you do it. When you jump out at around 15,000 feet it's two degrees above zero out there. Very cool, very beautiful. Sky-diving is the best fun there is. People ask me what do you feel up there, going down. Well, it's a feeling of its own. You don't feel like you're going down in an elevator. You feel—well, free. Sometimes another sky-diver and I, we'll play catch with oranges. The real fun is the free-fall—that's the whole turkey. The parachute part is just a matter of getting down safely.[7]

Surfers try to explain their mystique:

> The bigger the wave, the bigger the challenge, all alone, you blend for a moment with immense power. You feel close to God.[8]

> Surfing has become a state of mind, a wild, uninhibited existence that revolves around the sun, the surf, and the sand. The first time you stand on a surfboard, you'll know why surfing will never die out. Plummeting down a hill of water sometimes as high as a four-story building and being able to move your board to the right or left side and to somewhat control its speed, brings about a feeling similar to flying. Not only are you moving but the force you've harnessed is also moving. That is, until you take a wipeout. Surfers will attempt to surf any large body of ocean that seems surfable, and if it is not surfable they will attempt to ride it anyway in an exhausting effort to get their "charge" (optimum level of adventure). The life of the surfer has a definite rhythm and beat to it. This is the beat of the surf and the beat of the wild driving music he listens to when he is not at the beach surfing. Personally, this music plays a very important part of my life.[9]

Such statements approach cultism when they convey an idea of a special experience that is the best there is—higher than ordinary life, setting a person apart but closer to those who share the mystique, doing something important to change his conception of himself and center his life. Overcoming fear and weakness by an ordeal is often part of the mystique. Most daring and thrilling sports provide at least three important payoffs to identity: (1) intense encounter with "reality"; (2) discovery, proof, transcendence of self; and (3) an audience before whom to shine, not the least of which is a circle of devoted hero worshipers. Is it any surprise, then, that cults easily grow up in sports which provide such identity payoffs?

Chapter 13

POWER

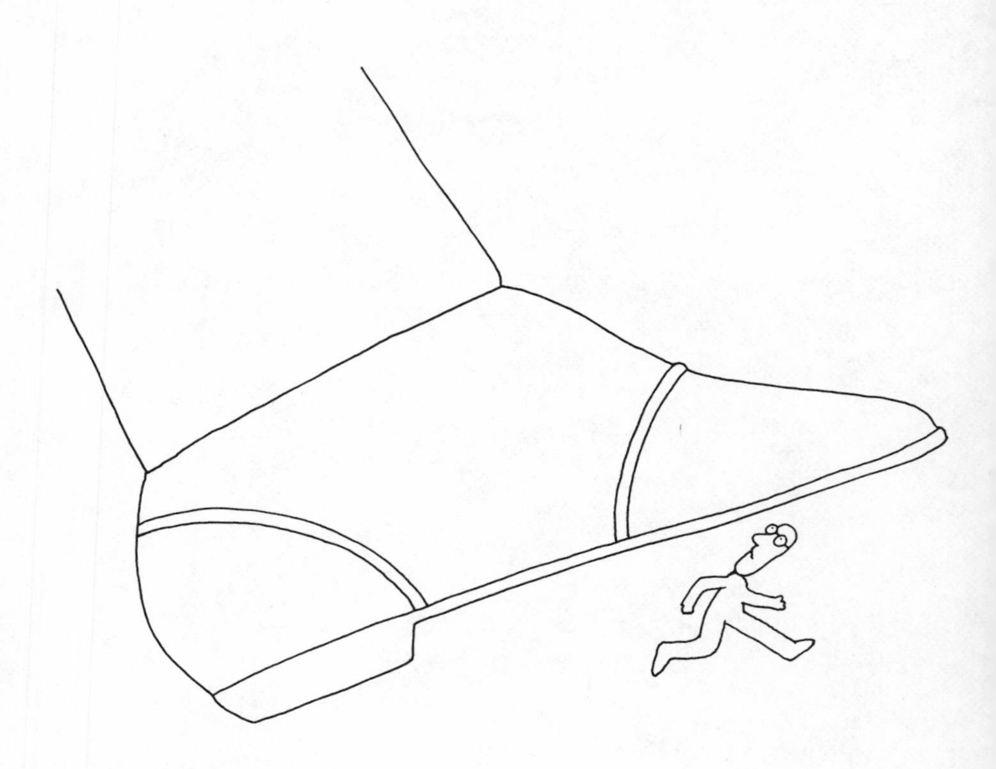

In every society people experience the realities of power. These realities are first experienced in the context of the family and other narrow spheres of a child's life. Sooner or later, however, every person encounters the larger structures of power of his society, especially—but by no means exclusively—in his contacts with the formal political system. The question of power has been the subject of extensive controversy in recent American sociology as well as political science and economics (not to mention political debates outside the field of social science).

Max Weber once more provided some of the key categories that sociologists have used in their analyses of power. His definition of power itself has been mentioned previously. Two other categories of far-reaching importance are those of authority and legitimacy. Power is not just a one-time exercise; it becomes institutionalized in on-going structures. It is these that Weber refers to under the category of "authority." The fundamental problem of authority is not just whether the resources of power are available but whether the conditions for their continuing application exist. Legitimacy is the belief that a particular system of power not only exists but has the right to exist and, as such, legitimacy is crucial for any continued and long-lasting exercise of power.

A persistent problem in the sociological analysis of power has been the relationship of power elites and other groups in society. A classical formulation of this was undertaken by the Italian sociologist Vilfredo Pareto (1848–1923). Pareto argues that real power in any society is always exercised by a rather small group of people. More importantly though, he argues that no such group (for example, no political elite) can maintain itself indefinitely, but that sooner or later, because of built-in stresses in any power system, one elite replaces another.

Within American sociology, the major debates have concerned the relationship of the formal political system, as expressed in the institutions of representative democracy, and formally unrecognized power elites (notably those of the business community). Again there has been widespread disagreement among sociologists as to the facts of the matter. Beginning in the 1950s with the work of C. Wright Mills, there emerged a school of "radical" political sociology, which has tended to debunk the claims of democratic ideology and has paid special attention to the "hidden" power of corporations, the "military industrial complex," and the like. Other political sociologists, while generally acknowledging the reality of these forces, have interpreted their relationship to the formal democratic institutions in a manner more consonant with the "official" legitimations of American politics.

The first selection, William Kornhauser's "'Power Elite' or 'Veto Groups'?" gives a summary of current positions taken regarding power and the powerful. One aspect of the current controversy is the relationship of the formal political system as expressed in the democratic institutions of American society, and the "hidden" power of corporations. The selection from Andrew Hacker's "A Country Called Corporate America" deals with this question. Similar problems are discussed in the selection from Seymour Martin Lipset's *Political Man,* entitled "Classes and Parties in American Politics."

WILLIAM KORNHAUSER

"POWER ELITE" OR "VETO GROUPS"?

RECENTLY two books appeared purporting to describe the structure of power in present-day America. They reached opposite conclusions: where C. Wright Mills found a "power elite," David Riesman found "veto groups." Both books have enjoyed a wide response, which has tended to divide along ideological lines. *The Power Elite* has been most favorably received by radical intellectuals, while *The Lonely Crowd* has found its main response among liberals. Mills and Riesman have not been oblivious to their differences. Mills is quite explicit on the matter: Riesman is a "romantic pluralist" who refuses to see the forest of American power inequalities from the trees of short-run and discrete balances of power among diverse groups.[1] Riesman has been less explicitly polemical, but he might have had Mills in mind when he spoke of those intellectuals "who feel themselves very much out of power and who are frightened of those who they think have the power," and who "prefer to be scared by the power structures they conjure up than to face the possibility that the power structure they believe exists has largely evaporated." [2]

I wish to intervene in this controversy just long enough to do two things: 1) locate as precisely as possible the issues upon which Riesman and Mills disagree; and 2) formulate certain underlying problems in the analysis of power which have to be considered before such specific disagreemens as those between Riesman and Mills can profitably be resolved.

1

We may compare Mills and Riesman on power in America along five dimensions:

Reprinted from *Culture and Social Character,* ed. Seymour Martin Lipset and Leo Lowenthal (Glencoe, Ill.: The Free Press of Glencoe, 1961). Reprinted by permission of The Macmillan Company.

1. structure of power: how power is distributed among the major segments of present-day American society;

2. changes in the structure of power: how the distribution of power has changed in the course of American history;

3. operation of the structure of power: the means whereby power is exercised in American society;

4. bases of the structure of power: how social and psychological factors shape and sustain the existing distribution of power;

5. consequences of the structure of power: how the existing distribution of power affects American society.

I) STRUCTURE OF POWER

It is symptomatic of their underlying differences that Mills entitles his major consideration of power simply "the power elite," whereas Riesman has entitled one of his discussions "Who has the power?" Mills is quite certain about the location of power, and so indicates by the assertive form of his title. Riesman perceives a much more amorphous and indeterminate power situation, and conveys this view in the interrogative form of his title. These contrasting images of American power may be diagrammed as two different pyramids of power. Mills' pyramid of power contains three levels:

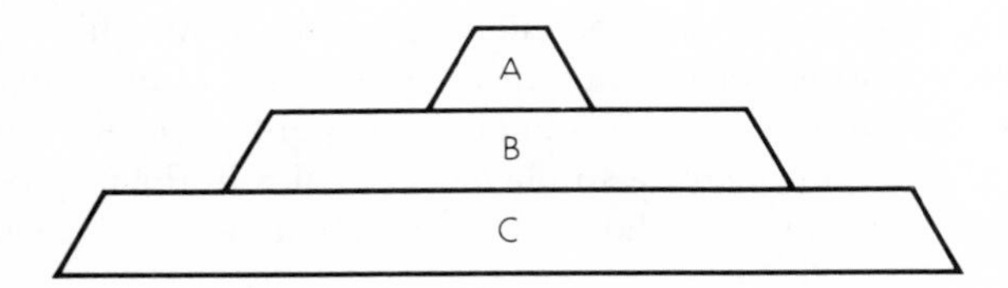

The apex of the pyramid ("A") is the "power elite": a unified power group composed of the top government executives, military officials, and corporation directors. The second level ("B") comprises the "middle levels of power": a diversified and balanced plurality of interest groups, perhaps most visibly at work in the halls of Congress. The third level ("C") is the "mass society": the powerless mass of unorganized and atomized people who are controlled from above.

Riesman's pyramid of power contains only two major levels:

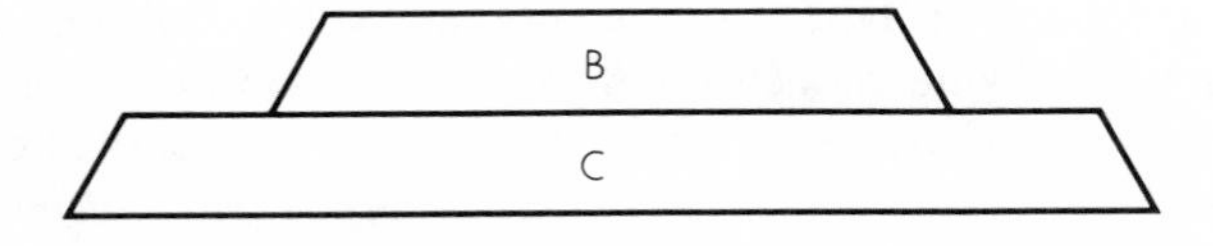

The two levels roughly correspond to Mills' second and third levels, and have been labeled accordingly. The obvious difference between the two pyramids is the presence of a peak in the one case and its absence in the other. Riesman sees no "power elite," in the sense of a single unified power group at the top of the structure, and this in the simplest terms contrasts his image of power in America with that of Mills. The upper level of Riesman's pyramid ("B") consists of "veto groups": a diversified and balanced plurality of interest groups, each of which is primarily concerned with protecting its jurisdiction by blocking actions of other groups which seem to threaten that jurisdiction. There is no decisive ruling group here, but rather an amorphous structure of power centering in the interplay among interest groups. The lower level of the pyramid ("C") comprises the more or less unorganized public, which is sought as an ally (rather than dominated) by the interest groups in their maneuvers against actual or threatened encroachments on the jurisdiction each claims for itself.

2) CHANGES IN THE STRUCTURE OF POWER

Riesman and Mills agree that the American power structure has gone through four major epochs. They disagree on the present and prospective future in the following historical terms: Mills judges the present to represent a fifth epoch, whereas Riesman judges it to be a continuation of the fourth.

The first period, according to Mills and Riesman, extended roughly from the founding of the republic to the Jacksonian era. During this period, Riesman believes America possessed a clearly demarcated ruling group, composed of a "landed-gentry and mercantilist-money leadership." [3] According to Mills, "the important fact about these early days is that social life, economic institutions, military establishment, and political order coincided, and men who were high politicians also played key roles in the economy and with their families, were among those of the reputable who made up local society." [4]

The second period extended roughly from the decline of Federalist leadership to the Civil War. During this period power became more widely dispersed, and it was no longer possible to identify a sharply defined ruling group. "In this society," Mills writes, "the 'elite' became a plurality of top groups, each in turn quite loosely made up." [5] Riesman notes that farmers and artisan groups became influential, and "occasionally, as with Jackson, moved into a more positive command." [6]

The third period began after the Civil War and extended through McKinley's administration in Riesman's view,[7] and until the New Deal according to Mills.[8] They agree that the era of McKinley marked the high point of the unilateral supremacy of corporate economic power. During this period, power once more became concentrated, but unlike the Federalist period and also unlike subsequent periods, the higher circles of economic institutions were dominant.

The fourth period took definite shape in the 1930's. In Riesman's view this period marked the ascendancy of the "veto groups," and rule by coalitions rather than by a unified power group. Mills judges it to have been so only in the early and middle Roosevelt administrations: "In these years, the New Deal as a system of power was essentially a balance of pressure groups and interest blocs." [9]

Up to World War II, then, Mills and Riesman view the historical development of power relations in America along strikingly similar lines. Their sharply contrasting portrayal of present-day American power relations begins with their diverging assessments of the period beginning about 1940. Mills envisions World War II and its aftermath as marking a new era in American power relations. With war as the major problem, there arises a new power group composed of corporate, governmental, and military directors.

> The formation of the power elite, as we may now know it, occurred during World War II and its aftermath. In the course of the organization of the nation for that war, and the consequent stabilization of the warlike posture, certain types of man have been selected and formed, and in the course of these institutional and psychological developments, new opportunities and intentions have arisen among them.[10]

Where Mills sees the ascendancy of a power elite, Riesman sees the opposite tendency toward the dispersal of power among a plurality of organized interests:

> There has been in the last fifty years a change in the configuration of power in America, in which a single hierarchy with a ruling class at its head has been replaced by a number of "veto groups" among which power is dispersed (239). The shifting nature of the lobby provides us with an important clue as to the difference between the present American political scene and that of the age of McKinley. The ruling class of businessmen could relatively easily (though perhaps mistakenly) decide where their interests lay and what editors, lawyers, and legislators might be paid to advance them. The lobby ministered to the clear leadership, privilege, and imperative of the business ruling class. Today we have substituted for that leadership a series of groups, each of which has struggled for and finally attained a power to stop things conceivably inimical to its interests and, within far narrower limits, to start things.[11]

We may conclude that both Mills and Riesman view the current scene as constituting an important break with the past; but where one finds a hitherto unknown *concentration* of power, the other finds an emerging *indeterminacy* of power.

3) OPERATION OF THE STRUCTURE OF POWER

Mills believes the power elite sets all important public policies, especially foreign policy. Riesman, on the other hand, does not believe that the same group or coalition of groups sets all major policies, but rather that the question of who exercises power varies with the issue at stake: most groups are inoperative on most issues, and all groups are operative primarily on those issues which vitally impinge on their central interests. This is to say that there are as many power structures as there are distinct spheres of policy.[12]

As to the modes of operation, both Mills and Riesman point to increasing *manipulation* rather than command or persuasion as the favored form of power play. Mills emphasizes the secrecy behind which important policy-determination occurs. Riesman stresses not so much manipulation under the guise of secrecy as manipulation under the guise of mutual tolerance for one another's interests and beliefs. Manipulation occurs, according to Riesman, because each group is trying to hide its concern with power in order not to antagonize other groups. Power relations tend to take the form of "monopolistic competition," "rules of fairness and fellowship [rather than the impersonal forces of competition] dictate how far one can go." [13] Thus both believe the play of power takes place to a considerable extent backstage; but Mills judges this power play to be under the direction of one group, while Riesman sees it as controlled by a mood and structure of accommodation among many groups.

Mills maintains that the mass media of communication are important instruments of manipulation: the media lull people to sleep, so to speak, by suppressing political topics and by emphasizing entertainment. Riesman alleges that the mass media give more attention to politics and problems of public policy than their audiences actually want, and thereby convey the false impression that there is more interest in public affairs than really exists in America at the present time. Where Mills judges the mass media of communication to be powerful political instruments in American society,[14] Riesman argues that they have relatively less significance in this respect.[15]

4) BASES OF THE STRUCTURE OF POWER

Power tends to be patterned according to the structure of interests in a society. Power is shared among those whose interests converge, and divides along lines where interests diverge. To Mills, the power elite is a reflection and solidification of a *coincidence of interests* among the ascendant institutional orders. The power elite rests on the "many interconnections and points of coinciding interests" of the corporations, political institutions, and military services.[16] For Riesman, on the other hand, there is an amorphous power structure which reflects a *diversity of interests* among the major organized groups. The power structure of veto groups rests on the divergent interests of political parties, business groups, labor organizations, farm blocs, and a myriad of other organized groups.[17]

But power is not a simple reflex of interests alone. It also rests on the capabilities and opportunities for cooperation among those who have similar interests, and for confrontation among those with opposing interests. Mills argues in some detail that the power elite rests not merely on the coincidence of interests among major institutions but also on the "psychological similarity and social intermingling" of their higher circles.[18] By virtue of similar social origins (old family, upper-class background), religious affiliations (Episcopalian and Presbyterian), education (Ivy League college or military academy), and the like, those who head up the major institutions share codes and values as well as material interests. This makes for easy communication, especially when many of these people already know one another, or at least know many people in common. They share a common way of life, and therefore possess both the will and the opportunity to integrate their lines of action as representatives of key institutions. At times this integration involves "explicit coordination," as during war.[19] So much for the bases of power at the apex of the structure.

At the middle and lower levels of power, Mills emphasizes the lack of independence and concerted purpose among those who occupy similar social positions. In his book on the middle classes,[20] Mills purports to show the weakness of white-collar people that results from their lack of economic independence and political direction. The white-collar worker simply follows the more powerful group of the moment. In his book on labor leaders,[21] Mills locates the alleged political impotence of organized labor in its dependence on government. Finally, the public is construed as composed of atomized and submissive individuals who are incapable of engaging in effective communication and political action.[22]

Riesman believes that power "is founded, in large measure, on interpersonal expectations and attitudes." [23] He asserts that in addition to the diversity of interest underlying the pattern of power in America there are widespread feelings of weakness and dependence at the top as well as at the bottom of the power structure: "if businessmen feel weak and dependent they do in actuality become weaker and more dependent, no matter what material resources may be ascribed to them." [24] In other words, the amorphousness of power in America rests in part on widespread feelings of weakness and dependence. These feelings are found among those whose position in the social structure provides resources which they could exploit, as well as among those whose position provides less access to the means of power. In fact, Riesman is concerned with showing that people at all levels of the social structure tend to feel weaker than their objective position warrants.

The theory of types of conformity that provides the foundation of so much of Riesman's writings enters into his analysis of power at this point. The "other-directed" orientation in culture and character helps to sustain the amorphousness of power.

The other-directed person in politics is the "inside-dopester," the person who possesses political competence but avoids political commitment. This is the dominant type in the veto groups, since other-direction is prevalent in the strata from which their leaders are drawn.

> Both within the [veto] groups and in the situation created by their presence, the political mood tends to become one of other-directed tolerance.[25]

However, Riesman does not make the basis of power solely psychological:

> This does not mean, however, that the veto groups are formed along the lines of character structure. As in a business corporation there is room for extreme inner-

directed and other-directed types, and all mixtures between, so in a veto group there can exist complex "symbiotic" relationships among people of different political styles. . . . Despite these complications I think it fair to say that the veto groups, even when they are set up to protect a clearcut moralizing interest, are generally forced to adopt the political manners of the other-directed.[26]

Riesman and Mills agree that there is widespread apathy in American society, but they disagree on the social distribution of political apathy. Mills locates the apathetic primarily among the lower social strata, whereas Riesman finds extensive apathy among people of higher as well as lower status. Part of the difference may rest on what criteria of apathy are used. Mills conceives of apathy as the lack of political meaning in one's life, the failure to think of personal interests in political terms, so that what happens in politics does not appear to be related to personal troubles.[27] Riesman extends the notion of apathy to include the politically uninformed as well as the politically uncommitted.[28] Thus political indignation undisciplined by political understanding is not a genuine political orientation. Riesman judges political apathy to be an important *basis* for amorphous power relations. Mills, on the other hand, treats political apathy primarily as a *result* of the concentration of power.

5) CONSEQUENCES OF THE STRUCTURE OF POWER

Four parallel sets of consequences of the structure of power for American society may be inferred from the writings of Mills and Riesman. The first concerns the impact of the power structure on the interests of certain groups or classes in American society. Mills asserts that the existing power arrangements enhance the interests of the major institutions whose leaders constitute the power elite.[29] Riesman asserts the contrary: no one group or class is decisively favored over others by the cumulated decisions on public issues.[30]

The second set of consequences concerns the impact of the structure of power on the quality of politics in American society. Here Mills and Riesman are in closer agreement. Mills maintains that the concentration of power in a small circle, and the use of manipulation as the favored manner of exercising power, lead to the decline of politics as public debate. People are decreasingly capable of grasping political issues, and of relating them to personal interests.[31] Riesman also believes that politics has declined in meaning for large numbers of people. This is not due simply to the ascendancy of "veto groups," although they do foster "the tolerant mood of other-direction and hasten the retreat of the inner-directed indignants." [32] More important, the increasing complexity and remoteness of politics make political self-interest obscure and aggravate feelings of impotence even when self-interest is clear.[33]

The third set of consequences of the American power structure concerns its impact on the quality of power relations themselves. Mills contends that the concentration of power has taken place without a corresponding shift in the bases of legitimacy of power: power is still supposed to reside in the public and its elected representatives, whereas in reality it resides in the hands of those who direct the key bureaucracies. As a consequence, men of power are neither responsible nor accountable for their power.[34] Riesman also implies that there is a growing discrepancy between the facts of power and the images of power, but for a reason opposite to Mills': power is more widely dispersed than is generally believed.[35]

Finally, a fourth set of consequences concerns the impact of the power structure on democratic leadership. If power tends to be lodged in a small group which is not accountable for its power, and if politics no longer involves genuine public debate, then there will be a *severe weakening of democratic institutions,* if not of leadership (the power elite exercises leadership in one sense of the term in that it makes decisions on basic policy for the nation). Mills claims that power in America has become so concentrated that it increasingly resembles the Soviet system of power:

> Official commentators like to contrast the ascendancy in totalitarian countries of a tightly organized clique with the American system of power. Such comments, however, are easier to sustain if one compares mid-twentieth-century Russia with mid-nineteenth-century America, which is what is often done by Tocqueville, quoting Americans making the contrast. But that was an America of a century ago, and in the century that has passed, the American elite have not remained as patrioteer essayists have described them to us. The "loose cliques" now head institutions of a scale and power not then existing and, especially since World War I, the loose cliques have tightened up.[36]

If, on the other hand, power tends to be dispersed among groups which are primarily concerned to protect and defend their interests, rather than to advance general policies and their own leadership, and if at the same time politics has declined as a sphere of duty and self-interest, then there will be a *severe weakening of leadership.* Thus Riesman believes that "power in America seems to [be] situational and mercurial; it resists attempts to locate it." [37] This "indeterminacy and amorphousness" of power inhibits the development of leadership:

> Where the issue involves the country as a whole, no individual or group leadership is likely to be very effective, because the entrenched veto groups cannot be budged. . . . Veto groups exist as defense groups, not as leadership groups.[38]

Yet Riesman does not claim that the decline of leadership directly threatens American democracy at least in the short run: the dispersion of power among a diversity of balancing "veto groups" operates to support democratic institutions even as it inhibits effective leadership. The long-run prospects of a leaderless democracy are of course not promising.

In the second part of this paper, I wish to raise certain critical questions about Riesman's and Mills' views of power. One set of questions seeks to probe more deeply the basic area of disagreement in their views. A second set of questions concerns their major areas of agreement.

Power usually is analyzed according to its distribution among the several units of a system. Most power analysts construe the structure of power as a *hierarchy*—a rank-order of units according to their amount of power. The assumption often is made that there is only one such structure, and that all units may be ranked vis-à-vis one another. Units higher in the hierarchy have power over units lower in the structure, so there is a one-way flow of power. Mills tends to adopt this approach to the structure of power.

Riesman rejects this conception of the power structure as mere hierarchy:

> The determination of who [has more power] has to be made all over again for our time: we cannot be satisfied with the answers given by Marx, Mosca, Michels, Pareto, Weber, Veblen, or Burnham.[39]
>
> The image of power in the contemporary America presented [in *The Lonely Crowd*] departs from current discussions of power which are usually based on a search for a ruling class.[40]

Riesman is not just denying the existence of a power elite in contemporary American society; he is also affirming the need to consider other aspects of power than only its unequal distribution. He is especially eager to analyze common responses to power:

> If the leaders have lost the power, why have the led not gained it? What is there about the other-directed man and his life situation which prevents the transfer? In terms of situation, it seems that the pattern of monopolistic competition of the veto groups resists individual attempts at power aggrandizement. In terms of character, the other-directed man simply does not seek power; perhaps, rather, he avoids and evades it.[41]

Whereas Mills emphasizes the *differences* between units according to their power, Riesman emphasizes their *similarities* in this respect. In the first view, some units are seen as dominated by other units, while in the second view, all units are seen as subject to constraints which shape and limit their use of power *in similar directions*.

The problem of power is not simply the differential capacity to make decisions, so that those who have power bind those who do not. Constraints also operate on those who are in decision-making positions, for if there are the places where acts of great consequence occur so are they the foci for social pressures. These pressures become translated into restrictions on the alternatives among which decision-makers can choose. Power may be meaningfully measured by ascertaining the range of alternatives which decision-makers can realistically consider. To identify those who make decisions is not to say how many lines of action are open to them, or how much freedom of choice they enjoy.

A major advance in the study of power is made by going beyond a formal conception of power, in which those who have the authority to make decisions are assumed to possess the effective means of power and the will to use it. Nor can it be assumed that those not in authority lack the power to determine public policy. The identification of effective sources of power requires analysis of how *decision-makers are themselves subject to various kinds of*

Two Portraits of the American Power Structure

POWER STRUCTURE	MILLS	RIESMAN
Levels	a) unified power elite b) diversified and balanced plurality of interest groups c) mass of unorganized people who have no power over elite	a) no dominant power elite b) diversified and balanced plurality of interest groups c) mass of unorganized people who have some power over interest groups
Changes	a) increasing concentration of power	a) increasing dispersion of power
Operation	a) one group determines all major policies b) manipulation of people at the bottom by group at the top	a) who determines policy shifts with the issue b) monopolistic competition among organized groups
Bases	a) coincidence of interests among major institutions (economic, military, governmental) b) social similarities and psychological affinities among those who direct major institutions	a) diversity of interests among major organized groups b) sense of weakness and dependence among those in higher as well as lower status
Consequences	a) enhancement of interests of corporations, armed forces, and executive branch of government b) decline of politics as public debate c) decline of responsible and accountable power—loss of democracy	a) no one group or class is favored significantly over others b) decline of politics as duty and self-interest c) decline of effective leadership

constraint. Major sources of constraint include 1) opposing elites and publics, and 2) cultural values and corresponding psychological receptivities and resistances to power. A comparison of Mills and Riesman with respect to these categories of constraint reveals the major area of disagreement between them.

Mills implies that both sources of constraint are inoperative on the highest levels of power. 1) There is little opposition among the top power-holders. Since they are not in opposition they do not constrain one another. Instead, they are unified and mutually supportive. Furthermore, there are few publics to constrain the elite. Groups capable of effective participation in broad policy determination have been replaced by atomized masses that are powerless to affect policy since they lack the bases for association and communication. Instead, people in large numbers are manipulated through organizations and media controlled by the elite. 2) Older values and codes no longer grip elites nor have they been replaced by new values and codes which could regulate the exercise of power. Top men of power are not constrained either by an inner moral sense or by feelings of dependence on others. The widespread permissiveness toward the use of expedient means to achieve success produces "the higher immorality," that is to say, the irresponsible exercise of power.

In sharp contrast to Mills, Riesman attaches great importance to constraints on decision-makers. 1) There is a plethora of organized groups, "each of which has struggled for and finally attained a power to stop things conceivably inimical to its interests." [42] Furthermore, there is extensive opportunity for large numbers of people to influence elites, because the latter are constrained by their competitive relations with one another to bid for support in the electoral arena and more diffusely in the realm of public relations. 2) The cultural emphasis on "mutual tolerance" and social conformity places a premium on "getting along" with others at the expense of taking strong stands. People are disposed to avoid long-term commitments as a result of their strong feelings of dependence on their immediate peers. "Other-directed" persons seek to maximize approval rather than power.

In general, the decisive consideration in respect to the restraint of power is the presence of multiple centers of power. Where there are many power groups, not only are they mutually constrained; they also are dependent on popular support, and therefore responsive to public demands. Now, there are many readily observable cases of regularized constraint among power groups in American society. Organized labor is one of many kinds of "countervailing power" in the market place.[43] In the political sphere, there is a strong two-party system and more or less stable factionalism within both parties, opposition among interest blocs in state and national legislatures, rivalry among executive agencies of government and the military services, and so forth.

Mills relegates these conflicting groups to the middle levels of power. Political parties and interest groups both inside and outside of government are not important units in the structure of power, according to Mills. It would seem that he takes this position primarily with an eye to the sphere of foreign policy, where only a few people finally make the big decisions. But he fails to put his argument to a decisive or meaningful test: he does not examine the pattern of decisions to show that foreign policy not only is made *by* a few people, but that it is made *for their particular interests.* Mills' major premise seems to be that all decisions are taken by and for special interests; there is no action oriented toward the general interests of the whole community. Furthermore, Mills seems to argue that because only a very few people occupy key decision-making *positions,* they are free to decide on whatever best suits their particular interests. But the degree of *autonomy* of decision-makers cannot be inferred from the *number* of decision-makers, nor from the *scope* of their decisions. It also is determined by the character of decision-making, especially *the dependence of decision-makers on certain kinds of procedure and support.*

Just as Mills is presenting a distorted image of power in America when he fails to consider the pressures on those in high positions, so Riesman presents a biased picture by not giving sufficient attention to *power differentials* among the various groups in society. When Riesman implies that if power is dispersed, then it must be relatively equal among groups and interests, with no points of concentration, he is making an unwarranted inference. The following statement conjures up an image of power in America that is as misleading on its side as anything Mills has written in defense of his idea of a power elite.

> One might ask whether one would not find, over a long period of time, that decisions in America favored one group or class . . . over others. Does not wealth exert its pull in the long run? In the past this has been so; for the future I doubt it. The future seems to be in the hands of the small business and professional men who control Congress, such as realtors, lawyers, car salesmen, undertakers, and so on; of the military men who control defense and, in part, foreign policy; of the big business managers and their lawyers, finance-committee men, and other counselors who decide on plant investment and influence the rate of technological change; of the labor leaders who control worker productivity and worker votes; of the black belt whites who have the greatest stake in southern politics; of the Poles, Italians, Jews, and Irishmen who have stakes in foreign policy, city jobs, and ethnic, religious and cultural organizations; of the editorializers and storytellers who help socialize the young, tease and train the adult, and amuse and annoy the aged; of the farmers—themselves a warring congeries of cattlemen, corn men, dairymen, cotton men, and so on—who control key departments and committees and who, as the living representatives of our inner-directed past, control many of our memories; of the Russians and, to a lesser degree, other foreign powers who control much of our agenda of attention; and so on.[44]

It appears that Riesman is asking us to believe that power differentials do not exist, but only differences in the spheres within which groups exercise control.

If Riesman greatly exaggerates the extent to which organized interests possess equal power, nevertheless he poses an important problem that Mills brushes aside. For Riesman goes beyond merely noting the existence of opposition among "veto groups" to suggest that they operate to smother one another's initiative and leadership. It is one thing for interest groups to constrain one another; it is something else again when they produce stalemate. Riesman has pointed to a critical problem for pluralist society: the danger that power may become fragmented among so many competing groups that effective general leadership cannot emerge.

On Mills' side, it is indisputable that American political institutions have undergone extensive centralization and bureaucratization. This is above all an *institutional* change wrought by the greatly expanded scale of events and decisions in the contemporary world. But centralization cannot be equated with a power elite! There can be highly centralized institutions and at the same time a fragmentation of power among a multiplicity of relatively independent public and private groups. Thus Riesman would appear to be correct that the substance of power resides in many large organizations, and that these organizations are not unified or coordinated in any firm fashion. If they were, surely Mills would have been able to identify the major mechanisms that could produce this result. That he has failed to do so is the most convincing evidence for their nonexistence.

To complete this analysis, we need only remind ourselves of the fundamental area of agreement between our two critics of American power relations. Both stress *the absence of effective political action* at all levels of the political order, in particular among the citizenry. For all of their differences, Mills and Riesman agree that there has been a decline in effective political participation, or at least a failure of political participation to measure up to the requirements of contemporary events and decisions. This failure has not been compensated by an increase in effective political action at the center: certainly Riesman's "veto groups" are not capable of defining and realizing the community's general aspirations; nor is Mills' "power elite" such a political agency. Both are asserting the inadequacy of political institutions, including public opinion, party leadership, Congress, and the Presidency, even as they see the slippage of power in different directions. In consequence, neither is sanguine about the capacity of the American political system to provide responsible leadership, especially in international affairs.

If there is truth in this indictment, it also may have its source in the very images of power that pervade Mills' and Riesman's thought. They are both inclined toward a negative response to power; and neither shows a willingness to confront the idea of a political system and the ends of power in it. Riesman reflects the liberal suspicion of power, as when he writes "we have come to realize that men who compete primarily for wealth are relatively harmless as compared with men who compete primarily for power." That such assertions as this may very well be true is beside the point. For certainly negative consequences of power can subsist along with positive ones. At times Riesman seems to recognize the need for people to seek and use power if they as individuals and the society as a whole are to develop to the fullest of their capacities. But his dominant orientation towards power remains highly individualistic and negative.

Mills is more extreme than Riesman on this matter, as he never asks what the community requires in the way of resources of power and uses of power. He is instead preoccupied with the magnitude of those resources and the allegedly destructive expropriation of them by and for the higher circles of major institutions. It is a very limited notion of power that construes it only in terms of coercion and conflict among particular interests. Societies require arrangements whereby resources of power can be effectively used and supplemented for public goals. This is a requirement for government, but the use of this term should not obscure the fact that government possesses power—or lacks effectiveness. Mills does not concern himself with the *ends* of power, nor with the conditions for their attainment. He has no conception of the bases of political order, and no theory of the functions of government and politics. He suggests nothing that could prevent his "power elite" from developing into a full-blown totalitarianism. The logic of Mills' position finally reduces to a contest between anarchy and tyranny.[45]

The problem of power seems to bring out the clinician in each of us. We quickly fasten on the pathology of power, whether we label the symptoms as "inside-dopesterism" (Riesman) or as "the higher immorality" (Mills). As a result, we often lose sight of the ends of power in the political system under review. It is important to understand that pivotal decisions increasingly are made at the national level, and that this poses genuine difficulties for the maintenance of democratic control. It is also important to understand that a multiplicity of relatively autonomous public and private agencies increasingly pressure decision-makers, and that this poses genuine difficulties for the maintenance of effective political leadership. But the fact remains that there are many cases of increasingly centralized decision-making *and* democratic control, and of multiple constraints on power *and* effective leadership. There is no simple relationship between the extent to which power is equally distributed and the efficacy of democratic order. For a modern democratic society requires strong government as well as a dispersal of power among diverse groups. Unless current tendencies are measured against both sets of needs, there will be little progress in understanding how either one is frustrated or fulfilled. Finally, in the absence of more disciplined historical and comparative research, we shall continue to lack a solid empirical basis for evaluating such widely divergent diagnoses of political malaise as those given us by Mills and Riesman.

ANDREW HACKER

A COUNTRY CALLED CORPORATE AMERICA

PROBLEMS like poverty, civil rights and juvenile delinquency may have been "discovered" only in the past few years, but such can hardly be said about the issue of bigness in American business. On and off, for the last three-quarters of a century, the question has been raised whether the nation's large corporations have reached the point where they can cut a swath through society without having to account for the consequences of their actions.

Allusions to "the trusts," "robber barons" and even "Wall Street" may have an archaic ring. Nevertheless, the frequency and magnitude of recent corporate mergers, the high level of profits despite the persistence of poverty and the latest furor over safety in the country's leading industry are bringing renewed life to a debate that has as much importance for 1966 as it did for 1896, 1912, and 1932.

Our large corporations are very large indeed. General Motors, for example, employs more than 600,000 people, a figure exceeding the combined payrolls of the state Governments of New York, California, Illinois, Pennsylvania, Texas and Ohio. The annual sales of Standard Oil of New Jersey are over $10 billion, more than the total tax collections of Wisconsin, Connecticut and Massachusetts, in addition to the six states just mentioned. In fact, our 50 largest companies have almost three times as many people working for them as our 50 states, and their combined sales are over five times greater than the taxes the states collect.

Yet here, as elsewhere, statistics can be made to tell several stories. For example, is big business getting bigger? Between 1957 and 1965, non-agricultural employment in the United States rose by about 10 percent. But during that same period the number of persons employed by the nation's largest industrial companies went up by 15 percent. Measured in this way, the big corporations seem to be taking three steps for every two taken by the economy as a whole.

At the same time it must be acknowledged that corporate America is by no means the fastest-growing sector in the country. Government employment, especially at the local level, is increasing at a higher rate; from 1957 to 1965 the public payroll, excluding the military, rose by 25 percent. Even higher was the percentage increase in service industries. Enterprises like boatyards, car washes and carry-out restaurants—many of them small and locally based—have come to constitute the most vital area of economic growth.

Moreover, the advent of automated processes in large-scale production has actually cut down corporate employment in several dominant industries. At the outset of 1965, for instance, such companies as General Electric and Gulf Oil and United States Steel actually had *fewer* people working for them than they had eight years earlier. While these firms are not yet typical, they may be harbingers of things to come—the apparent ability of corporations to increase their sales, production and profits with a decreasing work force.

If corporate size has a variety of yardsticks, corporate power is beyond precise measurement. It is not an overstatement to say that we know too much about the economics of big business and not nearly enough about the social impact of these institutions. Professional economists tend to focus on the freedom of large firms to set or manage prices, with the result that attention is deflected from the broader but less tangible role played by corporations in the society as a whole.

By the same token it is all too easy to expose egregious defects in consumer products or advertising or packaging. Congressional hearings make good forums for periodic charges of "irresponsibility," whether the target of the year happens to be automobiles or pharmaceuticals or cigarettes. It is true that the buyer is often stung—and sometimes laid to rest—by the products of even the most prestigeful of corporations. But the quality of merchandise, like the ability to fix prices, is only a secondary aspect of corporate power.

What calls for a good deal more thought and discussion is the general and pervasive influence of the large corporate entity in and on the society. For the decisions made by these huge companies guide and govern, directly and indirectly, all of our lives.

The large corporations shape the material contours of the nation's life. While original ideas for new products may come from a variety of sources, it is the big companies that have the resources to bring these goods to the public. The argument that the consumer has "free will," deciding what he will and will not buy, can be taken just so far. (Too much can be made of the poor old Edsel.) For in actual fact we *do* buy much or even most of what the large corporations put on the shelves or in the showrooms for us.

To be sure, companies are not unsophisticated and have a fair idea of what the consumer will be willing to purchase. But the general rule, with fewer exceptions than we would like to think, is that if they make it we will buy it. Thus we air-condition our bedrooms, watch color television in our living rooms, brush our teeth electrically in the bathroom and cook at eyelevel in the kitchen. It is time for frankness on this score: the American consumer is not notable for his imagination and does not know what he "wants." Thus he waits for corporate America to develop new products and, on hearing of them, discovers a long-felt "need" he never knew he had.

And more than any other single force in society, the large corporations govern the character and quality of the nation's labor market. The most visible example of this process has been the decision of companies to introduce computers into the world of work, bringing in train an unmistakable message to those who must earn a living. Millions of Americans are told, in so many words, what skills they will have to possess if they are to fill the jobs that will be available. A company has the freedom to decide *how* it will produce its goods and services, whether its product happens to be power mowers or life insurance or air transportation. And having made this decision, it establishes its recruiting patterns accordingly. Individuals, in short, must tailor themselves to the job if they want to work at all. Most of us and all of our children, will find ourselves adjusting to new styles of work whether we want to or not.

The impact of corporate organization and technology on the American educational system deserves far closer attention than it has been given. Whether we are talking of a vocational high school in Los Angeles or an engineering college in Milwaukee or a law school in New Haven, the shape of the curriculum is most largely determined by the job needs of our corporate enterprises. The message goes out that certain kinds of people having certain kinds of knowledge are needed. All American education, in a significant sense, is vocational. Liberal-arts students may enjoy a period of insulation but they are well aware that they will eventually have to find niches for themselves in offices or laboratories.

While many college graduates go into non-corporate or non-business employment, the fact remains that much of their educational tune is still being determined by corporate overtures. Even the liberal-arts college in which I teach has recently voted to establish within its precincts a department of "computer science." It is abundantly clear that while I.B.M. and Sperry Rand did not command Cornell to set up such a department, the university cannot afford to be insensitive to the changing character of the job market.

Our large firms both have and exercise the power to decide where they will build their new factories and offices. And these decisions, in their turn, determine which regions of the country will prosper and which will stagnate. The new face of the South is, in largest measure, the result of corporate choices to open new facilities in what was hitherto a blighted area. Not only has this brought new money to the region, but new kinds of jobs and new styles of work have served to transform the Southern mentality. The transition to the 20th century has been most rapid in the communities where national corporations have settled. You cannot remain an unrepentant Confederate and expect to get on in Du Pont.

By the same token the regions which have not prospered in postwar years have been those where corporations have opted not to situate. Too much can be made of the New England "ghost towns." Actually corporations have "pulled out" of very few places; more critical has been their failure to establish or expand facilities in selected parts of the country. Thus patterns of migration—from the countryside to the city and from the city to the suburb—are reflections of corporation decisions on plant and office location. If men adjust to machines, they also move their bodies to where the jobs are.

Related to this have been the corporate decisions to rear their headquarters in the center of our largest cities, especially the East Side of New York. Leaving aside the architectural transformation and the esthetic investment with which we will have to live for many years, the very existence of these prestige-palaces has had the effect of drawing hundreds of thousands of people into metropolitan areas not equipped to handle them. Thus not only the traffic snarls and the commuter crush, but also the burgeoning of suburbs for the young-marrieds of management and the thin-walled apartments for others in their twenties, fifties and sixties.

Much—perhaps too much—has been made of ours being an age of "organization men." Yet there is more than a germ of truth in this depiction of the new white-collar class which is rapidly becoming the largest segment of the American population. The great corporations created this type of individual, and the habits and style of life of corporate employment continue to play a key role in setting values and aspirations for the population as a whole.

Working for a large organization has a subtle but no less inevitable effect on a person's character. It calls for the virtues of adaptability, sociability, and that certain caution necessary when one knows one is forever being judged.

The types of success represented by the man who has become a senior engineer at Western Electric or a branch manager for Metropolitan Life are now models for millions. Not only does the prestige of the corporation rub off on the employe, but he seems to be affixed to an escalator that can only move in an upward direction. Too much has been made of the alleged "repudiation" of business and the corporate life by the current generation of college students. This may be the case at Swarthmore, Oberlin and in certain Ivied circles. But in actual fact, the great majority of undergraduates, who are after all at places like Penn State and Purdue, would like nothing better than a good berth in Ford or Texaco. Indeed, they are even now priming themselves to become the sort of person that those companies will want them to be.

The pervasive influence of the large corporations, in these and other areas, derives less from how many people they employ and far more from their possession of great wealth. Our largest firms are very well-off indeed, and they have a good deal of spare cash to spend as and where they like. These companies make profits almost automatically every year, and they find it necessary to give only a fraction of those earnings back to their stockholders in the form of dividends.

(If the largest companies are "competitive" it is only really in the sense that we all are: all of us have to keep working at our jobs if we are to survive as viable members of the society. Quite clearly the biggest corporations stand no risk of going out of business. Of the firms ranking among the top 40 a dozen years ago all but two are still in preeminent positions. And the pair that slipped—Douglas Aircraft and Wilson meat-packing—continue to remain in the top 100.)

Thus the big firms have had the money to create millions of new white-collar jobs. Department heads in the large companies ask for and are assigned additional assistants, coordinators, planners and programmers who fill up new acres of office space every year. What is ironic, considering that this is the business world, is that attempts are hardly ever made to discover whether these desk-occupiers actually enhance the profitability or the productivity of the company. But everyone keeps busy enough: attending meetings and conferences, flying around the country, and writing and reading and amending memoranda.

White-collar featherbedding is endemic in the large corporation, and the spacious amenities accompanying such employment make work an altogether pleasant experience. The travel and the transfers and the credit-card way of life turn work into half-play and bring with them membership in a cosmopolitan world. That a large proportion of these employes are not necessary was illustrated about 10 years ago when the Chrysler Corporation had its back to the wall and was forced to take the unprecedented step of firing one-third of its white-collar force. Yet the wholesale departure of these clerks and executives, as it turned out, had no effect on the company's production and sales. Nevertheless, Chrysler was not one to show that an empire could function half-clothed, and it hired back the office workers it did not need just as soon as the cash was again available.

If all this sounds a bit Alice-in-Wonderland, it would be well to ponder on what the consequences would be were all of our major corporations to cut their white-collar staffs to only those who were actually needed. Could the nation bear the resulting unemployment, especially involving so many people who have been conditioned to believe that they possess special talents and qualities of character?

Corporate wealth, then, is spent as a corporation wishes. If General Motors wants to tear down the Savoy-Plaza and erect a corporate headquarters for itself at Fifth Avenue and 59th Street, it will go ahead and do so. Quite obviously an office building could, at a quarter of the cost, have been located on Eleventh Avenue and 17th Street. But why should cost be the prime consideration? After all, the stockholders have been paid their dividends, new production facilities have been put into operation, and there is still plenty of money left over. Nor is such a superfluity of spare cash limited to the very largest concerns. Ford, which is generally thought of as General Motors' poor sister, was sufficiently well-heeled to drop a quarter of a billion dollars on its Edsel and still not miss a dividend.

If our large corporations are using their power to reshape American society, indeed to reconstruct the American personality, the general public's thinking about such concentrated influence still remains ambiguous.

There persists, for example, the ideology of anti-trust and the fond place in American hearts still occupied by small business. Thus politicians can count on striking a resonant chord when they call for more vigorous prosecutions under the Sherman Law and for greater appropriations for the Small Business Administration. Most Americans, from time to time, do agree that our largest companies are too big and should somehow or other be broken up into smaller units. But just how strong or enduring this sentiment is is hard to say. No one really expects that Mobil Oil or Bethlehem Steel can or will be "busted" into 10 or a dozen entirely new and independent companies. Thus, if the ideology that bigness equals badness lingers on, there is no serious impetus to translate that outlook into action.

Part of the problem is that if Americans are suspicious of bigness, they are not really clear about just what it is about large corporations that troubles them. Despite the periodic exposures of defective brake cylinders or profiteering on polio vaccine, the big story is not really one of callous exploitation or crass irresponsibility. Given the American system of values, it is difficult to mount a thoroughgoing critique of capitalism or to be "anti-business" in an unequivocal way. The result is that our commentaries in this area are piecemeal and sporadic in character. We have the vocabularies for criticizing both "big government" and "big labor" but the image of the large corporation is a hazy one, and despite its everyday presence in our midst our reaction to its very existence is uncertain.

Take the question of who owns our big enterprises. In terms of legal title the owners are the stockholders, and management is accountable to that amorphous group. But it is well known that in most cases a company's shares are so widely dispersed that the managers of a corporation can run the firm pretty well as they please. Yet even assuming that the executives are acting with the tacit consent of their company's theoretical owners, it is worth inquiring just who these stockholders are.

Interestingly, a rising proportion of the stockholders are not people at all but rather investing institutions. Among these non-people are pension funds, insurance companies, brokerage houses, foundations and universities. Thus some of the most significant "voters" at the annual meetings of the big companies are the Rockefeller Foundation, Prudential Life and Princeton University. And these institutions, out of habit and prudence, automatically ratify management decisions.

It is instructive that the corporations' own public-relations departments have just about given up trying to persuade us that these stockholder gatherings are just another version of the local town meeting. The last report I saw that did this was filled with photographs showing average-citizen stockholders rising to question the board of directors on all manner of company policies. "A sizable number of shareholders participated in the lively discussion periods," the reader is told. "Many more spoke individually with directors and other executives about the affairs of the company." However, in small type in the back of the report is an accounting of the five votes that were actually taken at the meeting. In no case did the management receive less than 96 percent of the ballots (i.e., shares) that were cast.

From these observations at least one answer is possible: yes, there is a "power élite" presiding over corporate America. Yet the problem with this term is that the "élite" in question consists not so much of identifiable personalities—how many of the presidents of our 20 largest corporations can any of us name?—but rather of the chairs in the top offices.

The typical corporation head stays at his desk for only about seven years. The power he exercises is less discretionary than we would like to believe, and the range of decisions that can be called uniquely his own is severely limited. (It is only in the small companies on the way up, such as the Romney days at American Motors, that the top men impress their personalities on the enterprise.) John Kenneth Galbraith once noted that when a corporation president retires and his successor is named, the price of the company's stock, presumably a barometer of informed opinion, does not experience a perceptible change.

Unfortunately it is far easier to think in terms of actual individuals than of impersonal institutions. Therefore it must be underlined that the so-called "élite" consists not of Frederic Donner and Frederick Kappel and Fred Borch but rather of *whatever* person happens to be sitting in the top seat at General Motors and A.T.&T. and General Electric. We are reaching the point where corporate power is a force in its own right, for all intents and purposes independent of the men who in its name make the decisions.

The modern corporation is not and cannot be expected to be a "responsible" institution in our society. For all the self-congratulatory handouts depicting the large firm as a "good citizen," the fact remains that a business enterprise exists purely and simply to make more profits—a large proportion of which it proceeds to pour back into itself. (True, the big companies do not seek to "maximize" their profits: their toleration of make-work and high living is enough evidence for this.)

But corporations, like all businesses whether large or small, are in the primary business of making money; indeed, they do not even exist to produce certain goods or services that may prove useful or necessary to society. If Eli Lilly or Searle and the other drug companies discovered that they could chalk up larger profits by getting out of vaccines and manufacturing frozen orange juice instead, they would have no qualms or hesitation about taking such a step.

A corporation, then, cannot be expected to shoulder the aristocratic mantle. No one should be surprised that in the areas of civil rights and civil liberties our large companies have failed to take any significant initiative. The men who preside over them are not philosopher-kings, and no expectation should be held out that they may become so. At best they can be counted on to give some well-publicized dollars to local community chests and university scholarships. But after those checks are written (and the handing-over of them has been photographed) it is time to get back to business.

And this is as it should be. Corporate power is great—in fact, far more impressive than corporation executives are willing to admit—and were large corporations to become "social-minded," their impact would be a very mixed blessing. For then the rest of us would have to let corporate management define just what constitutes "good citizenship," and we would have to accept such benefactions without an excuse for comment or criticism.

Therefore, when corporations, in the course of doing their business, create social dislocations there is no point in chiding or exhorting them to more enlightened ways. It would be wrong, of course, to lay the blame for all of our social ills at the doorsteps of the large firms. If the drug companies manufacture cheap and effective birth control pills it is a trifle presumptuous to take them to task for whatever promiscuity occurs as a consequence.

Nevertheless, the American corporation, in the course of creating and marketing new merchandise, presents us with temptations—ranging from fast cars to color television—to which we sooner or later succumb. There is nothing intrinsically wrong with color television. It is, rather, that the money we spend for a new set is money that can no longer be put aside for the college education of our children. (Thus, no one should be surprised when, 15 years from now, there is a demand for full Federal scholarships for college students. Not the least reason for such a demand will be that we were buying color TV back in 1966.)

Specific questions can be framed easily enough. It is the answers that are far from clear. We have unemployment: how far is it because corporations have not been willing or able to create enough jobs for the sorts of people who need them? We have a civil rights problem: how far is it because corporations have been reluctant to hire and train Negroes as they have whites? We have a shortage of nurses: how far is it because corporations outbid and undercut the hospitals by offering girls secretarial jobs at higher pay for less work? We have whole waves of unwanted and unneeded immigrants pouring into our large cities: how far is it because corporations have decided to locate in Ventura County in California rather than Woodruff County in Arkansas?

Questions like these may suggest differing answers but they do add up to the fact that a good measure of laissez-faire continues to exist in our corporate economy. For all their ritual protestations over Government intervention and regulation, our large companies are still remarkably free: free to make and sell what they want, free to hire the people they want for the jobs they have created, free to locate where they choose, free to dispose of their earnings as they like—and free to compel the society to provide the raw materials, human and otherwise, necessary for their ongoing needs.

The task of picking up the pieces left by the wayside belongs to Government. This is the ancient and implicit contract of a society committed to freedom of enterprise. But whether the agencies of Government have the resources or the public support to smooth out the dislocations that have been caused to our economy and society is not at all clear. Negro unemployment, the pollution of the Great Lakes, the architectural massacre of Park Avenue and the wasteland of television seem to be beyond the power and imagination of a Government that has traditionally understood its secondary and complementary role.

Corporate America, with its double-edged benefactions and its unplanned disruptions, is in fact creating new problems at a rate faster than our Governmental bureaus can possibly cope with them. Given that the articulate segments of the American public seem at times to show more confidence in United States Steel than in the United States Senate, the prognosis must be that the effective majority today prefers a mild but apparently bearable chaos to the prospect of serious Government allocation and planning.

The American commitment to private property means, at least for the foreseeable future, that we will be living with the large corporation. On the whole, Americans seem vaguely contented with this development, unanticipated as it may have been. In light of this stolidity the order of the day—to reverse Karl Marx's dictum—is to understand our world rather than change it; to identify, with as much clarity and precision as is possible, the extent to which a hundred or so giant firms are shaping the contours of our contemporary and future society. Only if we engage in such an enterprise will we be able to make any kind of considered judgment concerning the kind of nation in which we wish to live and the sort of people we want to be.

SEYMOUR MARTIN LIPSET

CLASSES AND PARTIES IN AMERICAN POLITICS

PARTY SUPPORTERS' SOCIAL CLASS POSITION

It often comes as a shock, especially to Europeans, to be reminded that the first political parties in history with "labor" or "working-man" in their names developed in America in the 1820s and 1830s. The emphasis on "classlessness" in American political ideology has led many European and American political commentators to conclude that party divisions in America are less related to class cleavages than they are in other Western countries. Polling studies, however, belie this conclusion, showing that in every American election since 1936 (studies of the question were not made before then), the proportion voting Democratic increases sharply as one moves down the occupational or income ladder. In 1948 almost 80 percent of the workers voted Democratic, a percentage which is higher than has ever been reported for left-wing parties in such countries as Britain, France, Italy, and Germany. Each year the lower-paid and less skilled workers are the most Democratic; even in 1952, two thirds of the unskilled workers were for Stevenson, though the proportion of all manual workers backing the Democrats dropped to 55 percent in that year—a drop-off which was in large measure a result of Eisenhower's personal "above the parties" appeal rather than a basic swing away from the Democratic party by the lower strata.[1]

In general, the bulk of the workers, even many who voted for Eisenhower in 1952 and 1956, still regard themselves as Democrats, and the results of the 1954 and 1958 congressional elections show that there has been no shift of the traditional Democratic voting base to the Republicans. Two thirds of

TABLE 17

Percent Republican Voting or Voting Preference among Occupational Groups and Trade-Union Members *

	1940	1948	1952	1954	1956
Business and Professional	64	77	64	61	68
White-collar Workers	52	48	60	52	63
Manual Workers (skilled and unskilled)	35	22	45	35	50
Farmers	46	32	67	56	54
Trade-union Members	28	13	39	27	43

* The 1940 figures represent pre-election voting preferences, and are recomputed from Hadley Cantril, *Public Opinion, 1935–1946* (Princeton: Princeton University Press, 1951), p. 602. Since the number of cases is not given in the table, estimates from census data on the relative proportion of persons in a given occupational category were made to facilitate the combination of several of them. The 1948 figures represent the actual voting reported by a national random sample, and are taken from Angus Campbell, Gerald Gurin, and Warren E. Miller, *The Voter Decides* (Evanston: Row, Peterson and Co., 1954), pp. 72–73. The remaining data may be found in an American Institute of Public Opinion news release, October 12, 1958, and also represent actual voting results.

the workers polled by Gallup in 1958 voted for a Democrat for Congress.

The same relationship between class, considered now as a very general differentiating factor, and party support exists within the middle and upper classes. The Democrats have been in a minority among the nonmanual strata, and, except among the intellectual professions, the Democratic proportion of the nonmanually occupied electorate declines inexorably with income and occupational status to the point where, according to one study, only 6 percent of the heads of corporations with more than 10,000 employees are Democrats. Perhaps the best single example of the pervasiveness of status differences as a factor in American politics is the political allegiances of the chief executives of major American corporations. This study, done in 1955 by the Massachusetts Institute of Technology's Center for International Studies, and based on interviews with a systematic sample of one thousand such men, found that even within this upper economic group, the larger the company of which a man was an officer, the greater the likelihood that he was a Republican (see Table 18).

TABLE 18

Relationship between Size of Firm and Political Party Allegiances of Corporation Executives—1955 *

SIZE OF FIRM	REPUBLICAN	DEMOCRATIC	INDEPENDENT
More than 10,000 workers	84%	6%	10%
1,000–9,999	80	8	12
100–999	69	12	19

* Data supplied to author through the courtesy of the Center for International Studies of the Massachusetts Institute of Technology.

Consistent with these findings are the popular images of typical supporters of each party. The Gallup Poll, shortly before the 1958 congressional elections, asked a nation-wide sample what their picture of the typical Democrat was, and received these answers most frequently: "middle class . . . common people . . . a friend . . . an ordinary person . . . works for his wages . . . average person . . . someone who thinks of everybody." The typical Republican, in contrast, is "better class . . . well-to-do . . . big businessman . . . money voter . . . well-off financially . . . wealthy . . . higher class." [2]

All in all, public opinion poll evidence confirms the conclusion reached by the historian Charles Beard in 1917 that "the center of gravity of wealth is on the Republican side while the center of gravity of poverty is on the Democratic side." [3] Beard's conclusions were based on an inspection of the characteristics of various geographical areas, and more recent studies using this ecological approach report similar findings. Thus the Harvard political scientist Arthur Holcombe found that among urban congressional districts, "the partisan pattern is the same. The only districts which have been consistently Republican for a considerable period of time are those with the highest rents. . . . The districts which have been most consistently Democratic are those with the lowest rents. . . . The districts with a preponderance of intermediate rents are the districts which have been most doubtful from the viewpoint of the major parties." [4] A detailed survey of party registrations in 1934 in the then strongly Republican Santa Clara County (suburban San Francisco) found a strong correlation between high occupational position and being a registered Republican. About 75 percent of plant superintendents, bankers, brokers, and managers of business firms identified publicly with the G.O.P., as contrasted with 35 percent of the cannery and other unskilled workers. Within each broad occupational group, property owners were much more likely to be registered Republicans than those who did not own property.[5]

Although most generalizations about the relationship of American parties to class differences are based on the variations in the backgrounds of their respective electorates, there is some as yet skimpy evidence that the same differences exist on the leadership level, particularly in the local community. A study of the backgrounds of candidates for nomination for county office in local primaries in three counties in Indiana indicates a close correspondence between the characteristics of leaders and voters. While 76 percent of those seeking Republican nominations were in professional or business-managerial occupations, 42 percent of the Democratic aspirants were manual workers. In Milwaukee, Wisconsin, 54 percent of the officers of the local Democratic party were manual workers or in sales and clerical positions. By contrast these groups represented only 10 percent among the Republicans, whose leaders were largely professionals or ran business firms.[6]

The relationship of socioeconomic position to political behavior in America as elsewhere is reinforced by religious and ethnic factors. Surveys indicate that, among the Christian denominations, the higher the average income of the membership of a given church group, the more likely its members are to vote Republican. If Christian religious groups in the United States are ranked according to the average socioeconomic status of their membership, they are, reading from high to low, Congregational, Presbyterian, Episcopal, Methodist, Lutheran, Baptist, and Catholic—and this rank order is identical to the one produced when the denominations are ranked by propensity to vote Republican. This suggests that socioeconomic status, rather than religious ideas, is the prime determinant of political values among different denominations. The fact that the Jews, who are one of the wealthiest religious groups in America, are shown by survey data to be most Democratic is probably due to their sensitivity to ethnic discrimination and their lack of effective social intercourse with the upper-status groups in America. But religious beliefs or loyalties, and the political values associated with them, nevertheless seem to have some independent effect on voting behavior. Working-class Protestants belonging to the Congregational or Presbyterian churches are more likely to be Republicans than workers who are Baptist or Catholic. Conversely, wealthy Baptists or Catholics are more apt to be Democrats than equally rich Congregationalists or Episcopalians are.[7]

Roughly speaking, the same differences appear between ethnic groups. Anglo-Saxons are more likely to be Republican than other Americans in the same class position who have a more recent immigrant background. Thus if an individual is middle class, Anglo-Saxon, and Protestant, he is very likely to be a Republican, whereas if he is working class, Catholic, and of recent immigrant stock, he will probably be a Democrat.

Even before the development of the two-party system in its present form, the political issues dividing the society tended to have a class character. Free public schools, for example, did not emerge naturally and logically from the structure and values of American society. Rather, as one historian of American education, Ellwood P. Cubberley, has pointed out: "Excepting for the battle for the abolition of slavery, perhaps no question has ever been before the American people for settlement which caused so much feeling or aroused such bitter antagonism." [8] In large part it was a struggle between liberals and conservatives in the modern sense of the term, although religious issues also played a strong role. "The friends of free schools were at first commonly regarded as fanatics, dangerous to the States, and the opponents of free schools were considered by them as old line conservatives or as selfish members of society." [9] Among the arguments presented for free education was that "a common state school, equally open to all, would prevent that class differentiation so dangerous in a Republic"; while opponents of such schools argued that they "will make education too common, and will educate people out of ther proper station in society . . . [and] would break down long-established and very desirable social barriers." [10] On one side of the issue were the poorer classes; on the other, "the old aristocratic class . . . the conservatives of society . . . taxpayers." [11]

The relationship between status or class position (as indicated by the three criteria of economic position, religion, and ethnic background) and party loyalty is thus not a new development in American history. Studies of the social bases of the Federalists, America's first conservative party, and the Jeffersonian Democrats in the late eighteenth and early nineteenth centuries indicate that they corresponded closely to the bases of the modern Republicans and Democrats, respectively. The Federalists were backed by the well-to-do farmers, urban merchants, persons of English extraction, and members of such high-status churches as the Congregationalists and the Episcopalians.[12] The Democrats were supported by urban workers, poorer farmers, persons of non-English background such as the Scotch-Irish, and members of the (then) poorer churches like the Presbyterians and the Catholics. The second conservative party, the Whigs, who fought the Democrats from 1836 to 1852, derived their strength from the same group as the Federalists, while the Democrats retained the groups which had backed Jefferson, and added most of the great wave of European immigrants.

Although the Republican party is often thought of as a newly created antislavery party, the research on the pre-Civil War period suggests that it inherited both the support and leadership of the northern Whigs. A detailed study of voting behavior in New York State before the Civil War shows that the Democrats kept their urban lower-class, Catholic, and immigrant support.[13]

The evidence compiled by various social scientists indicates that the men of wealth and economic power in America have *never* given more than minority support to the Democrats. . . .

A recent study by Mabel Newcomer of the political views of large business executives in 1900, 1925, and 1950 reports that in all three periods about three quarters of this group were Republicans. Even in 1925, a period not normally considered to be one of political class conflict in America, only 19 percent of the executives were Democrats. These data certainly underestimate the Republican majority among business executives, since they are based on public party enrollment rather than voting preference and include many registered as southern Democrats who would be Republicans if they were not living in a one-party region.[14] . . .

PARTY POLICIES DETERMINED BY PARTY SUPPORTERS

This division of Americans into supporters of one of two parties, one historically based on those who are poorer and the other on the more well to do, does not mean that the parties have always divided ideologically along traditional "left-right" political lines. Such issues did separate the parties in Jefferson's and Jackson's day, and also—for the most part—from 1896 to the present, although there were some significant exceptions, like the elections of 1904 and the 1920s. However, even when the parties did not present opposing positions on conventional left-right lines, there have almost always been issues between them which reflected the differences in their social bases.

For example, the Federalist-Whig-Republican party was less receptive to immigration in the nineteenth century than the Democratic party, and a Republican administration enacted the restrictive immigration legislation of the early 1920s. In general, the various nativist and anti-Catholic movements which have arisen at various periods in American history have been identified with the conservative parties on a local if not a national level.[15]

Even the controversy over slavery reflected differences in class. The Northern urban lower-class groups before the Civil War tended to be anti-Negro and were uninterested in the struggle for abolition. In New York State the conservatives supported the right of free Negroes to vote in the state constitutional conventions of 1820 and 1846, while the major Democratic spokesmen either opposed, or were uninterested in, the extension of the suffrage to Negroes. The free Negroes, in turn, were supporters of the Federalist and Whig parties before 1850, and the freed slaves and their descendants remained loyal backers of the Republican party until the election of Roosevelt produced for the first time a Democratic administration which showed an interest in their problems.[16] The Wilson Administration of 1913–21, although liberal on other issues, reflected southern attitudes in its race-relations policies. The movements reflecting Protestant middle-class morality, such as those designed to prohibit liquor and gambling, or those concerned with the elimination of corruption in government, have also made headway largely through the conservative parties. In the controversies over prohibition in the twentieth century, the Democrats in the North were the "wet" party, while the Republicans were "dry." And in the prosperous 1920s the Democrats as the party of the lower strata and the Catholics ran the campaign of 1928 largely on the platform of repeal of prohibition.[17]

The differences in the ethnic composition of their social bases have also been reflected in the foreign policy positions of the two parties. The one Democratic administration between 1861 and 1913—that of Grover Cleveland in the 1880s—opposed Great Britain on a number of issues and sympathized with the cause of Irish freedom. A recent study of the British immigrant in America shows that the British, though not viewed as a separate or alien ethnic group like arrivals from other countries, organized British clubs in the late nineteenth century as a means of fighting the political power of the Irish Democrats. These British associations gravitated toward the Republican party.[18] Even during World War I, such differences affected American policy. Although Wilson was personally sympathetic to the cause of Britain and the Allies, the bulk of Americans of non-Anglo-Saxon background were hostile either to Britain or to Czarist Russia, and it was the Republican party, based on middle-class Anglo-Saxons, which advocated greater help to the Allies. Wilson, it should be remembered, fought the election of 1916 on an *anti*-war platform and won or held the support of the Irish, Jews, and Germans for the Democratic party.

The position of the two parties on foreign policy has not only reflected their ethnic bases, but on occasion has alienated part of them. Millions of Americans of Irish and German extraction clearly resented American entry into World War I.[19] The Germans in particular suffered heavily as a result of social and economic discrimination during and after the war. Some analysts have suggested (although the statistical work to prove this has not been done) that the great Republican victory of 1920, in which Harding secured a larger percentage of the vote than any other Republican since the founding of the party, was in part at least a result of a shift away from the Democrats by members of ethnic groups who felt "betrayed" by Wilson's taking the country into war.

It is now largely forgotten that, in his early years of office, Franklin D. Roosevelt was an "isolationist," and that the Democratic party leadership in Congress acted as if it believed that America had been tricked into entering World War I by British propaganda and the manipulation of Wall Street bankers. The neutrality act passed in Roosevelt's first term of office by an overwhelmingly Democratic Congress reflected isolationist and anti-British attitudes. In this respect the Democrats returned to their traditional role of representing the major ethnic groups.[20]

The outbreak of World War II placed Roosevelt in the same dilemma that had faced Wilson earlier. He knew that he had to aid the Allies, but he also wanted to be re-elected. The fall of France left him with no alternative other than giving "all aid short of war"; but in the 1940 presidential campaign he still promised that the country would *not* go to war. This time, however, unlike 1916, the Republicans took the isolationist and pacifist position, largely as an election maneuver. Wendell Willkie was even more favorable to intervention than Roosevelt, yet he and his advisers apparently felt that their one hope of victory was to entice the Irish, German, and Italian voters, who were against intervention because of their national identification, away from the Democrats. Opinion-poll data for that year show that the Republicans were somewhat successful, since the Democratic vote did drop off greatly among these three groups. It was probably counterbalanced, at least in part, by a swing to Roosevelt among middle-class "Anglo-Saxons."

Ethnic reactions have also affected the handling of the Communist issue in the last decade. It should not be forgotten that Senator McCarthy was Irish and represented a state in which the influence of German-Americans is high. McCarthy, in his charges of Communist infiltration in the State Department, stressed that America was "betrayed" by men of upper-class Anglo-Saxon backgrounds, by the graduates of Harvard and other Ivy League schools. He was saying, in effect, to the isolationist ethnic groups which had been exposed to charges of "disloyalty" in the previous two wars, that the people who had really been disloyal were the upper-class Anglo-Saxons who had manipulated the United States into fighting the wrong enemy, and who had "lost the peace" to the Russians.[21] In 1952 the Republicans made a strong foreign policy appeal to the ethnic groups, especially the Catholics and Germans. Eisenhower, like Willkie before him, allowed electoral expediency to modify his public statements. And the election returns indicate that he made strong headway among middle-class Catholics, Germans, and Irish.[22] . . .

UPPER-CLASS LIBERALISM

. . . The strong strain of Tory radicalism [social and political, but not economic, liberalism in members of the upper class in politics, such as Franklin Roosevelt and Nelson Rockefeller] which has reappeared at crucial points in American history has served to reduce the tensions inherent in class and sectional cleavages. The Tory radicals, to use Richard Hofstadter's words describing the corporation lawyers who became leaders of the Progressive movement, have never wanted "a sharp change in the social structure, but rather the formation of a responsible elite which was to take charge of the popular impulse toward change, and direct it into moderate, and, as they would have said, 'constructive channels,' a leadership occupying, as Brandeis aptly put it, a position of independence between the wealthy (self-interested businessmen) and the people, prepared to curb the excesses of either." [23] From the standpoint of political stability, Tory radicalism has served to retain the loyalties of both the underprivileged out-groups who gain from needed reforms and the conservative strata who are outraged by the same measures. The participation of upper-class persons in liberal politics may also be seen as enlightened self-interest, since they are able to achieve needed reforms, exercise restraint, and, according to E. Digby Baltzell in *Philadelphia Gentleman,* "perpetuate upper-class influence on the functional class system as a whole by the very fact that they hold important positions within the new avenues to power." [24] At the same time, their presence serves to blur the class lines separating the parties.

THE EFFECT OF ONE-PARTY STATES

Regionalism—the Democratic control of the South and the traditional Republican domination of many northern states—represents one important deviation, but a disappearing one, from the class basis of American politics.[25] With the exception of the South, it is difficult to estimate how much of the voting behavior of the country remains a reflection of traditional regional attachments to a party which are independent of other factors. Republican regional strength outside of the South is based mainly on majorities in certain areas who have "Republican" social characteristics: rural and small-town middle-class Protestants. Traditionally Republican New England no longer includes Massachusetts, Connecticut, and Rhode Island, now highly urbanized, in which the Democrats more than hold their own. In Maine, Vermont, and New Hampshire, where rural and small-town votes play a more important role and the proportion of Catholics has grown more slowly, one-party Republican rule characterized state politics until the 1950s and is still far from dead in the last two states, although in 1958 Vermont elected a Democratic congressman for the first time in 106 years. The same pattern is true for a number of Midwest farm states: the absence of major urban centers has meant that many of these states remain Republican. Perhaps the best evidence that Republican strength in predominantly Republican states has not represented a purely regional attachment to a party is the fact that the working class in the cities of these states has been largely Democratic, and that, in fact, the principal urban centers have Democratic majorities. For example, the two largest cities in Vermont and New Hampshire, Burlington and Manchester, are old Democratic strongholds. In one Vermont industrial town, Winooski, the Democrats secure up to 90 percent of the vote. . . .

THE NATIONALIZATION OF POLITICS

The political pattern in the southern states is a major deviation from the normal picture of political diversity reflecting class, ethnic, and religious differences. With the exception of a brief period in the 1890s when the Populists had strength, the South has been solidly Democratic in presidential elections from the end of Reconstruction down to 1928. Since 1928 some southern states have moved out of the Democratic camp in four elections—1928, 1948, 1952 and 1956—although all of them remained Democratic on the state and local level.[26] The 1928 defection came from the resentment of a section of the Protestant, rural and small-town South to the Democratic nomination of a Catholic and anti-prohibitionist, Al Smith. The three more recent deviations illustrate the consequences of the nationalization of American politics under Roosevelt and Truman: the more conservative sections of the South are rebelling against the ideological liberalism of the national party, which is for equal rights for Negroes, supports trade-unions, and is hostile to big business.

The nationalization of American politics is one of the major consequences of the New Deal-Fair Deal revolution, and of the urbanization and industrialization of the entire country. Parochial issues which do not involve left-right or foreign policy differences have been reduced in importance. And one finds that the effects of traditional religious, ethnic, and regional factors are also diminishing, although they still play a role when foreign policy issues are salient. In 1952 Eisenhower carried middle-class residential districts in the urban South with about the same majority that he took in comparable districts in northern urban centers.[27] The majority of middle-class Catholics voted Republican in 1952 (a shift facilitated by the foreign policy issues discussed earlier), while the bulk of working-class Catholics, especially those belonging to trade-unions, remained Democratic.[28] Such factors as occupational status, income, and the class character of the district in which people live probably distinguish the support of the two major parties more clearly now than at any other period in American history since the Civil War.

The importance of class factors should not, however, cause us to overlook the fact that in the United States, as in every other democratic country, a large minority of the workers and a smaller section of the middle classes deviate from the dominant class tendency. One of the necessary conditions for a viable two-party system is that both parties hover around the 50 percent mark. Hence in every industrialized country the conservative parties must win working-class support; and in the United States and Britain the Republican and Conservative parties have necessarily accepted many of the reforms enacted by their adversaries. Actually the conservatives' gradual shift to the left is endemic in the sheer demography of democratic politics—the poor will always be in the majority.

The preponderance of "poorer" people also means that the conservatives must always attempt to reduce the saliency of class issues in politics. It is clearly to the advantage of the parties of the left for people to vote consciously in terms of class. Consequently, the Republicans always seek to emphasize nonclass issues, such as military defense, foreign policy, corruption, and so forth.

The way in which class and nonclass factors operate to affect electoral behavior was shown in a survey of the sources of the Eisenhower appeal. The political sociologist Oscar Glantz found that 1952 Eisenhower voters differed greatly among themselves in their occupational positions and traditional party loyalties. The more well to do a group, the more likely its members were to account for their vote for Eisenhower by identification with the Republican party rather than with Eisenhower; to be strongly favorable to business; and conservative on various domestic political issues. Conversely, the "converts" to Eisenhower from the Democratic ranks who were less well to do than traditional Republicans were less friendly to business and more liberal on domestic matters, and cited factors such as Eisenhower's leadership qualities and their dissatisfaction with the way the Democrats had handled noneconomic issues like the Korean War as their reasons for voting Republican in 1952. Glantz suggests, in line with the analysis here, that the interest basis of political behavior was not weakened by Eisenhower's victory, but rather that 1952 illustrates the way in which loyalties linked to class issues may be suppressed when the stimulus of a noneconomic factor is great.[29]

The current spirit of "moderation" in politics seems at first glance to belie the thesis that class factors have become more significant in distinguishing between the supporters of the two parties, but in reality it does not. There are two basic underlying processes which account for the present shift toward the center on the part of both parties. One, which has been stressed by many political commentators, is the effect of the prolonged period of prosperity the country has enjoyed and the resultant increase in the entire population's standard of living. It has been argued that the lower-class base to which the Democratic party appeals is declining numerically, and therefore the party cannot advocate reform measures if it desires to win. But at least as important as the change in economic conditions is the effect on the ideological "face" of both parties of being in or out of office. The policies of the "out" party are largely set by its representatives in the House and Senate. Thus the Democrats [when] in opposition are led by conservative or centrist southerners, while the Republican leadership in Congress veers most sharply to the right when the Democrats hold the presidency.

A Democratic President is invariably to the left of the Democratic congressional leadership, since he is basically elected by the large urban industrial states where trade-unions and minority groups constitute the backbone of the party, while the southerners continue to sway the congressional Democratic contingent. Similarly, a Republican President under current conditions must remain to the left of his congressional supporters, since he, too, must be oriented toward carrying or retaining the support of the industrial, urban, and therefore more liberal sections of the country, while most Republican Congressmen are elected in "safe" conservative districts. So when the Republicans hold the presidency, they move to the left as compared to their position in opposition, while the Democrats, shifting from presidential incumbency to congressional opposition, move to the right. This shift produces a situation in which the policies of the two parties often appear almost indistinguishable.

One result in this century has been that the opposition party often faces an ideological crisis when nominating a President. From 1940 to 1952, every Republican convention witnessed a fight between a right-wing congressional candidate—Taft or Bricker—and a candidate of the liberal Governors—Willkie, Dewey, and Eisenhower. The liberal candidate has always secured the nomination. The Democrats in 1952 and 1956 faced a similar problem, but seemingly resolved it differently by nominating a relatively conservative candidate, Stevenson, and rejecting more outspoken liberals like Kefauver and Harriman, much as in 1944 they chose the then moderate Truman against the liberal Wallace for Vice-President. In large part the difference in the reaction of the professional politicians in the two parties reflects their judgment of the character of the marginal voters whom they must win to be certain of a majority. The Republicans can be sure of the votes of almost all conservatives outside of the South, particularly the "villagers"; they must increase their vote among the more liberal strata concentrated in the cities. The Democrats have the solid vote of the working class and the liberals; they can easily lose the moderates in the center. The choice of the professionals in both parties is, therefore, a man of the center.

The differences between the social bases of the two major parties which have held up for more than a century and a half suggest that those who believe in the Tweedledee-Tweedledum theory of American politics have been taken in by campaign rhetoric and miss the underlying basis of the cleavage. It is especially ironic that the Marxist critics of American politics who pride themselves on differentiating between substructure and ideology have mistaken the ideology for the substructure. American politicians, particularly the conservative ones, have always sought to suppress any overt emphasis on the real differences between the parties: the fact that they have been based on and represented different classes, ethnic groups, and religions. To acknowledge and accentuate openly these differences would seemingly lose votes. Each party wants to gain votes from those who belong to groups linked to the other faction. Hence conservative party ideologists from Clay to Taft have been denied presidential nominations by the party professionals. The Whig party could only defeat the party of the "demagogues," as they described the Democrats, in 1840 and 1848, on both of which occasions they nominated military heroes whose political opinions were almost unknown. In 1848, Whig leaders started to nominate Zachary Taylor, the hero of the Mexican War, before "it was . . . known to what party he professed to belong," a practice repeated a hundred years later by their Republican descendants.[30] The Democrats, appealing to the more numerous lower strata, have been less inhibited about making explicit appeals to interests, but even for them electoral expediency has often inhibited the expression of class consciousness. The continuing receptivity of the Democrats to the lower strata, particularly urban workers, and minority ethnic groups, explains why all efforts to create a new party based on these groups has always failed: the Democratic party has been the expression of the political consciousness of lower-status groups, much as the Republicans have fulfilled a similar function for the more privileged. The problem of electoral majorities, which in a two-party system presses both parties to appeal to the center in their electoral tactics, clearly does not negate Tocqueville's still valid observation that the two parties in America reflect the ever-present conflict in all societies between aristocracy and democracy, that is to say, between the more and the less well to do. . . .

Chapter 14

DEVIANCE

EVERY society has norms of conduct. Every society also has groups and individuals that deviate from these norms. Deviance is thus a moral category in everyday life: it designates people who don't behave the way they *should*. The sociology of deviance has tried in various ways to go beyond the moralizing aspect of the phenomenon. This has been particularly important as sociologists have become aware of relativity in the social definitions of what is normal—and, therefore, of what is deviant. Also, contemporary society has been changing very rapidly in a number of areas where the definition of deviance used to be quite stable and unambiguous. Sexuality is one such area. Earlier sociologists tended to believe that deviance (especially deviance that leads to criminal acts) could be explained in biological terms. Very few sociologists take this position today. The tendency more recently has been to ascribe deviance to peculiarities in the social-psychological development of individuals or to the distinctive traits of one's social environment.

One school in this area, which is headed by Robert Merton, has tried to explain deviance in terms of the structural tensions and the changes in functionality between the various sectors of a highly differentiated modern society. In recent years, a school based on "labeling theory" has argued that deviance is mostly a matter of arbitrary definition—that is, that society *creates* deviance by affixing that label to particular complexes of conduct. These two views are not necessarily contradictory, but phenomena of deviance tend to appear very differently when viewed from these two perspectives. Stated simply, structural-functionalists (Merton's school) tend to see deviance from the angle of society's "management," while labeling theorists tend to identify with the deviant's own viewpoint and thus to debunk the "official" definitions of deviance.

The first selection in this chapter is Michael E. Brown's "The Condemnation and Persecution of Hippies," which deals with the treatment of "hippies" as a deviant group in American society. Sociologists have dealt with the phenomenon of deviance under different headings for a long time. In recent years there has been much debate as to whether deviant conduct is to be understood in terms of objective qualities belonging to it or simply in terms of "labels" arbitrarily affixed by society. The selected essay by Ronald L. Akers, "Problems in the Sociology of Deviance: Social Definitions and Behavior," summarizes the issues involved in this debate. Delinquency and crime have always occupied an important place in sociological work on deviance. The final selection, Walter B. Miller's "Lower Class Culture as a Generating Milieu of Gang Delinquency," deals with the relationship between delinquency and specific class features.

MICHAEL E. BROWN

THE CONDEMNATION AND PERSECUTION OF HIPPIES

THIS article is about persecution and terror. It speaks of the Hippie and the temptations of intimacy that the myth of Hippie has made poignant, and it does this to discuss the institutionalization of repression in the United States.

When people are attacked as a group, they change. Individuals in the group may or may not change, but the organization and expression of their collective life will be transformed. When the members of a gathering believe that there is a grave danger imminent and that opportunities for escape are rapidly diminishing, the group loses its organizational quality. It becomes transformed in panic. This type of change can also occur outside a situation of strict urgency: When opportunities for mobility or access to needed resources are cut off, people may engage in desperate collective actions. In both cases, the conversion of social form occurs when members of a collectivity are about to be hopelessly locked into undesired and undesirable positions.

The process is not, however, automatic. The essential ingredient for conversion is social control exercised by external agents on the collectivity itself. The result can be benign, as a panic mob can be converted into a crowd that makes an orderly exit from danger. Or it can be cruel.

The transformation of groups under pressure is of general interest; but there are special cases that are morally critical to any epoch. Such critical cases occur when pressure is persecution, and transformation is destruction. The growth of repressive mechanisms and institutions is a key concern in this time of administrative cruelty. Such is the justification for the present study.

SOCIAL CONTROL AS TERROR

Four aspects of repressive social control such as that experienced by Hippies are important. First, the administration of control is suspicious. It projects a dangerous future and guards against it. It also refuses the risk of inadequate coverage by enlarging the controlled population to include all who might be active in any capacity. Control may or may not be administered with a heavy hand, but it is always a generalization applied to specific instances. It is a rule and thus ends by pulling many fringe innocents into its bailiwick; it creates as it destroys.

Published by permission of *TRANS-action,* Inc., from *TRANS-action,* Vol. 6, Sept. 1969.

Second, the administration of control is a technical problem which depending on its site and object, requires the bringing together of many different agencies that are ordinarily dissociated or mutually hostile. A conglomerate of educational, legal, social welfare, and police organization is highly efficient. The German case demonstrates that. Even more important, it is virtually impossible to oppose control administered under the auspices of such a conglomerate since it includes the countervailing institutions ordinarily available. When this happens control is not only efficient and widespread, but also legitimate, commanding a practical, moral and ideological realm that is truly "one-dimensional."

Third, as time passes, control is applied to a wider and wider range of details, ultimately blanketing its objects' lives. At that point, as Hilberg suggests in his *The Destruction of the European Jews,* the extermination of the forms of lives leads easily to the extermination of the lives themselves. The line between persecution and terror is thin. For the oppressed, life is purged of personal style as every act becomes inexpressive, part of the struggle for survival. The options of a life-style are eliminated at the same time that its proponents are locked into it.

Fourth, control is relentless. It develops momentum as organization accumulates, as audiences develop, and as unofficial collaborators assume the definition of tasks, expression and ideology. This, according to W. A. Westley's "The Escalation of Violence Through Legitimation," is the culture of control. It not only limits the behaviors, styles, individuals and groups toward whom it is directed, it suppresses all unsanctioned efforts. As struggle itself is destroyed, motivation vanishes or is turned inward.

These are the effects of repressive control. We may contrast them with the criminal law, which merely prohibits the performance of specific acts (with the exception, of course, of the "crime without victims"—homosexuality, abortion, and drug use). Repression converts or destroys an entire social form, whether that form is embodied in a group, a style or an idea. In this sense, it is terror.

These general principles are especially relevant to our understanding of tendencies that are ripening in the United States day by day. Stated in terms that magnify it so that it can be seen despite ourselves, this is the persecution of the Hippies, a particularly vulnerable group of people who are the cultural wing of a way of life recently emerged from its quiet and individualistic quarters. Theodore Roszak, describing the Hippies in terms of their relationship to the culture and politics of dissent, notes that "the underlying unity of youthful dissent consists . . . in the effort of beat-hip bohemianism to work out the personality structure, the total life-style that follows from New Left social criticism." This life-style is currently bearing the brunt of the assault on what Roszak calls a "counter-culture"; it is an assault that is becoming more concentrated and savage every day. There are lessons for the American future to be drawn from this story.

PERSECUTION

Near Boulder, Colorado, a restaurant sign says "Hippies not served here." Large billboards in upstate New York carry slogans like "Keep America Clean: Take a Bath," and "Keep America Clean: Get a Haircut." These would be as amusing as ethnic jokes if they did not represent a more systematic repression.

The street sweeps so common in San Francisco and Berkeley in 1968 and 1969 were one of the first lines of attack. People were brutally scattered by club-wielding policemen who first closed exits from the assaulted area and then began systematically to beat and arrest those who were trapped. This form of place terror, like surveillance in Negro areas and defoliation in Vietnam, curbs freedom and forces people to fight or submit to minute inspection by hostile forces. There have also been one-shot neighborhood pogroms, such as the police assault on the Tompkins Square Park gathering in New York's Lower East Side on Memorial Day, 1967: "Sadistic glee was written on the faces of several officers," wrote the *East Village Other*. Some women became hysterical. The police slugged Frank Wise, and dragged him off, handcuffed and bloody, crying, "My God, my God, where is this happening? Is this America?" The police also plowed into a group of Hippies, Yippies, and straights at the April, 1968, "Yip-in" at Grand Central Station. The brutality was as clear in this action as it had been in the Tompkins Square bust. In both cases, the major newspapers editorialized against the police tactics, and in the first the Mayor apologized for the "free wielding of nightsticks." But by the summer of 1968, street sweeps and busts and the continuous presence of New York's Tactical Police Force had given the Lower East Side an ominous atmosphere. Arrests were regularly accompanied by beatings and charges of "resistance to arrest." It became clear that arrests rather than subsequent procedures were the way in which control was to be exercised. The summer lost its street theaters, the relaxed circulation of people in the neighborhood and the easy park gatherings.

Official action legitimizes nonofficial action. Private citizens take up the cudgel of law and order newly freed from the boundaries of due process and respect. After Tompkins Square, rapes and assaults became common as local toughs assumed the role, with the police, of defender of the faith. In Cambridge, Massachusetts, following a virulent attack on Hippies by the Mayor, *Newsweek* reported that vigilantes attacked Hippie neighborhoods in force.

Ultimately more damaging are the attacks on centers of security. Police raids on "Hippie pads," crash pads, churches and movement centers have become daily occurrences in New York and California over the past two and a half years. The usual excuses for raids are drugs, runaways and housing violations, but many incidents of unlawful entry by police and the expressions of a more generalized hostility by the responsible officials suggests that something deeper is involved. The Chief of Police in San Francisco put it bluntly; quoted in *The New York Times Magazine* in May, 1967, he said:

> Hippies are no asset to the community. These people do not have the courage to face the reality of life. They are trying to escape. Nobody should let their young children take part in this hippy thing.

The Director of Health for San Francisco gave teeth to this counsel when he sent a task force of inspectors on a door-to-door sweep of the Haight-Ashbury—"a two-day blitz" that ended with a strange result, again according to *The Times:* Very few of the Hippies were guilty of housing violations.

Harassment arrests and calculated degradation have been two of the most effective devices for introducing uncertainty to the day-to-day lives of the Hippies. Cambridge's Mayor's attack on the "hipbos" (the suffix stands for body odor) included, said *Newsweek* of Oct. 30, 1967, a raid on a "hippie pad" by the Mayor and "a platoon of television cameramen." They "seized a pile of diaries and personal letters and flushed a partially clad girl from the closet." In Wyoming, *The Times* reported that two "pacifists" were "jailed and shaved" for hitchhiking. This is a fairly common hazard, though Wyoming officials are perhaps more sadistic than most. A young couple whom I interviewed were also arrested in Wyoming during the summer of 1968. They were placed in solitary confinement for a week during which they were not permitted to place phone calls and were not told when or whether they would be charged or released. These are not exceptional cases. During the summer of 1968, I interviewed young hitchhikers throughout the country; most of them had similar stories to tell.

In the East Village of New York, one hears countless stories of apartment destruction by police (occasionally reported in the newspapers), insults from the police when rapes or robberies are reported, and cruel speeches and even crueler bails set by judges for arrested Hippies.

In the light of this, San Francisco writer Mark Harris' indictment of the Hippies as paranoid seems peculiar. In the September 1967 issue of *The Atlantic,* he wrote,

> The most obvious failure of perception was the hippies failure to discriminate among elements of the Establishment, whether in the Haight-Ashbury or in San Francisco in general. Their paranoia was the paranoia of all youthful heretics. . . .

This is like the demand of some white liberals that Negroes acknowledge that they (the liberals) are not the power structure, or that black people must distinguish between the good and the bad whites despite the fact that the black experience of white people in the United States has been, as the President's Commission on Civil Disorder suggested, fairly monolithic and racist.

Most journalists reviewing the "Hippie scene" with any sympathy at all seem to agree with *Newsweek* that "the hippies do seem natural prey for publicity-hungry politicians—if not overzealous police," and that they have been subjected to varieties of cruelty that ought to be intolerable. This tactic was later elaborated in the massive para-military assault on Berkeley residents and students during a demonstration in support of Telegraph Avenue's street people and their People's Park. The terror of police violence, a constant in the lives of street people everywhere, in California carries the additional threat of martial law under a still-active state of extreme emergency. The whole structure of repression was given legitimacy and reluctant support by University of California officials. Step by step, they became allies of Reagan's "dogs of war." Roger W. Heyns, chancellor of the Berkeley campus, found himself belatedly reasserting the university's property in the lot. It was the law and the rights of university that trapped the chancellor in the network of control and performed the vital function of providing justification and legitimacy for Sheriff Madigan and the National Guard. Heyns said: "We will have to put up a fence to re-establish the conveniently forgotten fact that this field is indeed the university's and to exclude unauthorized personnel from the site. . . . The fence will give us time to plan and consult. We tried to get this time some other way and failed—hence the fence." And hence "Bloody Thursday" and the new regime.

And what of the Hippies? They have come far since those balmy days of 1966–67, days of flowers, street-cleaning, free stores, decoration and love. Many have fled to the hills of Northern California to join their brethren who had set up camps there several years ago. Others have fled to communes outside the large cities and in the Middle West. After the Tompkins Square assault, many of the East Village Hippies refused to follow the lead of those who were more political. They refused to develop organizations of defense and to accept a hostile relationship with the police and neighborhood. Instead, they discussed at meeting after meeting, how they could show their attackers love. Many of those spirits have fled; others have been beaten or jailed too many times; and still others have modified their outlook in ways that reflect the struggle. Guerrilla theater, Up Against the Wall Mother Fucker, the Yippies, the urban communes; these are some of the more recent manifestations of the alternative culture. One could see these trends growing in the demonstrations mounted by Hippies against arrests of runaways or pot smokers, the community organizations, such as grew in Berkeley for self-defense and politics, and the beginnings of the will to fight back when trapped in street sweeps.

It is my impression that the Hippie culture is growing as it recedes from the eye of the media. As a consequence of the destruction of their urban places, there has been a redistribution of types. The flower people have left for the hills and become more communal; those who remained in the city were better adapted to terror, secretive or confrontative. The Hippie culture is one of the forms radicalism can take in this society. The youngsters, 5,000 of them, who came to Washington to counter-demonstrate against the Nixon inaugural showed the growing amalgamation of the New Left and its cultural wing. The Yippies who went to Chicago for guerrilla theater and learned about "pigs" were the multi-generational expression of the new wave. A UAWMF (Up Against the Wall Mother Fucker) drama, played at Lincoln

Center during the New York City garbage strike—they carted garbage from the neglected Lower East Side and dumped it at the spic 'n' span cultural center—reflected another interpretation of the struggle, one that could include the politically militant as well as the culturally defiant. Many Hippies have gone underground—in an older sense of the word. They have shaved their beards, cut their hair, and taken straight jobs, like the secret Jews of Spain; but unlike those Jews, they are consciously an underground, a resistance.

What is most interesting and, I believe, a direct effect of the persecution, is the enormous divergence of forms that are still recognizable by the outsider as Hippie and are still experienced as a shared identity. "The Yippies," says Abbie Hoffman, "are like Hippies, only fiercer and more fun." The "hippie types" described in newspaper accounts of drug raids on colleges turn out, in many cases, to be New Leftists.

The dimensions by which these various forms are classified are quite conventional: religious-political, visible-secret, urban-hill, communal-individualistic. As their struggle intensifies, there will be more efforts for unity and more militant approaches to the society that gave birth to a real alternative only to turn against it with a mindless savagery. Yippie leader Jerry Rubin, in an "emergency letter to my brothers and sisters in the movement" summed up:

> Huey Newton is in prison.
> Eldridge Cleaver is in exile.
> Oakland Seven are accused of conspiracy.
> Tim Leary is up for 30 years and how many of our brothers are in court and jail for getting high?
> . . .
> Camp activists are expelled and arrested.
> War resisters are behind bars.
> Add it up!

RONALD L. AKERS

PROBLEMS IN THE SOCIOLOGY OF DEVIANCE: SOCIAL DEFINITIONS AND BEHAVIOR

THE conflict criminologist, George Vold, reminded us some years ago that the phenomenon of crime involves two major dimensions—the behavioral and the definitional:

> There is, therefore, always a dual problem of explanation—that of accounting for the behavior *as behavior,* and equally important, accounting *for the definitions* by which specific behavior comes to be considered crime or non-crime.[1]

A growing number of criminologists have become aware of the two-sided nature of the problem and suggest that interest be turned increasingly to the study of and accounting for the contents of the criminal law. They have suggested that the legal norms defining certain behavior as crime be subjected to analysis and explanations sought for why some acts are defined as crime and others are not.[2] One has even suggested that this constitutes *the* problem in the sociology of crime.[3]

While directly traceable to a conflict orientation,[4] this newer emphasis on studying the law itself has also received major impetus from the theoretical issues raised by an interest in white-collar crime. The very differences between the set of laws regulating occupational behavior and other statutes embodying legal proscriptions and sanctions have raised questions about how and why they were enacted—not just why they have been violated.[5] The study of white-collar crime, as Newman says, calls for the investigator ". . . to cast his analysis not only in the framework of those who *break* laws, but in the context of those who *make* laws as well." [6]

In a recent article, Gibbs notes that one of the major questions left unanswered by the "new" labelling conception of deviance is whether the ultimate goal is "to explain deviant behavior or to explain reactions to deviations?" [7] A re-reading of the literature expounding this approach would suggest that the goal is not to account for *either* the behavior *or* the reaction, but *both.* Thus, in a sense, the labellers have illuminated the twofold problem of expla-

Reprinted from Ronald L. Akers, "Problems in the Sociology of Deviance: Social Definitions and Behavior," *Social Forces* 46 (June 1968): 455–65 by permission of the University of North Carolina Press.

nation in the broader study of deviance just as conflict and white-collar crime perspectives have done in the narrower field of criminology.

These, then, are the two basic problems facing the sociology of deviance: How and/or why certain kinds of behavior and people get defined and labelled as deviant. How and/or why some people engage in deviant acts. Our research must provide data on and our theories should explain *both* social definitions and behavior. The purpose of this paper is to explore the meaning of these two problems, the *nature* of theories and research revolving around them, and their implications for the future direction of the sociology of deviance.

THE BEHAVIORAL QUESTIONS

The explanation of deviant behavior must address itself to two interrelated problems: (a) accounting for the group and structural variations in rates of deviancy, and (b) describing and explaining the process by which individuals come to commit acts labelled deviant. Cressey, following Sutherland, refers to them as the problems of epidemiology and individual conduct.[8] Other terms could be used, but I prefer to talk about structural and processual questions. Some want to speak of them as the sociological and psychological "levels" of explanations.[9] Reference to the two problems of behavioral explanation in these terms continues the old Durkheimian polemic designed to assure sociologists that they have a unique discipline. This polemic in intellectual imperialism is probably no longer meaningful or necessary; but if it is, it is not relevant to this distinction between structural and processual explanations of deviant behavior, for both can be sociological.[10]

The theoretical emphasis in the sociology of deviance has been on structural explanations, however, and we have some fairly sophisticated notions about the kinds of structures and environments which produce certain kinds of deviancy. But this emphasis should be balanced by an increased concern with the process by which these environments do so. Although there are excellent miniature theories of embezzlement, drug addiction, marihuana use, and check forgery, little recently has been done to reconstruct the process of coming to commit various kinds of deviant behavior.[11] Moreover, there have been virtually no efforts to locate the common elements in these separate explanations toward the end of testing or constructing a general processual theory.

Sutherland's differential association theory is notable for standing nearly alone as a general processual theory of criminal behavior.[12] Sutherland recognized that while theories addressed to structural and processual questions may be different, the two must be consistent.[13] Not only should they be consistent, but as Cressey notes, they should be integrated.[14] This integration is possible because in the final analysis both kinds of theory propose answers to the *same overall* question of why some people come to commit deviant acts and others do not. The structural theories contend that more people in certain groups, located in certain positions in, or encountering particular pressures created by the social structure, will engage in deviancy than those in other groups and locations.[15] In a sense, they explain between-group variations. But in so doing, they implicitly or explicitly posit processes by which these structural conditions produce higher probabilities of deviancy. Processual theories explain within-group variation. They say that the individual commits deviancy because he has encountered a particular life history. But in so doing, they are also saying something about the deviancy-producing groups, structures, and circumstances he must encounter to increase the probability of his becoming and remaining deviant.[16]

The differential association-reinforcement theory formulated by Burgess and Akers avoids some of the problems of Sutherland's original formulation and describes the general process (consistently and integrally with a broader theory of behavior) of deviant behavior.[17] It is capable of identifying the common elements in the separate processual theories and provides the groundwork for integrating structural and processual explanations. By conceptualizing groups and social structure as learning environments which structure the patterns of associations and reinforcement, a long step is taken in the direction of bringing the two together. Differential association-reinforcement spells out the *mechanisms* by which environmental stimuli produce and maintain behavior and the structural theories explicate the type of *environments* most likely to sustain norm and law-violating behavior.

But before we will learn which theory or combination of theories turns out to be our best explanation of deviance, we must broaden our data-gathering technology beyond the traditional research strategies. Compared to those in other sociological specialties, students of deviancy have relatively ready access to regularly compiled data on behavior that has come to the attention of official agencies. Nevertheless, we still do not know very much about even the official distribution and variations in rates of some kinds of deviancy and are practically ignorant of the true distribution of nearly every type of deviant behavior. To test theories adequately, we need to utilize as broad a range of research techniques and as representative data as possible. This means that we must be more imaginative in ferreting out other sources of data beyond the usual official statistics, records, and populations of apprehended offenders to get a better idea of the distribution of rates of deviant behavior. More precise knowledge of the behavioral process in deviancy not only awaits more systematic and extensive use of case histories and analytic inductive techniques, but requires greater utilization of laboratory experimentation.

Reliable and valid techniques of studying delinquency in representative samples of the general adolescent population are being developed, and these could be extended into other fields.[18] Of course, there is much yet to be done even in the restricted area of delinquency, and it could be argued that these techniques are simply inappropriate to some types of deviancy. But the spirit, if not the substance, of the developing technology of unofficial measures of delinquency could be applied to some adult violations. This could be combined with the utilization of official and semi-official compilations. We have not yet realized the full implications of the pioneering work of Sutherland, Clinard and Hartung on white-collar offenses.[19] There is a whole class of research sites of which we have thus far little availed ourselves—the files and records of private police, detective, security and similar agencies, insurance and management consultant firms, and business and occupational control boards and commissions.[20] Also, we are just beginning to recognize the importance and potential impact on sociology of the very effective experimental technology developing in the operant conditioning tradition.[21]

SOCIAL DEFINITIONS

In the broadest sense, the problem of explaining social definitions is that of accounting for why people come to have the values and norms they do. But more specifically the problem is to learn how the prevailing conceptions of what behavior constitutes the major forms of deviancy in society have come to be established. This entails examining two related processes: (a) establishing the rules, definitions, norms, and laws the infraction of which constitutes deviance, and (b) reacting to people who have or are believed to have violated the norms by applying negative sanctions and labels to them (and applying deviant labels to others not because of their actions but because they possess some physical characteristic or disability). We are infinitely more knowledgeable about both the behavioral questions than either of these two aspects of the defining and labelling process.

Those criminologists who have given thought to the problem of why certain behavior is defined as criminal offer explanations that converge on a group-conflict theme—what becomes defined as illegal is related to the ability of certain groups in society to have their values, norms, and interests incorporated into or reflected in the law and its administration even against the interests of other groups and the general public.[22] Thus, they have been led to recognize the political nature of crime and a well-documented observation concerning politics in democratic society: the passage and content of nearly all laws, the formulation of nearly all public policy, and nearly every public decision, including court decisions, are in some measure the outcome of the direct or indirect political influence of competing interest groups.[23]

Investigation of the political process by which the criminal label is established and applied remains a neglected area, however. What literature there is on this point has been largely suggestive and programmatic, with little in the way of research.[24] We will have to scrutinize more carefully the process by which the criminal law is formed and enforced in a search for those variables which determine what of the total range of behavior becomes prohibited and which of the total range of norms become a part of the law. Certainly these variables would include not only the activities of political-interest groups but also a range of factors in the changing social, economic, political, and normative structures of society, its historical development, legal traditions, governmental forms, and the general political process.

It is the politically dominant subunits of society that at any given time can see to it that public policy reflects their interests, whether these be whole classes or segments, major vested interests, more specialized private groups, or even agencies of the government itself which exert influence in the law-making and enforcing processes. However, the law reflects not only the interests of particular segments but also the changing needs and functions of the whole society. The extent to which public policy is the result of the victories, compromises, and resolutions of group conflict is an empirical question that must be answered for the specific form of deviancy in question. Empirical research on this question may take the form of longitudinal or current study of some policy-in-the-making or historical reconstruction of the way the policy came into being. Such research is already available on theft law,[25] vagrancy law,[26] public policy on drugs,[27] prohibition,[28] and legal regulation of certain professional practices.[29] But the surface has just been scratched. The field is wide open for sociological research in the process of legislation, the content of laws, and the operation of administrative agencies, courts, police,[30] and other agents of formal social control as well as the operation and function of private police and detectives as more informal enforcement agencies.

Conflict explanations and research into the degree to which group interests form a part of the total political process is only a start. If differences in the power of interest groups account for some differences in the laws and public policy, what makes for power? [31] What differentials in organizational and other group properties account for differential political influence? Sociological theory has been nearly mute and systematic research almost nonexistent on this problem.[32]

This type of research should be complemented by research into the variables operative in the defining and reacting to deviance by more informal social audiences than those connected with the formulation and implementation of the law. The significant theoretical problem here is the relationship between prevailing moral sentiments of the society and the normative stances incorporated into public policy. This concern with the overlap of current public opinion and the contents of the law is part of the old question of whether the law always flows from or can induce changes in the folkways and mores. The reasonable answer, of course, is that law is both an independent and dependent variable in society, but research and theoretical perspectives are needed which define when and under what conditions it is one or the other.[33] Some research has already been reported which utilizes survey data on public views of certain kinds of deviancy and public policy,[34] and "impact" studies have been conducted or are underway in the sociology of law.[35] This sort of research could be combined with studies of the similarities in the norms of specialized groups and particular legal norms.[36] Research of this nature may not tell us anything about the basic process by which norms are formed and social control is exercised, however. We may have to have recourse to historical, cross-cultural, and small-group experimental studies to get at this question.[37]

LABELLING AND DEVIANT BEHAVIOR

The professors of the labelling perspective contend that the important questions concerning deviancy are: Who applies the deviant label to whom? Whose rules shall prevail? Under what circumstances is the label successfully and unsuccessfully applied? How does a community decide what forms of conduct should be singled out for this kind of attention?[38] These kinds of questions do not exclude a concern with deviant behavior, but they do give it secondary importance. One would expect, then, that this approach offers a valuable balance to exclusive concern with the causes of deviation and the characteristics of deviants. Indeed, some of the research alluded to above has been generated by this perspective. However, the theoretical contribution of this approach to the problem of social definitions has not been as great as its promise.

Today only the most unregenerate biological or constitutional determinist would quibble with the basic contention of this school that the deviant nature of acts resides not in the acts or the person committing them, but rather in group definitions and reactions. Certainly, it has been a long time since sociologists have said otherwise. It is true that in the past we have sometimes forgotten the basically social nature of deviance, and in an effort to untangle the etiology, we have become overly concerned with the conditions and characteristics of the deviants themselves. But this is just another way of saying that we have devoted most of our energy to the behavioral question and have implicitly accepted, as given, the established norms defining various kinds of behavior as deviant. Nonetheless, this in no way implies that sociologists thought that there was something inherently evil or deviant about the behavior itself. Not since Garofalo have we attempted to erect universal or natural categories of inherently criminal behavior, and certainly since Durkheim we have been cognizant of the centrality of social definitions to the conception of criminal and deviant behavior. Yet, much of the effort expended by the writers in the labelling tradition has been to exorcise this nonexistent fallacy.

The labelling approach does rightly emphasize, neither wholly originally nor uniquely, the importance of studying social definitions and the process by which acts and people get labelled as deviant. But when labelling theorists have attempted to answer questions about social definitions, they say little more than what conflict theorists have been saying for some time, i.e., the dominant groups in society will have their norms and values prevail, will successfully apply their conceptions of who are the deviants and become more or less official definers of deviancy.[39] In fact, the most sophisticated statement (going much beyond this) about the determinants of one type of labelling, "criminalization," is in a recent article written from an avowedly conflict perspective, not a stigma or labelling perspective.[40] Rather, although those of this school come dangerously close to saying that the actual behavior is unimportant, their contribution to the study of deviancy comes precisely in their conception of the impact of labelling on behavior. One sometimes gets the impression from reading this literature that people go about minding their own business, and then—"wham"—bad society comes along and slaps them with a stigmatized label. Forced into the role of deviant the individual has little choice but to be deviant. This is an exaggeration, of course, but such an image can be gained easily from an overemphasis on the impact of labelling. However, it is exactly this image, toned down and made reasonable, which is the central contribution of the labelling school to the sociology of deviance.

Thanks to this image, we are now more appreciative of the impact of norm enforcement in the furtherance of deviancy. Societal reaction may deter individuals from engaging in further deviant behavior, but it may not effectively reduce the behavior it was designed to combat. In fact, it may play a role in setting up conditions conducive to subsequent and other deviancy. The stigmatization of deviance may have an impact such that the deviant comes to view himself as irrevocably deviant, becomes more committed to a deviant role, or becomes involved in deviant groups; this influences his future deviance and may force him to participate in various kinds of secondary deviance.[41] This perspective has also generated some ideas about what kinds of deviancy are likely to be affected in this way.[42]

When carried too far, however, this insight serves as a blinder. The labelling creates the deviance, yes, and often operates to increase the probability that certain stigmatized persons will commit future deviancy, and to promote deviant behavior that might not have occurred otherwise. But the label does not create the behavior in the *first place*. People can and do commit deviant acts because of the particular contingencies and circumstances in their lives, quite apart from or in combination with the labels others apply to them.[43] The labelling process is not completely arbitrary and unrelated to the behavior of those detected and labelled. Although errors are made and criteria extraneous to behavior are used, we do not react to others as homosexuals unless they exhibit behavior believed to be indicative of homosexuality, and the courts do not stigmatize with the label of criminal until it has been legally determined that criminal acts have been committed. There were addicts loose in the land long before Anslinger and the Narcotics Bureau were let loose on the addicts. Obviously, the behavior is not itself deviant; it is only because others have defined it so. But once defined, aside from questions of secondary deviations, the behavior is prior to the labelling reaction. One may say that, in this sense, the behavior creates the label.

Such statements do not go far enough, however. Neither behavior nor its social definitions occur in a vacuum; they are mutually influencing. A given class of behaviors and those performing them may carry no particular stigma or at best be mildly disapproved by the norm-enforcing and conventional segments of society until such time as they become more widespread, visible, and offensive to extant notions of propriety. There is certainly some question about the deviancy of taking *LSD-25* and similar drugs until some and then many began this practice. Who, just a few months ago, would have condemned the smoking of the baked parts of the inside of banana skins? There surely were those who made regular use of certain drugs before that use became proscribed. But that use began to violate other norms regarding drug "abuse," became offensive to large parts of the population, partly because it was by those already considered deviant on other counts, and was stigmatized to some degree. Once the behavior was proscribed and deviant labels attached to those engaging in this behavior, these users became more committed deviants, subcultural support for the behavior evolved, and various forms of secondary deviancy emerged. This then evoked stronger prohibitions and more stigmatized and even criminal labels, which then turned some off and others on the drugs. The theoretical problem, then, becomes one of specifying the ways and under what conditions such an interactive process involving both behavior and definitions will take place. What determines when a behavior pattern not previously specifically stigmatized will become defined as deviant and when labelling increases or decreases the probability of further involvement in that behavior pattern?

CONCLUSIONS

The two major problems of accounting for behavior and definitions are of equal importance, but both criminology and the broader field of the sociology of deviance have tended to give short shrift to explaining social definitions of deviancy. We have tended to see our job nearly exclusively as the study of those who break the laws and violate the norms. The time is here for increased theoretical and research interest in the making and enforcing of the laws and norms. Conflict theorists, those criminologists interested in white-collar crime, and those in the labelling school of deviance have been more alert to the problem than have those operating from other perspectives. But we are in the beginning stage of mapping out the parameters and still know very little about the determining variables in the social defining and reacting process.

To give more complete answers to the questions about deviant behavior, increased attention needs to be turned to explicating the process by which individuals come to commit deviant acts, and eventually we must integrate processual and structural explanations. We are not likely to come up with a complete explanation of deviant behavior in general until we tie it into a general theory of social behavior. By the same token, the study of societal control of deviance must be tied to the larger study and theory of conflict, power, and norm formation and enforcement.

Finally, the ultimate goal should be to specify the interaction and integrate explanations of behavior and social definitions. Jeffery has attempted something similar to this regarding crime and the law.[44] His formulation is not entirely satisfactory as a broader solution, however, mainly because it is too narrowly concerned with these rather than the whole problem of behavior and social definitions. It should be said, by no means originally and at the risk of sounding trite and mouthing platitudes, that the solution will need to be general enough that the sociology of deviance becomes simply sociology. Patterned social behavior, both conforming and nonconforming, is the stuff of which society is made, and social definitions is just another term for normative and value patterns, i.e., culture. The study of society *and* culture, of course, is the stuff of which sociology is made.

WALTER B. MILLER

LOWER CLASS CULTURE AS A GENERATING MILIEU OF GANG DELINQUENCY

THE etiology of delinquency has long been a controversial issue and is particularly so at present. As new frames of reference for explaining human behavior have been added to traditional theories, some authors have adopted the practice of citing the major postulates of each school of thought as they pertain to delinquency, and of going on to state that causality must be conceived in terms of the dynamic interaction of a complex combination of variables on many levels. The major sets of etiological factors currently adduced to explain delinquency are, in simplified terms, the physiological (delinquency results from organic pathology), the psychodynamic (delinquency is a "behavioral disorder" resulting primarily from emotional disturbance generated by a defective mother-child relationship), and the environmental (delinquency is the product of disruptive forces, "disorganization," in the actor's physical or social environment).

This paper selects one particular kind of "delinquency" [1]—law-violating acts committed by members of adolescent street corner groups in lower class communities—and attempts to show that the dominant component of motivation underlying these acts consists in a directed attempt by the actor to ad-

Reprinted from *The Journal of Social Issues* 14 (1958): 5–19 by permission of the publisher.

here to forms of behavior, and to achieve standards of value, as they are defined within that community. It takes as a premise that the motivation of behavior in this situation can be approached most productively by attempting to understand the nature of cultural forces impinging on the acting individual as they are perceived *by the actor himself*—although by no means only that segment of these forces of which the actor is consciously aware—rather than as they are perceived and evaluated from the reference position of another cultural system. In the case of "gang" delinquency, the cultural system which exerts the most direct influence on behavior is that of the lower class community itself—a long-established, distinctively patterned tradition with an integrity of its own—rather than a so-called "delinquent subculture" which has arisen through conflict with middle class culture and is oriented to the deliberate violation of middle class norms.

The bulk of the substantive data on which the following material is based was collected in connection with a service-research project in the control of gang delinquency. During the service aspect of the project, which lasted for three years, seven trained social workers maintained contact with twenty-one corner group units in a "slum" district of a large eastern city for periods of time ranging from ten to thirty months. Groups were Negro and white, male and female, and in early, middle, and late adolescence. Over eight thousand pages of direct observational data on behavior patterns of group members and other community residents were collected; almost daily contact was maintained for a total time period of about thirteen worker years. Data include workers' contact reports, participant observation reports by the writer—a cultural anthropologist—and direct tape recordings of group activities and discussions.[2]

FOCAL CONCERNS OF LOWER CLASS CULTURE

There is a substantial segment of present-day American society whose way of life, values, and characteristic patterns of behavior are the product of a distinctive cultural system which may be termed "lower class." Evidence indicates that this cultural system is becoming increasingly distinctive, and that the size of the group which shares this tradition is increasing.[3] The lower class way of life, in common with that of all distinctive cultural groups, is characterized by a set of focal concerns—areas or issues which command widespread and persistent attention and a high degree of emotional involvement. The specific concerns cited here, while by no means confined to the American lower classes, constitute a distinctive *patterning* of concerns which differs significantly, both in rank order and weighting, from that of American middle class culture. Table 19 presents a highly schematic and simplified listing of six of the major concerns of lower class culture. Each is conceived as a "dimension" within which a fairly wide and varied range of alternative behavior patterns may be followed by different individuals under different situations. They are listed roughly in order of the degree of *explicit* attention

TABLE 19

Focal Concerns of Lower Class Culture

AREA	PERCEIVED ALTERNATIVES (STATE, QUALITY, CONDITION)	
1. Trouble:	law-abiding behavior	law-violating behavior
2. Toughness:	physical prowess, skill; "masculinity"; fearlessness, bravery; daring	weakness, ineptitude; effeminacy; timidity, cowardice; caution
3. Smartness:	ability to outsmart, dupe, "con"; gaining money by "wits"; shrewdness, adroitness in repartee	gullibility, "con-ability"; gaining money by hard work; slowness, dull-wittedness, verbal maladroitness
4. Excitement:	thrill; risk, danger; change, activity	boredom; "deadness," safeness; sameness, passivity
5. Fate:	favored by fortune, being "lucky"	ill-omened, being "unlucky"
6. Autonomy:	freedom from external constraint; freedom from superordinate authority; independence	presence of external constraint; presence of strong authority; dependency, being "cared for"

accorded each and, in this sense, represent a weighted ranking of concerns. The "perceived alternatives" represent polar positions which define certain parameters within each dimension. As will be explained in more detail, it is necessary in relating the influence of these "concerns" to the motivation of delinquent behavior to specify *which* of its aspects is oriented to, whether orientation is *overt* or *covert, positive* (conforming to or seeking the aspect) or *negative* (rejecting or seeking to avoid the aspect).

The concept "focal concern" is used here in preference to the concept "value" for several interrelated reasons: (1) It is more readily derivable from direct field observation. (2) It is descriptively neutral—permitting independent consideration of positive and negative valences as varying under different conditions, whereas "value" carries a built-in positive valence. (3) It makes possible more refined analysis of subcultural differences, since it reflects actual behavior, whereas "value" tends to wash out intracultural differences since it is colored by notions of the "official" ideal.

Trouble. Concern over "trouble" is a dominant feature of lower class culture. The concept has various shades of meaning; "trouble" in one of its aspects represents a situation or a kind of behavior which results in unwelcome or complicating involvement with official authorities or agencies of middle class society. "Getting into trouble" and "staying out of trouble" represent major issues for male and female, adults and children. For men, "trouble" frequently involves fighting or sexual adventures while drinking; for women, sexual involvement with disadvantageous consequences. Expressed desire to avoid behavior which violates moral or legal norms is often based less on an explicit commitment to "official" moral or legal standards than on a desire to avoid "getting into trouble," e.g., the complicating consequences of the action.

The dominant concern over "trouble" involves a distinction of critical importance for the lower class community—that between "law-abiding" and "non-law-abiding" behavior. There is a high degree of sensitivity as to where each person stands in relation to these two classes of activity. Whereas in the middle class community a major dimension for evaluating a person's status is "achievement" and its external symbols, in the lower class personal status is very frequently gauged along the law-abiding–non-law-abiding dimension. A mother will evaluate the suitability of her daughter's boyfriend less on the basis of his achievement potential than on the basis of his innate "trouble" potential. This sensitive awareness of the opposition of "trouble-producing" and "non-trouble-producing" behavior represents both a major basis for deriving status distinctions and an internalized conflict potential for the individual.

As in the case of other focal concerns, which of two perceived alternatives —"law-abiding" or "non-law-abiding"—is valued varies according to the individual and the circumstances; in many instances there is an overt commitment to the "law-abiding" alternative, but a covert commitment to the "non-law-abiding." In certain situations, "getting into trouble" is overtly recognized as prestige-conferring; for example, membership in certain adult and adolescent primary groupings ("gangs") is contingent on having demonstrated an explicit commitment to the law-violating alternative. It is most important to note that the choice between "law-abiding" and "non-law-abiding" behavior is still a choice *within* lower class culture; the distinction between the policeman and the criminal, the outlaw and the sheriff, involves primarily this one dimension; in other respects they have a high community of interests. Not infrequently brothers raised in an identical cultural milieu will become police and criminals respectively.

For a substantial segment of the lower class population "getting into trouble" is not in itself overtly defined as prestige-conferring, but is implicitly recognized as a means to other valued ends, e.g., the covertly valued desire to be "cared for" and subject to external constraint, or the overtly valued state of excitement or risk. Very frequently "getting into trouble" is multifunctional and achieves several sets of valued ends.

Toughness. The concept of "toughness" in lower class culture represents a compound combination of qualities or states. Among its most important components are physical prowess, evidenced both by demonstrated possession of strength and endurance and by athletic skill; "masculinity," symbolized by a distinctive complex of acts and avoidances (bodily tattooing, absence of sentimentality, non-concern with "art," "literature," conceptualization of women as conquest objects, etc.); and bravery in the face of physical threat. The model for the "tough guy"—hard, fearless, undemonstrative, skilled in physical combat—is represented by the movie gangster of the thirties, the "private eye," and the movie cowboy.

The genesis of the intense concern over "toughness" in lower class culture is probably related to the fact that a significant proportion of lower class males are reared in a predominantly female household and lack a consistently present male figure with whom to identify and from whom to learn essential components of a "male" role. Since women serve as a primary object of identification during pre-adolescent years, the almost obsessive lower class concern with "masculinity" probably resembles a type of compulsive reaction-formation. A concern over homosexuality runs like a persistent thread through lower class culture. This is manifested by the institutionalized practice of baiting "queers," often accompanied by violent physical attacks, an expressed contempt for "softness" or frills, and the use of the local term for "homosexual" as a generalized pejorative epithet (e.g., higher class individuals or upwardly mobile peers are frequently characterized as "fags" or "queers"). The distinction between "overt" and "covert" orientation to aspects of an area of concern is especially important in regard to "toughness." A positive overt evaluation of behavior defined as "effeminate" would be out of the question for a lower class male; however, built into lower class culture is a range of devices which permit men to adopt behaviors and concerns which in other cultural milieux fall within the province of women, and at the same time to be defined as "tough" and manly. For example, lower class men can be professional short-order cooks in a diner and still be regarded as "tough." The highly intimate circumstances of the street corner gang involve the recurrent expression of strongly affectionate feelings towards other men. Such expressions, however, are disguised as their opposite, taking the form of ostensibly aggressive verbal and physical interaction (kidding, "ranking," roughhousing, etc.).

Smartness. "Smartness," as conceptualized in lower class culture, involves the capacity to outsmart, outfox, outwit, dupe, "take," "con" another or others and the concomitant capacity to avoid being outwitted, "taken," or duped oneself. In its essence, smartness involves the capacity to achieve a valued entity—material goods, personal status—through a maximum use of mental agility and a minimum use of physical effort. This capacity has an extremely long tradition in lower class culture and is highly valued. Lower class culture can be characterized as "non-intellectual" only if intellectualism is defined specifically in terms of control over a particular body of formally learned knowledge involving "culture" (art, literature, "good" music, etc.), a generalized perspective on the past and present conditions of our own and other societies, and other areas of knowledge imparted by formal educational institutions. This particular type of mental attainment is, in general, overtly disvalued and frequently associated with effeminacy; "smartness" in the lower class sense, however, is highly valued.

The lower class child learns and practices the use of this skill in the street corner situation. Individuals continually practice duping and outwitting one another through recurrent card games and other forms of gambling, mutual exchanges of insults, and "testing" for mutual "con-ability." Those who demonstrate competence in this skill are accorded considerable prestige. Leadership roles in the corner group are frequently allocated according to demonstrated capacity in the two areas of "smartness" and "toughness"; the ideal leader combines both, but the "smart" leader is often accorded more prestige than the "tough" one—reflecting a general lower class respect for "brains" in the "smartness" sense.[4]

The model of the "smart" person is represented in popular media by the card shark, the professional gambler, the "con" artist, the promoter. A conceptual distinction is made between two kinds of people: "suckers," easy marks, "lushes," dupes, who work for their money and are legitimate targets of exploitation; and sharp operators, the "brainy" ones, who live by their wits and "getting" from the suckers by mental adroitness.

Involved in the syndrome of capacities related to "smartness" is a dominant emphasis in lower class culture on ingenious aggressive repartee. This skill, learned and practiced in the context of the corner group, ranges in form from the widely prevalent semi-ritualized teasing, kidding, razzing, "ranking," so characteristic of male peer group interaction, to the highly ritualized type of mutual insult interchange known as "the dirty dozens," "the dozens," "playing house," and other terms. This highly patterned cultural form is practiced on its most advanced level in adult male Negro society, but less polished variants are found throughout lower class culture—practiced, for example, by white children, male and female, as young as four or five. In essence, "doin' the dozens" involves two antagonists who vie with each other in the exchange of increasingly inflammatory insults, with incestuous and perverted sexual relations with the mother a dominant theme. In this form of insult interchange, as well as on other less ritualized occasions for joking, semi-serious, and serious mutual invective, a very high premium is placed on ingenuity, hair-trigger responsiveness, inventiveness, and the acute exercise of mental faculties.

Excitement. For many lower class individuals the rhythm of life fluctuates between periods of relatively routine or repetitive activity and sought situations of great emotional stimulation. Many of the most characteristic features of lower class life are related to the search for excitement or "thrill." Involved here are the highly prevalent use of alcohol by both sexes and the widespread use of gambling of all kinds—playing the numbers, betting on horse races, dice, cards. The quest for excitement finds what is perhaps its most vivid expression in the highly patterned practice of the recurrent "night on the town." This practice, designated by various terms in different areas ("honky-tonkin' "; "goin' out on the town"; "bar hoppin' "), involves a patterned set of activities in which alcohol, music, and sexual adventuring are major components. A group or individual sets out to "make the rounds" of various bars or night clubs. Drinking continues progressively throughout the evening. Men seek to "pick up" women, and women play the risky game of entertaining sexual advances. Fights between men involving women, gambling, and claims of physical prowess, in various combinations, are frequent consequences of a night of making the rounds. The explosive potential of this type of adventuring with sex and aggression, frequently leading to "trouble," is semi-explicitly sought by the individual. Since there is always a good likelihood that being out on the town will eventuate in fights, etc., the practice involves elements of sought risk and desired danger.

Counterbalancing the "flirting with danger" aspect of the "excitement" concern is the prevalence in lower class culture of other well-established patterns of activity which involve long periods of relative inaction or passivity. The term "hanging out" in lower class culture refers to extended periods of standing around, often with peer mates, doing what is defined as "nothing," "shooting the breeze," etc. A definite periodicity exists in the pattern of activity relating to the two aspects of the "excitement" dimension. For many lower class individuals the venture into the high risk world of alcohol, sex, and fighting occurs regularly once a week, with interim periods devoted to accommodating to possible consequences of these periods, along with recurrent resolves not to become so involved again.

Fate. Related to the quest for excitement is the concern with fate, fortune, or luck. Here also a distinction is made between two states—being "lucky" or "in luck" and being unlucky or jinxed. Many lower class individuals feel that their lives are subject to a set of forces over which they have relatively little control. These are not directly equated with the supernatural forces of formally organized religion, but relate more to a concept of "destiny," or man as a pawn of magical powers. Not infrequently this often implicit world view is associated with a conception of the ultimate futility of directed effort towards a goal: if the cards are right, or the dice good to you, or if your lucky number comes up, things will go your way; if luck is against you it's not worth trying. The concept of performing semi-magical rituals so that one's "luck will change" is prevalent; one hopes as a result to move from the state of being "unlucky" to that of being "lucky." The element of fantasy plays an important part in this area. Related to and complementing the notion that "only suckers work" (Smartness) is the idea that once things start going your way, relatively independent of your own effort, all good things will come to you. Achieving great material rewards (big cars, big houses, a roll of cash to flash in a fancy night club), valued in lower class as well as in other parts of American culture, is a recurrent theme in lower class fantasy and folk lore; the cocaine dreams of Willie the Weeper or Minnie the Moocher present the components of this fantasy in vivid detail.

The prevalence in the lower class community of many forms of gambling, mentioned in connection with the "excitement" dimension, is also relevant here. Through cards and pool which involve skill, and thus both "toughness" and "smartness"; or through race horse betting, involving "smartness"; or through playing the numbers, involving predominantly "luck," one may make a big killing with a minimum of directed and persistent effort within conventional occupational channels. Gambling in its many forms illustrates the fact that many of the persistent features of lower class culture are multi-functional—serving a range of desired ends at the same time. Describing some of the incentives behind gambling has involved mention of all of the focal concerns cited so far—Toughness, Smartness, and Excitement, in addition to Fate.

Autonomy. The extent and nature of control over the behavior of the individual—an important concern in most cultures—has a special significance and is distinctively patterned in lower class culture. The discrepancy between what is overtly valued and what is covertly sought is particularly striking in this area. On the overt level there is a strong and frequently expressed resentment of the idea of external controls, restrictions on behavior, and unjust or coercive authority. "No one's gonna push *me* around," or "I'm gonna tell him he can take the job and shove it. . . ." are commonly expressed sentiments. Similar explicit attitudes are maintained to systems of behavior-restricting rules, insofar as these are perceived as representing the injunctions and bearing the sanctions of superordinate authority. In addition, in lower class culture a close conceptual connection is made between "authority" and "nurturance." To be restrictively or firmly controlled is to be cared for. Thus the overtly negative evaluation of superordinate authority frequently extends as well to nurturance, care, or protection. The desire for personal independence is often expressed in such terms as "I don't need *nobody* to take care of me. I can take care of myself!" Actual patterns of behavior, however, reveal a marked discrepancy between expressed sentiment and what is covertly valued. Many lower class people appear to seek out highly restrictive social environments wherein stringent external controls are maintained over their behavior. Such institutions as the armed forces, the mental hospital, the disciplinary school, the prison or correctional institution, provide environments which incorporate a strict and detailed set of rules, defining and limiting behavior and enforced by an authority system which controls and applies coercive sanctions for deviance from these rules. While under the jurisdiction of such systems, the lower class person generally expresses to his peers continual resentment of the coercive, unjust, and arbitrary exercise of authority. Having been released, or having escaped from these milieux, however, he will often act in such a way as to insure recommitment, or choose recommitment voluntarily after a temporary period of "freedom."

Lower class patients in mental hospitals will exercise considerable ingenuity to insure continued commitment while voicing the desire to get out; delinquent boys will frequently "run" from a correctional institution to activate efforts to return them; to be caught and returned means that one is cared for. (Since "being controlled" is equated with "being cared for," attempts are frequently made to "test" the severity or strictness of superordinate authority to see if it remains firm. If intended or executed rebellion produces swift and firm punitive sanctions, the individual is reassured, at the same time that he is complaining bitterly at the injustice of being caught and punished. Some environmental milieux, having been tested in this fashion for the "firmness" of their coercive sanctions, are rejected, ostensibly for being too strict, actually for not being strict enough. This is frequently so in the case of "problematic" behavior by lower class youngsters in the public schools which generally cannot command the coercive controls implicitly sought by the individual.

A similar discrepancy between what is overtly and covertly desired is found in the area of dependence-independence. The pose of tough rebellious independence often assumed by the lower class person frequently conceals powerful dependency cravings. These are manifested primarily by obliquely expressed resentment when "care" is not forthcoming rather than by expressed satisfaction when it is. The concern over autonomy-dependency is related both to "trouble" and "fate." Insofar as the lower class individual feels that his behavior is controlled by forces which often propel him into "trouble" in the face of an explicit determination to avoid it, there is an implied appeal to "save me from myself." A solution appears to lie in arranging things so that his behavior will be coercively restricted by an externally imposed set of controls strong enough to forcibly restrain his inexplicable inclination to get into trouble. The periodicity observed in connection with the "excitement" dimension is also relevant here; after involvement in trouble-producing behavior (assault, sexual adventure, a "drunk"), the individual will actively seek a locus of imposed control (his wife, prison, a restrictive job); after a given period of subjection to this control, resentment against it mounts, leading to a "break away" and a search for involvement in further "trouble."

FOCAL CONCERNS OF THE LOWER CLASS ADOLESCENT STREET CORNER GROUP

The one-sex peer group is a highly prevalent and significant structural form in the lower class community. There is a strong probability that the prevalence and stability of this type of unit is directly related to the prevalence of a stabilized type of lower class child-rearing unit—the "female-based" household. This is a nuclear kin unit in which a male parent is either absent from the household, present only sporadically, or, when present, only minimally or inconsistently involved in the support and rearing of children. This unit usually consists of one or more females of child-bearing age and their offspring. The females are frequently related to one another by blood or marriage ties, and the unit often includes two or more generations of women, e.g., the mother and/or aunt of the principal child-bearing female.

The nature of social groupings in the lower class community may be clarified if we make the assumption that it is the *one-sex peer unit* rather than the two-parent family unit which represents the most significant relational unit for both sexes in lower class communities. Lower class society may be pictured as comprising a set of age-graded one-sex groups which constitute the major psychic focus and reference group for those over twelve or thirteen. Men and women of mating age leave these groups periodically to form temporary marital alliances, but these lack stability, and after varying periods of "trying out" the two-sex family arrangement, they gravitate back to the more "comfortable" one-sex grouping, whose members exert strong pressure on the individual *not* to disrupt the group by adopting a two-sex household pattern of life.[5] Membership in a stable and solidary peer unit is vital to the lower class individual precisely to the extent to which a range of essential functions —psychological, educational, and others—are not provided by the "family" unit.

The adolescent street corner group represents the adolescent variant of this lower class structural form. What has been called the "delinquent gang" is one subtype of this form, defined on the basis of frequency of participation in law-violating activity; this subtype should not be considered a legitimate unit of study per se, but rather as one particular variant of the adolescent street corner group. The "hanging" peer group is a unit of particular importance for the adolescent male. In many cases it is the most stable and solidary primary group he has ever belonged to; for boys reared in female-based households the corner group provides the first real opportunity to learn essential aspects of the male role in the context of peers facing similar problems of sex-role identification.

The form and functions of the adolescent corner group operate as a selective mechanism in recruiting members. Toughness (physical prowess, bravery, skill in athletics and games such as pool and cards), Smartness (skill in repartee, capacity to "dupe" fellow group members), and the like. The term "ranking," used to refer to the pattern of intragroup aggressive repartee, indicates awareness of the fact that this is one device for establishing the intragroup status hierarchy.

The concern over status in the adolescent corner group involves in particular the component of "adultness," the intense desire to be seen as "grown up," and a corresponding aversion to "kid stuff." "Adult" status is defined less in terms of the assumption of "adult" responsibility than in terms of certain external symbols of adult status—a car, ready cash, and, in particular, a perceived "freedom" to drink, smoke, and gamble as one wishes and to come and go without external restrictions. The desire to be seen as "adult" is often a more significant component of much involvement in illegal drinking, gambling, and automobile driving than the explicit enjoyment of these acts as such.

The intensity of the corner group member's desire to be seen as "adult" is sufficiently great that he feels called upon to demonstrate qualities associated with adultness (Toughness, Smartness, Autonomy) to a much greater degree than a lower class adult. This means that he will seek out and utilize those avenues to these qualities which he perceives as available with greater intensity than an adult and less regard for their "legitimacy." In this sense the adolescent variant of lower class culture represents a maximization or an intensified manifestation of many of its most characteristic features.

Concern over status is also manifested in reference to other street corner groups. The term "rep" used in this regard is especially significant and has broad connotations. In its most frequent and explicit connotation, "rep" refers to the "toughness" of the corner group as a whole relative to that of other groups; a "pecking order" also exists among the several corner groups in a given interactional area, and there is a common perception that the safety or security of the group and all its members depends on maintaining a solid "rep" for toughness vis-a-vis other groups. This motive is most frequently advanced as a reason for involvement in gang fights: "We *can't* chicken out on this fight; our rep would be shot!"; this implies that the group would be relegated to the bottom of the status ladder and become a helpless and recurrent target of external attack.

On the other hand, there is implicit in the concept of "rep" the recognition that "rep" has or may have a dual basis—corresponding to the two aspects of the "trouble" dimension. It is recognized that group as well as individual status can be based on both "law-abiding" and "law-violating" behavior. The situational resolution of the persisting conflict between the "law-abiding" and "law-violating" bases of status comprises a vital set of dynamics in determining whether a "delinquent" mode of behavior will be adopted by a group, under what circumstances, and how persistently. The determinants of this choice are evidently highly complex and fluid, and rest on a range of factors including the presence and perceptual immediacy of different community reference-group loci (e.g., professional criminals, police, clergy, teachers, settlement house workers), the personality structures and "needs" of group members, the presence in the community of social work, recreation, or educational programs which can facilitate utilization of the "law-abiding" basis of status, and so on.

What remains constant is the critical importance of "status" both for the members of the group as individuals and for the group as a whole insofar as members perceive their individual destinies as linked to the destiny of the group, and the fact that action geared to attain status is much more acutely oriented to the fact of status itself than to the legality or illegality, morality or immorality of the means used to achieve it.

LOWER CLASS CULTURE AND THE MOTIVATION OF DELINQUENT BEHAVIOR

The customary set of activities of the adolescent street corner group includes activities which are in violation of laws and ordinances of the legal code. Most of these center around assault and theft of various types (the gang fight; auto theft; assault on an individual; petty pilfering and shoplifting; "mugging"; pocketbook theft). Members of street corner gangs are well aware of the law-violating nature of these acts; they are not psychopaths, or physically or mentally "defective"; in fact, since the corner group supports and enforces a rigorous set of standards which demand a high degree of fitness and personal competence, it tends to recruit from the most "able" members of the community.

Why, then, is the commission of crimes a customary feature of gang activity? The most general answer is that the commission of crimes by members of adolescent street corner groups is motivated primarily by the attempt to achieve ends, states, or conditions which are valued and to avoid those that are disvalued within their most meaningful cultural milieu, through those culturally available avenues which appear as the most feasible means of attaining those ends.

The operation of these influences is well illustrated by the gang fight—a prevalent and characteristic type of corner group delinquency. This type of activity comprises a highly stylized and culturally patterned set of sequences. Although details vary under different circumstances, the following events are generally included. A member or several members of group A "trespass" on the claimed territory of group B. While there they commit an act or acts which group B defines as a violation of their rightful privileges, an affront to their honor, or a challenge to their "rep." Frequently this act involves advances to a girl associated with group B; it may occur at a dance or party; sometimes the mere act of "trespass" is seen as deliberate provocation. Members of group B then assault members of group A, if they are caught while still in B's territory. Assaulted members of group A return to their "home" territory and recount to members of their group details of the incident, stressing the insufficient nature of the provocation ("I just *looked* at her! Hardly even said anything!"), and the unfair circumstances of the assault ("About *twenty* guys jumped just the *two* of us!"). The highly colored account is acutely inflammatory; group A, perceiving its honor violated and its "rep" threatened, feels obligated to retaliate in force. Sessions of detailed planning now occur; allies are recruited if the size of group A and its potential allies appears to necessitate larger numbers; strategy is plotted, and messengers dispatched. Since the prospect of a gang fight is frightening to even the "toughest" group members, a constant rehearsal of the provocative incident or incidents and declamations of the essentially evil nature of the opponents accompany the planning process to bolster possibly weakening motivation to fight. The excursion into "enemy" territory sometimes results in a full scale fight; more often group B cannot be found, or the police appear and stop the fight, "tipped off" by an anonymous informant. When this occurs, group members express disgust and disappointment; secretly there is much relief; their honor has been avenged without incurring injury; often the anonymous tipster is a member of one of the involved groups.

The basic elements of this type of delinquency are sufficiently stabilized and recurrent as to constitute an essentially ritualized pattern, resembling both in structure and expressed motives for action classic forms such as the European "duel," the American Indian tribal war, and the Celtic clan feud. Although the arousing and "acting out" of individual aggressive emotions are inevitably involved in the gang fight, neither its form nor motivational dynamics can be adequately handled within a predominantly personality-focused frame of reference.

It would be possible to develop in considerable detail the processes by which the commission of a range of illegal acts is either explicitly supported by, implicitly demanded by, or not materially inhibited by factors relating to the focal concerns of lower class culture. In place of such a development, the following three statements condense in general terms the operation of these processes:

1. Following cultural practices which comprise essential elements of the total life pattern of lower class culture automatically violates certain legal norms.
2. In instances where alternate avenues to similar objectives are available, the non-law-abiding avenue frequently provides a relatively greater and more immediate return for a relatively smaller investment of energy.
3. The "demanded" response to certain situations recurrently engendered within lower class culture involves the commission of illegal acts.

The primary thesis of this paper is that the dominant component of the motivation of "delinquent" behavior engaged in by members of lower class corner groups involves a positive effort to achieve states, conditions, or qualities valued within the actor's most significant cultural milieu. If "conformity to immediate reference group values" is the major component of motivation of "delinquent" behavior by gang members, why is such behavior frequently referred to as negativistic, malicious, or rebellious? Albert Cohen, for example, in *Delinquent Boys* (Glencoe, Ill.: Free Press, 1955) describes behavior which violates school rules as comprising elements of "active spite and malice, contempt and ridicule, challenge and defiance." He ascribes to the gang "keen delight in terrorizing 'good' children, and in general making themselves obnoxious to the virtuous." A recent national conference on social work with "hard-to-reach" groups characterized lower class corner groups as "youth groups in conflict with the culture of their (*sic*) communities." Such characterizations are obviously the result of taking the middle class community and its institutions as an implicit point of reference.

A large body of systematically interrelated attitudes, practices, behaviors, and values characteristic of lower class culture are designed to support and maintain the basic features of the lower class way of life. In areas where these differ from features of middle class culture, action oriented to the achievement and maintenance of the lower class system may violate norms of middle class culture and be perceived as deliberately non-conforming or malicious by an observer strongly cathected to middle class norms. This does not mean, however, that violation of the middle class norm is the dominant component of motivation; it is a by-product of action primarily oriented to the lower class system. The standards of lower class culture cannot be seen merely as a reverse function of middle class culture—as middle class standards "turned upside down"; lower class culture is a distinctive tradition many centuries old with an integrity of its own.

From the viewpoint of the acting individual, functioning within a field of well-structured cultural forces, the relative impact of "conforming" and "rejective" elements in the motivation of gang delinquency is weighted preponderantly on the conforming side. Rejective or rebellious elements are inevitably involved, but their influence during the actual commission of delinquent acts is relatively small compared to the influence of pressures to achieve what is valued by the actor's most immediate reference groups. Expressed awareness by the actor of the element of rebellion often represents only that aspect of motivation of which he is explicitly conscious; the deepest and most compelling components of motivation—adherence to highly meaningful group standards of Toughness, Smartness, Excitement, etc.—are often unconsciously patterned. No cultural pattern as well established as the practice of illegal acts by members of lower class corner groups could persist if buttressed primarily by negative, hostile, or rejective motives; its principal motivational support, as in the case of any persisting cultural tradition, derives from a positive effort to achieve what is valued within that tradition, and to conform to its explicit and implicit norms.

Chapter 15

CHANGE

SOCIAL change is a universal human experience. Under the conditions of modern industrial society this experience has been vastly intensified and accelerated. From the beginning, sociology has been influenced by very different conceptualizations of change. August Comte continued the Enlightenment idea of progress, which had a long lasting hold on French and American sociology in particular. The idea of evolution also played an important part in sociological theories of change. The Marxian idea of history as a struggle between classes has been a direct inspiration for some sociologists, and a stimulus to construct contrary theories to others. Along with the general question "What is social change?" sociologists have always been preoccupied with the question "What are the peculiar changes that have led to modern society?"

In contemporary sociology an important division can be made between those sociologists who stand in the tradition of what may be called the classical sociological paradigm—particularly as it was shaped by Durkheim and Weber—and those sociologists who have tried to revitalize the Marxian paradigm. The former group (the structural-functionalists belong here as well as others) will try to characterize contemporary society by such broad categories as "rationalization," "differentiation," or "modernization." The terms used to describe contemporary society as a whole will reflect this—"modern society," "advanced industrial society," "developed society," or even "postmodern" or "postindustrial" society. Sociologists adhering to the Marxian paradigm will understand contemporary society primarily in terms of the dynamics of the capitalist system and the resulting class conflicts. Accordingly, their favorite term is "late capitalist society." Although there are clear contradictions between these two theoretical viewpoints, there are various phenomena where a convergence perspective is possible.

The first selection is from one of the most famous community studies in American sociology, that of Muncie, Indiana, which Robert and Helen Lynd examined first in the 1920s and then again after the Great Depression. The passage selected is from their second study, *Middletown in Transition,* in which the Lynds discuss the profound changes undergone by this community in the wake of the Depression. The Marxian view is presented in a selection from *The Communist Manifesto* (1848), by Karl Marx and Friedrich Engels. "Classical" sociology was at least partly concerned with the search for interpretations of social change alternative to the Marxian one. In the selection from Emile Durkheim's *Division of Labor in Society,* social change is viewed in terms of the progressive differentiation of society. The selected essay by Max Weber discusses "rationalization," by which Weber meant the specifically modern process through which rational modes of conduct and thinking are imposed upon ever wider segments of social life. The final selection, "The Blueing of America" by Peter L. and Brigitte Berger, discusses some questions of cultural and social change in contemporary America.

ROBERT S. LYND AND HELEN M. LYND

MIDDLETOWN FACES BOTH WAYS

[EARLIER WE] have sought to make explicit the elements of permanence and of change in Middletown as the city has met with four types of experience peculiarly conducive to cultural change: sudden and great strain on its institutions, widespread dislocation of individual habits, pressure for change from the larger culture surrounding it, and at some points the actual implementing from without of a changed line of action. These ten years of boom and depression might be expected to leave permanent marks on the culture.

The boom experiences were not essentially different in kind from those Middletown had known before: optimism, growth, making money—these things are in the city's main stream of tradition. Such an experience as climbing to the very verge of the long-expected population of 50,000 contained elements of novelty and has, despite the depression, left a permanent deposit in the city in the form of increased self-regard. The prosperity of the fat years, while sharpening the disappointments of the depression, also remains today in Middletown in the form of enhanced personal goals and glimpsed new psychological standards of living for many of its citizens. The fact that Middletown does not regard the depression as in any sense "its own fault," or even the fault of the economy by which it lives, makes it easy for the city to think of the confusion following 1929 as "just a bad bump in the road," one of those inevitable occurrences that spoil things temporarily but do not last. The gold-rush scramble back to confidence which the research staff witnessed in 1935 was the inevitable result of such a rationale of the depression. Middletown was in effect saying, albeit soberly and decidedly anxiously: "It's all over, thank God! And now we'll get after all those things we were planning for ourselves in 1928–29!" In a culture built on money, the experience of better homes, better cars, winter vacations in Florida, and better educated children dies hard; and while some people's hopes, especially among the working class, have been mashed out permanently by the depression, the influential business group who determine the wave length of Middletown's articulate hopes are today busily broadcasting the good news that everything is all right again.

The depression experiences contained more outright novelty than did the years 1925–29:

A city exultantly preoccupied with the question, "How fast can we make even more money?" was startled by being forced to shift its central concern for a period of years to the stark question, "Can we manage to keep alive?"

A city living excitedly *at* a future which all signs promised would be golden lived for a while *in* the present with its exigent demands.

A city living by the faith that everyone can and should support himself lived through a period of years in which it had to confess that at least temporarily a quarter of its population could not get work.

A city intensely opposed to society's caring for able-bodied people has taxed itself to support for an indefinitely long period one in every four of its families.

A city that has chronically done without many manifestly needed civic improvements, on the philosophy that it does no good to hunt up and plan desirable things to do because there isn't any money to pay for them, has lived for a time in a world in which not money but ability to plan and carry out progress was the limiting factor.

A city built around the theory of local autonomy has lived in a world experiencing rapid centralization of administrative authority and marked innovations in the interference by these centralized agencies in local affairs.

A city that lives by the thought that it is one big cooperating family has had the experience of a wholesale effort by its working class to organize against its business class under sponsorship from Washington.

A city committed to faith in education as the key to its children's future has had to see many of its college-trained sons and daughters idle, and to face the question as to what education is really "worth."

A city devoted to the doctrine that "Work comes first," to an extent that has made many of its citizens scarcely able to play, has faced the presence of enforced leisure and heard people talk of "the new leisure." Civicly, the community has begun to state positively the problem of the leisure of the mass of its people, and to make wider provision for popular leisure pursuits.

A city still accustomed to having its young assume largely the values of their parents has had to listen to an increasing number of its young speak of the world of their parents as a botched mess.

A city in which the "future" has always been painted in terms of its gayer-hued hopes has been forced to add to its pigments the somber dark tones of its fears.

Experiences such as these partake in their cumulative effect of the crisis quality of a serious illness, when life's customary busy immediacies drop away and one lies helplessly confronting oneself, reviewing the past, and asking abrupt questions of the future. What has Middletown learned from its crisis and partial convalescence?

[Earlier we] stated some of the larger questions of this sort which the research staff took to Middletown in June, 1935. The broad answer to these questions is that basically the texture of Middletown's culture has not changed. Those members of the research staff who had expected to find sharp differences in group alignments within the city, in ways of thinking, or feeling, or carrying on the multifarious daily necessities of life, found little to support their hypotheses. Middletown is overwhelmingly living by the values by which it lived in 1925; and the chief additions are defensive, negative elaborations of already existing values, such as, among the business class, intense suspicion of centralizing tendencies in government, of the interference of social legislation with business, of labor troubles, and of radicalism. Among the working class, tenuous and confused new positive values are apparent in such a thing as the aroused conception of the possible role of government in bolstering the exposed position of labor by social legislation, including direct relief for the unemployed. But, aside from these, no major new symbols or ideologies of a positive sort have developed as conspicuous rallying points. Leadership in the community has not shifted in kind, but has become more concentrated in the same central group observed in 1925. The different rates of change pointed out in the earlier study as occurring in the different areas of living have not altered materially: [1] economic activities have set the pace and determined the cadence of these years, though the changes have not differed in kind over these ten years anything like so sharply as during the thirty-five-year period covered in the earlier study; in terms of actual rate of change and radical quality of innovation, the institutions concerned with care for the unable leaped into the lead during the depression, although Middletown likes to regard the changes in this area as "purely emergency and temporary" in character; education, leisure, and the relations among family members have exhibited some changes; while the city's local government and religion have remained as before most resistant of all its institutions to change.

With the exception of the widespread innovations in caring for the unemployed, which by 1936 were already contracting their scope, a map of Middletown's culture shows today much the same contours as before; no wholly new hills and valleys appear save in this "temporary" provision for the unemployed and the resulting new public works; the configuration is the same. Even the fault lines which appear today and show signs of developing into major fissures within the community were faintly visible in 1925. In the main, a Rip Van Winkle, fallen asleep in 1925 while addressing Rotary or the Central Labor Union, could have awakened in 1935 and gone right on with his interrupted address to the same people with much the same ideas.

Such changes as are going forward in Middletown are disguised by the thick blubber of custom that envelops the city's life. The city is uneasily conscious of many twinges down under the surface, but it resembles the person who insists on denying and disregarding unpleasant physical symptoms on the theory that everything *must* be all right, and that if anything really is wrong it may cure itself without leading to a major operation. The conflicts under the surface in Middletown are not so much new as more insistent, more difficult to avoid, harder to smooth over. Many of these latent conflicts, aggravated by the depression and now working themselves toward the surface of the city's life, have been pointed out in the preceding pages: conflicts among values hitherto held as compatible; conflicts among institutions—economic and political, economic and educational and religious, economic and familial; conflicts among groups in the community breaking through the symbols of the unified city; conflicts between deep-rooted ideas of individual and collective responsibility; conflicts, above all, between symbols and present reality.

The physical and personal continuities of life are relatively great in the small community, and the average dweller in such a community probably has a sense of "belonging" that is qualitatively somewhat different from that of the big-city dweller. The institutions in the small city tend to be familiar and, with the help of many assumptions of long standing as to how they are linked together and operate, a quality of simplicity is imparted to them in the minds of local people. By assuming continuities and similarities, this simplicity is interpreted outward to include "American life" and "American institutions."

One of the major elements of conflict imparted by the depression to Middletown has been the injection of a new sense of the inescapable complexity of this assumedly simple world. As indicated earlier, the more alert Middletown people met the depression with an earnest desire to "understand" it—only to be thrown back later, in many instances, with a sense that it was "too big" for them and that all they could do was to try to stick to their jobs and save their own skins. One suspects that for the first time in their lives many Middletown people have awakened, in the depression, from a sense of being at home in a familiar world to the shock of living as an atom in a universe dangerously too big and blindly out of hand. With the falling away of literal belief in the teachings of religion in recent decades, many Middletown folk have met a similar shock, as the simpler universe of fifty years ago has broken up into a vastly complicated physical order; but, there, they have been able to retain the shadowy sense of their universe's being in beneficent control by the common expedient of believing themselves to live in a world of unresolved duality, in which one goes about one's daily affairs without thought of religion but relies vaguely on the ultimates in life being somehow divinely "in hand." In the economic order, however, it is harder for Middletown to brush aside the shock by living thus on two largely unconnected levels, for the economic out-of-handness is too urgently threatening to daily living.

So Middletown tries to forget and to disregard the growing disparities in the midst of which it lives. Its adult population has, through its socially gay youth and busy adult life, resisted the patient scrutiny of problems and the teasing out of their less obvious antecedents and implications. As a local man remarked in 1924 in commenting on the pressure of modern living, "We've lost the ability to ponder over life. We're too busy." And, if in the boom days Middletown was "too busy" to ponder, it was too worried to do so in the depression. It is quite characteristic, for instance, that, as one woman remarked in 1935, "We never get down to talking about things like the coming of fascism. The only time we ever talk about any of those things is when we comment on a radio program." Rather than ponder such things, Middletown prefers either to sloganize or to personalize its problems. And the more the disparities have forced themselves to attention, the more things have seemed "too big" and "out of hand," the more Middletown has inclined to heed the wisdom of sticking to one's private business [2] and letting the uncomfortable "big problems" alone save for a few encompassing familiar slogans. Where Middletown cannot avoid these big problems and must on occasion present at least the semblance of a balance in this system of nonbalancing intellectual bookkeeping, it is resorting increasingly to the suppression of detailed entries and to the presentation of only the alleged totals.

One frequently gets a sense of people's being afraid to let their opinions become sharp. They believe in "peace, but—." They believe in "fairness to labor, but—." In "freedom of speech, but—." In "democracy, but—." In "freedom of the press, but—." This is in part related to the increased apprehensiveness that one feels everywhere in Middletown: fear on the part of teachers of the D.A.R. and the Chamber of Commerce; fear by businessmen of high taxes and public ownership of utilities and of the Roosevelt administration; fear by laborers of joining unions lest they lose their jobs; fear by office-holders wanting honest government of being framed by the politicians; fear by everyone to show one's hand, or to speak out.

But this process of avoiding issues goes on less and less fluently. With a widening gap between symbol and practice in the most immediate concerns of living, there are more forced choices as to where one's emphasis is to be placed. Middletown wants to be adventurous and to embrace new ideas and practices, but it also desperately needs security, and in this conflict both businessmen and workingmen appear to be clinging largely to tried sources of security rather than venturing out into the untried. Middletown people want to be kind, friendly, expansive, loyal to each other, to make real the idea of a friendly city working together for common ends; but, in a business world where one is struggling for self-preservation, or for power and prestige as a supposed means to self-preservation, warm personal relations, like the more fastidious sorts of integrity, may tend to become a luxury and be crowded to the wall. If necessary, one dispenses with affection. People want to continue to live hopefully and adventurously into the future, but if the future becomes too hazardous they look steadily toward the known past.[3]

On the surface, then, Middletown is meeting such present issues and present situations as it cannot escape by attempting to revert to the old formulas: we must always believe that things are good and that they will be better, and we must stress their hopeful rather than their pessimistic aspects. This leads to the stating of such social problems as may arise defensively and negatively [4]—rather than to engaging in a positive program for social analysis and reconstruction. It is still true in 1936 that, to Middletown, such things as poverty or a depression are simply exceptions to a normally good state of affairs; and anything that goes wrong is the fault of some individuals (or, collectively, of "human nature") rather than anything amiss with the organization and functioning of the culture. The system is fundamentally right and only the persons wrong; the cures must be changes in personal attitudes, not in the institutions themselves. Among these personal cures for its social woes are the following six basic qualities needed for a better world outlined in a local address: "faith, service, cooperation, the Golden Rule, optimism, and character." "The typical citizen," says an editorial approvingly, "discounts the benefits of the political and economic New Deal and says that common sense is the answer to the depression. . . . He thinks hard work is the depression cure." Or again, "If profits are low, it is still possible to get a good deal of enjoyment by doing the best possible under adverse circumstances and by taking pride in our work."

KARL MARX AND FRIEDRICH ENGELS

BOURGEOIS AND PROLETARIANS

THE history of all hitherto existing society is the history of class struggles.

Freeman and slave, patrician and plebeian, lord and serf, guildmaster and journeyman, in a word, oppressor and oppressed, stood in constant opposition to one another, carried on an uninterrupted, now hidden, now open fight, that each time ended, either in the revolutionary reconstitution of society at large, or in the common ruin of the contending classes.

In the earlier epochs of history we find almost everywhere a complicated arrangement of society into various orders, a manifold gradation of social rank. In ancient Rome we have patricians, knights, plebeians, slaves; in the middle ages, feudal lords, vassals, guildmasters, journeymen, apprentices, serfs; in almost all of these classes, again, subordinate gradations.

The modern bourgeois society that has sprouted from the ruins of feudal society, has not done away with class antagonisms. It has but established new classes, new conditions of oppression, new forms of struggle in place of the old ones. Our epoch, the epoch of the bourgeois, possesses, however, this distinctive feature: it has simplified the class antagonisms. Society as a whole is

Reprinted from *The Communist Manifesto* by Karl Marx and Friedrich Engels (Chicago: Charles H. Kerr, 1888).

more and more splitting up into two great hostile camps, into two great classes directly facing each other: Bourgeoisie and Proletariat.

From the serfs of the middle ages sprang the chartered burghers of the earliest towns. From these burgesses the first elements of the bourgeoisie were developed.

The discovery of America, the rounding of the Cape, opened up fresh ground for the rising bourgeoisie. The East Indian and Chinese markets, the colonization of America, trade with the colonies, the increase in the means of exchange and in commodities generally, gave to commerce, to navigation, to industry, an impulse never before known, and thereby, to the revolutionary element in the tottering feudal society, a rapid development.

The feudal system of industry, under which industrial production was monopolized by closed guilds, now no longer sufficed for the growing wants of the new markets. The manufacturing system took its place. The guild masters were pushed on one side by the manufacturing middle class; division of labor between the different corporate guilds vanished in the face of division of labor in each single workshop.

Meantime the markets kept ever growing, the demand ever rising. Even manufacture no longer sufficed. Thereupon steam and machinery revolutionized industrial production. The place of manufacture was taken by the giant, Modern Industry, the place of the industrial middle class, by industrial millionaires, the leaders of whole industrial armies, the modern bourgeois.

Modern industry has established the world's market, for which the discovery of America paved the way. The market has given an immense development to commerce, to navigation, to communication by land. This development has, in its turn, reacted on the extension of industry; and in proportion as industry, commerce, navigation and railways extended, in the same proportion the bourgeoisie developed, increased its capital, and pushed into the background every class handed down from the middle ages.

We see, therefore, how the modern bourgeoisie is itself the product of a long course of development, of a series of revolutions in the modes of production and of exchange.

Each step in the development of the bourgeoisie was accompanied by a corresponding political advance of that class. An oppressed class under the sway of the feudal nobility, an armed and self-governing association in the mediæval commune, here independent urban republic (as in Italy and Germany), there taxable "third estate" of the monarchy (as in France), afterwards, in the period of manufacture proper, serving either the semi-feudal or the absolute monarchy as a counterpoise against the nobility, and, in fact, corner-stone of the great monarchies in general, the bourgeoisie has at last, since the establishment of Modern Industry and of the world's market, conquered for itself, in the modern representative State, exclusive political sway. The executive of the modern State is but a committee for managing the common affairs of the whole bourgeoisie.

The bourgeoisie, historically, has played a most revolutionary part.

The bourgeoisie, wherever it has got the upper hand, has put an end to all feudal, patriarchal, idyllic relations. It has pitilessly torn asunder the motley feudal ties that bound man to his "natural superiors," and has left remaining no other nexus between man and man than naked self-interest, callous "cash payment." It has drowned the most heavenly ecstasies of religious fervor, of chivalrous enthusiasm, of philistine sentimentalism, in the icy water of egotistical calculation. It has resolved personal worth into exchange value, and in place of the numberless indefeasible chartered freedoms, has set up that single, unconscionable freedom—Free Trade. In one word, for exploitation, veiled by religious and political illusions, it has substituted naked, shameless, direct, brutal exploitation.

The bourgeoisie has stripped of its halo every occupation hitherto honored and looked up to with reverent awe. It has converted the physician, the lawyer, the priest, the poet, the man of science, into its paid wage laborers.

The bourgeoisie has torn away from the family its sentimental veil, and has reduced the family relation to a mere money relation.

The bourgeoisie has disclosed how it came to pass that the brutal display of vigor in the middle ages, which Reactionists so much admire, found its fitting complement in the most slothful indolence. It has been the first to show what man's activity can bring about. It has accomplished wonders far surpassing Egyptian pyramids, Roman aqueducts, and Gothic cathedrals; it has conducted expeditions that put in the shade all former Exoduses of nations and crusades.

The bourgeoisie cannot exist without constantly revolutionizing the instruments of production, and thereby the relations of production, and with them the whole relations of society. Conservation of the old modes of production in unaltered forms, was, on the contrary, the first condition of existence for all earlier industrial classes. Constant revolutionizing of production, uninterrupted disturbance of all social conditions, everlasting uncertainty and agitation, distinguish the bourgeois epoch from all earlier ones. All fixed, fast-frozen relations, with their train of ancient and venerable prejudices and opinions, are swept away; all new-formed ones become antiquated before they can ossify. All that is solid melts into air, all that is holy is profaned, and man is at last compelled to face with sober senses his real conditions of life and his relations with his kind.

The need of a constantly expanding market for its products chases the bourgeoisie over the whole surface of the globe. It must nestle everywhere, settle everywhere, establish connections everywhere.

The bourgeoisie has through its exploitation of the world's market given a cosmopolitan character to production and consumption in every country. To the great chagrin of Reactionists, it has drawn from under the feet of industry the national ground on which it stood. All old-established national industries have been destroyed or are daily being destroyed. They are dislodged by new industries, whose introduction becomes a life and death question for all civilized nations, by industries that no longer work up indigenous raw material, but raw material drawn from the remotest zones, industries whose products are consumed, not only at home, but in every quarter of the globe. In place of the old wants, satisfied by the productions of the country, we find new wants, requiring for their satisfaction the products of distant lands and climes. In place of the old local and national seclusion and self-sufficiency, we have intercourse in every direction, universal inter-dependence of nations. And as in material, so also in intellectual production. The intellectual creations of individual nations become common property. National one-sidedness and narrow-mindedness become more and more impossible, and from the numerous national and local literatures, there arises a world literature.

The bourgeoisie, by the rapid improvement of all instruments of production, by the immensely facilitated means of communication, draws all, even the most barbarian, nations into civilization. The cheap prices of its commodities are the heavy artillery with which it batters down all Chinese walls, with which it forces the barbarians' intensely obstinate hatred of foreigners to capitulate. It compels all nations, on pain of extinction, to adopt the bourgeois mode of production; it compels them to introduce what it calls civilization into their midst, *i.e.,* to become bourgeois themselves. In one word, it creates a world after its own image.

The bourgeoisie has subjected the country to the rule of the towns. It has created enormous cities, has greatly increased the urban population as compared with the rural, and has thus rescued a considerable part of the population from the idiocy of rural life. Just as it has made the country dependent on the towns, so it has made barbarian and semi-barbarian countries dependent on the civilized ones, nations of peasants on nations of bourgeois, the East on the West.

The bourgeoisie keeps more and more doing away with the scattered state of the population, of the means of production, and of property. It has agglomerated population, centralized means of production, and has concentrated property in a few hands. The necessary consequence of this was political centralization. Independent, or but loosely connected provinces, with separate interests, laws, governments and systems of taxation, became lumped together into one nation, with one government, one code of laws, one national class interest, one frontier, and one customs tariff.

The bourgeoisie, during its rule of scarce one hundred years, has created more massive and more colossal productive forces than have all preceding generations together. Subjection of Nature's forces to man, machinery, application of chemistry to industry and agriculture, steam navigation, railways, electric telegraphs, clearing of whole continents for cultivation, canalization of rivers, whole populations conjured out of the ground—what earlier century had even a presentiment that such productive forces slumbered in the lap of social labor?

We see then: the means of production and of exchange on whose foundation the bourgeoisie built itself up, were generated in feudal society. At a certain stage in the development of these means of production and of exchange, the conditions under which feudal society produced and exchanged, the feudal organization of agriculture and manufacturing industry, in one word, the feudal relations of property, became no longer compatible with the already developed productive forces; they became so many fetters. They had to be burst asunder.

Into their place stepped free competition, accompanied by a social and political constitution adapted to it, and by the economical and political sway of the bourgeois class.

A similar movement is going on before our own eyes. Modern bourgeois society with its relations of production, of exchange, and of property, a society that has conjured up such gigantic means of production and of exchange, is like the sorcerer, who is no longer able to control the powers of the nether world whom he has called up by his spells. For many a decade past the history of industry and commerce is but the history of the revolt of modern productive forces against modern conditions of production, against the property relations that are the conditions for the existence of the bourgeoisie and of its rule. It is enough to mention the commercial crises that by their periodical return put on its trial, each time more threateningly, the existence of the bourgeois society. In these crises a great part not only of the existing products, but also of the previously created productive forces, is periodically destroyed. In these crises there breaks out an epidemic that, in all earlier epochs, would have seemed an absurdity—the epidemic of overproduction. Society suddenly finds itself put back into a state of momentary barbarism; it appears as if a famine, a universal war of devastation had cut off the supply of every means of subsistence; industry and commerce seem to be destroyed; and why? because there is too much civilization, too much means of subsistence, too much industry, too much commerce. The productive forces at the disposal of society no longer tend to further the development of the conditions of bourgeois property; on the contrary, they have become too powerful for these conditions, by which they are fettered, and so soon as they overcome these fetters, they bring disorder into the whole of bourgeois society, endanger the existence of bourgeois property. The conditions of bourgeois society are too narrow to comprise the wealth created by them. And how does the bourgeoisie get over these crises? On the one hand by enforced destruction of a mass of productive forces; on the other, by the conquest of new markets, and by the more thorough exploitation of the old ones. That is to say, by paving the way for more extensive and more destructive crises, and by diminishing the means whereby crises are prevented.

The weapons with which the bourgeoisie felled feudalism to the ground are now turned against the bourgeoisie itself.

But not only has the bourgeoisie forged the weapons that bring death to itself; it has also called into existence the men who are to wield those weapons—the modern working class—the proletarians.

In proportion as the bourgeoisie, *i.e.,* capital, is developed, in the same proportion is the proletariat, the modern working class, developed; a class of laborers, who live only so long as they find work, and who find work only so long as their labor increases capital. These laborers, who must sell themselves piecemeal, are a commodity, like every other article of commerce, and are consequently exposed to all the vicissitudes of competition, to all the fluctuations of the market.

Owing to the extensive use of machinery and to division of labor, the work of the proletarians has lost all individual character, and, consequently, all charm for the workman. He becomes an appendage of the machine, and it is only the most simple, most monotonous, and most easily acquired knack, that is required of him. Hence, the cost of production of a workman is restricted almost entirely to the means of subsistence that he requires for his maintenance, and for the propagation of his race. But the price of a commodity, and therefore also of labor, is equal, in the long run, to its cost of production. In proportion, therefore, as the repulsiveness of the work increases, the wage decreases. Nay, more, in proportion as the use of machinery and division of labor increase, in the same proportion the burden of toil also increases, whether by prolongation of the working hours, by increase of the work exacted in a given time, or by increased speed of the machinery, etc.

Modern industry has converted the little workshop of the patriarchal master into the great factory of the industrial capitalist. Masses of laborers, crowded into the factory, are organized like soldiers. As privates of the industrial army they are placed under the command of a perfect hierarchy of officers and sergeants. Not only are they slaves of the bourgeois class, and of the bourgeois State, they are daily and hourly enslaved by the machine, by the over-seer, and, above all, by the individual bourgeois manufacturer himself. The more openly this despotism proclaims gain to be its end and aim, the more petty, the more hateful and the more embittering it is.

The less skill and exertion of strength is implied in manual labor, in other words, the more modern industry becomes developed, the more is the labor of men superseded by that of women. Differences of age and sex have no longer any distinctive social validity for the working class. All are instruments of labor, more or less expensive to use, according to age and sex.

No sooner is the exploitation of the laborer by the manufacturer so far at an end that he receives his wages in cash, than he is set upon by the other portions of the bourgeoisie, the landlord, the shopkeeper, the pawnbroker, etc.

The lower strata of the middle class—the small tradespeople, shopkeepers, and retired tradesmen generally, the handicraftsmen and peasants—all these sink gradually into the proletariat, partly because their diminutive capital does not suffice for the scale on which modern industry is carried on, and is swamped in the competition with the large capitalists, partly because their specialized skill is rendered worthless by new methods of production. Thus the proletariat is recruited from all classes of the population.

The proletariat goes through various stages of development. With its birth begins its struggle with the bourgeoisie. At first the contest is carried on by individual laborers, then by the workpeople of a factory, then by the operatives of one trade, in one locality, against the individual bourgeois who directly exploits them. They direct their attacks not against the bourgeois conditions of production, but against the instruments of production themselves; they destroy imported wares that compete with their labor, they smash to pieces machinery, they set factories ablaze, they seek to restore by force the vanished status of the workman of the middle ages.

At this stage the laborers still form an incoherent mass scattered over the whole country, and broken up by their mutual competition. If anywhere they unite to form more compact bodies, this is not yet the consequence of their own active union, but of the union of the bourgeoisie, which class, in order to attain its own political ends, is compelled to set the whole proletariat in motion, and is moreover yet, for a time, able to do so. At this stage, therefore, the proletarians do not fight their enemies, but the enemies of their enemies, the remnants of absolute monarchy, and land owners, the nonindustrial bourgeois, the petty bourgeoisie. Thus the whole historical movement is concentrated in the hands of the bourgeoisie; every victory so obtained is a victory for the bourgeoisie.

But with the development of industry the proletariat not only increases in number; it becomes concentrated in greater masses, its strength grows and it feels that strength more. The various interests and conditions of life within the ranks of the proletariat are more and more equalized, in proportion as machinery obliterates all distinctions of labor, and nearly everywhere reduces wages to the same low level. The growing competition among the bourgeois, and the resulting commercial crises, make the wages of the workers ever more fluctuating. The unceasing improvement of machinery, ever more rapidly developing, makes their livelihood more and more precarious; the collisions between individual workman and individual bourgeois take more and more the character of collisions between two classes. Thereupon the workers begin to form combinations (Trades' Unions) against the bourgeois; they club together in order to keep up the rate of wages; they found permanent associations in order to make provision beforehand for these occasional revolts. Here and there the contest breaks out into riots.

Now and then the workers are victorious, but only for a time. The real fruit of their battles lies not in the immediate result but in the ever improved means of communication that are created in modern industry and that place the workers of different localities in contact with one another. It was just this contact that was needed to centralize the numerous local struggles, all of the same character, into one national struggle between classes. But every class struggle is a political struggle. And that union, to attain which the burghers of the middle ages, with their miserable highways, required centuries, the modern proletarians, thanks to railways, achieve in a few years.

This organization of the proletarians into a class and consequently into a political party, is continually being upset again by the competition between the workers themselves. But it ever rises up again; stronger, firmer, mightier. It compels legislative recognition of particular interests of the workers, by taking advantage of the divisions among the bourgeoisie itself. Thus the ten-hours' bill in England was carried.

Altogether collisions between the classes of the old society further, in many ways, the course of the development of the proletariat. The bourgeoisie finds itself involved in a constant battle. At first with the aristocracy; later on, with those portions of the bourgeoisie itself whose interests have become antagonistic to the progress of industry; at all times with the bourgeoisie of foreign countries. In all these countries it sees itself compelled to appeal to the proletariat, to ask for its help, and thus to drag it into the political arena. The bourgeoisie itself, therefore, supplies the proletariat with weapons for fighting the bourgeoisie.

Further, as we have already seen, entire sections of the ruling classes are, by the advance of industry, precipitated into the proletariat, or are at least threatened in their conditions of existence. These also supply the proletariat with fresh elements of enlightenment and progress.

Finally, in times when the class struggle nears the decisive hour, the process of dissolution going on within the ruling class, in fact within the whole range of old society, assumes such a violent, glaring character, that a small section of the ruling class cuts itself adrift, and joins the revolutionary class, the class that holds the future in its hands. Just as, therefore, at an earlier period, a section of the nobility went over to the bourgeoisie, so now a portion of the bourgeoisie goes over to the proletariat, and in particular, a portion of the bourgeois ideologists, who have raised themselves to the level of comprehending theoretically the historical movement as a whole.

Of all the classes that stand face to face with the bourgeoisie to-day, the proletariat alone is a really revolutionary class. The other classes decay and finally disappear in the face of modern industry; the proletariat is its special and essential product.

The lower middle class, the small manufacturer, the shopkeeper, the artisan, the peasant, all these fight against the bourgeoisie to save from extinction their existence as fractions of the middle class. They are therefore not revolutionary, but conservative. Nay, more, they are reactionary, for they try to roll back the wheel of history. If by chance they are revolutionary, they are so only in view of their impending transfer into the proletariat; they thus defend not their present, but their future interests, they desert their own standpoint to place themselves at that of the proletariat.

The "dangerous class," the social scum, that passively rotting class thrown off by the lowest layers of old society, may, here and there, be swept into the movement by a proletarian revolution; its conditions of life, however, prepare it far more for the part of a bribed tool of reactionary intrigue.

In the conditions of the proletariat, those of old society at large are already virtually swamped. The proletarian is without property; his relation to his wife and children has no longer anything in common with the bourgeois family relations; modern industrial labor, modern subjection to capital, the same in England as in France, in America as in Germany, has stripped him of every trace of national character. Law, morality, religion, are to him so many bourgeois prejudices, behind which lurk in ambush just as many bourgeois interests.

All the preceding classes that got the upper hand sought to fortify their already acquired status by subjecting society at large to their conditions of appropriation. The proletarians cannot become masters of the productive forces of society, except by abolishing their own previous mode of appropriation, and thereby also every other previous mode of appropriation. They have nothing of their own to secure and to fortify; their mission is to destroy all previous securities for, and insurances of, individual property.

All previous historical movements were movements of minorities, or in the interest of minorities. The proletarian movement is the self-conscious, independent movement of the immense majority, in the interest of the immense majority. The proletariat, the lowest stratum of our present society, cannot stir, cannot raise itself up, without the whole superincumbent strata of official society being sprung into the air.

Though not in substance, yet in form, the struggle of the proletariat with the bourgeoisie is at first a national struggle. The proletariat of each country must, of course, first of all settle matters with its own bourgeoisie.

In depicting the most general phases of the development of the proletariat, we traced the more or less veiled civil war, raging within existing society, up to the point where that war breaks out into open revolution, and where the violent overthrow of the bourgeoisie lays the foundation for the sway of the proletariat.

Hitherto every form of society has been based, as we have already seen, on the antagonism of oppressing and oppressed classes. But in order to oppress a class certain conditions must be assured to it under which it can, at least, continue its slavish existence. The serf, in the period of serfdom, raised himself to membership in the commune, just as the petty bourgeois, under the yoke of feudal absolutism, managed to develop into a bourgeois. The modern laborer, on the contrary, instead of rising with the progress of industry, sinks deeper and deeper below the conditions of existence of his own class. He becomes a pauper, and pauperism develops more rapidly than population and wealth. And here it becomes evident that the bourgeoisie is unfit any longer to be the ruling class in society and to impose its conditions of existence upon society as an over-riding law. It is unfit to rule because it is incompetent to assure an existence to its slave within his slavery, because it cannot help letting him sink into such a state that it has to feed him instead of being fed by him. Society can no longer live under this bourgeoisie; in other words, its existence is no longer compatible with society.

The essential condition for the existence, and for the sway of the bourgeois class, is the formation and augmentation of capital; the condition for capital is wage-labor. Wage-labor rests exclusively on competition between the laborers. The advance of industry, whose involuntary promoter is the bourgeoisie, replaces the isolation of the laborers, due to competition, by their revolutionary combination, due to association. The development of modern industry, therefore, cuts from under its feet the very foundation on which the bourgeoisie produces and appropriates products. What the bourgeoisie therefore produces, above all, are its own grave diggers. Its fall and the victory of the proletariat are equally inevitable.

EMILE DURKHEIM

THE CAUSES FOR THE PROGRESS OF THE DIVISION OF LABOR

WE must, then, look for the causes explaining the progress of the division of labor in certain variations of the social scene. The results of the preceding book enable us to infer at once what these variations are.

We saw how the organized structure, and, thus, the division of labor, develop as the segmental structure disappears. Hence, either this disappearance is the cause of the development, or the development is the cause of the disappearance. The latter hypothesis is inadmissible, for we know that the segmental arrangement is an insurmountable obstacle to the division of labor, and must have disappeared at least partially for the division of labor to appear. The latter can appear only in proportion to the disappearance of the segmental structure. To be sure, once the division of labor appears, it can contribute towards the hastening of the other's regression, but it is in evidence only after the regression has begun. The effect reacts upon the cause, but never loses its quality of effect. The reaction it exercises is, consequently, secondary. The growth of the division of labor is thus brought about by the social segments

Reprinted from *The Division of Labor in Society* by Emile Durkheim (Glencoe, Ill.: Free Press, 1947), pp. 256–62, by permission of The Macmillan Company.

losing their individuality, the divisions becoming more permeable. In short, a coalescence takes place which makes new combinations possible in the social substance.

But the disappearance of this type can have this consequence for only one reason. That is because it gives rise to a relationship betweeen individuals who were separated, or, at least, a more intimate relationship than there was. Consequently, there is an exchange of movements between parts of the social mass which, until then, had no effect upon one another. The greater the development of the cellular system, the more are our relations enclosed within the limits of the cell to which we belong. There are, as it were, moral gaps between the various segments. On the contrary, these gaps are filled in as the system is leveled out. Social life, instead of being concentrated in a multitude of little centres, distinctive and alike, is generalized. Social relations,—more exactly, intra-social—consequently become more numerous, since they extend, on all sides, beyond their original limits. The division of labor develops, therefore, as there are more individuals sufficiently in contact to be able to act and react upon one another. If we agree to call this relation and the active commerce resulting from it dynamic or moral density, we can say that the progress of the division of labor is in direct ratio to the moral or dynamic density of society.

But this moral relationship can only produce its effect if the real distance between individuals has itself diminished in some way. Moral density cannot grow unless material density grows at the same time, and the latter can be used to measure the former. It is useless to try to find out which has determined the other; they are inseparable.

The progressive condensation of societies in historical development is produced in three principal ways:

1. Whereas lower societies are spread over immense areas according to population, with more advanced people population always tends to concentrate. As Spencer suggests, if we oppose the rate of population in regions inhabited by savage tribes to that of regions of the same extent in Europe; or again, if we oppose the density of the population in England under the Heptarchy to its present density, we shall recognize that the growth produced by the union of groups is also accompanied by interstitial growth.[1] The changes brought about in the industrial life of nations prove the universality of this transformation. The industry of nomads, hunters, or shepherds implies the absence of all concentration, dispersion over the largest possible surface. Agriculture, since it necessitates a sedentary life, presupposes a certain tightening of the social fibre, but it is still incomplete, for there are stretches of land between families.[2] In the city, although the condensation was greater, the houses were not contiguous, for joint property was no part of the Roman law.[3] It grew up on our soil, and is proof that the social web has become tighter.[4] On the other hand, from their origins, European societies have witnessed a continuous growth in their density in spite of exceptions of short-lived regressions.[5]

2. The formation of cities and their development is an even more characteristic symptom of the same phenomenon. The increase in average density may be due to the material increase of the birth-rate, and, consequently, can be reconciled with a very feeble concentration, a marked maintenance of the segmental type. But cities always result from the need of individuals to put themselves in very intimate contact with others. They are so many points where the social mass is contracted more strongly than elsewhere. They can multiply and extend only if the moral density is raised. We shall see, moreover, that they receive recruits especially by immigration. This is only possible when the fusion of social segments is advanced.

As long as social organization is essentially segmental, the city does not exist. There are none in lower societies. They did not exist among the Iroquois nor among the ancient Germans.[6] It was the same with the primitive populations of Italy. "The peoples of Italy," says Marquardt, "originally did not live in cities, but in familial communities or villages (*pagi*) over which farms (*vici*, οἶκοι) were spread." [7] But in a rather short time the city made its appearance. Athens and Rome are or become cities, and the same transformation is made in all Italy. In our Christian societies, the city is in evidence from the beginning, for those left by the Roman empire did not disappear with it. Since then, they have increased and multiplied. The tendency of the country to stream into the city, so general in the civilized world,[8] is only a consequence of this movement. It is not of recent origin; from the seventeenth century, statesmen were preoccupied with it.[9]

Because societies generally begin with an agricultural period there has sometimes been the temptation to regard the development of urban centres as a sign of old age and decadence.[10] But we must not lose sight of the fact that this agricultural phase is as short as societies are elevated. Whereas in Germany, among the Indians of America, and with all primitive peoples, it lasts as long as the people themselves, in Rome and Athens, it ends rather soon, and, with us, we can say that it has never existed alone. On the other hand, urban life commences sooner, and consequently extends further. The regularly more rapid acceleration of this development proves that, far from constituting a sort of pathological phenomenon, it comes from the very nature of higher social species. The supposition that this movement has attained alarming proportions in our societies today, which perhaps no longer have sufficient suppleness to adapt themselves, will not prevent this movement from continuing either within our societies or after them, and the social types which will be formed after ours will likely be distinguished by a still more complete and rapid regression of agricultural civilization.

3. Finally, there are the number and rapidity of ways of communication and transportation. By suppressing or diminishing the gaps separating social segments, they increase the density of society. It is not necessary to prove that they are as numerous and perfected as societies are of a more elevated type.

Since this visible and measurable symbol reflects the variations of what we have called moral density,[11] we can substitute it for this latter in the formula we have proposed. Moreover, we must repeat here what we said before. If society, in concentrating, determines the development of the division of labor, the latter, in its turn, increases the concentration of society. But no matter, for the division of labor remains the derived fact, and, consequently, the advances which it has made are due to parallel advances of social density, whatever may be the causes of the latter. That is all we wish to prove.

But this factor is not the only one.

If condensation of society produces this result, it is because it multiplies intra-social relations. But these will be still more numerous, if, in addition, the total number of members of society becomes more considerable. If it comprises more individuals at the same time as they are more intimately in contact, the effect will necessarily be re-enforced. Social volume, then, has the same influence as density upon the division of labor.

In fact, societies are generally as voluminous as they are more advanced, and consequently as labor is more divided. Societies, as living organisms, in Spencer's words, begin in the form of a bud, sprouting extremely tenuous bodies, compared to those they finally become. The greatest societies, as he says, have emerged from little wandering hordes, such as those of lower races. This is a conclusion which Spencer finds cannot be denied.[12] What we have said of the segmental constitution makes this an indisputable truth. We know, indeed, that societies are formed by a certain number of segments of unequal extent which mutually envelop one another. These moulds are not artificial creations, especially in origin, and even when they have become conventional, they imitate and reproduce, as far as possible, the forms of the natural arrangement which has preceded. There are a great many old societies maintained in this form. The most vast among these subdivisions, those comprising the others, correspond to the nearest inferior social type. Indeed, among the segments of which they are in turn composed, the most extensive are vestiges of the type which comes directly below the preceding, and so on. There are found traces of the most primitive social organization among the most advanced peoples.[13] Thus, the tribe is formed of an aggregate of hordes or clans. The nation (the Jewish nation, for example) and the city are formed of an aggregate of tribes; the city, in turn, with the villages subordinate to it, enters as an element of the most complex societies, etc. Thus, the social volume cannot fail to increase, since each species is constituted by a repetition of societies of the immediately anterior species.

There are exceptions, however. The Jewish nation, before the conquest, was probably more voluminous than the Roman city of the fourth century. Nevertheless, it was of an inferior species. China and Russia are a great deal more populous than the most civilized nations of Europe. With these people, consequently, the division of labor is not developed in proportion to the social volume. That is because the increase of volume is not necessarily a mark of superiority if the density does not increase at the same time and in the same relation, for a society can attain great dimensions because it comprises a very great number of segments, whatever may be the nature of the latter. If, then, even the most vast among them reproduce only societies of very inferior type, the segmental structure will remain very pronounced, and, consequently, social organization little elevated. Even an immense aggregate of clans is below the smallest organized society, since the latter has run through stages of evolution within which the other has remained. In the same way, if the number of social units has influence on the division of labor, it is not through itself and necessarily, but it is because the number of social relations generally increases with that of individuals. But, for this result to be attained, it is not enough that society take in a great many people, but they must be, in addition, intimately enough in contact to act and react on one another. If they are, on the contrary, separated by opaque milieux, they can only be bound by rare and weak relations, and it is as if they had small populations. The increase of social volume does not, then, always accelerate the advances of the division of labor, but only when the mass is contracted at the same time and to the same extent. Consequently, it is only an additional factor, but when it is joined to the first, it amplifies its effects by action peculiar to it, and therefore is to be distinguished from that.

We can then formulate the following proposition: *The division of labor varies in direct ratio with the volume and density of societies, and, if it progresses in a continuous manner in the course of social development, it is because societies become regularly denser and generally more voluminous.*

At all times, it is true, it has been well understood that there was a relation between these two orders of fact, for, in order that functions be more specialized, there must be more co-operators, and they must be related to co-operate. But, ordinarily, this state of societies is seen only as the means by which the division of labor develops, and not as the cause of its development. The latter is made to depend upon individual aspirations toward well-being and happiness, which can be satisfied so much better as societies are more extensive and more condensed. The law we have just established is quite otherwise. We say, not that the growth and condensation of societies *permit,* but that they necessitate a greater division of labor. It is not an instrument by which the latter is realized; it is its determining cause.[14]

MAX WEBER

THE "RATIONALIZATION" OF EDUCATION AND TRAINING

WE cannot here analyze the far-reaching and general cultural effects that the advance of the rational bureaucratic structure of domination, as such, develops quite independently of the areas in which it takes hold. Naturally, bureaucracy promotes a 'rationalist' way of life, but the concept of rationalism allows for widely differing contents. Quite generally, one can only say that the bureaucratization of all domination very strongly furthers the development of 'rational matter-of-factness' and the personality type of the professional expert. This has far-reaching ramifications, but only one important element of the process can be briefly indicated here: its effect upon the nature of training and education.

Educational institutions on the European continent, especially the institutions of higher learning—the universities, as well as technical academies, business colleges, gymnasiums, and other middle schools—are dominated and influenced by the need for the kind of 'education' that produces a system of special examinations and the trained expertness that is increasingly indispensable for modern bureaucracy.

The 'special examination,' in the present sense, was and is found also outside of bureaucratic structures proper; thus, today it is found in the 'free' professions of medicine and law and in the guild-organized trades. Expert examinations are neither indispensable to nor concomitant phenomena of bureaucratization. The French, English, and American bureaucracies have for a long time foregone such examinations entirely or to a large extent, for training and service in party organizations have made up for them.

'Democracy' also takes an ambivalent stand in the face of specialized examinations, as it does in the face of all the phenomena of bureaucracy—although democracy itself promotes these developments. Special examinations, on the one hand, mean or appear to mean a 'selection' of those who qualify from all social strata rather than a rule by notables. On the other hand, democracy fears that a merit system and educational certificates will result in a privileged 'caste.' Hence, democracy fights against the special-examination system.

Reprinted from *Max Weber: Essays in Sociology,* edited and translated by H. H. Gerth and C. Wright Mills (New York: Oxford University Press, Galaxy Books, 1958), pp. 240–44.

The special examination is found even in pre-bureaucratic or semi-bureaucratic epochs. Indeed, the regular and earliest locus of special examinations is among prebendally organized dominions. Expectancies of prebends, first of church prebends—as in the Islamite Orient and in the Occidental Middle Ages—then, as was especially the case in China, secular prebends, are the typical prizes for which people study and are examined. These examinations, however, have in truth only a partially specialized and expert character.

The modern development of full bureaucratization brings the system of rational, specialized, and expert examinations irresistibly to the fore. The civil-service reform gradually imports expert training and specialized examinations into the United States. In all other countries this system also advances, stemming from its main breeding place, Germany. The increasing bureaucratization of administration enhances the importance of the specialized examination in England. In China, the attempt to replace the semi-patrimonial and ancient bureaucracy by a modern bureaucracy brought the expert examination; it took the place of a former and quite differently structured system of examinations. The bureaucratization of capitalism, with its demand for expertly trained technicians, clerks, et cetera, carries such examinations all over the world. Above all, the development is greatly furthered by the social prestige of the educational certificates acquired through such specialized examinations. This is all the more the case as the educational patent is turned to economic advantage. Today, the certificate of education becomes what the test for ancestors has been in the past, at least where the nobility has remained powerful: a prerequisite for equality of birth, a qualification for a canonship, and for state office.

The development of the diploma from universities, and business and engineering colleges, and the universal clamor for the creation of educational certificates in all fields make for the formation of a privileged stratum in bureaus and in offices. Such certificates support their holders' claims for intermarriages with notable families (in business offices people naturally hope for preferment with regard to the chief's daughter), claims to be admitted into the circles that adhere to 'codes of honor,' claims for a 'respectable' remuneration rather than remuneration for work done, claims for assured advancement and old-age insurance, and, above all, claims to monopolize socially and economically advantageous positions. When we hear from all sides the demand for an introduction of regular curricula and special examinations, the reason behind it is, of course, not a suddenly awakened 'thirst for education' but the desire for restricting the supply for these positions and their monopolization by the owners of educational certificates. Today, the 'examination' is the universal means of this monopolization, and therefore examinations irresistibly advance. As the education prerequisite to the acquisition of the educational certificate requires considerable expense and a period of waiting for full remuneration, this striving means a setback for talent (charisma) in favor of property. For the 'intellectual' costs of educational certificates are always low, and with the increasing volume of such certificates, their intellectual costs do not increase, but rather decrease.

The requirement of a chivalrous style of life in the old qualification for fiefs in Germany is replaced by the necessity of participating in its present rudimental form as represented by the dueling corps of the universities which also distribute the educational certificates. In Anglo-Saxon countries, athletic and social clubs fulfil the same function. The bureaucracy, on the other hand, strives everywhere for a 'right to the office' by the establishment of a regular disciplinary procedure and by removal of the completely arbitrary disposition of the 'chief' over the subordinate official. The bureaucracy seeks to secure the official position, the orderly advancement, and the provision for old age. In this, the bureaucracy is supported by the 'democratic' sentiment of the governed, which demands that domination be minimized. Those who hold this attitude believe themselves able to discern a weakening of the master's prerogatives in early weakening of the arbitrary disposition of the master over the officials. To this extent, bureaucracy, both in business offices and in public service, is a carrier of a specific 'status' development, as have been the quite differently structured officeholders of the past. We have already pointed out that these status characteristics are usually also exploited, and that by their nature they contribute to the technical usefulness of the bureaucracy in fulfilling its specific tasks.

'Democracy' reacts precisely against the unavoidable 'status' character of bureaucracy. Democracy seeks to put the election of officials for short terms in the place of appointed officials; it seeks to substitute the removal of officals by election for a regulated procedure of discipline. Thus, democracy seeks to replace the arbitrary disposition of the hierarchically superordinate 'master' by the equally arbitrary disposition of the governed and the party chiefs dominating them.

Social prestige based upon the advantage of special education and training as such is by no means specific to bureaucracy. On the contrary! But educational prestige in other structures of domination rests upon substantially different foundations.

Expressed in slogan-like fashion, the 'cultivated man,' rather than the 'specialist,' has been the end sought by education and has formed the basis of social esteem in such various systems as the feudal, theocratic, and patrimonial structures of dominion: in the English notable administration, in the old Chinese patrimonial bureaucracy, as well as under the rule of demagogues in the so-called Hellenic democracy.

The term 'cultivated man' is used here in a completely value-neutral sense; it is understood to mean solely that the goal of education consists in the quality of a man's bearing in life which was *considered* 'cultivated,' rather than in a specialized training for expertness. The 'cultivated' personality formed the educational ideal, which was stamped by the structure of domination and by the social condition for membership in the ruling stratum. Such education aimed at a chivalrous or an ascetic type; or, at a literary type, as in China; a gymnastic-humanist type, as in Hellas; or it aimed at a conventional type, as in the case of the Anglo-Saxon gentleman. The qualification of the ruling stratum as such rested upon the possession of 'more' cultural quality (in the absolutely changeable, value-neutral sense in which we use the term here), rather than upon 'more' expert knowledge. Special military, theological, and juridical ability was of course intensely practiced; but the point of gravity in Hellenic, in medieval, as well as in Chinese education, has rested upon educational elements that were entirely different from what was 'useful' in one's specialty.

Behind all the present discussions of the foundations of the educational system, the struggle of the 'specialist type of man' against the older type of 'cultivated man' is hidden at some decisive point. This fight is determined by the irresistibly expanding bureaucratization of all public and private relations of authority and by the ever-increasing importance of expert and specialized knowledge. This fight intrudes into all intimate cultural questions.

During its advance, bureaucratic organization has had to overcome those essentially negative obstacles that have stood in the way of the leveling process necessary for bureaucracy. In addition, administrative structures based on different principles intersect with bureaucratic organizations. Since these have been touched upon above, only some especially important structural *principles* will be briefly discussed here in a very simplified schema. We would be led too far afield were we to discuss all the actually existing types. We shall proceed by asking the following questions:

1. How far are administrative structures subject to economic determination? Or, how far are opportunities for development created by other circumstances, for instance, the purely political? Or, finally, how far are developments created by an 'autonomous' logic that is solely of the technical structure as such?
2. We shall ask whether or not these structural principles, in turn, release specific economic effects, and if so, what effects. In doing this, one of course from the beginning has to keep his eye on the fluidity and the overlapping transitions of all these organizational principles. Their 'pure' types, after all, are to be considered merely as border cases which are especially valuable and indispensable for analysis. Historical realities, which almost always appear in mixed forms, have moved and still move between such pure types.

The bureaucratic structure is everywhere a late product of development. The further back we trace our steps, the more typical is the absence of bureaucracy and officialdom in the structure of domination. Bureaucracy has a 'rational' character: rules, means, ends, and matter-of-factness dominate its bearing. Everywhere its origin and its diffusion have therefore had 'revolutionary' results, in a special sense, which has still to be discussed. This is the same influence which the advance of *rationalism* in general has had. The march of bureaucracy has destroyed structures of domination which had no rational character, in the special sense of the term. Hence, we may ask: What were these structures?

PETER L. BERGER AND BRIGITTE BERGER

THE BLUEING OF AMERICA

A sizable segment of the American intelligentsia has been on a kick of revolution talk for the last few years. Only very recently this talk was carried on in a predominantly Left mood, generating fantasies of political revolution colored red or black. The mood appears to have shifted somewhat. Now the talk has shifted to cultural revolution. Gentle grass is pushing up through the cement. It is "the kids," hair and all, who will be our salvation. But what the two types of revolution talk have in common is a sovereign disregard for the realities of technological society in general, and for the realities of class and power in America.

Only the most religious readers of leftist publications could ever believe that a political revolution from the Left had the slightest prospects in America. The so-called black revolution is at a dividing fork, of which we shall speak in a moment. But as to the putatively green revolution, we think that the following will be its most probable result: It will accelerate social mobility in America, giving new opportunities for upward movement of lower-middle-class and working-class people, and in the process will change the ethnic and religious composition of the higher classes. Put differently: far from "greening" America, the alleged cultural revolution will serve to strengthen the vitality of the technological society against which it is directed, and will further the interests of precisely those social strata that are least touched by its currently celebrated transformations of consciousness.

The cultural revolution is not taking place in a social vacuum, but has a specific location in a society that is organized in terms of classes. The cadres of the revolution, not exclusively but predominantly, are the college-educated children of the upper-middle class. Ethnically, they tend to be Wasps and Jews. Religiously, the former tend to belong to the main-line Protestant denominations, rather than to the more fundamentalist or sectarian groups. The natural focus of the revolution is the campus (more precisely, the type of

Reprinted from *The New Republic,* April 3, 1971, pp. 20–23, by permission of *The New Republic.*

campus attended by this population), and such satellite communities as have been springing up on its fringes. In other words, the revolution is taking place, or minimally has its center, in a subculture of upper-middle-class youth.

The revolution has not created this subculture. Youth, as we know it today, is a product of technological and economic forces intimately tied to the dynamics of modern industrialism, as is the educational system within which the bulk of contemporary youth is concentrated for ever-longer periods of life. What is true in the current interpretations is that some quite dramatic transformations of consciousness have been taking place in this sociocultural ambience. These changes are too recent, and too much affected by distortive mass-media coverage, to allow for definitive description. It is difficult to say which manifestations are only transitory and which are intrinsic features likely to persist over time. Drugs are a case in point. So is the remarkable upsurge of interest in religion and the occult. However, one statement can be made with fair assurance: the cultural revolution has defined itself in diametric opposition to some of the basic values of bourgeois society, those values that since Max Weber have commonly been referred to as the "Protestant ethic"—discipline, achievement and faith in the onward-and-upward thrust of technological society. These same values are now perceived as "repression" and "hypocrisy," and the very promises of technological society are rejected as illusionary or downright immoral. A hedonistic ethic is proclaimed in opposition to the "Protestant" one, designed to "liberate" the individual from the bourgeois inhibitions in all areas of life, from sexuality through aesthetic experience to the manner in which careers are planned. Achievement is perceived as futility and "alienation," its ethos as "uptight" and, in the final analysis, inimical to life. Implied in all this is a radical aversion to capitalism and the class society that it has engendered, thus rendering the subculture open to leftist ideology of one kind or another.

Its radicalism, though, is much more far-reaching than that of ordinary, politically defined leftism. It is not simply in opposition to the particular form of technological society embodied in bourgeois capitalism but to the very idea of technological society. The rhetoric is Rousseauean rather than Jacobin, the imagery of salvation is intensely bucolic, the troops of the revolution are not the toiling masses of the Marxist prophecy but naked children of nature dancing to the tune of primitive drums.

When people produce a utopia of childhood it is a good idea to ask what their own childhood has been like. In this instance, the answer is not difficult. As Philippe Ariès has brilliantly shown, one of the major cultural accomplishments of the bourgeoisie has been the dramatic transformation of the structure of childhood, in theory as well as in practice. Coupled with the steep decline in child mortality and morbidity that has been brought about by modern medicine and nutrition, this transformation is one of the fundamental facts of modern society. A new childhood has come into being, probably happier than any previous one in human society. Its impact, however, must be seen in conjunction with another fundamental fact of modern society—namely, the increasing bureaucratization of all areas of social life. We would see the turmoil of youth today as being rooted in the clash between these two facts—paraphrasing Max Weber, in the clash between the new "spirit of childhood" and the "spirit of bureaucracy." However one may wish to judge the merits of either fact, both are probably here to stay. Logically enough, the clash almost invariably erupts when the graduates of the new childhood first encounter bureaucracy in their own life—to wit, in the educational system.

We cannot develop this explanation any further here, though we would like to point out that it is almost exactly the opposite of the Freudian interpretations of the same clash provided, for example, by Lewis Feuer or Bruno Bettelheim: Rebellious youth is not fighting against any fathers; on the contrary, it is outraged by the *absence* of parental figures and familial warmth in the bureaucratic institutions that envelop it. The point to stress, though, is that the transformation of childhood, born of the bourgeoisie, today affects nearly all classes in American society—*but it does not affect them equally*. As, for example, the work of John Seeley and Herbert Gans has demonstrated, there exist far-reaching differences between the childrearing practices of different classes. The transformation, and with it the new "spirit of childhood," developed most fully and most dramatically in the upper-middle class—that is, in the same social context that is presently evincing the manifestations of "greening."

To say this is in no way to engage in value judgments. If value judgments are called for, we would suggest calibrated ones. Very few human cultures (or subcultures) are either wholly admirable or wholly execrable, and the intellectuals who extoll this particular one are as much *terribles simplificateurs* as the politicians who anathematize it. In any case, our present purpose is to inquire into the probable consequences of the cultural changes in question.

The matrix of the green revolution has been a class-specific youth culture. By definition, this constitutes a biographical way station. Long-haired or not, *everyone,* alas, gets older. This indubitable biological fact has been used by exasperated over-thirty observers to support their hope that the new youth culture may be but a noisier version of the old American pattern of sowing wild oats. Very probably this is true for many young rebels, especially those who indulge in the external paraphernalia and gestures of the youth culture without fully entering into its new consciousness. But there is evidence that for an as yet unknown number, the way station is becoming a place of permanent settlement. For an apparently growing number there is a movement *from youth culture to counter-culture*. These are the ones who drop out permanently. For yet others, passage through the youth culture leaves, at any rate, certain permanent effects, not only in their private lives but in their occupational careers. As with the Puritanism that gave birth to the bourgeois culture of America, this movement too has its fully accredited saints and those who only venture upon a *halfway covenant*. The former, in grim righteousness, become sandal makers in Isla Vista. The latter at least repudiate the more obviously devilish careers within "the system"—namely, those in scientific technology, business and government that lead to positions of status and privilege in the society. They do not drop out, but at least they shift their majors—in the main, to the humanities and the social sciences, as we have recently seen in academic statistics.

The overall effects of all this will, obviously, depend on the magnitude of these changes. To gauge the effect, however, one will have to relate them to the class and occupational structures of the society. For those who become permanent residents of the counter-culture, and most probably for their children, the effect is one of downward social mobility. This need not be the case for the halfway greeners (at least as long as the society is ready to subsidize, in one way or another, poets, T-group leaders and humanistic sociologists). But they too will have been deflected from those occupational careers (in business, government, technology and science) that continue to lead to the higher positions in a modern society.

What we must keep in mind is that whatever cultural changes may be going on in this or that group, the personnel requirements of a technological society not only continue but actually expand. The notion that as a result of automation fewer and fewer people will be required to keep the technological society going, thus allowing the others to do their own thing and nevertheless enjoy the blessings of electricity, is in contradiction to all the known facts. Automation has resulted in changes in the occupational structure, displacing various categories of lower-skilled labor, but it has in no way reduced the number of people required to keep the society going. On the contrary, it has increased the requirements for scientific, technological and (last but not least) bureaucratic personnel. (The recent decline in science and engineering jobs is due to recession, and does not affect the long-term needs of the society.) The positions disdained by the aforementioned upper-middle-class individuals will therefore have to be filled by someone else. The upshot is simple: *There will be new "room at the top."*

Who is most likely to benefit from this sociological windfall? It will be the newly college-educated children of the lower-middle and working classes. To say this, we need not assume that they remain untouched by their contact with the youth culture during their school years. Their sexual mores, their aesthetic tastes, even their political opinions might become permanently altered as compared with those of their parents. We do assume, though, that they will, now as before, reject the anti-achievement ethos of the cultural revolution. They may take positions in intercourse that are frowned upon by Thomas Aquinas, they may continue to listen to hard rock on their hi-fi's and they may have fewer racial prejudices. But all these cultural acquisitions are, as it were, functionally irrelevant to making it in the technocracy. Very few of them will become sandal makers or farmers on communes in Vermont. We suspect that not too many more will become humanistic sociologists.

Precisely those classes that remain most untouched by what is considered to be the revolutionary tide in contemporary America face *new prospects of upward social mobility*. Thus, the "revolution" (hardly the word) is not at all where it seems to be, which should not surprise anyone. The very word *avant-garde* suggests that one ought to look behind it for what is to follow—and there is no point asking the *avant-gardistes,* whose eyes are steadfastly looking forward. Not even the Jacobins paid attention to the grubby tradesmen waiting to climb up over their shoulders. A technological society, given a climate of reasonable tolerance (mainly a function of affluence), can afford a sizable number of sandal makers. Its "knowledge industry" (to use Fritz Machlup's term) has a large "software" division, which can employ considerable quantities of English majors. And, of course, the educational system provides a major source of employment for nontechnocratic personnel. To this may be added the expanding fields of entertainment and therapy, in all their forms. All the same, quite different people are needed to occupy the society's command posts and to keep its engines running. These people will have to retain the essentials of the old "Protestant ethic"—discipline, achievement orientation, and also a measure of freedom from gnawing self-doubt. If such people are no longer available in one population reservoir, another resevoir will have to be tapped.

There is no reason to think that "the system" will be unable to make the necessary accommodations. If Yale should become hopelessly greened, Wall Street will get used to recruits from Fordham or Wichita State. Italians will have no trouble running the RAND Corporation, Baptists the space program. Political personnel will change in the wake of social mobility. It is quite possible that the White House may soon have its first Polish occupant (or, for that matter, its first Greek). Far from weakening the class system, these changes will greatly strengthen it, moving new talent upward and preventing rigidity at the top (though, probably, having little effect at the *very* top). Nor will either the mechanics or the rewards of social mobility change in any significant degree. A name on the door will still rate a Bigelow on the floor; only there will be fewer Wasp and fewer Jewish names. Whatever other troubles "the system" may face, from pollution to Russian ICBMs, it will not have to worry about its being brought to a standstill by the cultural revolution.

It is, of course, possible to conceive of such economic or political shocks to "the system" that technological society, as we have known it in America, might collapse, or at least seriously deteriorate. Ecological catastrophe on a broad scale, massive malfunction of the capitalist economy, or an escalation of terrorism and counter-terror would be cases in point. Despite the currently fashionable prophecies of doom for American society, we regard these eventualities as very unlikely. If any of them should take place after all, it goes without saying that the class system would stop operating in its present form. But whatever else would then be happening in America, it would *not* be the green revolution. In the even remoter eventuality of a socialist society in this country, we would know where to look for our greeners—in "rehabilitation camps," along the lines of Castro's Isle of Pines.

We have been assuming that the children of the lower-middle and working classes remain relatively unbitten by the "greening" bug—at least sufficiently unbitten so as not to interfere with their aspirations of mobility. If they too should drop out, there would be literally no one left to mind the technological store. But it is not very easy to envisage this. America falling back to the status of an underdeveloped society? Grass growing over the computers? A totalitarian society, in which the few remaining "uptight" people run the technocracy, while the rest just groove? Or could it be Mongolian ponies grazing on the White House lawn? Even if the great bulk of Americans were to become "beautiful people," however, the rest of the world is most unlikely to follow suit. So far in history, the uglies have regularly won out over the "beautiful people." They probably would again this time.

The evidence does not point in this direction. The data we have on the dynamics of class in a number of European countries would suggest that the American case may not be all that unique. Both England and western Germany have been undergoing changes in their class structures very similar to those projected by us, with new reservoirs of lower-middle-class and working-class populations supplying the personnel requirements of a technological society no longer served adequately by the old elites.

What we have described as a plausible scenario is not terribly dramatic, at least compared with the revolutionary visions that intellectuals so often thrive on. Nor are we dealing with a process unique in history. Vilfredo Pareto called this type of process the "circulation of elites." Pareto emphasized (rightly, we think) that such circulation is essential if a society is going to survive. In a Paretian perspective, much of the green revolution would have to be seen in terms of decadence (which, let us remark in passing, is not necessarily a value judgment—some very impressive flowerings of human creativity have been decadent in the same sociological sense).

But even Marx may, in a paradoxical manner, be proven right in the end. It may be the blue-collar masses that are, at last, coming into their own. "Power to the people!"—nothing less than that. The "class struggle" may be approaching a new phase, with the children of the working class victorious. These days we can see their banner all over the place. It is the American flag. In that perspective, the peace emblem is the old bourgeoisie, declining in the face of a more robust adversary. Robustness here refers, above all, to consciousness—not only to a continuing achievement ethos, but to a self-confidence not unduly worried by unending self-examination and by a basically intact faith in the possibilities of engineering reality. Again, it would not be the first time in history that a declining class leaned toward pacifism, as to the "beautiful things" of aesthetic experience. Seen by that class, of course, the blue-collar masses moving in suffer from considerable aesthetic deficiences.

"Revolutionary" America? Perhaps, in a way. We may be on the eve of its blueing.

Chapter 16

OLD AGE, ILLNESS, AND DEATH

EVERY individual sooner or later experiences the limits of the human condition in the events of aging, illness, and death. If a society is to function as a viable framework for the individual's biography it must provide values and institutional forms that enable the individual to cope with these experiences both in his own life and in that of others. As a result of modern conditions, especially modern medicine, there has been a sharp rise in life expectancy and thus in the rise of the aged as a group in the population.

The aged in contemporary society are increasingly segregated from the rest of the population, largely as a result of the transformations in the structure of the family mentioned earlier. This fact has produced a severe social problem that has attracted the attention of an increasing number of sociologists. With the rise of the hospital as the major institutional setting in which illness takes place, the seriously sick also have come to be segregated from ordinary social life. Sociologists have studied the various aspects of the "sick role" and the institutional process of the hospital as well as the professional organizations of medical care. The approach of labeling theory has been used very effectively in this area. A small group of sociologists has been concerned with the "management of death" in contemporary society.

The first selection, from Jules Henry's *Culture against Man,* shows features of the segregation of the aged in specialized institutional settings. In the next selection, Julius A. Roth discusses some aspects of the professional-patient interaction in a tuberculosis hospital in his article "The Treatment of Tuberculosis as a Bargaining Process." Death is the experience that finally and dramatically puts in question all social patterns of ordinary life. The manner in which a society handles death reveals much about its values. The last selection, from Robert Blauner's "Death and the Social Structure," deals with the handling of death in contemporary society.

JULES HENRY

HUMAN OBSOLESCENCE

ROSEMONT is a private institution run for profit by Mrs. Dis. She is genial, cooperative, and always one legal step ahead of the Health Department. This report starts with an overview of a typical ward.[1]

From *Culture against Man,* by Jules Henry (New York: Random House, 1963).

As I entered the ward a few of the men turned their heads in my direction. Others paid no attention or were asleep. Most of them were dressed in street clothes. A few were in pajamas and robes. The clothing looked old and poorly fitting and some of it was torn at the elbows. The ward smelled strongly of urine even though the windows were open. The beds were so close together that often there was room only for a chair between them. All walls were lined with beds and there were some in the center of the room. The mattresses were thin, the beds sagged in the middle, the sheets were dingy and some of them were smeared with dried feces. The beds with no assigned occupants were covered by a thin grayish cover; others had a faded blue, red, green, or brown blanket folded at the foot or spread over the bed. The upper half of the windows had dark curtains. On the walls were a picture of George Washington, one of Jesus, another of the Madonna and Child, and a religious calendar. A couple of men had clocks at the head of their beds. The dark floor was dotted with wet spots, and I noticed several men spitting on it. Most of them were staring into space and they did not talk to one another.

I could not but notice the contrast between the attractive flooring and the drabness of the rest. The floor is tiled in colored squares and at one place there is a crest set in tile bearing the letter R. The floor is clean and waxed but two walls are dingy and have soil spots. A third has unpainted areas where remodeling has been done. The fourth wall is a flimsy partition between the two sides of the division. A picture of George Washington hangs askew on the wall above the negro patients, and there are two other pictures.

There was an odor of urine. My general impression of the patients was one of apathy and depression. Most of them were sitting slumped over, heads bowed, hands folded. The few who were moving did so slowly and without animation. . . . While I was standing near the center of the ward Mr. Nathan, a large man in a dirty green shirt walked slowly over to me. . . . He talked about how hungry he was, saying that this was true of all the patients. He had never had a large appetite, he said, but even he was hungry on the food they got here. While he was talking several patients walked over and looked at a clock on the south wall. Mr. Nathan explained, "You see, it's getting near lunch time and they're all hungry. That's what everyone does from eleven o'clock until lunch time—they look at the clock."

There was some activity at the east end of the division, so I walked over. Mr. Quilby and Mr. Segram, two dirty, thin, gray-haired little men were in the same bed. One was talking loudly and the other was paying no attention. The bed had no linens, and the mattress, which was slit from end to end, had several wet spots. A second bed was empty and it too had no linens. The empty bed had a large wet spot in the center. In the third bed Mr. Quert, a patient who seemed more oriented, was sitting on the side of his bed apparently keeping the two patients from getting into his bed. He explained that someone had to watch or they would hurt themselves. Mr. Quert seemed to be keeping the two men in the bed by putting a bedside table in front of them whenever they tried to get out. I noted that Mr. Quert had a puddle of urine under his bed too. . . . The two men were the most depressing sight I have ever seen. They were only partially dressed and neither of them had shoes on. They bumped into each other as they constantly moved back and forth in the bed. One of them was kicking or scratching the other. One tried constantly to get out of the bed, first on one side and then on the other, but was always prevented by the table. As he turned from side to side he would bump into the other man in the bed and would lift his legs high to avoid bumping him. Horrified, I stood watching for some time. I tried to speak to the men but they seemed not to hear. Mr. Ansmot (a patient) shouted at one to get into his own bed, but got no results. . . . From the way he moved, one of the men must have been blind, for he always felt around with his feet or hands before he moved in any direction.

When I had seen enough of this I walked out. As I went through the north division I saw a white and a colored aide sitting in chairs, and a white aide was calling loudly and sternly to a patient who had had an incontinent stool and had feces smeared all over himself and his bed, "Sam, you get that sheet up over you."

I watched the patient in the second bed in the center for a few moments. He had feces all over himself and the bed. I failed to find out what his name was. He did not reply when I spoke to him. I left the division feeling completely depressed and contaminated.

I noticed that Mr. Link and Mr. Scope were both incontinent and that the odor was especially bad on this side of the division. As I walked down this aisle I looked down and noticed I was standing in a puddle of urine about an inch deep. I jumped over it and looking back saw that the urine had collected in the center aisle and ran almost all the way from the east to the west end of the division. It started from the beds of four patients, Link, Scope, Yankton and Merchant. I walked down the aisle carefully avoiding the stream of urine.

JULIUS A. ROTH

THE TREATMENT OF TUBERCULOSIS AS A BARGAINING PROCESS

THE person in a skilled service occupation tries to keep the initiative in his relationship to his client.[1] He uses his superior specialized knowledge, his

Reprinted from *Human Behavior and Social Processes: An Interactionist Approach* (ed. Arnold M. Rose). (Boston: Houghton Mifflin, 1962), pp. 575–88. Reprinted by permission of the publishers.

special frame of reference, and the lingo of his trade to assert his right to make decisions for the client. Even in non-professional occupations this desire to keep control is evident. The auto repair man does not want the motorist telling him "his business"; the carpenter knows better what kind of cabinets the home owner wants or needs than does the home owner himself; janitors know better when the drafts of a furnace should be opened and closed than do the residents of a building.

The ideal of a professional group is to control its own standards of service. The practitioner is to be judged only by his colleagues; clients are to accept his advice and direction as that of an expert whose competence in the area of service is far superior to their own and that of the client's family and acquaintances. The closer a professional group comes to this ideal, the more autonomous and "successful" it is.

Such autonomy can be reduced by the nature of the relationship with other professional groups (for example, the subordination of professional nurses to physicians). It can also be reduced by certain aspects of the professional person's relationship to his client. It is the latter aspect which I shall discuss in this essay.

For one thing, the goals of the professional and the client are never entirely the same. The doctor may want to keep the patient in the hospital for a long period to increase the chance that he will not have a relapse, while the patient is more willing to take his chances on leaving earlier in order to get back to his family, job, or other activities. To the physician, the patient's apparent neglect of his health will appear irrational; to the patient, the doctor's apparent refusal to consider the patient's family relationships, career, and the chance to live a freer kind of life will appear inhuman. Stated more generally, the goals of the professional in his relationship to the client tend to be highly specialized (for example, the clinician is concerned primarily with arresting the infection in the patient's lungs), whereas the goals of the client include goals generated by all of his roles in addition to that of client of a given professional person.

In many lines of work the client must be educated to get him to do what the professional person thinks is best for him. But with better education, the client is also better able to judge the service and to escape from professional control. Thus, in tuberculosis hospitals the staff is faced with the constant dilemma of how to give the patient enough education about his disease and its treatment so that he can "cooperate" in that treatment, but not so much education that he thinks he knows as much as the doctor.

Even if the professional person carefully weighs the information he gives to the client, however, he may not be able to control many other sources of information to the client. Information control is the key to much of the control of decision-making and evaluation of service because the client can evaluate the service if he knows (or thinks he knows) the basis for given decisions and actions. Tuberculosis hospital patients are relatively well educated about the service they are receiving. They are surrounded by "experts"—their fellow patients—who readily pool their observations made over periods of months about their experiences with the same disease and the same forms of treatment given by the same doctors and nurses. The patient can see for himself how long patients have to wait before getting a pass, how active they can be without having a relapse, what condition they are in following a bronchoscopy, how many of what kinds of pills you should get with each meal, and so on.

With the help of such information and with goals which contradict those of the staff, patients can (and do) formulate plans for resisting the control the hospital staff tries to maintain over their lives. Again, such resistance by the clients is not unusual in professional-client relationships. All professionals experience a greater or lesser degree of such resistance to control from their clients and must in part yield to it if the relationship is to be continued. The relationship could, of course, be terminated. A patient can leave the hospital against advice in order to escape completely the control of the hospital staff, or the staff may discharge a patient whom they consider to be too "uncooperative." Most often, however, both sides succeed in continuing the relationship by compromising their goals and yielding in part to the pressure of the other party.

Thus, the treatment relationship may be conceived of as a conflict for the control of the patients' behavior, a conflict usually resolved by bargaining. The tuberculosis hospital as it exists today is the product of such long-term bargaining.

One aspect of a bargaining relationship is the relative power of the parties concerned. At one extreme are the prison and concentration camp where—at first sight—the "client" is largely helpless before those who prescribe his behavior. Even at this extreme, however, the underdog is not completely crushed and is often able to gain his goals to some degree—usually by subtle and surrepitious means. . . . At the other extreme are cases where the professional persons feel forced to comply with the "unreasonable" demands of the clients, for example in the case of jazz musicians playing for a "square" audience. Here again, however, the professionals develop subtle means of controlling the audience without seeming to do so. . . . We also have the limiting case of the client approaching colleague status, which seems to happen to some extent in the research ward reported by Fox. . . . In this context the tuberculosis hospital is closer to the concentration camp, with the professionals holding most of the formal power and the clients having to rely mainly on indirect means to gain ends conflicting with those of the professionals.

Generally speaking, the patients want more and earlier privileges, more pass time, earlier discharge, and greater freedom from restrictions, and the staff want fewer and later privileges, fewer and shorter passes, longer hospitalization, and more restrictions for a longer time (for the patients). In addition, there are frequently differences between patients and staff about the need for given medicines or surgery or other forms of treatment.

A patient may believe that further hospitalization is not necessary in his case, but he is unable to leave against advice because he knows the staff can cut him off from outpatient treatment or prevent him from getting a job or refuse welfare aid to himself and his family. Such official power does not mean that the patient is helpless, however. When a patient threatens to leave against advice, the doctor is faced with the possibility that the patient's condition may deteriorate if cut off from treatment or that the patient may be a source of infection to his family and other associates. The doctor may be willing to give the patient more pass time or other privileges if these will serve to hold him in the hospital for a longer period. Thus, the patient will have used his very weakness and disability as a counter with which to bargain for greater control over his own behavior. The very regulations which are intended to strengthen the doctor's hand—cutting off the patient who leaves against advice from all treatment for a specified number of months—usually three) may thus actually cause the more conscientious doctor to give in to many of the patients' demands rather than run the risk of having the patients derprived of treatment. In fact, the more "irresponsible" patient is likely to be in a stronger position than the more conforming one because the staff member believes the former will actually carry out his threat if privileges are withheld, while the latter is more likely to be thought of as bluffing.[2]

It would seem at first that nurses can control the patients with regard to medicines, diet, and other aspects of the treatment carried out on the ward. Here again, however, the patients are not completely helpless. Through their observations and pooling of information patients learn which nurses are most likely to give certain drugs and nursing services and the patients ask these nurses whenever possible. Thus, a patient may get an aspirin or laxative from one nurse after being refused by another. He may get food outside of his diet from other patients. He may have visitors sneak in contraband medicine or food. He may fake or exaggerate symptoms to convince nurses that they should give him medicine he thinks he needs. Persistent complaining and demands over a period of time are often sufficient in themselves to get nurses to "give in" and give the patient what he wants to keep him quiet. The nurses, who usually believe that patients take too much medicine and should be discouraged from taking more,[3] make use of such tactics as stalling the patient off, trying to talk him out of his symptoms, or giving innocuous placebos. They defend their action among themselves by claiming that the patient's symptoms are "all in his head," that he might become addicted or suffer untoward reactions if given the medicine he asks for, or that he is deliberately exaggerating his complaints. At the same time, the nurse is restrained in withholding medicine and using placebos by the fact that the patient might "go over her head" and complain to the doctor that his distress is not being relieved. The nurse's bargaining position is to some degree dependent upon the readiness with which the doctor accedes to the patients' demands.[4]

Surgery is another area in which considerable bargaining goes on between the patients and staff. Patients frequently refuse surgery recommended by the physicians. Physicians may try to convince the patient of the possible dire consequences if he does not take surgery. The patient may point to the possible dire consequences (as he has observed them in other patients getting surgery) if he *does* have surgery. He may also argue that surgery is unnecessary by pointing to other patients whose cases appear similar to his and who have made a good recovery without surgery. On the other hand, some patients regard surgery as the quickest way out of the hospital or the "sure cure" and insist on having it even when the physicians are reluctant. In some cases, surgery may be considered by the physicians only because the patient has demanded it. Patients who say they want surgery and want it as soon as possible are more likely to be considered for surgery, and, if recommended, are likely to get it at an earlier time. A patient who says openly that he is against surgery and does not want it is likely to have consideration of the possibility of surgery postponed by the medical staff and is less likely to have it recommended if it *is* brought up for consideration. A threat by the patient that he will leave against advice by a certain date if not given surgery may sometimes prompt the medical staff to move toward surgery faster than they otherwise would have.

Even an unconscious patient can bargain with a surgeon. Thus, if a pulmonary resection [5] has been planned, but the surgeon finds on entering the chest cavity that a thoracoplasty [6] seems more suitable, he may refrain from doing a thoracoplasty and simply sew the patient up again if the patient has insisted strenuously before the operation that he did not want a thoracoplasty under any circumstances.

Sometimes explicit "deals" can be made, similar to those of union-management bargaining. Patients *are* sometimes given regular and frequent passes to induce them to remain in the hospital. Patients who are threatening to walk out are sometimes offered a pass to think things over if they promise not to leave for good at this time. Patients are sometimes promised earlier conferences to reduce the chance of their leaving against advice. Patients who have left against advice have been permitted to return to the hospital after a few days, despite a ninety-day exclusion rule, on the condition that they sign up for surgery. Patients have been promised discharges within a specified period of time if they agree to take another round of gastric cultures, bronchoscopy, planigrams, and other diagnostic procedures. Patients are sometimes given an extended leave of absence to take care of what they insist is an urgent personal or family problem if they promise to return after it is over and spend at least a certain minimum period of time in the hospital.

More often, the "deal" is never explicitly stated by anyone, but is gradually worked out over a period of time as a result of pushes and pulls from either side. Thus, in a hospital where the patients had come to think of an "average" stay as about one year, the physician decided that one patient with extensively diseased lungs should stay for at least eighteen months after the patient had refused recommended surgery. The patient, however, felt that he had been promised a discharge within a year and simply would not go along with the idea of staying a year and a half. The doctor backed down to some extent under pressure from the patient. To insist on holding him the full eighteen months, the doctor thought, might only lead to his leaving against advice immediately. The patient, at the same time, wanted to avoid leaving against advice and thus being cut off from follow-up treatment if he could manage to get a sanctioned discharge without undue delay. First the doctor thought of keeping the patient in Class 3 [7] until he had been in the hospital a year and then holding him six weeks in each of Classes 4 and 5. In this way, he would get in about fifteen months of treatment, in itself a compromise over the doctor's original intentions, but yet substantially longer than the patient wanted to stay. The patient, however, wanted to be promoted immediately and spend only a month each in Classes 4 and 5 so that he could get out of the hospital in about a year. The doctor made no definite promises, but thought he would try stalling off the patient as much as possible on each promotion. The patient's continued demands and protests pushed the doctor into promoting him to Class 4 before his first year was up, but the doctor got the patient to serve most of his six-week periods in Classes 4 and 5 before he discharged him. Thus, the doctor did not keep the patient quite the full fifteen months he had hoped for after his initial compromise and the patient did not get out in the year which he had considered the "proper" length of hospitalization. On the other hand, the mutual compromise allowed the post-hospital follow-up examinations and treatment to be continued, something which both doctor and patient desired.

A bargaining ploy may have a delayed and/or disguised effect so that the person who applies the pressure cannot know for sure just how successful he has been. To the patient who assails the physician with arguments, threats, and (in the case of women) tears in an effort to get his therapy conferences, privileges, passes, and discharge sooner than the doctors want to give them, it may look as if the physician is unyielding. Yet, back in the conference room the medical staff take the patient's "beefs" into consideration when making their decisions about managing his case. They may well give the next privilege somewhat earlier or more readily give a discharge in a borderline case than they would have if the patient had applied no pressure. This process seems analogous to what very likely happens in baseball when a player protests an umpire's decision. Umpires never change their decisions on balls and strikes, so it may at first seem that the player is wasting his breath when he protests. If one of the subsequent pitches is doubtful, however, the umpire is more likely to call it a ball if the batter has vigorously protested a previous strike call than he would if the batter had said nothing. In this situation, if the catcher and pitcher wish to uphold *their* bargaining position, they in turn must protest a close call against them in order to place the umpire under obligation to even things up with *them*.

In most of the illustrations I have given thus far, the bargaining process has been rather open, even though the effects have been at times more or less hidden. Actually, if bargaining situations could be quantified, such obvious examples would almost certainly be in a small minority. I have relied heavily on such obvious examples simply because the bargaining process could be more easily described with them.

Most bargaining, I would guess, is a product not of overt demands and pressures, but rather of the *anticipation* by the parties involved of the likely or possible consequences of certain behavior on their part. Take the case of patients who would like to get more time out on pass than they are presently getting. The "standard" pass may, in a given hospital, be regarded as three days once a month. Some patients may push the limits somewhat by asking for a pass three weeks after the previous one or by requesting five days instead of the customary three. However, if they want as much pass time as they can get, why don't they ask for another pass after only one week or why don't they ask for ten days instead of only five? The fact is that patients believe that requesting just a little more time may result in success, but making an "unreasonable" request will only earn them the anger of the physician and perhaps a flat refusal to grant any pass at all. (Of course, how to define the limits of "reasonable" and "unreasonable" is an important question in itself.) They are in much the same position as a labor union whose members are getting two dollars an hour and want to bargain for more. They will ask for an additional thirty cents an hour (with the hope of getting fifteen, just as patients may increase their pass request from three to five days and hope to get four), but they would not think of asking for an additional two dollars an hour. Such a request would be considered grossly unreasonable by everyone concerned and would weaken their bargaining position by making negotiation impossible and alienating public support. The union in such a case is faced with the problem of deciding what are the outside limits of a "reasonable" demand which will give them a basis for negotiation. Tuberculosis patients operate in the same way when they discuss with one another how much pass time they can "reasonably" request for an occasion such as Christmas. Physicians, on the other hand, will *anticipate* whether a given patient will be unduly upset or will leave against advice if refused a pass. They will *anticipate* whether giving a "special pass" to one patient will subject them to an outburst of such requests for "special passes" from other patients. Their anticipation of such results will affect their decision about whether or not to give a pass even though no threats have been made and no overt pressure has been brought to bear.

Or a physician may *anticipate* that a given patient will "take advantage" of a promotion in activity classification and assume that he is cured and need no longer observe any rest rules. In such a case the doctor may hold up the patient's promotion beyond the time that he would for most patients in a similar condition as a means of holding down the patient's activity level. "I know

Abrams is already taking more activity than most patients in Class 3 even though he's only in Class 2. He's already been in Class 2 longer than most of the other guys and he's ready for a promotion so far as his condition is concerned. But if I put him in Class 3 now, he won't stay in bed at all any more, he won't wear a mask off the ward, he'll do leather work twelve hours a day instead of only six, he'll think he's cured. The only way I can show him that he still needs some rest is to hold him in Class 2." At the same time, patients who believe they have discovered the doctor's line of reasoning may try to improve their bargaining position by making a point of appearing "cooperative"—telling the doctor how anxious they are to follow his advice, carefully hiding their violations of activity restrictions from the nurses, and so forth—on the assumption that a "good" reputation will enable them to win earlier privileges, more passes, and earlier discharge.

Patients do not actually have to threaten to leave against advice in order to get concessions toward speeding up their treatment timetable. (In fact, if the threat occurs in a covert or indirect manner, it is likely to be more successful because it does not challenge the physician's authority.) A doctor, as a result of an off-the-cuff personality diagnosis and reports from the nurses about the patient's activities, may conclude that a given patient "can't stand confinement." In order to hold on to his patient, the doctor may from time to time give him a somewhat earlier privilege to keep him feeling that he is progressing. The doctor may not originally have wanted to discharge the patient earlier than most others and may believe that he needs just as much hospital treatment as the average. But even without overt demands from the patient, the doctor may find that the logic of the timetable requires an earlier discharge in keeping with the earlier privileges granted, unless there is a very clear-cut opposing consideration, such as recent positive sputum tests or obvious changes in recent chest X-rays. "I don't think he should be released just yet, but I can't think of any good reason for holding him."

The anticipations of others' reactions are, of course, often mistaken. The staff member or patient may find that his allowances for another's reactions did not have the effect he had hoped or that allowances on one matter had unexpected effects on other matters. The doctor whom the patient decided was a "soft touch" for another pass may just have launched a "crackdown" to stem the creeping increase in pass time. The patient who seemed to accept his fate so passively may become violently angry and walk out of the hospital when the staff decides to hold him another three months. However, such unsought and unexpected effects themselves become part of the information which the doctor and patient use to decide what to do next.

Anticipation and allowances on both sides influence each other and thus form a dialectic of mutual pushes and pulls operating over periods of time. Some of the pushes and pulls are outspoken demands from staff or patients, some are talked over only within the colleague group, some operate only within the minds of the individuals. Such a dialectic cannot be meaningfully analyzed as an event or a single interchange, but only as a continuing process with an arbitrary beginning and end, as in the following oversimplified illustration.

Some patients may press for the somewhat earlier granting of a given privilege, let us say "outside privileges." [8] The doctor may resist such pressure for a time. But on one occasion a patient offers a particularly compelling argument and is granted outside privileges a month earlier than he otherwise would have been. (The doctor's anticipation of the effect on the patient of his agreement to or refusal of this request enters into his decision.) Now it becomes more difficult for the doctor to refuse similar requests from other patients because he is faced with the question of whether it is "fair" to grant the earlier privilege to one and not to others. He therefore finds himself granting the same earlier outside privileges to others with "less compelling" arguments. The more conforming patient, who would not have thought of pressing for earlier outside privileges before, now cannot see why he should not get what others get. Patients are no longer asking to be granted an "exception," but simply to be given what "everybody else" is getting.

At this point the earlier outside privileges may become stabilized at a new point in the timetable. They may even have the effect of moving forward some other privileges since the patients, and even the physician in his own mind, can argue: "If outside privileges can be taken safely a month earlier, why can't [other privileges] be taken earlier?" The doctor may feel retrospectively that he has been tricked, but he may not consider the issue important enough to fight about and lets the matter stand. On the other hand, he may decide that the changed state of affairs is not in keeping with a proper treatment regimen, or that the patients will take advantage of his leniency in this case to make inroads on the restrictions in other areas, or that the shift in outside privileges to an earlier point on the timetable damages the logic of the graduated activity program and thus threatens to destroy the program as a whole. In such a case he may wipe out the earlier outside privileges altogether, or at least firmly refuse to grant any more. The patients call such action a "crackdown." There will be a period of ill feeling while the patients accuse the doctor of reneging on his bargain and the doctor accuses the patients of having tried to put a fast deal over on him. (Such accusations may pass openly between doctor and patients or they may be passed behind each other's backs.)

For a time the patients will let up on their demands and their pressure against the limits in all directions because they believe the doctor is "not in the mood" to tolerate or give in to pressure for more freedom of action. They "lie low" in the same way as does a whorehouse operator who suspends activities temporarily when the district attorney announces a drive to "wipe out vice." The doctor, in the meantime, is keeping a sharp eye open to see that he is not "tricked" again. However, in hospitals, as in politics, such campaigns blow over. The patients dissect the doctor's words and actions, as well as pick up any information they can from the ward personnel and others, to find out when the doctor has "cooled off." Finally, a few of the more venturesome patients will tentatively renew their demands. The first efforts may be sharply rejected, proving that the patients' timing was wrong. But the time will come when the traumatic effects of the previous incident will wear off and the doctor will once again grapple with himself about just what the "fair" and "humane" way is to deal with the "needs" of the patients. (Concepts of "fairness," humanity, decency, and the like themselves become counters in the bargaining process and may bolster the bargaining position of one side or the other.) Again, he will begin to make concessions to the pressures and anticipated pressures of the patients and the cycle begins again—although things are never *quite* the same way after any cycle as they were before.

Thus, much (probably most) bargaining between persons or groups is not a matter of open threats and the deployment of power positions. It tends rather to be more a matter of A anticipating B's reaction to A's potential behavior and modifying his own behavior in an effort to control B's reaction. A's behavior then creates an image which calls for certain anticipations on B's part when B decides what response to make. This process of anticipation and modification of behavior continues so long as A and B are in communication with one another. When A or B becomes aware of this process (and there is probably always some degree of awareness of it), he may deliberately try to manipulate his own image (as seen by the other party) as a means of gaining a desired response.

ROBERT BLAUNER

DEATH AND THE SOCIAL STRUCTURE

BUREAUCRATIZATION OF MODERN DEATH CONTROL

Since there is no death without a body—except in mystery thrillers—the corpse is another consequence of mortality that contributes to its disruptiveness, tending to produce fear, generalized anxiety, and disgust.[1] Since families and work groups must eventually return to some kind of normal life, the time they are exposed to corpses must be limited. Some form of disposal (earth or sea burial, cremation, exposure to the elements) is the core of mortuary institutions everywhere. A disaster that brings about massive and unregulated ex-

Reprinted from *Psychiatry* 29 (1966): 378–94 by permission of the author and the William Alanson White Foundation, Inc.

posure to the dead, such as that experienced by the survivors of Hiroshima and also at various times by survivors of great plagues, famines, and death-camps, appears to produce a profound identification with the dead and a consequent depressive state.[2]

The disruptive impact of a death is greater to the extent that its consequences spill over onto the larger social territory and affect large numbers of people. This depends not only on the frequency and massiveness of mortality, but also on the physical and social settings of death. These vary in different societies, as does also the specialization of responsibility for the care of the dying and the preparation of the body for disposal. In premodern societies, many deaths take place amid the hubbub of life, in the central social territory of the tribe, clan, or other familial group. In modern societies, where the majority of deaths are now predictably in the older age brackets, disengagement from family and economic function has permitted the segregation of death settings from the more workaday social territory. Probably in small towns and rural communities, more people die at home than do so in urban areas. But the proportion of people who die at home, on the job, and in public places must have declined consistently over the past generations with the growing importance of specialized dying institutions—hospitals, old people's homes, and nursing homes.[3]

Modern societies control death through bureaucratization, our characteristic form of social structure. Max Weber has described how bureaucratization in the West proceeded by removing social functions from the family and the household and implanting them in specialized institutions autonomous of kinship considerations. Early manufacturing and entrepreneurship took place in or close to the home; modern industry and corporate bureaucracies are based on the separation of the workplace from the household.[4] Similarly, only a few generations ago most people in the United States either died at home, or were brought into the home if they had died elsewhere. It was the responsibility of the family to lay out the corpse—that is, to prepare the body for the funeral.[5] Today, of course, the hospital cares for the terminally ill and manages the crisis of dying; the mortuary industry (whose establishments are usually called "homes" in deference to past tradition) prepares the body for burial and makes many of the funeral arrangements. A study in Philadelphia found that about ninety per cent of funerals started out from the funeral parlor, rather than from the home, as was customary in the past.[6] This separation of the handling of illness and death from the family minimizes the average person's exposure to death and death's disruption of the social process. When the dying are segregated among specialists for whom contact with death has become routine and even somewhat impersonal, neither their presence while alive nor as corpses interferes greatly with the mainstream of life.

Another principle of bureaucracy is the ordering of regularly occurring as well as extraordinary events into predictable and routinized procedures. In addition to treating the ill and isolating them from the rest of society, the modern hospital as an organization is committed to the routinization of the handling of death. Its distinctive competence is to contain through isolation, and reduce through orderly procedures, the disturbance and disruption that are associated with the death crisis. The decline in the authority of religion as well as shifts in the functions of the family underlies this fact. With the growth of the secular and rational outlook, hegemony in the affairs of death has been transferred from the church to science and its representatives, the medical profession and the rationally organized hospital.

Death in the modern hospital has been the subject of two recent sociological studies: Sudnow has focused on the handling of death and the dead in a county hospital catering to charity patients; and Glaser and Strauss have concentrated on the dying situation in a number of hospitals of varying status.[7] The county hospital well illustrates various trends in modern death. Three quarters of its patients are over 60 years old. Of the 250 deaths Sudnow observed, only a handful involved people younger than 40.[8] This hospital is a setting for the concentration of death. There are 1,000 deaths a year; thus approximately three die daily, of the 330 patients typically in residence. But death is even more concentrated in the four wards of the critically ill; here roughly 75 per cent of all mortality occurs, and one in 25 persons will die each day.[9]

Hospitals are organized to hide the facts of dying and death from patients as well as visitors. Sudnow quotes a major text in hospital administration:

> "The hospital morgue is best located on the ground floor and placed in an area inaccessible to the general public. It is important that the unit have a suitable exit leading onto a private loading platform which is concealed from hospital patients and the public." [10]

Personnel in the high-mortality wards use a number of techniques to render death invisible. To protect relatives, bodies are not to be removed during visiting hours. To protect other inmates, the patient is moved to a private room when the end is foreseen. But some deaths are unexpected and may be noticed by roommates before the hospital staff is aware of them. These are considered troublesome because elaborate procedures are required to remove the corpse without offending the living.

The rationalization of death in the hospital takes place through standard procedures of covering the corpse, removing the body, identifying the deceased, informing relatives, and completing the death certificate and autopsy permit. Within the value hierarchy of the hospital, handling the corpse is "dirty work"; and when possible attendants will leave a body to be processed by the next work shift. As with so many of the unpleasant jobs in our society, hospital morgue attendants and orderlies are often Negroes. Personnel become routinized to death and are easily able to pass from mention of the daily toll to other topics; new staff members stop counting after the first half-dozen deaths witnessed.[11]

Standard operating procedures have even routinized the most charismatic and personal of relations, that between the priest and the dying patient. It is not that the church neglects charity patients. The chaplain at the county hospital daily goes through a file of the critically ill for the names of all known Catholic patients, then enters their rooms and administers extreme unction. After completing his round on each ward, he stamps the index card of the patient with a rubber stamp which reads: "Last Rites Administered. Date ——— clergyman —————." Each day he consults the files to see if new patients have been admitted or put on the critical list. As Sudnow notes, this rubber stamp prevents him from performing the rites twice on the same patient.[12] This example highlights the trend toward the depersonalization of modern death, and is certainly the antithesis of the historic Catholic notion of "the good death."

In the hospitals studied by Glaser and Strauss, depersonalization is less advanced. Fewer of the dying are comatose; and as paying patients with higher social status they are in a better position to negotiate certain aspects of their terminal situation. Yet nurses and doctors view death as an inconvenience, and manage interaction so as to minimize emotional reactions and fuss. They attempt to avoid announcing unexpected deaths because relatives break down too emotionally; they prefer to let the family members know that the patient has taken "a turn for the worse," so that they will be able to modulate their response in keeping with the hospital's need for order.[13] And drugs are sometimes administered to a dying patient to minimize the disruptiveness of his passing—even when there is no reason for this in terms of treatment or the reduction of pain.

The dying patient in the hospital is subject to the kinds of alienation experienced by persons in other situations in bureaucratic organizations. Because doctors avoid the terminally ill, and nurses and relatives are rarely able to talk about death, he suffers psychic isolation.[14] He experiences a sense of meaninglessness because he is typically kept unaware of the course of his disease and his impending fate, and is not in a position to understand the medical and other routines carried out in his behalf.[15] He is powerless in that the medical staff and the hospital organization tend to program his death in keeping with their organizational and professional needs; control over one's death seems to be even more difficult to achieve than control over one's life in our society.[16] Thus the modern hospital, devoted to the preservation of life and the reduction of pain, tends to become a "mass reduction" system, undermining the subjecthood of its dying patients.

The rationalization of modern death control cannot be fully achieved, however, because of an inevitable tension between death—as an event, a crisis, an experience laden with great emotionality—and bureaucracy, which must deal with routines rather than events and is committed to the smoothing out of affect and emotion. Although there was almost no interaction between dying patients and the staff in the county hospital studied by Sudnow, many nurses in the other hospitals became personally involved with their patients and experienced grief when they died. Despite these limits to the general trend, our society has gone far in containing the disruptive possibilities of mortality through its bureaucratized death control.

THE DECLINE OF THE FUNERAL IN MODERN SOCIETY

Death creates a further problem because of the contradiction between society's need to push the dead away, and its need "to keep the dead alive." [17] The social distance between the living and the dead must be increased after death, so that the group first, and the most affected grievers later, can reestablish their normal activity without a paralyzing attachment to the corpse. Yet the deceased cannot simply be buried as a dead body: The prospect of total exclusion from the social world would be too anxiety laden for the living, aware of their own eventual fate. The need to keep the dead alive directs societies to construct rituals that celebrate and insure a transition to a new social status, that of spirit, a being now believed to participate in a different realm.[18] Thus, a funeral that combines this status transformation with the act of physical disposal is universal to all societies, and has justly been considered one of the crucial *rites de passage*.[19]

Because the funeral has been typically employed to handle death's manifold disruptions, its character, importance, and frequency may be viewed as indicators of the place of mortality in society. The contrasting impact of death in primitive and modern societies, and the diversity in their modes of control, are suggested by the striking difference in the centrality of mortuary ceremonies in the collective life. Because death is so disruptive in simple societies, much "work" must be done to restore the social system's functioning. Funerals are not "mere rituals," but significant adaptive structures, as can be seen by considering the tasks that make up the funeral work among the LoDagaa of West Africa. The dead body must be buried with the appropriate ritual so as to give the dead man a new status that separates him from the living; he must be given the material goods and symbolic invocations that will help guarantee his safe journey to the final destination and at the same time protect the survivors against his potentially dangerous intervention in their affairs (such as appearing in dreams, "walking," or attempting to drag others with him); his qualities, lifework, and accomplishments must be summed up and given appropriate recognition; his property, roles, rights, and privileges must be distributed so that social and economic life can continue; and, finally the social units—family, clan, and community as a whole—whose very existence and functioning his death has threatened, must have a chance to vigorously reaffirm their identity and solidarity through participation in ritual ceremony.[20]

Such complicated readjustments take time; and therefore the death of a mature person in many primitive societies is followed by not one, but a series of funerals (usually two or three) that may take place over a period ranging from a few months to two years, and in which the entire society, rather than just relatives and friends, participates.[21] The duration of the funeral and the fine elaboration of its ceremonies suggest the great destructive possibilities of death in these societies. Mortuary institutions loom large in the daily life of the community; and the frequent occurrence of funerals may be no small element in maintaining societal continuity under the precarious conditions of high mortality.[22]

In Western antiquity and the middle ages, funerals were important events in the life of city-states and rural communities.[23] Though not so central as in high-mortality and sacred primitive cultures (reductions in mortality rates and secularism both antedate the industrial revolution in the West), they were still frequent and meaningful ceremonies in the life of small-town, agrarian America several generations ago. But in the modern context they have become relatively unimportant events for the life of the larger society. Formal mortuary observances are completed in a short time. Because of the segregation and disengagement of the aged and the gap between generations, much of the social distance to which funerals generally contribute has already been created before death. The deceased rarely have important roles or rights that the society must be concerned about allocating; and the transfer of property has become the responsibility of individuals, in cooperation with legal functionaries. With the weakening of beliefs in the existence and malignancy of ghosts, the absence of "realistic" concern about the dead man's trials in his initiation to spirithood, and the lowered intensity of conventional beliefs in an afterlife, there is less demand for both magical precautions and religious ritual. In a society where disbelief or doubt is more common than a firm acceptance of the reality of a life after death,[24] the funeral's classic function of status transformation becomes attenuated.

The recent attacks on modern funeral practices by social critics focus on alleged commercial exploitation by the mortuary industry and the vulgar ostentatiousness of its service. But at bottom this criticism reflects this crisis in the function of the funeral as a social institution. On the one hand, the religious and ritual meanings of the ceremony have lost significance for many people. But the crisis is not only due to the erosion of the sacred spirit by rational, scientific world views.[25] The social substructure of the funeral is weakened when those who die tend to be irrelevant for the ongoing social life of the community, and when the disruptive potentials of death are already controlled by compartmentalization into isolated spheres where bureaucratic routinization is the rule. Thus participation and interest in funerals are restricted to family members and friends rather than involving the larger community, unless an important leader has died.[26] Since only individuals and families are affected, adaptation and bereavement have become their private responsibility, and there is little need for a transition period to permit society as a whole to adjust to the fact of a single death. Karl Marx was proved wrong about "the withering away of the state," but with the near disappearance of death as a public event in modern society, the withering away of the funeral may become a reality.

In modern societies, the bereaved person suffers from a paucity of ritualistic conventions in the mourning period. He experiences grief less frequently, but more intensely, since his emotional involvements are not diffused over an entire community, but are usually concentrated on one or a few people.[27] Since mourning and a sense of loss are not widely shared, as in premodern communities, the individualization and deritualization of bereavement make for serious problems in adjustment. There are many who never fully recover and "get back to normal," in contrast to the frequently observed capacity of the bereaved in primitive societies to smile, laugh, and go about their ordinary pursuits the moment the official mourning period is ended.[28] The lack of conventionalized stages in the mourning process results in an ambiguity as to when the bereaved person has grieved enough and thus can legitimately and guiltlessly feel free for new attachments and interests. Thus at the same time that death becomes less disruptive to the society, its prospects and consequences become more serious for the bereaved individual.

Chapter **17**

VALUES AND ULTIMATE MEANINGS

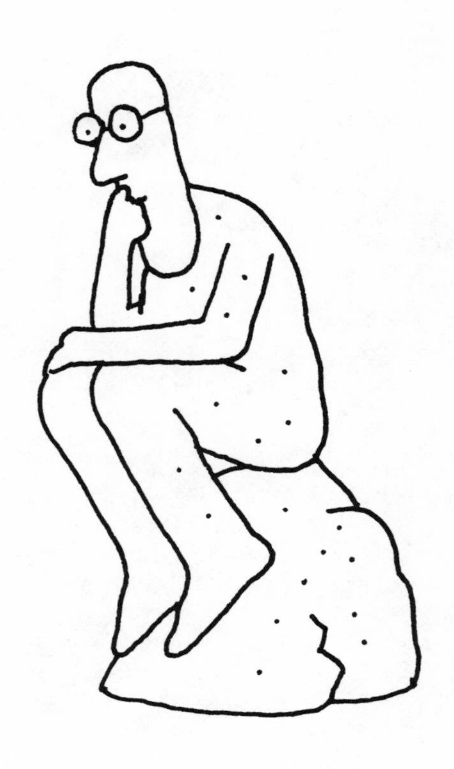

ONE of the major contributions of Emile Durkheim to sociological thought was the insight that no society can exist without some overarching beliefs that serve to integrate and solidify discrepant groups and group interests. Such overarching beliefs may be religious in the conventional sense of the word—that is, they may refer to God, gods, or other supernatural entities—but they may also refer to values and interpretations of existence that are not religious in the customary sense of the word.

All social institutions are permeated by values. To understand a society, one must understand its values, particularly those that provide ultimate meanings for individual life and for the society as a whole. It is because of this that religion occupies a very large place in the works of almost all classical sociologists. Durkheim developed a theory of society in which religion (though admittedly in a very broad sense) is understood as the root social fact. Weber interpreted the rise of modern society as being at least partially determined by transformations in the religious outlook of Western man. A continuing question on sociological theory has been the relationship of values and ideas to the "hard" interests of social groups.

In the Marxian paradigm of society, values and ideas are interpreted as essentially nothing but expressions of such interests, especially, of course, of class interests. Weber and other sociologists of the classical period tried to develop a more complex view of this relationship, with values and ideas not only expressing interests, but also shaping them. The sociology of religion has been one manifestion of this continuing concern of sociologists. Another has been the so-called sociology of knowledge, which has tried to understand the general relationship between social structure and human thought.

The contemporary situation has been widely interpreted as being the result of the decline of traditional religious beliefs and values (this decline is commonly called secularization). Thus, contemporary society exhibits a pronounced instability and insecurity in the area of values. Indeed, some sociologists have argued that this is one of the most fundamental facts to understand with regard to modern society.

The first selection, W. Lloyd Warner's "An American Sacred Ceremony," provides an excellent illustration of a political ritual in contemporary America that expresses the basic Durkheimian theme in pure form. The passage from Warner deals with the common values of the society that have often been called the American "civil religion." The second selection, Thomas Luckmann's "Belief, Unbelief, and Religion," discusses the problems of belief under the impact of secularization. There have been a number of events in recent years, particularly in America, that put in question the notion that secularization is a progressive and irreversible process. Resurgences of religion have cropped up in some very unexpected places. The third selection, Andrew Greeley's "There's a New-Time Religion on Campus," deals with some of these.

W. LLOYD WARNER

AN AMERICAN SACRED CEREMONY

MEMORIAL DAY AND SYMBOLIC BEHAVIOR

Every year in the springtime when the flowers are in bloom and the trees and shrubs are most beautiful, citizens of the Union celebrate Memorial Day. Over most of the United States it is a legal holiday. Being both sacred and secular, it is a holy day as well as a holiday and is accordingly celebrated.

For some it is part of a long holiday of pleasure, extended outings, and great athletic events; for others it is a sacred day when the dead are mourned and sacred ceremonies are held to express their sorrow; but for most Americans, especially in the smaller cities, it is both sacred and secular. They feel the sacred importance of the day when they, or members of their family, participate in the ceremonies; but they also enjoy going for an automobile trip or seeing or reading about some important athletic event staged on Memorial Day. This chapter will be devoted to the analysis and interpretation of Memorial Day to learn its meanings as an American sacred ceremony, a rite that evolved in this country and is native to it.

Memorial Day originated in the North shortly after the end of the Civil War as a sacred day to show respect for the Union soldiers who were killed in the War between the States. Only since the last two wars has it become a day for all who died for their country. In the South only now are they beginning to use it to express southern respect and obligation to the nation's soldier dead.

Memorial Day is an important occasion in the American ceremonial calendar and as such is a unit of this larger ceremonial system of symbols. Close examination discloses that it, too, is a symbol system in its own right existing within the complexities of the larger one.

Symbols include such familiar things as written and spoken words, religious beliefs and practices, including creeds and ceremonies, the several arts, such familiar signs as the cross and the flag, and countless other objects and acts which stand for something more than that which they are. The red, white, and blue cloth and the crossed sticks in themselves and as objects mean very little, but the sacred meanings which they evoke are of such deep significance to some that millions of men have sacrificed their lives for the first as the Stars and Stripes and for the second as the Christian Cross.

Reprinted from *American Life: Dream and Reality* by W. Lloyd Warner (Chicago: University of Chicago Press, 1953), pp. 5–19, by permission of the publisher and Mildred Warner.

Symbols are substitutes for all known real and imaginary actions, things, and the relations among them. They stand for and express feelings and beliefs about men and what they do, about the world and what happens in it. What they stand for may or may not exist. What they stand for may or may not be true, for what they express may be no more than a feeling, an illusion, a myth, or a vague sensation falsely interpreted. On the other hand, that for which they stand may be as real and objectively verifiable as the Rock of Gibraltar.

The ceremonial calendar of American society, this yearly round of holidays and holy days, partly sacred and partly secular, but more sacred than secular, is a symbol system used by all Americans. Christmas and Thanksgiving, Memorial Day and the Fourth of July, are days in our ceremonial calendar which allow Americans to express common sentiments about themselves and share their feelings with others on set days pre-established by the society for this very purpose. This calendar functions to draw all people together to emphasize their similarities and common heritage; to minimize their differences; and to contribute to their thinking, feeling, and acting alike. All societies, simple or complex, possess some form of ceremonial calendar, if it be no more than the seasonal alternation of secular and ceremonial periods, such as that used by the Australian aborigines in their yearly cycle.

The integration and smooth functioning of the social life of a modern community are very difficult because of its complexity. American communities are filled with churches, each claiming great authority and each with its separate sacred symbol system. Many of them are in conflict, and all of them in opposition to one another. Many associations, such as the Masons, the Odd Fellows, and the like, have sacred symbol systems which partly separate them from the whole community. The traditions of foreign-born groups contribute to the diversity of symbolic life. The evidence is clear for the conflict among these systems.

It is the thesis of this chapter that the Memorial Day ceremonies and subsidiary rites (such as those of Armistice or Veterans' Day) of today, yesterday, and tomorrow are rituals of a sacred symbol system which functions periodically to unify the whole community, with its conflicting symbols and its opposing, autonomous churches and associations. It is contended here that in the Memorial Day ceremonies the anxieties which man has about death are confronted with a system of sacred beliefs about death which gives the individuals involved and the collectivity of individuals a feeling of well-being. Further, the feeling of triumph over death by collective action in the Memorial Day parade is made possible by re-creating the feeling of well-being and the sense of group strength and individual strength in the group power, which is felt so intensely during the wars, when the veterans' associations are created and when the feeling so necessary for the Memorial Day's symbol system is originally experienced.

Memorial Day is a cult of the dead which organizes and integrates the various faiths and national and class groups into a sacred unity. It is a cult of the dead organized around the community cemeteries. Its principal themes are those of the sacrifice of the soldier dead for the living and the obligation of the living to sacrifice their individual purposes for the good of the group, so that they, too, can perform their spiritual obligations.

MEMORIAL DAY CEREMONIES

We shall first examine the Memorial Day ceremony of an American town for evidence. The sacred symbolic behavior of Memorial Day, in which scores of the town's organizations are involved, is ordinarily divided into four periods. During the year separate rituals are held by many of the associations for their dead, and many of these activities are connected with later Memorial Day events. In the second phase, preparations are made during the last three or four weeks for the ceremony itself, and some of the associations perform public rituals. The third phase consists of the scores of rituals held in all the cemeteries, churches, and halls of the associations. These rituals consist of speeches and highly ceremonialized behavior. They last for two days and are climaxed by the fourth and last phase, in which all the separate celebrants gather in the center of the business district on the afternoon of Memorial Day. The separate organizations, with their members in uniform or with fitting insignia, march through the town, visit the shrines and monuments of the hero dead, and, finally, enter the cemetery. Here dozens of ceremonies are held, most of them highly symbolic and formalized. Let us examine the actual ritual behavior in these several phases of the ceremony.

The two or three weeks before the Memorial Day ceremonies are usually filled with elaborate preparations by each participating group. Meetings are held, and patriotic pronouncements are sent to the local paper by the various organizations which announce what part each organization is to play in the ceremony. Some of the associations have Memorial Day processions, memorial services are conducted, the schools have patriotic programs, and the cemeteries are cleaned and repaired. Graves are decorated by families and associations and new gravestones purchased and erected. The merchants put up flags before their establishments, and residents place flags above their houses.

All these events are recorded in the local paper, and most of them are discussed by the town. The preparation of public opinion for an awareness of the importance of Memorial Day and the rehearsal of what is expected from each section of the community are done fully and in great detail. The latent sentiments of each individual, each family, each church, school, and association for its own dead are thereby stimulated and related to the sentiments for the dead of the nation.

One of the important events observed in the preparatory phase in the community studied occurred several days before Memorial Day, when the man who had been the war mayor wrote an open letter to the commander of the American Legion. It was published in the local paper. He had a city-wide reputation for patriotism. He was an honorary member of the American Legion. The letter read: "Dear Commander: The approaching Poppy Day [when Legion supporters sold poppies in the town] brings to my mind a visit to the war zone in France on Memorial Day, 1925, reaching Belleau Wood at about 11 o'clock. On this sacred spot we left floral tributes in memory of our town's boys—Jonathan Dexter and John Smith, who here had made the supreme sacrifice, that the principle that 'might makes right' should not prevail."

Three days later the paper in a front-page editorial told its readers: "Next Saturday is the annual Poppy Day of the American Legion. Everybody should wear a poppy on Poppy Day. Think back to those terrible days when the red poppy on Flanders Field symbolized the blood of our boys slaughtered for democracy." The editor here explicitly states the symbolism involved.

Through the early preparatory period of the ceremony, through all its phases and in every rite, the emphasis in all communities is always on sacrifice—the sacrifice of the lives of the soldiers of the city, willingly given for democracy and for their country. The theme is always that the gift of their lives was voluntary; that it was freely given and therefore above selfishness or thought of self-preservation; and, finally, that the "sacrifice on the altars of their country" was done for everyone. The red poppy became a separate symbol from McCrae's poem "In Flanders Fields." The poem expressed and symbolized the sentiments experienced by the soldiers and people of the country who went through the first war. The editor makes the poppy refer directly to the "blood of the boys slaughtered." In ritual language he then recites the names of some of the city's "sacrificed dead," and "the altars" (battles) where they were killed. "Remember Dexter and Smith killed at Belleau Wood," he says. "Remember O'Flaherty killed near Château-Thierry, Stulavitz killed in the Bois d'Ormont, Kelley killed at Côte de Châtillon, Jones near the Bois de Montrebeaux, Kilnikap in the Saint-Mihiel offensive, and the other brave boys who died in camp or on stricken fields. Remember the living boys of the Legion on Saturday."

The names selected by the editor covered most of the ethnic and religious groups of the community. They included Polish, Russian, Irish, French-Canadian, and Yankee names. The use of such names in this context emphasized the fact that the voluntary sacrifice of a citizen's life was equalitarian. They covered the top, middle, and bottom of the several classes. The newspapers throughout the country each year print similar lists, and their editorials stress the equality of sacrifice by all classes and creeds.

The topic for the morning services of the churches on the Sunday before Memorial Day ordinarily is the meaning of Memorial Day to the town and to the people as Christians. All the churches participate. Because of space limitations, we shall quote from only a few sermons from one Memorial Day to show the main themes; but observations of Memorial Day behavior since the

Second World War show no difference in the principal themes expressed before and after the war started. Indeed, some of the words are almost interchangeable. The Rev. Hugh McKellar chose as his text, "Be thou faithful until death." He said:

> "Memorial Day is a day of sentiment and when it loses that, it loses all its value. We are all conscious of the danger of losing that sentiment. What we need today is more sacrifice, for there can be no achievement without sacrifice. There are too many out today preaching selfishness. Sacrifice is necessary to a noble living. In the words of our Lord, 'Whosoever shall save his life shall lose it and whosoever shall lose his life in My name shall save it.' It is only those who sacrifice personal gain and will to power and personal ambition who ever accomplish anything for their nation. Those who expect to save the nation will not get wealth and power for themselves.
>
> "Memorial Day is a religious day. It is a day when we get a vision of the unbreakable brotherhood and unity of spirit which exists and still exists, no matter what race or creed or color, in the country where all men have equal rights."

The minister of the Congregational Church spoke with the voice of the Unknown Soldier to emphasize his message of sacrifice:

> "If the spirit of that Unknown Soldier should speak, what would be his message? What would be the message of a youth I knew myself who might be one of the unknown dead? I believe he would speak as follows: 'It is well to remember us today, who gave our lives that democracy might live, we know something of sacrifice.' "

The two ministers in different language expressed the same theme of the sacrifice of the individual for national and democratic principles. One introduces divine sanction for this sacrificial belief and thereby succeeds in emphasizing the theme that the loss of an individual's life rewards him with life eternal. The other uses one of our greatest and most sacred symbols of democracy and the only very powerful one that came out of the First World War—the Unknown Soldier. The American Unknown Soldier is Everyman; he is the perfect symbol of equalitarianism.

There were many more Memorial Day sermons, most of which had this same theme. Many of them added the point that the Christian God had given his life for all. That afternoon during the same ceremony the cemeteries, memorial squares named for the town's dead, the lodge halls, and the churches had a large number of rituals. Among them was the "vacant chair." A row of chairs decorated with flags and wreaths, each with the name of a veteran who had died in the last year, was the center of this ceremony held in a church. Most of the institutions were represented in the ritual. We shall give only a small selection from the principal speech:

> "Now we come to pay tribute to these men whose chairs are vacant, not because they were eminent men, as many soldiers were not, but the tribute we pay is to their attachment to the great cause. We are living in the most magnificent country on the face of the globe, a country planted and fertilized by a Great Power, a power not political or economic but religious and educational, especially

in the North. In the South they had settlers who were there in pursuit of gold, in search of El Dorado, but the North was settled by people seeking religious principles and education."

In a large city park, before a tablet filled with the names of war dead, one of our field workers shortly after the vacant-chair rite heard a speaker in the memorial ritual eulogize the two great symbols of American unity—Washington and Lincoln. The orator said:

"No character except the Carpenter of Nazareth has ever been honored the way Washington and Lincoln have been in New England. Virtue, freedom from sin, and righteousness were qualities possessed by Washington and Lincoln, and in possessing these characteristics both were true Americans, and we would do well to emulate them. Let us first be true Americans. From these our friends beneath the sod we receive their message, 'Carry on.' Though your speaker will die, the fire and spark will carry on. Thou are not conqueror, death, and thy pale flag is not advancing."

In all the other services the same themes were used in the speeches, most of which were in ritualized, oratorical language, or were expressed in the ceremonials themselves. Washington, the father of his country, first in war and peace, had devoted his life not to himself but to his country. Lincoln had given his own life, sacrificed on the altar of his country. Most of the speeches implied or explicitly stated that divine guidance was involved and that these mundane affairs had supernatural implications. They stated that the revered dead had given the last ounce of devotion in following the ideals of Washington and Lincoln and the Unknown Soldier and declared that these same principles must guide us, the living. The beliefs and values of which they spoke referred to a world beyond the natural. Their references were to the supernatural.

On Memorial Day morning the separate rituals, publicly performed, continued. The parade formed in the early afternoon in the business district. Hundreds of people, dressed in their best, gathered to watch the various uniformed groups march in the parade. Crowds collected along the entire route. The cemeteries, carefully prepared for the event, and the graves of kindred, covered with flowers and flags and wreaths, looked almost gay.

The parade marched through the town to the cemeteries. The various organizations spread throughout the several parts of the graveyards, and rites were performed. In the Greek quarter ceremonies were held; others were performed in the Polish and Russian sections; the Boy Scouts held a memorial rite for their departed; the Sons and Daughters of Union Veterans went through a ritual, as did the other men's and women's organizations. All this was part of the parade in which everyone from all parts of the community could and did participate.

Near the end of the day all the men's and women's organizations assembled about the roped-off grave of General Fredericks. The Legion band played. A minister uttered a prayer. The ceremonial speaker said:

"We meet to honor those who fought, but in so doing we honor ourselves. From them we learn a lesson of sacrifice and devotion and of accountability to God and honor. We have an inspiration for the future today—our character is strengthened—this day speaks of a better and greater devotion to our country and to all that our flag represents."

After the several ceremonies in the Elm Hill Cemetery, the parade re-formed and started the march back to town, where it broke up. The firing squad of the American Legion fired three salutes, and a bugler sounded the "Last Post" at the cemetery entrance as they departed. This, they said, was a "general salute for all the dead in the cemetery."

Here we see people who are Protestant, Catholic, Jewish, and Greek Orthodox involved in a common ritual in a graveyard with their common dead. Their sense of separateness was present and expressed in the different ceremonies, but the parade and the unity gained by doing everything at one time emphasized the oneness of the total group. Each ritual also stressed the fact that the war was an experience where everyone sacrificed and some died, not as members of a separate group, but as citizens of a whole community.

The full significance of the unifying and integrative character of the Memorial Day ceremony—the increasing convergence of the multiple and diverse events through the several stages into a single unit in which the many become the one and all the living participants unite in the one community of the dead—is best seen in Figure 1. It will be noticed that the horizontal extension at the top of the figure represents space; and the vertical dimension, time. The four stages of the ceremony are listed on the left-hand side, the arrows at the bottom converging and ending in the cemetery. The longer and wider area at the top with the several well-spread rectangles represents the time and space diversities of stage 1; the interconnected circles in stage 3 show the closer integration that has been achieved by this time.

During stage 1 it will be recalled that there is no synchronization of rituals. They occur in each association without any reference to one another. All are separate and diverse in time and space. The symbolic references of the ceremonies emphasize their separateness. In general, this stage is characterized by high diversity, and there is little unity in purpose, time, or space.

Although the ceremonies of the organizations in stage 2 are still separate, they are felt to be within the bounds of the general community organization. There is still the symbolic expression of diversity, but now diversity in a larger unity (see Figure 1). In stage 3 there are still separate ceremonies but the time during which they are held is the same. Inspection of the chart will show that time and space have been greatly limited since the period of stage 1.

The ceremonies in stage 4 become one in time and one in space. The representatives of all groups are unified into one procession. Thereby, organizational diversity is symbolically integrated into a unified whole. This is not necessarily known to those who participate, but certainly it is felt by them.

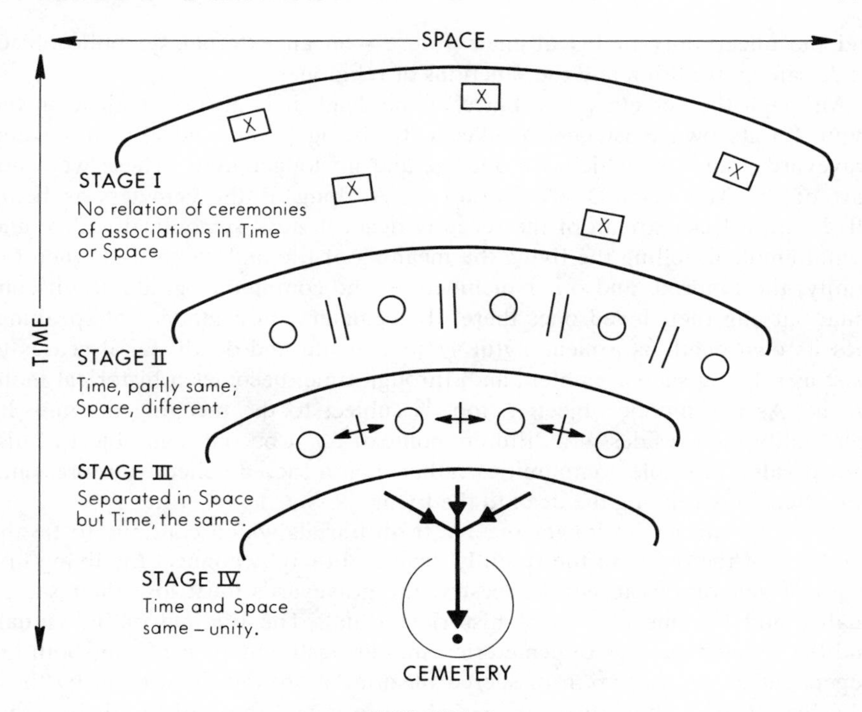

FIGURE 1
Progress of the Memorial Day Ceremony

The chart is designed to symbolize the progressive integration and symbolic unification of the group.

Moreover, at the conclusion of the ceremony, when the entire collectivity moves physically from diversity and extension, spread thinly throughout the city, into the inclosed, confined, consecrated unity of the receptacle (the cemetery as depicted in the chart), the celebrants themselves and their great ceremony symbolically incorporate the full spiritual power of the cemetery as a sacred symbol system.

Yankee City cemeteries are themselves collective representations which reflect and express many of the community's basic beliefs and values about what kind of society it is, what the persons of men are, and where each fits into the secular world of the living and the spiritual society of the dead. Whenever the living think about the deaths of others, they necessarily express some of their own concern about their own extinction. The cemetery provides them with enduring visible symbols which help them to contemplate man's fate and their own separate destinies. The cemetery and its gravestones are the hard, enduring signs which anchor each man's projections of his innermost fantasies and private fears about the certainty of his own death—

and the uncertainty of his ultimate future—on an external symbolic object made safe by tradition and the sanctions of religion.

Although the cemetery is a place of the dead, it is also dependent on the living for its own existence. Yankee City, being a very old city, has many graveyards, some of which are ancient and no longer used. These were not part of the Memorial Day celebrations. As long as the cemetery is being filled with a fresh stream of the recently dead, it stays symbolically alive and a vital emblem, telling the living the meaning of life and death. But when the family, the kindred, and other members of the community gradually discontinue burying their loved ones there, the cemetery, in a manner of speaking, dies its own death as a meaningful symbol of life and death, for it ceases to exist as a living sacred emblem and, through time, becomes a historical monument. As a symbolic object it, too, is subject to the meaning of time. Its spirituality then resides in a different context, for it becomes an object of historical value in stable communities rather than a sacred collective representation effectively relating the dead to the living.

When cemeteries no longer receive fresh burials which continue to tie the emotions of the living to the recently dead and thereby connect the living in a chain of generations to early ancestry, the graveyards must lose their sacred quality and become objects of historical ritual. The lifetime of individuals and the living meanings of cemeteries are curiously independent, for both are dependent on an ascription of sacred meaning bestowed upon them by those who live. The symbols of death say what life is and those of life define what death must be. The meanings of man's fate are forever what he makes them. . . .

THOMAS LUCKMANN

BELIEF, UNBELIEF, AND RELIGION

BELIEF and unbelief are historical phenomena, but they are not contingent. Their emergence as social facts depends on a particular constellation of conditions, some of which pertain to the social structure and some of which involve the conformation of the socially objectivated world view. The former

Reprinted from *The Culture of Unbelief,* ed. Rocco Caporale and Antonio Grumelli (Berkeley: University of California Press, 1971), pp. 23–31. Originally published by the University of California Press; reprinted by permission of the Regents of the University of California.

may be subsumed under the heading of institutional specialization of religion; the latter consist in the segregation of symbolic representations in the form of a sacred cosmos.

In the analysis of religion in complex societies—especially in modern industrial societies—it may be permissible to treat institutional specialization of religion and the segregation of a sacred cosmos as interdependent. It should be noted, however, that the formation of specialized religious institutions always presupposes a fairly distinct articulation of a sacred cosmos while, on the contrary, the segregation of religious representations in the world view does not necessarily require a specialized institutional basis in society.

Segregation of religious representations means that two major domains become polarized in the world view. A corresponding bipolarity marks the culturally determined patterns of subjective experience. The relation between everyday life and the sacred becomes indirect. Only the "ultimate" significance of ordinary, habitual experience and, of course, the "extraordinary" experiences that break the routines of everyday life are taken to refer to the sacred level of reality, a level that often receives the status of the supernatural. The ordinary meaning of conduct in everyday life, on the other hand, is less rigorously determined by the logic of the sacred cosmos. The world of everyday life thus gives rise to more immediately pragmatic systems of relevance. Although some representations are set apart in the world view as specifically religious, their significance—and the authority of the entire sacred cosmos—rests on what may be called the elementary religious function of the world view as a whole. This function consists in the socialization of human organisms into an order that transcends the individual. Socially objectivated world views integrate transcendences that are subjectively experienced on different levels: they connect the fleeting sensation of the organism to a past and future that are embedded in shared, social dimensions of time; they link the individual to a community of the living and the dead; they bridge the gap between the private world of dreams and the public world of everyday life.

It is questionable whether the world view as an undifferentiated whole in fact ever performed this elementary religious function. Perhaps such was the case in the earliest forms of human society. Data on archaic societies indicate that certain representations already tend to acquire a special status within the world view. Even before such representations coalesce into a segregated sacred cosmos, religious representations may visibly symbolize—and in part perform—the religious function of the world view as a whole. "Religious" is nonetheless still "diffused" throughout the world view. Only when the sacred cosmos is sharply segregated from other levels of reality is the elementary religious function of the world view taken over as the special and exclusive function of religious representations.

Neither the social objectivation nor the social maintenance and transmission of the sacred cosmos necessarily requires an institutional basis of its own. In archaic societies and (to a lesser extent) also in what, for lack of a better term, one may call traditional civilizations, religious representations pervade such institutions as kinship, the division of labor, and the regulation and exercise of power. In such societies the sacred cosmos legitimates conduct in the full range of social situations and bestows meaning on the entire course of an individual biography. There is, therefore, nothing—whether it be their ecology, economy, or systems of knowledge—that can be fully understood of such societies without reference to religion. Clearly it is otherwise in highly complex societies whose social structure consists of relatively independent subsystems.[1]

The very pervasiveness of religion has as its corollary that belief and unbelief are absent as social facts. There is no such thing as selective internalization—or refusal of internalization—of the sacred cosmos. For a number of reasons individual members of archaic societies may be socialized with different degrees of success into the social order but the notion of unbelief would be entirely misplaced in this connection. It would be equally beside the point to use the concept of belief to describe the subjective results of successful internalization of the world view.

Institutional specialization of religion radically altered the relation of the individual to the social order. It hardly needs to be stressed that institutional specialization of religion is merely one aspect of a global process of social change: the segmentation of the social structure in consequence of the differentiation and specialization of diverse institutional areas. Attempts to discern the essential features of archaic societies are well known.[2] The typical conformation of such societies was described in terms of the primitive level of the division of labor and by reference to kinship as the guiding principle of social organization. The prevalence of restitutive law and the absence of central political organization was noted, as was the peculiar form of solidarity and the pervasiveness of face-to-face social relations.[3]

The most general structural trait of this type of society, however, is emphasized in Redfield's concept of "primitive fusion." [4] Its meaning is well illustrated by the description of a Maya Indian peasant and his son working in the maize field. Their performance is neither an instance of kinship behavior nor an economic action nor religious conduct, but all of this at once, in an indissoluble unity of meaning. The concept thus describes a form of social organization where "bundles" of specific institutions are not tied together, that is, where institutions do not form functionally specialized domains. It is the "logic" of the sacred universe rather than the pragmatic logic of functional subsystems that integrate various institutionalized actions into meaningful wholes in the life of the individual and the community.

Our knowledge of the reasons for the transformation of certain archaic societies into the great ancient civilizations surely is not satisfactory. For all that, it is not difficult to see the common element in the various explanations which have been advanced. Increasing complexity in the division of labor, resulting from domestication of plants and leading to the production of a surplus over the subsistence minimum, the emergence of supracommunal and supratribal political organization for the management of hydraulic works, the differentiation of occupational roles and emergence of distinct social classes and a sharp rise in the inequality of the social distribution of knowledge may be all seen as processes of differentiation and partial specialization of institutional areas. While a certain differentiation of religious institutions seems to have accompanied this process everywhere it did not result in full specialization of these institutions. They remained closely intertwined with the structure of kinship and political organization.

The processes that led to the emergence of modern industrial society are somewhat better known and understood. One may say summarily that they were marked by an irregular sequence of phases in which not only economic and political but also religious institutions were increasingly more specialized in their functions. The social structure was segmented to a much higher degree than before into institutional domains. With the exception of religion, the norms of the several domains tended toward functional "rationality." After a long period of what may be called jurisdictional disputes between the domains, institutional norms in different fields achieved a remarkable autonomy. In other words, they became partly, if not fully independent from the sacred cosmos. The hierarchical structure of the world view was seriously weakened. "Ultimate" significance was no longer effectively diffused from the sacred level of reality into all reaches of the world view. A plurality of systems of knowledge came into existence, each developing an internal logic of its own and each having an institutional basis of its own. Not only technology and science and not only political ideologies such as, for example, nationalism but also the traditional sacred cosmos found a distinct, rather clearly identifiable location in the social structure. Various forms of competition and adaptation marked the relations between the various semiautonomous systems of knowledge and determined the degree of success with which they penetrated the world of everyday life.

This indicates the profound transformation that took place in the internal conformation of the social structure, in the basic relations of the social structure to the world view and in the relation of the individual member of society to the social order. In the case of religion it was this general change rather than an intrinsic trend toward functional "rationality" that led to the increasingly restricted jurisdiction of specifically religious norms.

Institutional specialization of religion exhibited, nevertheless, some typical parallels with the specialization of other institutional domains. Specifically religious social roles were soon defined as parts of an emergent occupational and status system. Their exclusive task was the administration of knowledge and regulation of performance expressly pertaining to the sacred universe. The interest of religious experts in controlling recruitment and training led to a development that conforms in general outline to the process that sociologists like to call "professionalization." The corollary of this process, the establishment of some kind of ecclesiastic organization, resulted both in the growth of specific institutional traditions and, eventually, in the emergence of ecclesiastic criteria of functional "rationality."

In consequence of a differentiated pattern of social stratification and an increasingly heterogeneous distribution of knowledge, religious representations, too, were unequally distributed in society. To be sure, everybody continued to participate in the sacred cosmos in some fashion. The function of the sacred cosmos was, of course, similar, *grosso modo,* for all members of society. But the typical, structurally determined differences in socialization that characterize complex societies eventually led to the articulation of different versions of the sacred cosmos among the social strata.

The simple and stable relationship between the world view and social structure that characterized archaic societies was thus radically changed. The sacred cosmos had been an integral part of a rather homogeneously distributed world view that was internalized in general processes of socialization. The reality and authority of the sacred cosmos was supported by the entire social structure and remained unproblematic. Now the sacred cosmos obtained a distinct and, in a manner of speaking, visible location in a restricted segment of the social structure. It was internalized in specific processes of socialization whose character varied in a manner that was determined by a complex social structure and a differentiated system of stratification. With some simplification one may say that the social structure produced its own internal "cultural contacts" and "culture conflicts." Here was a double task for the religious experts. The maintenance and transmission of the sacred cosmos now required the explicit formulation of an "official" model of religion. This involved standardization and systematization of religious representations and led to the emergence of "higher" forms of sacred knowledge, that is, of various theological disciplines. It also involved the interpretation of sacred knowledge in different modes that were adapted to the structurally determined needs of different groups and strata. This had not only "pastoral" but also serious political implications for the emergent forms of ecclesiastic organization. In this situation the identity of the elementary religious function of the world view with the articulated sacred cosmos and with the actual performance and the social effects of specialized religious institutions can no longer be taken for granted. In other words, "religion"—defined by its elementary function—and visible religion as an institutionally defined social fact are no longer necessarily the same.

Religious institutions may achieve different degrees of specialization. The process is of course not restricted to the history of Christendom in Western society. In various ancient civilizations the presence of a priesthood indicates that religious institutions were highly differentiated—but the various forms of theocratic government, divine kingship, and the like, also indicate that institutional specialization of religion, for various reasons, did not proceed very far. More highly specialized religious institutions may be found in traditional Islamic societies. But only the development of the Christian churches in Occidental history resulted in full specialization of religious institutions. To be sure, no single factor can account for this development. In the background there was the sharp segregation of a sacred cosmos in Israel. Then there were the conditions surrounding the emergence of the early Church such as the pluralism of world views and the differentiation of specifically religious communities in an empire whose political and economic institutions already had a relatively high degree of autonomy. Then came the long history of jurisdictional disputes between Church and state throughout the Middle Ages. With the Renaissance the rate of change in society accelerated: the growth of cities, the "discovery" of disembodied "classical" systems of values and of knowledge, the contact with alien civilizations, the peculiarly Occidental combination of science and technology and, finally, the rise of modern capitalism ushered in a new world. Only this intricate and unique combination of factors may perhaps explain why religious institutions were specialized to a degree not paralleled elsewhere.

In the emergent industrial societies specifically religious communities are linked to ecclesiastic organizations with a long tradition and with their own criteria of functional rationality. The traditional sacred cosmos is held in clear monopoly by religious experts. The sacred cosmos is highly systematized in the form of doctrine and there is a proliferation of "higher" forms of expert religious knowledge. The entire domain of specialized religion is both highly autonomous and clearly divorced from the other institutional components of the social structure.

An important consequence of institutional specialization in religion is that the population that is not expert in religious matters participates in the sacred cosmos in a somewhat indirect manner. Specialized institutions predefine for it the cognitive and behavioral norms of what is properly religious. These institutions control the expectations and sanction the performances in matters that are recognized by everybody as distinctly religious.

This is the decisive point at which one may begin to speak of belief and unbelief as social facts. It would have been entirely unprofitable to define belief in general terms as the subjective correlate of a system of knowledge considered to be hostile to "religion." In societies in which the sacred cosmos is internalized integrally in the overall process of socialization into a hierarchically structured world view there is no belief nor unbelief.

In this light, it may be helpful to present in a highly schematic manner three main ways in which the location of "religion" in society is objectively experienced:

1. All conduct is "religiously" relevant. All situations in everyday life as well as the entire biography are endowed with a significance that refers directly to sacred reality. There is no specifically religious socialization;

socialization *tout court* is essentially religious. There are neither belief nor unbelief, neither believers nor unbelievers.

2. Some kinds of conduct are "religiously" more relevant than others. Some behavior is determined by specifically religious norms, and the sacred cosmos is set apart from everyday life. But all conduct still rests on internalized general norms that are infused with sacred significance and all conduct is subject to general social controls.

3. Some patterns of behavior are generally recognized as specifically religious while other forms of conduct are guided by distinctly different normative expectations. Religious norms are internalized in specific processes of socialization that are directly and indirectly controlled by specialized institutions. Definition and interpretation of religious representations is held in monopoly by religious institutions which may also enforce "correct" religious performance. It is under such circumstances that belief and unbelief emerge as social facts.

ANDREW M. GREELEY

THERE'S A NEW-TIME RELIGION ON CAMPUS

DURING a recent unpleasantness between the University of Chicago and its Students for a Democratic Society the normal, decorous quiet of the Social Science Building was rent one fine afternoon by earpiercing sounds. Secretaries, research assistants and even a few faculty members dashed to their office doors to discover who was being murdered. Three young women dressed in shabby and tattered garments were standing in front of the Sociology Department office shrieking. "Fie on thee, Morris Janowitz! A hex on thy strategy!" WITCH (Women's International Terrorists Corps from Hell) had come to put a curse on the Sociology Department.

So far, nothing seems to have happened to Professor Janowitz or the Sociology Department. But if it does, there's going to be an awful lot of frightened people along the Midway. (I offered to sprinkle holy water on the departmental office; but, while social science seems ready for witchcraft, it is not yet ready for exorcism.)

Reprinted from *The New York Times Magazine,* June 1, 1969, by permission of the author and the publisher.

WITCH is only one manifestation—though a spectacular one—of a resurgence of interest in the occult on the college campuses of the country. Although some observers of WITCH's "hexing" dismiss it as a form of "guerrilla theater," the WITCHes themselves elaborate a quasi-scholarly explanation of how they continue a neolithic religion that worshipped the great earth mother goddess until it was replaced by Christianity. One suspects that the WITCHes are but first cousins of the California Druids who also claim to be carrying on a tradition from the neolithic underground—thus confounding those of us who thought that the only Druids left in the world were Irish Monsignors.

WITCH is a combination of the put-on and the serious, the deliberately comic and the profoundly agonized, of the bizarre and the holy. The same is true of the other manifestations of the neo-sacred now observable around the country:

Prof. Huston Smith of M.I.T. describes an experience with a seminar of some of the best students in the institution. "I cannot recall the exact progression of topics, but it went something like this: Beginning with Asian philosophy, it moved on to meditation, then yoga, then Zen, then Tibet, then successively to the 'Bardo Thodol,' tantra, the kundalini, the chakras, the *I CHING* [*ee-ching*, a book presenting an ancient Chinese divination device which enables one to make decisions—a sort of pre-I.B.M. computer], karate and aikido, the yang-yin macrobiotic (brown rice) diet, Gurdjieff, Maher Baba, astrology, astral bodies, auras, U.F.O.'s, tarot cards, parapsychology, witchcraft and magic. And, underlying everything, of course, the psychedelic drugs. Nor were the students dallying with these subjects. They were *on* the drugs; they were eating brown rice; they were meditating hours on end; they were making their decisions by *I Ching* divination, which one student designated the most important discovery of his life; they were constructing complicated electronic experiments to prove that their thoughts, via psychokinesis, could affect matter directly.

"And they weren't plebeians. Intellectually they were aristocrats with the highest average math scores in the land, Ivy League verbal scores, and two to three years of saturation in M.I.T. science."

A certain Catholic university discovered that it had a coven of warlocks on campus (warlocks, for the uninitiated, are male witches). As the dean of the institution put it, "We've really become progressive around here. A couple of hundred years ago we would have burned them at the stake. Twenty-five years ago I would have expelled them. Now we simply sent them all to psychiatrists."

At a Canadian university, the student body was given a chance to recommend courses of its own choosing to be included in the curriculum. The majority of the courses chosen had to do with astrology, Zen, sorcery and witchcraft.

In most of the élite universities in the country, horoscopes and the prediction of the future by the use of tarot cards are widespread. Not all the students, not even a majority, are engaging in such divination. But a minority is and the majority does not ridicule their efforts. On the contrary, one has the impression that the majority reacts the same way it reacts to the S.D.S.: "We understand why they want to do it, even if we are not yet ready to do it ourselves."

Catholic girls' colleges seem to be particularly disposed to producing groups of young women who make decisions by use of the *I Ching*.

In a number of colleges, particularly in California, semi-monastic cults have arisen composed of young people who subsist on vegetarian diets, take vows not to cut their hair and spend long hours in contemplation.

A thin network of students has formed a loose "community" to support one another through the stresses of graduate school, a community which does not take spatial separation to be a very serious problem in the providing of mutual support. One leader of the "community" describes quite bluntly what the "community" is about: "You might say we're forming a new religious order."

In the hills of Sonoma, Calif., there flourishes an institution called the "Six-Day School," composed largely of Berkeley dropouts, who learn about political pacifism, astrology, vegetarian dieting, mysticism and magic, during the course of their stay at the school—which usually exceeds six days. One group last year left Sonoma to proceed to Mount Shasta, there to await the end of the world.

A bookstore just off Harvard Square, clearly doing an excellent business, announces itself as "The Sphinx—Occult Books."

The White Brotherhood, a medieval Catharist sect made up of those in direct contact with "the spirit" (who has revealed to them that they are among the 144,000 white-robed martyrs of the Book of Revelation), is spreading at West Coast universities from Seattle to San Diego. Interestingly enough, the brotherhood is being spread by the same messengers who propagated it in the 13th century—wandering poets and minstrels, or, as they used to be called, troubadours and meistersingers.

I remarked in one of my classes that I had been able to locate almost every kind of offbeat religious behavior on our campus save for spiritualistic séances and wondered why someone hadn't thought of hunting up a medium. Several members of the class promptly assured their confreres that I hadn't looked very far; spiritualism was alive and well in Hyde Park.

Perhaps the most puzzling aspect of the new pursuit of the sacred is that it is so funny and yet so serious. Students cannot talk about it without laughing and yet they must interrupt their laughter to protest that they respect the goals of new devotees of the sacred. However, the puzzle is less difficult when one understands that the cultists are engaging in a form of *drama*—partly, one suspects, under the influence of their cousins, the hippies. Drama about the sacred is *liturgy;* and liturgy, as J. Huizinga pointed out in his famous book, *Homo Ludens,* is sacred play.

The sacred by its very nature has large components of the playful and the comic close to its core. Only with the Reformation did the idea that the sacred was grimly serious finally triumph in the Western world. Catholic clerics will probably admit, now that the venerable Solemn High Mass has fallen into disuse, that they frequently found it hard to keep a straight face during its complex ceremonies. The new manifestations of the sacred, like the Solemn High Mass, are simultaneously much in earnest and a hilarious put-on. To put a hex on the Sociology Department is comic; but it is also a tentative assertion that there are powers in heaven and on earth which may transcend sociology departments.

Let us, first of all, make all the proper qualifications. Only a minority of students is engaged in the pursuit of the bizarrely sacred. Such a pursuit is not new among young people but is a continuation of the interest in the occult and the mystical which has persisted for some time. It is a form of romanticism which has recurred in one fashion or another periodically in years gone by. It is experimental and does not indicate any return to the organized churches; as one student said to me, "Who in the world would expect to find anything sacred in the churches?"

The evidence for this resurgence of interest in the sacred is "impressionistic" and not yet based on the kind of "hard" empirical data that so delight the heart of the social scientist. Nevertheless, with all these qualifications, it still does seem there has been a very notable increase, however temporary, in interest in the sacred and particularly the bizarrely sacred among students on the college and university campuses in the last few years. Furthermore, the "return of the sacred" has happened exactly where one would least expect it —among the élite students at the best colleges and universities in the land, precisely those places where secularization would presumably have been most effective and most complete.

What the hell is going on? God is dead, but the devil lives?

One of the things that strikes an interviewer who talks to students about the "return of the sacred"—even though they may themselves not be involved in witchcraft, astrology, or the *I Ching*—is that they resolutely refuse to dismiss as foolish those who are so involved. The first reason that young people give for the "return of the sacred" is the failure of science. One graduate student said to me, "Let's face it, science is dead. While the newspapers and magazines were giving all the attention to the death of God, science was really the one that was dying."

The extent and the depth of the revolts against positivism come as a considerable shock to those like myself whose training in the positive sciences took place in a time when they were totally unquestioned at the great universities. During the last winter quarter I put a statistical table on the blackboard and proceeded to explain the implications. One of my students respectfully but pointedly observed, "Mr. Greeley, I think you're an empiricist. In fact, at times I even think you are a *naive* empiricist." The accusation didn't surprise me because I guess I am an empiricist, but the tone of it did, for it was a tone of voice that used to be reserved for the accusation of being a "clerical Fascist."

The student then went on to deliver a fierce harangue against "the epistemology of science," and to assert that the "imperialism" of science by which it claimed to be the only valid form of human knowledge and the only valid rationale for organizing society was completely unsatisfactory to his generation. A number of other students rose to offer vigorous support to this position.

After class I pondered the matter in some confusion and returned the following session to ask if there was anyone who disagreed. Would no one rise with the appropriate quote from Kaspar Naegele to defend empiricism, positivism and rationality? The class was completely silent, until one young woman remarked, "I think we all agree with what was said in the last class." At the beginning of the nineteen-sixties when I was in graduate school, such thoughts would have been "thinking the unthinkable."

The young people seem to be angry at science for its failures. A coed observed, "Science hasn't ended war, it hasn't ended injustices, and it doesn't respond to most of man's needs. Why should we take it seriously?" And another joined in: "Pure rationalism just isn't rational because man is more than reason and religion knows that even if positive science doesn't." And a third coed concluded, "Science was something that we had to work through our system. It only started with people like Darwin and it's not surprising that for a while everybody thought it was the only thing that mattered. It's just now that we've come to know better."

Other students explained the return to the sacred as a reaction to the failure of the university administrations and faculties to live up to their own rationalist and scientific principles. A young man put it this way: "When we see the utter incapacity of the rationalists to engage in rational discourse with us, we begin to rediscover the legitimacy of emotions. From these it is just a short step to the legitimacy of the sacred."

The rhetoric of the return to the sacred is not so very different from the rhetoric of the radical political movements. Words like "honesty," "integrity," "fidelity," "love," "openness" and "community" abound. The hippy culture with its emphasis on drugs stands midway between the two, bridging the gap between and pervading both other movements with its influence. Yet the movements are distinct. The hippies and the radicals may frequently use religious terminology and even respond to "religious needs," but their concerns still tend to be this-worldly. The neo-sacralists, on the other hand, are willing to accept as a working possibility a world which, if it does not completely transcend the present world, at least to some extent stands beyond it.

It is precisely this "standing beyond" which young people relate to the second reason for the return to the sacred: The sacred seems to provide an avenue for personal efficacy. As a male undergraduate described it, "Why use the *I Ching* in a world where you have the I.B.M. 360? The answer is easy. You can't understand the 360 and you don't have much control over it. The *I Ching* says that there are powers that stand beyond and are more powerful than the 360, powers with which in some way you can enter into a meaningful relationship when you can't do it with the 360." And one of his friends added, "Most of us realize that other people make our decisions for us quite arbitrarily. Whether I go to Vietnam or not, whether I get killed there or not doesn't depend at all on who I am or what I think. I'd sooner feel that my future was being shaped by the stars or by the turn of the cards because these would represent powers that would be more concerned about me than would either my draft board or the Pentagon."

I pressed these two young men to question whether they really did think that there was something beyond that made itself known through the movements of the stars or the turn of the cards. One of them shrugged uncomfortably. "I'm not sure," he said, "but I like to think so." And the other commented, "It's like the conclusion of Arthur Miller's 'Death of a Salesman.' In death, somebody did 'notice' Willy Loman. When someone turns the cards for you, you feel at least here you are being noticed."

The theme that religion is a response to alienation and to a feeling of unimportance against the larger society is widespread in the students' comments. "Religion makes you feel like you're a *person.*" A woman undergraduate told me, "It makes you feel that you are important and that what you do does matter and you can have influence on others." Students are further impressed by the enthusiasm and confidence of the cultists. "They really believe that what they say is *true,*" observed one young man to me. "They really believe that they do have the answer and that they do know what is ethically right and ethically wrong. It's hard to avoid being affected by their enthusiasm after you've been in a school that really isn't sure what is true or what is right or wrong."

Like the radicals and the hippies, the neo-sacralists are in desperate search for something to belong to. The religious groups are *communities,* places where you are more than just an I.B.M. file card. A young woman put it this way; "If you get into a group like that, you at least know that somebody will notice the difference if you're murdered. Around this university, you could be dead in your room for days and nobody would ever know the difference." And another commented, "We don't have to worry anymore, at least not very much, about where our food and housing is going to come from so we worry about ourselves and about finding ourselves. The only place where we are going to find ourselves is in deep relationships with others and that means either religion or sex and maybe both." The religious communities that grew up around the various cults of the sacred are felt to provide opportunities for meaningful intimate relationships over against the depersonalizing formalism of the academic and governmental bureaucracies. "You're a *person* in the group even if you don't want to be. You're forced to face yourself and discover who you are."

The quest for community in small groups makes the neo-sacralists quite conscious of the relevance to their quest of T-groups (T stands for training), encounter groups, and the whole bag of group-dynamics tricks. Just as for some students group dynamics or sensitivity training become almost a religion, so for others already involved in quasi-religious behavior, sensitivity training and its cousins become an important means of religious growth. One girl told me how delighted she was to be part of a T-group which included two people who were on drugs and two others who were making their major decisions by means of horoscopes. It was, she noted, a fascinating experience.

Underlying the other three explanations the young people offer for the return to the sacred is a fourth: the sacred provides meaning. "In one way," a charming young woman said to me, "the sacred is even better than drugs, because when you're on drugs the world looks beautiful to you only if you're on a trip. But religion has persuaded some people that the world is beautiful most of the time despite the ugliness we see. That's terribly important."

One of her male classmates chimed in, "What we're really concerned about is whether anything is real, I mean, whether it is *really real*. Is there something that is so powerful that it can even make *us* real?" And an older graduate student (a clergyman, I suspect, but nowadays it's hard to tell) pointed out, "And Mircea Eliade [professor of history of religions at the University of Chicago and one of the world's most distinguished experts on the sacred] tells us that this is exactly what the sacred is, the *really real*."

Some of those who have kept an eye on WITCH argue that its principal contribution is to give its members some sense of what it means to be a woman, even if it is a bizarre concept of womanhood. Full meaning involves not only understanding what the world is all about or at least understanding whether the world has anything in it that is "really real," but it also involves having some sense of what you are all about and whether there is a possibility that you are "really real." The religious experience in the final analysis is seen as "ecstatic," that is to say, that it, like sex, takes a person out of himself and brings him into contact not only with other human beings but with the "creative powers" which presumably underpin the cosmos.

My very unsystematic survey of student opinion on the neo-sacred leads me to conclude that what is going on is authentically, if perhaps transiently and bizarrely, religious. Personal efficacy, meaning, community, encounter with the ecstatic and the transcendental, and the refusal to believe that mere reason can explain either life or personhood—all of these have traditionally been considered religious postures. An anthropologist visiting the secular university campus from another planet could not help but conclude that there was a lot of very interesting religious behavior going on and he would probably feel that it was very primal and primordial, if not indeed primitive religious behavior.

Do these students believe in God? I have the impression that most of the young neo-sacralists would not understand the question or at least would find it premature. They don't believe in the God they left behind in their parish congregations. But they are frankly experimenting—as a part of a self-conscious "psychosocial moratorium"—with the "experience of the sacred" to see whether there is anything there which could add depth and richness to their lives. Most of them seem to hope, at times rather forlornly, that they will be able to find something or Something. But they are not ready to give it or It a name just yet.

The new religious enthusiasts clearly owe a major debt of gratitude to the hippies. Indeed, one might even consider them to be merely one wing of the hippie movement. Both emphasize the prerational if not the antirational. The quest for the spontaneous and the "natural" in the two dissenting groups is a protest against the "hang-ups" of a society that is viewed as overorganized and overrationalized but less than human. Both are a search for "experience" and for a specific kind of experience—one that "takes one out of oneself." Both have, as noted, a strong comic element about them—an irresistible urge to "put on" the rational society.

Both the neo-sacralists and the hippies are communitarian, seeking experience and vitality from intimate friendships, friendships which in many instances are strongly at odds with the conventions of the large society. The two groups are further linked by their longing for the mystical and the reflective, again because these activities are seen as a means of standing apart from the rest of society, which has so little time for anything else but activity. The ceremonies, the rituals, even the *vestments* of the two groups also represent a common revolt against the sober and somber garb of the suburban businessman and his daily schedule. (Long hair used to be important but now that even the suburban executive is wearing sideburns and maybe a goatee, we might expect the deviants to imitate Buddhist monks and shave their heads.)

So the new search for the sacred shares with the hippies an "acted out" rejection of the rationalized bourgeois society, a rejection that is also a put-on of that society. The hippies were the first to become concerned with the mystical and the occult. But many of the new religious enthusiasts are not hippies in the ordinary sense and most of them are not willing to go the "drug route" with the hippies. Some say quite frankly that they view religion as a substitute for drugs and one that is much less dangerous. One hesitates to say it, but the neo-sacralists appear to be much more "respectable" than the hippies.

But the important difference, I think, is that the religious cultists are seeking for something that hippies refuse to be hung up on; they seem to be looking for what the sociologists would call a "meaning system" or an "interpretative scheme." The hippies put on life because they think life is a put-on and ought not to be taken seriously. Those who are engaging in the quest for the sacred are, with a greater or lesser amount of explicit acknowledgment of the fact, looking for an explanation for life and for themselves. They're not sure they'll find it; they're not even sure that the search is anything more than a joke. But they'd like to think it just might be.

It might be pertinent to ask why we are so surprised about the return of the sacred. The non-rational has been with man a long time and so has the supernatural and even the superstitious. Was it not unduly naive of us to assume that it would disappear so quickly? Astrology has always been a rather successful industry. Superstition is widespread in the general population. More than two-fifths of the American population go to church every week as do almost half the college students in the country. The limited amount of longitudinal research done on religious beliefs and behavior shows very little change in the last two decades. If the sacred and the superstitious still permeate the larger society, why are we so surprised that they have been tenacious enough to reassert themselves on the college campus?

Students themselves will cheerfully admit that their lives are not at all free from superstitious behavior even if they don't take the sacred or the supernatural very seriously. As one girl said to me, "I always wear the same sweatshirt every time I take an exam and I know other people who simply refuse to go into an exam unless they've had a shower beforehand. When you ask us why we do these things, the only response we can come up with is, why not? Sure, it might not make any difference, but then again it might, and there's no point in taking any chances." One is reminded of the famous agnostic prayer which is addressed "To Whom it may concern."

My friend, Peter H. Rossi, chairman of the Department of Social Relations at Johns Hopkins University, summarized only half-facetiously the relationship between agnosticism and superstition: "I'm not sure that I believe in good spirits but I have the uncanny feeling that there might be evil spirits."

Will the interest in the sacred on the college campus survive? For some individual students, it is clearly nothing more than an experiment which is part of their youthful "psychosocial moratorium," a part of their quest for personal identity. If the data on graduate student church attendance are to be believed, the moratorium will end not so much with agnosticism or even a new form of religion for most students, but rather with a return to some form of traditional religion. (A recent survey of graduate students showed that about 40 per cent of the students at the 12 major arts and science graduate schools in the country were regular church-attenders.) Witchcraft, astrology and divination, if they lose at all, are likely to lose to the traditional religions.

Yet some of the present concern about the sacred is likely to continue influencing the traditional church system within—perhaps even leading to the formation of new religious sects. Like most everything else on campus, the return to the sacred, while it is communitarian, is profoundly anti-organizational. Whatever of the present commitment to the sacred survives is likely to be informal and casual but such groups as the Druids and WITCH could conceivably grow much larger.

The students, in any event, have little doubt that the sacred will continue to interest them and that it will continue to fascinate their successors on campus. As one undergraduate male argued, "The interest in the sacred is rooted in a kind of existentialist dissatisfaction with the way things are and since the way things are is not likely to change for a while, there is no reason to think the sacred is going to go away either."

Certainly the dissatisfaction with the failures of positive science does not seem to be reversible and one is inclined to suspect that it will be a fairly long time before the argument that religion or the supernatural or the sacred are not "scientific" will be persuasive. Not everybody will be religious; but neither will religion be in full retreat, and some of the more bizarre, primitive, and superstitious forms of the religious are likely to enjoy a respectability for a number of years to come.

Max Weber, the founder of modern social theory, anticipated the rise of new prophecy (or the resurgence of old ones) as long ago as the first decade of the present century and anticipated it precisely as a revolt against the rationalism of the "spirit of capitalism." He wrote: "In the field of its highest development, in the United States, the pursuit of wealth, stripped of its religious and ethical meaning, tends to become associated with purely mundane passions, which often actually give it the character of sport. No one knows who will live in this cage in the future, or whether at the end of this tremendous development entirely new prophets will arise, or there will be a great rebirth of old ideas and ideals, or, if neither, mechanized petrification, embellished with a sort of convulsive self-importance. For of the last stage of this cultural development, it might well be truly said: 'Specialists without spirit, sensualists without heart; this nullity imagines that it has attained a level of civilization never before achieved.' "

Most of the contemporary manifestations of the sacred on the college and university campus are a form of withdrawal from the larger society—if not positively destructive in their view of said society. Yet the constructive element is not completely absent. I remarked a few weeks ago to my seminar on the sociology of religion that I thought most of the new religious forms offered personal redemption but despaired of social redemption. One young woman raised her hand. "Mr. Greeley," she asked, "have you ever heard of a book called 'The Phenomenon of Man' by a man named Teilhard de Chardin?" I admitted that I had. "Well," she said, "I was deeply impressed by that book because even though there is so much wrong with the world right now, I think Teilhard is right when he says that we're on the verge of a great leap into a much better form of human life, that we are moving into the noosphere and that we are traveling toward an omega point. I think there's a lot of us that feel that we can have a faith in the sacred which will help us to create a better world."

I admitted to being something of a Teilhardist myself and conceded that there might be some of what she described in the student quest for the sacred, but insisted that I didn't see much of it and that I felt the "return of the sacred" assumed largely that the world was unredeemable and discriminating.

That was a mistake, for all kinds of hands rose up in the seminar room and all kinds of students rose up to assert their faith in the possibility of a Teilhardist-like vision of evolution toward the omega point.

I still can't quite figure out the meaning of that experience, but I must say I never expected to encounter a classroom of Teilhardists at the University of Chicago. Such people are necessarily a minority—a very small minority—of the whole student population, or indeed of the whole population of the world. There are not very many Teilhardists around. But it wouldn't take very many. . . .

POSTSCRIPT—WHY SOCIOLOGY?

SOCIOLOGY may be an interesting intellectual pastime for a small number of professionals. Does it have any broader human significance? The final selection, Peter L. Berger's "Sociology and Freedom," discusses this question.

PETER L. BERGER

SOCIOLOGY AND FREEDOM

SOCIOLOGY, greatly to the surprise of most of its older practitioners, has acquired the reputation of a liberating discipline. Sociology courses are crowded with students in search of the intellectual tools with which to demolish the hypocritical world of their elders and fashion for themselves, if not for society at large, a new authenticity and a new freedom. Even more astonishing expectations are directed toward sociology by students who adhere to the radical left. For them, sociology is nothing less than the theoretical arm of revolutionary praxis, that is, a liberating discipline in the literal sense of a radical transformation of the social order. It is sociology in this latter understanding that has been associated with the remarkable proportion of students of the field who are among leading activists of the New Left, both in America and in western Europe—to the point where there now are firms in Germany and in France screening job applicants in order to bar those who have taken sociology courses. Even in this country, where sociology is established more firmly in academia, there are places where the field has taken on a slightly disreputable flavor.

All this is very recent indeed. Only a few years ago most outsiders, if they thought of a sociologist at all, thought of him as a dry character, with an insatiable lust for statistics who at best might dig up some data of use to policy makers and at worst (in the words of one malevolent commentator) might spend ten thousand dollars to discover the local house of ill repute. It would have required a wild imagination to conceive of this unexciting type as an object of interest either for young seekers after salvation or for the FBI. It has happened all the same. Especially among younger members of the profession there are now serious aspirants to drastically different images of the sociologist. There is the image of the sociologist as one of several guru types within the youth culture, in close proximity to the evangelists of psychedelia, T-group mysticism, and other fashionable gospels. There is also the image of

Reprinted from Peter L. Berger, "Sociology and Freedom," *The American Sociologist,* 1971, Vol. 6 (February): 1–5, by permission of the publisher.

the sociologist as a carrier of revolutionary doctrine and, potentially at least, as a character throwing Molotov cocktails through the windows of the faculty club (in either direction, depending on circumstances). Both images have provoked dismay as well as enthusiasm. The former image is especially galling for psychologists, who suddenly find themselves challenged in what so recently was a monopoly in the treatment of the metaphysical afflictions of intellectuals. The latter image is a source of alarm not only to university administrators and law enforcement officers, but to orthodox Marxists, who describe the new radical sociologists in terms that could have been borrowed from Spiro Agnew.

The greatest dismay, naturally, comes from sociologists. Placid purveyors of Parsonian theory are suddenly confronted with demands to be "relevant" to the turbulent and constantly shifting commitments of the young. Graduates of the Bureau of Applied Social Research, collectors and producers of multiple correlations with impeccable margins of error, suddenly hear themselves denounced as academic hirelings of the military-industrial complex. This confrontation between the old and the new sociology, a yawning generation gap if there ever was one, could be fully observed at the 1969 meetings of the American Sociological Association in San Francisco. There were the various caucuses of radical leftists, black militants, and (perhaps most frightening of all) liberated or wanting-to-be-liberated women sociologists, each group doing its thing in the antiseptic corridors of the San Francisco Hilton. Amid this novel furor, the majority, almost furtively, went about its usual business of interviewing job candidates, drinking publishers' liquor, and reading papers in atrocious English.

Sociology should be an instrument for the existential liberation of the individual; it should be a weapon in the revolutionary struggle to liberate society. To anyone familiar with the history of the discipline, these notions are startling, if not ironic. In the origins of sociology, there was indeed a quasi-religious conception of it—the conception of Auguste Comte and his followers. Comte, however, envisaged sociology as an *anti*revolutionary doctrine, as the new church that was to restore order and progress in the wake of the havoc caused by the French Revolution. With few exceptions, however, the Comtian view of sociology as *Heilswissen* (to use Max Scheler's term) did not survive into the classic age of the discipline, the period roughly between 1890 and 1930. None of the classic sociologists would have been able to make much sense of the current notion of sociology as a vehicle of personal liberation.

As to understanding sociology to be a doctrine of revolutionary praxis, it is noteworthy that some of the greatest classic figures (such as Max Weber, Emile Durkheim and Vilfredo Pareto) invested a good deal of effort in what they considered to be refutations of Marxism. Most classic sociology in Europe was a counterrevolutionary and (at least implicitly) conservative doctrine. Early American sociology had a strong reformist animus, but this was more congenial to YMCA secretaries than to revolutionaries or preachers of spiritual salvation. Even this mild reformism became, at most, a submerged motif as "value-freedom" and technical proficiency became established as binding norms within the profession.

I have no satisfying explanation for the recent dramatic changes in the conception of sociology. One can point, of course, to certain intellectual sources—C. Wright Mills in this country, the so-called Frankfurt School in Germany, and Marxists-turned-sociologists, such as Henri Lefebvre, in France. This, though, does not explain why these individuals and their ideas have suddenly come to exert such a powerful influence. I strongly suspect that, as is often the case in the history of ideas, there is a strong element of chance in the new affinity between sociology and political radicalism. In any case, I don't intend to devote myself here to speculation about the reasons for this slightly bizarre marriage (not the least reason being that I doubt whether it will last long). Rather than to explore historical causes, I wish to look at the theoretical question at issue, to wit: *In what sense, if at all, can sociology be called a liberating discipline?*

I shall approach the question by way of two seemingly contradictory propositions: (1) sociology is subversive of established patterns of thought, and (2) sociology is conservative in its implications for the institutional order. I suggest that *both* propositions are correct, and that understanding this entails also grasping the relationship between sociology and freedom, at least on the level of politics. (I should add here that the epistemological problem of how an empirical science can or cannot deal with man's freedom is clearly outside the scope of this paper.)

Sociology is subversive of established patterns of thought. This, of course, is today a favorite notion of those who would marry sociology to radical politics. A few years ago most sociologists would have been shocked or honestly bewildered by the proposition. Then, it was those with a vested interest in established patterns of thought who (if the inelegant simile may be forgiven) smelled the rat before those who put it there. I recall a remark made to me in 1956 by a barber in the southern town where I had just started my first teaching job. After I told him what I was teaching, he paused (more pensively than hostilely) and remarked, "Oh, I know about sociologists. You're the guys who wrote all those footnotes in the Supreme Court decision on getting the colored into the schools." He was right, of course, in an extended sense, if not literally. I wonder how many of the sociologists who busily gathered all those data on the place of the Negro in America (some of them Southerners living quite comfortably in a segregated society) imagined that they were providing the legitimations for one of the great social transformations of our time. Put differently, I suggest that there is in sociology a subversive impulse that strives for expression regardless of the intentions of individual sociologists.

Every human society has assumptions that, most of the time, are neither challenged nor reflected upon. In other words, in every society there are patterns of thought that most people accept without question as being of the very nature of things. Alfred Schutz called the sum of these "the world-taken-for-granted," which provides the parameters and the basic programs for our everyday lives. Robert and Helen Lynd, in their classic studies of *Middletown,* pointed to the same phenomenon with their concept of "of course statements"—statements that people take for granted to such a degree that, if questioned about them, they preface their answers with "of course." These socially established patterns of thought provide the individual with what we may call his basic reality kit (paraphrasing Erving Goffman), that is, with the cognitive and normative tools to build a coherent universe to live in. It is difficult to see how social life would be possible without this. But specific institutions and specific vested interests are also legitimated by such taken-for-granted patterns of thought. Thus, a threat to the taken-for-granted quality of legitimating thought patterns can very quickly become a threat to the institutions being legitimated and to the individuals who have a stake in the institutional status quo.

Sociology, willy-nilly and by its own intrinsic logic, keeps generating such threats. Simply by doing its cognitive job, sociology puts the institutional order *and* its legitimating thought patterns under critical scrutiny. Sociology has a built-in debunking effect. It shows up the fallaciousness of socially established interpretations of reality by demonstrating that the facts do not gibe with the "official" view or, even more simply by relativizing the latter, that is, by showing that it is only one of several possible views of society. *That* is already dangerous enough and would provide sufficient grounds for sociologists to become what the Prussian authorities used to call *polizeibekannt*—of interest to the cognitive if not to the actual police—and, let me add, every society has its cognitive policemen who administer the "official" definitions of reality. But sociology, at least in certain situations, is more directly subversive. It unmasks vested interests and makes visible the manner in which the latter are served by social fictions. At least in certain situations, then, sociology can be political dynamite.

A favorite term of the New Left in Europe and Latin America is derived from the vocabulary of psychoanalysis—*Bewusstmachung* in German, *concientización* in Spanish—perhaps best translated as "bringing to consciousness." This is the process of social critique by which the mystifications of "false consciousness" are demolished and the way is prepared for the demystified consciousness necessary for revolutionary praxis. I shall return shortly to the question of revolutionary praxis. As to the first aspect of the term, the subversive effects of critical social analysis on consciousness, it must be admitted that it pertains to sociology in a very basic way. Anyone who pursues the sociological perspective to its logical consequences will find himself undergoing a transformation of his consciousness of society. At least potentially, this makes him unsafe in the eyes of the guardians of law and order. It also produces unsafety, sometimes with catastrophic effects, for his own peace of mind.

"Bringing to consciousness," in this sense, does indeed have a liberating quality. But the freedom to which it leads, quite apart from its possible political effects, can be a rather terrible thing. It is the freedom of ecstasy, in the literal sense of *ek-stasis*—stepping or standing outside the routine ways and assumptions of everyday life—and this, let us recall, also includes standing apart from routine comforts and routine security. Thus, if there is a relationship between "bringing to consciousness" and the ecstasy of liberation, there is also a relationship between that ecstasy and the possibility of desperation. Toward the end of his life Max Weber was asked by a friend to whom he had been explaining the very pessimistic conclusions of his sociological analysis, "But, if you think this way, why do you continue doing sociology?" Weber's reply is one of the most chilling statements I know in the history of western thought: "Because I want to know how much I can stand." Alfred Seidel, a student of Weber's who was also greatly influenced by Freud, came to an even more pessimistic conclusion in his little book appropriately entitled *Bewusstsein als Verhaengnis—Consciousness as Doom*. Seidel concluded that the combined critical consciousness of sociology and psychoanalysis was not only politically subversive but inimical to life itself. Whatever other motives there may have been, Seidel's suicide, as a young man in the 1920s, was an existential ratification of this view of the "bringing to consciousness" of sociology.

My purpose is not to suggest that sociologists, to be consistent, should all commit suicide. I have a somewhat more benign view of the existential possibilities of sociological consciousness. Rather, I want to point out that the relationship between sociology and freedom is not as simple, or as cheerful, as the radicals in the profession would have us believe. Yes, there is a liberating quality to the discipline of sociology. Yes, there are situations where sociological understanding can be liberating in a political and (at least in terms of my own values) morally significant sense—as in the service that sociology can render to the liberation of American blacks from racial oppression. But for individual sociologists, the discipline can bring to consciousness aspects of the world that are profoundly disturbing and a freedom that, in the extreme instance, evokes truly Kierkegaardian terrors.

Sociology is conservative in its implications for the institutional order. This second proposition, put differently, means that sociology, far from leading inevitably to revolutionary praxis, actually inhibits the latter in most cases. Put differently once more, fomenters of revolution have *as* good reason to be suspicious of sociology as policemen have. This point can be made economically by way of three imperatives which, in my opinion, sociological understanding can show to be present in every human community: the imperatives of order, of continuity, and of triviality. Each of these flies in the face of some of the fondest beliefs of the contemporary left.

After a recent lecture of mine on sociological theory, a perceptive student remarked to me, "You sure have a hangup on order, don't you?" I conceded the description, but I added that my "hangup" was not arbitrary or inadvertent. Behind it is the conviction that sociology leads to the understanding that order is *the* primary imperative of social life. There is the additional conviction (which I cannot develop here) that this fact is rooted in the fundamental constitution of man, that is, that not only sociology but philosophical anthropology must lead to a "hangup on order."

Society, in essence, is the imposition of order upon the flux of human experience. Most people will first think here of what American sociologists call "social control"—the imposition of coercive power upon deviant individuals or groups—and, of course, it is in this sense that radicals will understand, and disagree with, my "hangup on order." Coercion and external controls, however, are only incidental aspects of society's imposition of order. Beginning with language, *every* social institution, no matter how "nonrepressive" or "consensual," is an imposition of order. If this is understood, it will be clear that social life abhors disorder as nature abhors a vacuum. This has the directly political implication that, except for rare and invariably brief periods, the forces of order are always stronger than those of disorder and, further, there are fairly narrow limits to the toleration of disorder in any human society.

The left, by and large, understands that all social order is precarious. It generally fails to understand that *just because of this precariousness* societies will react with almost instinctive violence to any fundamental or long-lasting threat to their order. The idea of "permanent revolution" is an anthropologically absurd fantasy. Indeed, revolutionary movements can be successful only if they succeed, and succeed fairly rapidly, in establishing new structures of order within which people can settle down with some semblance of social *and* psychic safety. Mao Tse Tung's cultural revolution can serve as a textbook example of the grotesque failure in store for any revolutionary praxis that fails to grasp this point.

The imperative of continuity is closely related to, but not identical with, the imperative of order. I suppose that, finally, it is rooted in the simple fact that people have children. If one has children, one feels a necessity to explain the past to them and to relate the present to the past. If one loves one's children (and I take it that this is the case with most people who have them), one will want to project into the future whatever good things one has possessed in one's own life—and there are very few people, even among the most oppressed, who have possessed nothing good at all. Conversely, if one loves one's parents (the current "generation crisis" notwithstanding, I am inclined to think that this, too, is something of an anthropological constant), one will not finally want to disparage *everything* that constituted the parents' world—especially not if one comes to have children of one's own, who not only ask what will become of them but from where they come. *Children are our hostages to history.* Consequently, to be a parent means (however dimly and on whatever level of intellectual sophistication) to have a stake in the continuity of the social order. As a result, there are limits not only to social disorder but to social discontinuity. Enthusiasts for violent change (most of whom, I have noticed, don't have children) fail to recognize this. Successful revolutionaries find out about the limits of disorder, usually to their dismay, as they must settle down to govern the society over which they have gained control. The experiences of the Soviet regime with the institutions of the family and of religion are instructive in this regard.

The imperative of triviality is also, I suspect, rooted in some basic facts of the human condition—namely, the facts that man's attention span is limited and that man can tolerate only a limited amount of excitement. Perhaps the physiological foundation of this is the need for sleep. Be this as it may, social life would be psychologically intolerable if each of its moments required from us full attention, deliberate decision, and high emotional involvement. I would thus give the status of a sociological axiom to this proposition: *Triviality is one of the fundamental requirements of social life*. It is sociologically, anthropologically, and perhaps even biologically *necessary* that a goodly portion of social life take place in a state of dim awareness or semisleep. Precisely for this reason the institutional order "programs" the individual's activity. Put simply, society protects our sanity by preempting a large number of choices—not only choices of action but choices of thought. If we understand this (the understanding has been worked out systematically, by the way, in the theory of institutions by the contemporary German sociologist Arnold Gehlen), we shall see that there are limits not only to disorder and discontinuity but to the frequency of "significant events." We shall then take more seriously "meaningless rituals," "empty forms," or "mere routines" in social life—simply through recognizing that were social life in its entirety to be charged with profound meaning, we would all go out of our minds. The "meaninglessness" of so much of social life, currently decried as the source of so-called "alienation," is in fact a necessary condition for both individual and collective sanity. The currently fashionable left ideal of full participation in the sense that everybody will participate in every decision affecting his life, would, if realized, constitute a nightmare comparable to unending sleeplessness. Fortunately, it is anthropologically unrealizable, though the endless "discussion" that goes on in radical groups gives a certain approximation of the horror that its realization would signify. It is one of the mercies of human nature that, finally, all participants and all discussants must fall asleep.

I have tried to explicate the conservative bent of sociology by pointing to some basic imperatives of social life that should make the sociologist skeptical of notions of violent change and hesitant to commit himself to revolutionary praxis. I think that similar conclusions can be arrived at, by way of sociological or historical empirical analysis, for the actual processes of revolution. If all this adds up to a conservative propensity, it should be emphasized that the conservatism in question is of a peculiar kind. It is *not* a conservatism based on the conviction that the institutions of the status quo are sacred, inexorably right, or empirically inevitable. The aforementioned subversive impulse of sociology precludes this type of conservatism. Rather, it is a conservatism based on skepticism about the status quo in society *as well as* about various programs for new social orders. It is, if you wish, the conservatism of the pessimist. The seeming contradiction between our two propositions about the subversiveness and the conservatism of sociology thus resolves itself into a paradoxical but by no means irrational stance: *the stance of a man who thinks daringly but acts carefully.* This, of course, is exactly the kind of man whom our young revolutionaries will call a fink. So be it. It is probably one of the unavoidable blindnesses of youth to fail to see that acting carefully in society may, for some, be the simple result of wanting to preserve their little applecarts, but for others, motivated quite differently, it may reflect a carefully thought-through concern to avoid senseless pain and to protect the good things of ordinary life. There is some irony, though, in the fact that a generation that has made a culture hero out of Albert Camus should extol his *Rebel* at the expense of his hymns of praise to the ordinary pleasures of ordinary men on sun-drenched beaches.

Sociology, therefore, is a liberating discipline in a very specific way. There can be no doubt about its liberating effects on consciousness. At least potentially, sociology may be a prelude to liberation not only of thought but of action. At the same time, however, sociology points up the social limits of freedom—the very limits that, in turn, provide the social space for any empirically viable expression of freedom. This perspective, alas, is not simple. It requires intellectual effort and is not easily harnessed to political passions. I contend that the effort is worth it and that it will serve well precisely those political purposes that come from a concern for living men rather than for abstract doctrines of liberation.

So much for sociology as a discipline. What about the sociologist? A good case can be made that there is a crisis of freedom in the world today. What is to be the place of the sociologist in this crisis?

While the place of sociology and the place of the sociologist are not identical, they are interrelated. Perhaps the easiest way to explain the difference is in terms of so-called "value-freedom," that Weberian term that has become a sort of middle-echelon devil in the conceptual hell of the sociological left. The *discipline of sociology,* I insist as emphatically as I can, must be value-free—however difficult this may be in some situations. The moment the discipline ceases to be value-free in principle, it ceases to be a science and becomes nothing but ideology, propaganda, and a part of the instrumentarium of political manipulation. The *practitioner of the discipline,* the sociologist —a living human being,—must *not* be value-free. The moment he is, he betrays his humanity and (in an operation that can simultaneously be called "false consciousness" and "bad faith") transforms himself into a ghostly embodiment of abstract science. These two statements about value-freedom are made, of course, in discrete frames of reference. The statement about the value-freedom of sociology is a methodological one; the statement about the value-freedom of the sociologist is ethical. But perhaps it is appropriate to conclude these observations with a little homily.

We may return here to the two images of the sociologist that were conjured up earlier—that of the sociologist as the antiseptically neutral technician and that of the sociologist as the fiercely committed partisan. I think that the sociological left has been very largely right, ethically speaking, in its denunciations of the former type (even if it has been unfair in individual instances). In an age in which not only freedom but the very survival of man is in jeopardy, there is something obscene about the scientist who claims that he is not responsible for the uses to which his science is put. This is not to deny in any way the right of individuals to live the theoretical life or to abstain from political engagement. This right, however, can be exercised more acceptably by Byzantinologists than by most sociologists. Sociology is too much linked to the agonizing dilemmas of our time to permit most of its practitioners to pursue their theoretical interests in detachment from the struggles of their fellow-men. It is clear, beyond that, that the sociologist in the employ of politically relevant organizations cannot disclaim political repsonsibility for his work—a point that has been impressed on us very forcefully by the debate that followed the revelations about Project Camelot.

Because of these considerations, I emphasize my belief in the political partisanship of sociologists and concede that at times this partisanship may be quite fierce. For example, when it comes to the Pentagon's view of Latin America, my own political reactions tend to be of considerable ferociousness. It is equally important to stress, however, that the sociologist has no doctrine of redemption to bring into the political arena. What he has to contribute is the critical intelligence that is, or should be, the foundation of his discipline. This is a political *as well as* a methodological mandate. There are plenty of passions available, and the sociologist may well participate in some of them. His distinctive contribution to politics should be his consistent, unswerving application of critical intelligence—to the status quo, yes, and to any challengers of the status quo. Indeed, when a sociologist joins a revolutionary movement (an option I have indicated I would not normally prescribe), his most important political contribution to it will be his ongoing critique *of it*. Put differently, my principal objection to most of my radicalized colleagues is not that they are engaged in the business of "bringing to consciousness" but that they are not doing enough of it.

To whom will such a conception of the sociologist's role appeal? Evidently not to those who simply want a career in any kind of establishment—and not to those who see themselves as Messianic figures. It is all too clear that both such types are strongly represented in American sociology today. I have found, however, and not least among my students, that there are others—those who are still willing to commit themselves militantly to reason. And reason has its own seductiveness.

Notes

CHAPTER 2 THE DISCIPLINE OF SOCIOLOGY

THE STRANGER

1. Instead of mentioning individual outstanding contributions by American writers such as W. G. Sumner, W. I. Thomas, Florian Znaniecki, R. E. Park, H. A. Miller, E. V. Stonequist, E. S. Bogardus, and Kimball Young, and by German authors, especially Georg Simmel and Robert Michels, we refer to the valuable monograph by Margaret Mary Wood, *The Stranger: A Study in Social Relationship,* New York, 1934, and the bibliography quoted therein.

2. This insight seems to be the most important contribution of Max Weber's methodological writings to the problems of social science. Cf. the present writer's *Der sinnhafte Aufbau der sozialen Welt,* Vienna, 1932, 2nd ed. 1960.

3. John Dewey, *Logic, the Theory of Inquiry,* New York, 1938, Chap. iv.

4. For the distinction of these two kinds of knowledge cf. William James, *Principles of Psychology,* New York, 1890, Vol. I, pp. 221–22.

5. Max Scheler, "Probleme einer Soziologie des Wissens," *Die Wissensformen und die Gesellschaft,* Leipzig, 1926, pp. 58ff.; cf. Howard Becker and Helmuth Otto Dahlke, "Max Scheler's Sociology of Knowledge," *Philosophy and Phenomenological Research,* Vol. II, 1942, pp. 310–22, esp. p. 315.

6. Robert S. Lynd, *Middletown in Transition,* New York, 1937, Chap. xii, and *Knowledge for What?,* Princeton, 1939, pp. 58–63.

7. As one account showing how the American cultural pattern depicts itself as an "unquestionable" element within the scheme of interpretation of European intellectuals we refer to Martin Gumpert's humorous description in his book, *First Papers,* New York, 1941, pp. 8–9. Cf. also books like Jules Romains, *Visite chez les Américains,* Paris, 1930, and Jean Prevost Usonie, *Esquisse de la civilisation américaine,* Paris, 1939, pp. 245–66.

TWO STYLES OF RESEARCH IN CURRENT SOCIAL STUDIES

1. Paul F. Lazarsfeld, et al., *The People's Choice* (New York: Duell, Sloan and Pearce, 1944).

2. We can also consider it in relation to its users—the pragmatic dimension—which I am not here considering. These are the three dimensions of meaning which Charles M. Morris has systematized in his "Foundations of the Theory of Signs," *International Encyclopedia of Unified Science,* Volume I: Number 2. University of Chicago Press, 1938.

3. To sort out the dimensions of a macroscopic concept requires us to elaborate it syntactically, while keeping our eyes open for semantic indices for each implication so elaborated. To translate each of these points into molecular terms requires us to trace the hierarchy of inference down to single, clear-cut variables. In assertions using macroscopic concepts, we must watch for whether or not the assertion (1) states a proposition, or (2) unlocks an implication. The guide-rule is whether the statement involves one empirical factor or at least two. If it involves only one factor, then it simply "spells out" or specifies one of the conceptual implications of that one factor; its meaning is syntactical. If the assertion involves two factors, it may be a proposition, a statement of a relation which can be true or false; its meaning is semantical.

CHAPTER 3 BECOMING A MEMBER OF SOCIETY—SOCIALIZATION

THE EVERYDAY WORLD OF THE CHILD

1. A. A. Milne, *Winnie the Pooh* (New York: E. P. Dutton & Co., 1954), p. 75.

MIND, SELF AND SOCIETY

1. ["The Relation of Play to Education," *University of Chicago Record, I* (1886–97), 140 ff.]

2. It is possible for inanimate objects, no less than for other human organisms, to form parts of the generalized and organized—the completely socialized—other for any given human individual, in so far as he responds to such objects socially or in a social fashion (by means of the mechanism of thought, the internalized conversation of gestures). Any thing—any object or set of objects, whether animate or inanimate, human or animal, or merely physical—toward which he acts, or to which he responds, socially, is an element in what for him is the generalized other; by taking the attitudes of which toward himself he becomes conscious of himself as an object or individual, and thus develops a self or personality. Thus, for example, the cult, in its primitive form, is merely the social embodiment of the relation between the given social group or community and its physical environment—an organized social means, adopted by the individual members of that group or community, of entering into social relations with that environment, or (in a sense) of carrying on conversations with it; and in this way that environment becomes part of the total generalized other for each of the individual members of the given social group or community.

3. We have said that the internal conversation of the individual with himself in terms of words or significant gestures—the conversation which constitutes the process or activity of thinking—is carried on by the individual from the standpoint of the "generalized other." And the more abstract that conversation is, the more abstract thinking happens to be, the further removed is the generalized other from any connection with particular individuals. It is especially in abstract thinking, that is to say, that the conversation involved is carried on by the individual with the generalized other, rather than with any particular individuals. Thus it is, for example, that abstract concepts are concepts stated in terms of the attitudes of the entire social group or community; they are stated on the basis of the individual's consciousness of the attitudes of the generalized other toward them, as a result of his taking these attitudes of the generalized other and then responding to them. And thus it is also that abstract propositions are stated in a form which anyone—any other intelligent individual—will accept.

FUN MORALITY: AN ANALYSIS OF RECENT AMERICAN CHILD-TRAINING LITERATURE

1. My analysis is based on the six editions indicated. I was unable to obtain those of 1926 and 1940.
2. *Infant Care* (1914), p. 58.
3. *Ibid.*, p. 62.
4. *Ibid.*
5. *Ibid.*
6. *Ibid.*
7. *Ibid.*, p. 61.
8. *Ibid.*
9. *Ibid.* (1942), p. 60.
10. *Ibid.*
11. *Ibid.*
12. *Ibid.*, pp. 59–60.

13. *Ibid.*
14. *Ibid.* (1914), pp. 60–61.
15. *Ibid.* (1945), p. 52.
16. In a recent television advertisement, Angelo Patri is quoted as saying: "Youngsters today need television for their morale as much as they need fresh air and sunshine for their health."
17. *Infant Care* (1945), p. 95.
18. *Ibid.* (1914), pp. 60–61.
19. *Ibid.* (1945), p. 30.
20. *Ibid.* (1914), pp. 59–60.
21. *Ibid.*, p. 62.
22. *Ibid.* (1942), p. 41.
23. *Ibid.*
24. *Ibid.*
25. *Ibid.* (1914), p. 34.
26. *Ibid.* (1945), pp. 1, 29, 38, 62.

CHAPTER 4 WHAT IS AN INSTITUTION? THE CASE OF LANGUAGE

THE LORE AND LANGUAGE OF SCHOOLCHILDREN

1. Witches were supposed to say their prayers backwards, with awful effect, as is noted in Robert Greene's *A Qvip for an Vpstart Courtier,* 1592, sig. D3, and this method of raising the devil is also mentioned by Defoe in his *System of Magick,* 1727, pp. 259–60. References to schoolboys' possession of the secret occur in *Notes and Queries,* 1st ser., vol. iv, 1851, p. 53, and 3rd ser., vol. iv, 1863, p. 492; and we knew it ourselves in our schooldays, but never dared test it. For accounts of boys who did, see William Henderson, *Folk Lore of the Northern Counties,* 1866, p. 19, and *The Listener,* 3 January 1957, p. 10. A 13-year-old Tredegar, Monmouthshire, girl tells us it is believed there that if one runs round the church three times the devil will appear.
2. A correspondent hailing from Faversham informs us that the couplet was current in her childhood around 1910.

SELECTING AN IDENTIFICATION BY MEANS OF SOCIAL CATEGORIES

1. Schneider, D. *American Kinship: A Cultural Account.* Englewood Cliffs, N.J.: Prentice-Hall, 1968, pp. 30–31.

INTRODUCTION TO NANCY MITFORD'S *Noblesse Oblige*

1. The title "Honourable" is pronounced *'onorable.* "Hon" is pronounced as it is spelled.
2. The use of a hyphen between upper and middle is evidently non-U.
3. Professor Ross is a U scholar in a non-U university.
4. "As the co-founder, with my sister Jessica, of the Hons Club, I would like to point out that . . . the word Hon meant Hen in Honnish. . . . We were very fond of chickens and on the whole preferred their company to that of human beings. . . ." Deborah Devonshire (the Duchess of Devonshire) in a letter to *Encounter.*
5. So, according to a letter to *Encounter,* is the word *weekend.*
6. "Years of age" seems to me rather non-U. "Old" would be U, I should think.
7. It is non-U to bathe, but presumably it is downright vulgar not to.

CHAPTER 5 THE FAMILY

FAMILY TYPES AND THE URBAN VILLAGERS

1. Donald Pitkin, "Land Tenure and Farm Organization in an Italian Village," Unpublished Ph.D. Dissertation, Harvard University, 1954, p. 114.

2. Others have reported the lack of contact between the generations and the frequency of contact within them among second-generation Italians. See Philip Garigue and Raymond Firth, "Kinship Organization of Italianates in London," in Raymond Firth, ed., *Two Studies of Kinship in London,* London: Athlone Press, 1956, especially pp. 74, 82. Comparative studies of Italian and Irish populations have also reported this pattern, noting the difference in the Irish family, where old people are venerated and powerful. See M. K. Opler and J. L. Singer, "Ethnic Differences in Behavior and Psycopathology," *International Journal of Social Psychiatry,* vol. 2 (1956), pp. 11–22; and Ezra F. Vogel, "The Marital Relationships of Parents of Emotionally Disturbed Children," Unpublished Ph.D. Dissertation, Harvard University, 1958, Chap. 6.

3. The role of the "Mum" has been described in many studies of English working-class life. See, for example, Michael Young and Peter Willmott, *Family and Kinship in East London,* London: Routledge and Kegan Paul, 1957. . . .

4. These terms are taken from Elizabeth Bott, *Family and Social Network,* London: Tavistock Publications, 1957, pp. 53–54.

5. Bott, *op. cit.,* p. 53.

6. S. M. Miller and Frank Riessman have used similar terms—parent-centered and child-centered—to distinguish working-class families from middle-class ones in "The Working-Class Subculture: A New View," *Social Problems,* vol. 9 (1961), p. 92. For a different typology of family organization, using somewhat the same terms, see Bernard Farber, "Types of Family Organization: Child-Oriented, Home-Oriented, and Parent-Oriented," in Arnold Rose, ed., *Human Behavior and Social Processes,* Boston: Houghton Mifflin, 1962, pp. 285–306.

7. For an example of what I call adult-directed child-rearing, see J. Seeley, R. Sim, and E. Loosley, *Crestwood Heights,* New York: Basic Books, 1956, especially Chaps. 7–9.

IN DEFENSE OF THE NEGRO FAMILY

1. One should be very careful to specify what the presumed deficits are. It is not correct to talk about language deficiency as a general defect among low-income populations. It is more accurate to speak about deficiency in syntax and formal language. But in other aspects of language, such as the use of metaphor, rich adjectives, hip language, the connection between verbal and non-verbal communication, there is positive strength. A much more significant deficit of low-income populations is lack of school know-how and lack of system know-how. Similarly, in relation to the family, it is very important to specify what the supposed weaknesses are, and how these supposed weaknesses are translated into preventing the individual from functioning in the employment and education structures. The mechanism of this connection is often vague.

FAMILY STRUCTURE AND SEXUAL RELATIONS IN THE COMMUNAL FAMILY

1. We think, indeed, that there is a close relationship between the commune movement, on the one hand, and the complex of stirrings in the middle class which includes the encounter movement, swingers, sensitivity training, and the incipient gestures toward group marriage represented by "wife-swapping." Each represents an attempt to cope with similar problems (e.g., alienation, existential discontents with the prospects or the realities of middle-class life) by groups of people differently situated in the life-career cycle: the communards being mainly college dropouts in their twenties, the others

being mainly married couples in their thirties or forties with children and already well into their professional careers with which they may have become disenchanted.

MARRIAGE AND THE CONSTRUCTION OF REALITY

1. *Cf.* Talcott Parsons and Robert Bales, *Family Socialization and Interaction Process* (Glencoe, Ill.: Free Press, 1955), pp. 3–34, 353–396.
2. *Cf.* Philippe Ariès, *Centuries of Childhood* (New York, Knopf, 1962), pp. 339–410.
3. *Cf.* Kurt Wolff (ed.), *The Sociology of Georg Simmel* (Glencoe, Ill., Free Press, 1950), pp. 118–144.
4. *Cf.* Schutz, *Aufbau,* pp. 29–36, 149–153.
5. *Cf.* Schutz, *Aufbau,* pp. 186–192, 202–210.
6. David Riesman's well-known concept of "other-direction" would also be applicable here.

CHAPTER 6 THE COMMUNITY

THE GOLD COAST AND THE SLUM

1. *Chicago Herald and Examiner,* July, 1923.
2. United Charities of Chicago: *Sixty Years of Service.* In 1920–21 there were 90 contributions to the United Charities in less than a square mile on the Gold Coast, and 460 poverty cases in the square mile behind it.
3. Taking figures for five widely differing "communities" in Chicago, this fact is clearly brought out:

COMMUNITY	POPULATION	WHO'S WHO	PHYSICIANS	POLITICIANS	POVERTY CASES	SUICIDES
Back of the Yards *	39,908	1	28	4	185	8
Bridgeport †	64,875	0	44	12	180	3
Lawndale ‡	105,819	1	212	14	251	6
Woodlawn ¶	69,504	31	185	14	48	8
Near North	83,819	151	212	30	555	28

* Immigrant community back of the Stockyards.
† Polish "area of first settlement" on the Southwest Side.
‡ Jewish "area of second settlement" on the West Side.
¶ South Side residential community, surrounding the University of Chicago, containing many professional men and women.

4. A five-room house on Hill Street, the rooms in which are 9x12x10 feet high, has thirty occupants. Another nurse told the writer of being called on a case on Sedgwick Street and finding two couples living in one room. One couple worked days, the other nights; one couple went to bed when the other couple got up. Mrs. Louise De Koven Bowen (*Growing Up with a City*), reminiscing of her United Charities experiences, tells of a woman who for three years existed on the food she procured from garbage cans and from the samples of department store demonstration counters. She adds: "Sometimes fate seems to be relentless to the point of absurdity, as in one case I remember of an Italian family. . . . The man was riding on a street car and was suddenly assaulted by an irate passenger. . . . His nose was broken and he was badly disfigured. . . . A few days later, on his way home from a dispensary where he had gone to have his wound dressed, he fell off a sidewalk and broke his leg. The mother gave birth to a

child the same day. Another child died the following day, and the eldest girl, only fourteen years old, who had been sent out to look for work, was foully assaulted on the street." Such is the life of the slum!

5. *Chicago Evening American,* December 21, 1923.

URBANISM AND SUBURBANISM AS WAYS OF LIFE:
A RE-EVALUATION OF DEFINITIONS

1. Louis Wirth. "Urbanism as a Way of Life," *American Journal of Sociology,* 44 (1938), pp. 1–24. Reprinted in Paul Hatt and Albert J. Reiss, Jr. (eds.), *Cities and Society*. Glencoe, Ill.: The Free Press, 1957, pp. 46–64.

2. *Ibid.,* p. 50.

3. *Ibid.,* pp. 54–60.

4. Richard Dewey. "The Rural-Urban Continuum: Real but Relatively Unimportant," *American Journal of Sociology,* 66 (1960), pp. 60–66.

5. Morris Axelrod. "Urban Structure and Social Participation," *American Sociological Review,* 21 (1956), pp. 13–18; Dewey, *op. cit.;* William H. Form, *et al.,* "The Compatibility of Alternative Approaches to the Delimitation of Urban Sub-areas," *American Sociological Review,* 19 (1954), pp. 434–440; Herbert J. Gans, *The Urban Villagers: A Study of the Second Generation Italians in the West End of Boston.* Boston: Center for Community Studies, December 1959 (mimeographed); Scott Greer, "Urbanism Reconsidered: A Comparative Study of Local Areas in a Metropolis," *American Sociological Review,* 21 (1956), pp. 19–25; Scott Greer, and Ella Kube, "Urbanism and Social Structure: A Los Angeles Study," in Marvin B. Sussman (ed.), *Community Structure and Analysis.* New York: Thomas Y. Crowell Company, 1959, pp. 93–112; Morris Janowitz, *The Community Press in an Urban Setting.* Glencoe, Ill.: The Free Press, 1952; Albert J. Reiss, Jr., "An Analysis of Urban Phenomena," in Robert M. Fisher (ed.), *The Metropolis in Modern Life.* Garden City, N.Y.: Doubleday & Company, Inc., 1955, pp. 41–49; Albert J. Reiss, Jr., "Rural-Urban and Status Differences in Interpersonal Contacts," *American Journal of Sociology,* 65 (1959), pp. 182–195; John R. Seeley, "The Slum: Its Nature, Use and Users," *Journal of the American Institute of Planners,* 25 (1959), pp. 7–14; Joel Smith, William Form, and Gregory Stone. "Local Intimacy in a Middle-Sized City," *American Journal of Sociology,* 60 (1954), pp. 276–284; Gregory P. Stone, "City Shoppers and Urban Identification: Observations on the Social Psychology of City Life," *American Journal of Sociology,* 60 (1954), pp. 36–45; William F. Whyte, Jr., *Street Corner Society.* Chicago: The University of Chicago Press, 1955; Harold L. Wilensky, and Charles Lebeaux, *Industrial Society and Social Welfare.* New York: Russell Sage Foundation, 1958; Michael Young, and Peter Willmott. *Family and Kinship in East London.* London: Routledge & Kegan Paul, Ltd., 1957.

I shall not attempt to summarize these studies, for this task has already been performed by Dewey, Reiss, Wilensky, and others.

6. Wirth, *op. cit.,* p. 56.

7. *Ibid.,* p. 52.

8. *Ibid.,* pp. 60–62.

9. Greer and Kube, *op. cit.,* p. 112.

10. Wilensky, *op. cit.,* p. 121.

11. By the *inner city,* I mean the transient residential areas, the Gold Coasts and the slums that generally surround the central business district, although in some communities they may continue for miles beyond that district. The *outer city* includes the stable residential areas that house the working- and middle-class tenant and owner. The *suburbs* I conceive as the latest and most modern ring of the outer city, distinguished from it only by yet lower densities, and by the often irrelevant fact of the ring's location outside the city limits.

12. Louis Wirth. *The Ghetto,* Chicago: The University of Chicago Press, 1928.

13. Arnold M. Rose. "Living Arrangements of Unattached Persons," *American Sociological Review,* 12 (1947), pp. 429–435.

14. Gans, *op. cit.*

15. Seeley, *op. cit.*

16. *Idem;* The trapped are not very visible, but I suspect that they are a significant element in what Raymond Vernon has described as the "gray areas" of the city. See Raymond Vernon, *The Changing Economic Function of the Central City,* New York: Committee on Economic Development, Supplementary Paper No. 1, January 1959.

17. Greer, *op. cit.,* p. 283.

18. If the melting pot has resulted from propinquity and high density, one would have expected second-generation Italians, Irish, Jews, Greeks, Slavs, etc. to have developed a single "pan-ethnic culture," consisting of a synthesis of the cultural patterns of the propinquitous national groups.

19. The corporation transients (see Whyte, *Organization Man;* Wilensky and Lebeaux, *op. cit.*), who provide a new source of residential instability to the suburb, differ from city transients. Since they are raising families, they want to integrate themselves into neighborhood life, and are usually able to do so, mainly because they tend to move into similar types of communities wherever they go.

20. The negative social consequences of overcrowding are a result of high room and floor density, not of the land coverage of population density which Wirth discussed. Park Avenue residents live under conditions of high land density, but do not seem to suffer visibly from overcrowding.

21. Whether or not these social phenomena have the psychological consequences Wirth suggested depends on the people who live in the area. Those who are detached from the neighborhood by choice are probably immune, but those who depend on the neighborhood for their social relationships—the unattached individuals, for example—may suffer greatly from loneliness.

22. Needless to say, residential instability must ultimately be traced back to the fact that, as Wirth pointed out, the city and its economy attract transient—and, depending on the sources of outmigration, heterogeneous—people. However, this is a characteristic of urban-industrial society, not of the city specifically.

23. By neighborhoods or residential districts I mean areas demarcated from others by distinctive physical boundaries or by social characteristics, some of which may be perceived only by the residents. However, these areas are not necessarily socially self-sufficient or culturally distinctive.

24. Wirth, *op. cit.,* p. 56.

25. For the definition of *outer city,* see Footnote 11.

26. Wirth, *loc. cit.*

27. Because neighborly relations are not quite primary, and not quite secondary, they can also become *pseudo-primary;* that is, secondary ones disguised with false affect to make them appear primary. Critics have often described suburban life in this fashion, although the actual prevalence of pseudo-primary relationships has not been studied systematically in cities or suburbs.

28. Stone, *op. cit.*

29. These neighborhoods cannot, however, be considered as urban folk societies. People go out of the area for many of their friendships, and their allegiance to the neighborhood is neither intense nor all-encompassing. Janowitz has aptly described the relationship between resident and neighborhood as one of "limited liability." See Janowitz, *op. cit.,* chap. 7.

30. Were I not arguing that ecological concepts cannot double as sociological ones, this way of life might best be described as small-townish.

31. Arthur J. Vidich, and Joseph Bensman. *Small Town in Mass Society: Class, Power and Religion in a Rural Community.* Princeton, N.J.: Princeton University Press, 1958.

32. Harold Wattell. "Levittown: A Suburban Community," in William M. Dobriner (ed.), *The Suburban Community.* New York: G. P. Putnam's Sons, 1958, pp. 287–313.

33. Bennett Berger. *Working Class Suburb: A Study of Auto Workers in Suburbia.* Berkeley, Calif.: University of California Press, 1960; Vernon, *op. cit.*

34. Berger, *op. cit.*

35. They may, of course, be significant for the welfare of the total metropolitan area. Cf. Wattell, *op. cit.*

36. Otis Dudley Duncan, and Albert J. Reiss, Jr. *Social Characteristics of Rural and Urban Communities, 1950,* New York: John Wiley & Sons, 1956, p. 131.

37. Donald L. Foley. "The Use of Local Facilities in a Metropolis," in Paul Hatt and Albert J. Reiss, Jr. (eds.), *Cities and Society*. Glencoe, Ill.: The Free Press, 1957, pp. 237–247; and Christen T. Jonassen, *The Shopping Center Versus Downtown,* Columbus, Ohio: Bureau of Business Research, Ohio State University, 1955.

38. Jonassen, *op. cit.,* pp. 91–92.

39. A 1958 study of New York theater goers showed a median income of close to $10,000 and 35 per cent were reported as living in the suburbs. See John Enders, *Profile of the Theater Market*. New York: Playbill, undated and unpaged.

40. A. C. Spectorsky. *The Exurbanites,* Philadelphia: J. B. Lippincott, 1955.

41. I am thinking here of adults; teen-agers do suffer from the lack of informal meeting places within walking or bicycling distance.

42. Herbert J. Gans. "Planning and Social Life: An Evaluation of Friendship and Neighbor Relations in Suburban Communities," *Journal of the American Institute of Planners,* 27 (1961), pp. 134–140.

43. Christen T. Jonassen. *The Shopping Center Versus Downtown,* Columbus, Ohio: Bureau of Business Research, Ohio State University, 1955.

44. Thomas Ktsanes and Leonard Reissman. "Suburbia: New Homes for Old Values," *Social Problems,* vol. 7 (Winter 1959–60), pp. 187–194.

ETHNOGENESIS AND NEGRO-AMERICANS TODAY

1. For definitions see as examples W. L. Warner, "Formal Education and the Social Structure," *Journal of Educational Sociology,* vol. 19 (May 1936) pp. 524–31. W. L. Warner and A. Davis, "A Comparative Study of American Caste," in *Race Relations and the Race Problem,* ed. Thompson (Durham, N.C.: Duke University Press, 1939). A. Davis, B. B. and M. R. Gardner, *Deep South* (Chicago, Ill.: University of Chicago Press, 1941). G. Myrdal, *An American Dilemma* (New York: Harper & Bros., 1944).

2. See as an example O. C. Cox, *Caste, Class, and Race* (New York: Doubleday & Co., Inc., 1948).

3. This structural emphasis and static quality is especially the case with the caste approach.

4. J. W. Bennett and M. M. Tumin, *Social Life* (New York: Alfred A. Knopf, Inc., 1949), p. 140. Note that the authors go on to say, "For instance, in our society all people with dark skin may be considered as belonging to a social category."

5. It may be added that social categories cannot interact in any sociological sense. For example, when we speak of the interaction of men and women, it may sound as if we are referring to the interaction of categories. But actually we are referring to the cumulative interaction of individuals or, more likely, to typical aspects of interaction situations that involve individuals. Certainly such interaction situations are influenced by the beliefs and values attached to the several category definitions but, just as certainly, it is incorrect to say that the social categories are interacting.

6. For a definition of ethnic group see note 11. Interestingly enough, both Myrdal and Cox approached the issue of Negroes as an entity, but neither one developed the implications of his own suggestive comments. See Myrdal's references to "a separate community" and "a nation within a nation" (note 1, above), pp. 680, 785, 1003–04 and Cox's reference to a "quasi-society" (note 2, above), p. 503. See also L. Singer, "A Comparative Analysis of Selected Approaches to Negro-White Relations in the United States" (Doctor's dissertation, Columbia University, 1958), pp. 140, 231–32, 270–72.

7. This is a serious consideration when the investigator's concern is diachronic and developmental. But this is not to say that category concepts are useless, for they are precisely applicable to synchronic analysis.

8. See E. F. Frazier, *The Negro in the United States* (New York: The Macmillan Co., 1949), pp. 6–21. This may be contrasted, for example, with Brazilian slavery. See D. Pierson, *Negroes in Brazil* (Chicago: University of Chicago Press, 1942), especially pp. 38–45, IX, and X.

9. *Race and Culture* (New York: Free Press of Glencoe, Inc., 1950), p. 268.

10. See H. Aptheker, *American Negro Slave Revolts* (New York: Columbia University Press, 1943).

11. The term ethnic group is derived from the Greek word *ethnos* meaning "a people." Ethnic group is used here to mean a set of persons that may be distinguished from other such sets by virtue of: 1) a shared pattern of values, beliefs, norms, tastes, and so forth 2) an awareness of their own distinctiveness, partially reflected in a "we-feeling"; (These two distinctions taken together make up their *ethos*.) 3) some structure of relationships among them; and 4) the tendency to maintain generational continuity by marriage within the group. This is very close to E. K. Francis' definition with its emphasis on the *Gemeinschaft* quality of an ethnic group. See, for example, his "The Nature of an Ethnic Group," *American Journal of Sociology,* vol. 52 (March 1947) pp. 393–400 and "The Russian Mennonites: from Religious to Ethnic Group," *American Journal of Sociology,* vol. 54 (September 1948) pp. 101–07. See also the definition given by R. M. MacIver and C. H. Page which specifies both primary and secondary relationships in *Society* (New York: Rinehart & Co., Inc.), p. 387. Compare with R. M. Williams, Jr., *The Reduction of Intergroup Tensions* (New York: Social Science Research Council, 1947), p. 42.

12. This process seems to me to be one of several kinds of "group-forming" processes. Roscoe C. Hinkle, in a private communication, has suggested "socio-genesis" as the generic term.

13. The distinguished population segment may become dominant, although this is not usually the case. See, for example, MacIver and Page, note 11, p. 388.

14. In the case of the Africans in the English colonies this would mean, not slavery per se, but the qualities which slavery in America achieved by the end of the seventeenth century. It was only then that the European-Christian "slaves" and the African-heathen "slaves" became differentiated into servants—who would serve for a limited time—and chattel slaves—who would serve for the duration of their lives. See O. Handlin, *Race and Nationality in American Life* (New York: Doubleday Anchor Books, 1957), especially I.

15. See, for example, A. Rose, *The Negro's Morale* (Minneapolis: University of Minnesota Press, 1949), especially II.

16. From one point of view Reconstruction was "a prolonged race riot." G. B. Johnson, cited in Myrdal, note 1, p. 449.

17. See Myrdal, note 1, pp. 191–96. The bulk of the Negroes are still in the South, although the northward movement continues as well as the movement of Southern Negroes to Southern cities. It might be pointed out that Cox hypothesizes (note 2, p. xxxii) that, "In the future Negroes will probably become more highly urbanized than any other native-born population group in the country."

18. For a similar view of the social-psychological consequences of the Negro press see Myrdal, note 1, pp. 908–24 and A. Rose, note 15, pp. 102–08.

19. The "Freedom Riders" have a special significance in that whites are actively participating in the attempt to end discrimination in transportation. Both Myrdal and Cox hypothesize that Negro-Americans cannot achieve their aspirations without overt support from whites. The "sit-ins" are also especially significant because of the youth of the demonstrators.

20. C. M. Arensberg in a private communication suggests such instances as "Italian-Americans in the U.S.A.; . . . Africans and East Indians in the Carbibbean (cf. Trinidad); Hispanos (Puerto Ricans, and so forth) in New York (just started); Pennsylvania Dutch (1600–1800); 'white Southerners' paralleling Negroes in the U.S.A.; . . . In Israel, today, the Oriental Jews . . . are being welded into a self-conscious minority vis-à-vis the 'Askenazim' or 'Europeans' by all accounts. . . ."

21. Beyond the differences in prestige-ranking of field hands and house servants, the only common cultural element was the slave family. On the relative weakness of this structure see Frazier, note 8, II, especially 40–41.

22. This is the sort of process which E. K. Francis refers to: "Yet even on the ground of our limited knowledge it becomes clear that, generally speaking, the stages of development traversed by ethnic groups are: expansion—fission—new combination." (note 11, p. 398). What we have here called ethnogenesis is related to Francis' sequence

at two points. It is, on the one hand, temporally prior in that ethnic groups must have formed before they could expand. On the other hand, the last stage of the sequence is ethnogenesis. Consequently, the expanded sequence should be: ethnogenesis—expansion—fission—new combination (that is, ethnogenesis).

23. See, for example, A. Davis and M. and B. Gardner, especially the section, "Endogamy-Keystone of Caste," pp. 24–44. Note also the various state laws, and not only in the deep South, prohibiting intermarriage.

24. No slight is intended to Frederick Douglass by this comment. While it is true that he lived through the period and died in 1895, he is not typical of it.

CHAPTER 7 THE STRATIFIED COMMUNITY

SOCIAL CLASS IN AMERICA

1. See Chapter 15 [of *Social Class in America*] for a description of the several volumes of "Yankee City Series." New and poorly organized towns sometimes have class systems which have no old-family (upper-upper) class.

2. See W. Lloyd Warner and Paul S. Lunt, *The Social Life of a Modern Community,* Vol. I, "Yankee City Series" (New Haven: Yale University Press, 1941), pp. 58–72.

3. The evidence for the statements in the paragraph can be found in *The Social Life of a Modern Community,* pp. 287–300.

4. It is conceivable that in smaller communities there may be only three, or even two, classes present.

5. Allison Davis, Burleigh B. Gardner, and Mary R. Gardner, *Deep South* (Chicago: University of Chicago Press, 1941). Also read: John Dollard, *Caste and Class in a Southern Town* (New Haven: Yale University Press, 1937); Mozell Hill, "The All-Negro Society in Oklahoma" (Unpublished Ph.D. dissertation, University of Chicago, 1936); Harry J. Walker, "Changes in Race Accommodation in a Southern Community" (Unpublished Ph.D. dissertation, University of Chicago, 1945).

6. See St. Clair Drake and Horace R. Cayton, *Black Metropolis* (New York: Harcourt, Brace & Co., 1945), for studies of two contrasting caste orders; read the "Methodological Note" by Warner in *Black Metropolis* for an analysis of the difference between the two systems.

7. See W. Lloyd Warner and Leo Srole, *The Social Systems of American Ethnic Groups,* Vol. III, "Yankee City Series" (New Haven: Yale University Press, 1945). Chapter X discusses the similarities and differences and presents a table of predictability on their probable assimilation and gives the principles governing these phenomena.

8. Gunnar Myrdal, *An American Dilemma* (New York: Harper & Bros., 1944). For an early publication on color-caste, see W. Lloyd Warner, "American Caste and Class," *American Journal of Sociology,* XLII, No. 2 (September, 1936), 234–37, and "Formal Education and the Social Structure," *Journal of Educational Sociology, IX* (May, 1936), 524–531.

THE WORKING CLASS SUBCULTURE: A NEW VIEW

1. Allison Davis and Robert J. Havighurst, "Social Class and Color Differences in Child Rearing," *American Sociological Review,* 11 (December, 1946), pp. 698–710.

2. Cf. David Riesman in his introduction to Ely Chinoy's *American Workers and Their Dreams* (New York: Doubleday & Company, 1955).

3. Louis Schneider and Sverre Lysgaard, "The Deferred Gratification Pattern: A Preliminary Study," *American Sociological Review,* 18 (April, 1953), pp. 142–9.

4. Joseph A. Kahl, *The American Class Structure* (New York: Rinehart and Company, 1959), pp. 205 ff.

5. For the original report, see A. B. Hollingshead and Frederick C. Redlich, *Social*

Class and Mental Illness (New York: John Wiley and Sons, 1958). The point above is taken from S. M. Miller and Elliot G. Mishler, "Social Class, Mental Illness, and American Psychiatry," *Milbank Memorial Fund Quarterly,* XXXVII (April, 1959), pp. 174–99.

6. For a review of the relevant literature, see Frank Riessman, *Education and the Culturally Deprived Child* (New York: Harper and Brothers, 1961).

7. Riessman, *Education and the Culturally Deprived Child,* has a discussion of some of the relevant literature.

8. Charles H. Hession, S. M. Miller and Curwen Stoddart, *The Dynamics of the American Economy* (New York: Alfred A. Knopf, 1956), Chapter 11.

9. Ely Chinoy, *op. cit.,* and Charles R. Walker, *Steeltown* (New York: Harper and Brothers, 1950), have data showing the considerable reluctance of workers to become foremen.

10. The initial attraction of many working-class youth to engineering is partly due to the apparently concrete and clear nature of the work and the presumed definiteness of the education for a particular type of job. Motivating working-class youth to go to college may require an expansion and sharpening of working-class children's interpretation of the job market.

11. Lloyd Reynolds, *Labor Economics and Labor Relations* (Englewood Cliffs, N.J.: Prentice-Hall, Inc., 1949), pp. 7–23.

12. The discussion by Miller and Swanson on the "motoric" orientation of workers is one of the most suggestive in the literature. Daniel R. Miller and Guy E. Swanson, *Inner Conflict and Defense* (New York: Henry Holt and Company, 1960).

13. S. M. Miller, *Union Structure and Industrial Relations: A Case Study of a Local Labor Union,* unpublished Ph.D. thesis, Princeton University, 1951.

14. Melvin L. Kohn, "Social Class and the Exercise of Parental Authority," *American Sociological Review,* 24 (June, 1959), pp. 364–5.

15. The data to support this assertion can be computed from the two American studies detailed in the appendix to S. M. Miller, "Comparative Social Mobility," *Current Sociology,* 1961.

WHITE COLLAR: THE AMERICAN MIDDLE CLASSES

1. It is impossible to isolate the salaried foremen from the skilled urban wage-workers in these figures. If we could do so, the income of lower white-collar workers would be closer to that of semi-skilled workers.

2. According to our calculations, the proportions of women, 1940, in these groups are: farmers, 2.9%; businessmen, 20%; free professionals, 5.9%; managers, 7.1%; salaried professionals, 51.7%; salespeople, 27.5%; office workers, 51%; skilled workers, 3.2%; semi-skilled and unskilled, 29.8%; rural workers, 9.1%.

CHAPTER 8 THE STRATIFIED SOCIETY

SOCIAL MOBILITY AND PERSONAL IDENTITY

1. We will limit ourselves here to citing the principal sources to be used for a comparative analysis of these societies: Reinhold Bendix and Seymour Lipset (eds.), *Class, Status and Power* (Glencoe: Free Press, 1953); Arnold Rose (ed.), *The Institutions of Advanced Societies* (Minneapolis: University of Minnesota Press, 1958); Ralf Dahrendorf, *Class and Class Conflict in Industrial Society* (Stanford: Stanford University Press, 1959); *European Journal of Sociology,* I: 2 (1960) (Issue entitled "A la recherche des classes perdues"); Raymond Mack, Linton Freeman and Seymour Yellin, *Social Mobility, Thirty Years of Research and Theory* (Syracuse: Syracuse University Press, 1957); Seymour Lipset and Reinhold Bendix, *Social Mobility in Industrial Society* (Berkeley: University of California Press, 1959); David V. Glass (ed.), *Social Mobility in Britain*

(London: Routledge and Kegan Paul, 1954). For the American case, on which we have concentrated, *cf.* Milton Gordon, *Social Class in American Sociology* (Durham, N.C.: Duke University Press, 1958); Joseph Kahl, *The American Class Structure* (New York: Rinehart, 1959); Kurt B. Mayer, *Class and Society* (New York: Random House, 1955); Bernard Barber, *Social Stratification* (New York: Harcourt, Brace and Co., 1957).

2. We are using the concept of class in a broadly Weberian sense, that is, as an economically based stratum that entails common life chances and life styles. We are using the concept of status more loosely to refer to the location of the individual in a prestige system. We are aware of the debate on the precise meaning of these concepts but we do not wish to enter into the argument here. For a recent discussion of the contemporary applicability of these concepts *cf.* Dahrendorf, *op. cit.* On the applicability of the concept of class to Communist societies *cf.* Milovan Djilas, *The New Class* (New York: Praeger, 1957). For an instructive discussion of the impact of mobility aspirations on traditional preindustrial societies *cf.* Daniel Lerner, *The Passing of Traditional Society* (Glencoe: Free Press, 1958).

3. Parsons speaks of looseness, also of vagueness and indefiniteness of the class structure. *Cf.* Talcott Parsons, "An Analytical Approach to Social Stratification" in *Essays in Social Theory, Pure and Applied* (Glencoe: Free Press, 1949), pp. 182 and 183, and, "A Revised Analytical Approach to the Theory of Social Stratification," in Bendix and Lipset (eds.), *op. cit.*, pp. 106 and 122.

4. *Cf.* Barber, *op. cit.*, pp. 19–49. The problems for research on class which result from the multi-dimensionality of class criteria can be best seen in the work of W. Lloyd Warner and his associates in the five volumes of the "Yankee City Series" (New Haven: Yale University Press, 1941, 1959). Also, *cf.* the procedural hand book put out by this group of researchers—W. Lloyd Warner, Marchia Meeker and Kenneth Eells, *Social Class in America* (Chicago: Scientific Research Associates, 1949).

5. Emile Durkheim, *Suicide* (Glencoe: Free Press, 1951) pp. 246, 247, 257, 258.

6. Robert K. Merton, *Social Theory and Social Structure* (Glencoe: Free Press, 1957—revised and enlarged edition), chapters IV and V. *Cf.* also Mary Lystad, "Social Mobility among Selected Groups of Schizophrenic Patients," *ASR,* XXII (1957), 288–292; Jack Gibbs and Walter Martin, "A Theory of Status Integration and Its Relationship to Suicide," *ASR,* XXVIII (1958), 140–147; Elwin Powell, "Occupation, Status, and Suicide: Toward a Redefinition of Anomie," *ASR,* XXIII (1958), 131–139; Jack Gibbs and Walter Martin, "On Status Integration and Suicide Rates in Tulsa," *ASR,* XXIV (1959), 392–396; Austin Porterfield and Jack Gibbs, "Occupational Prestige and Social Mobility of Suicides in New Zealand," *AJS,* LXVI (1960), 147–152; David Matza and Gresham Sykes, "Juvenile Delinquency and Subterranean Values," *ASR,* XXVI (1961), 712–719; Albert Reiss, Jr., and Albert Rhodes, "The Distribution of Juvenile Delinquency in the Social Class Structure," *ASR,* XXVI (1961), 720–732; Elton Jackson, "Status Consistency and Symptoms of Stress," *ASR,* XXVII (1962), 469–480; Lee Robins, Harry Gyman and Patricia O'Neal, "The Interaction of Social Class and Deviant Behavior," *ASR,* XXVI (1962), 480–492; Warren Breed, "Occupational Mobility and Suicide," *ASR,* XXVIII (1963), 179–188; Robert Kleinen and Seymour Parker, "Goal-Striving, Social Status and Mental Disorder: A Research Review," *ASR,* XXVIII (1963), 189–203.

7. *Cf.* Gerhard Lenski, "Status Crystalization: A Non-Vertical Dimension of Social Status," *ASR,* XIX (1954), 12–18; Gerhard Lenski, "Social Participation and Status Crystallization," *ASR,* XXI (1956), 458–464; Irwin Goffman, "Status Consistency and Preference for Change in Power Distribution," *ASR,* XXII (1957), 275–281; Werner Landecker, "Class Crystallization and Its Urban Pattern," *Social Research,* XXVII (1960), 308–320; Raymond Murphy and Richard Morris, "Occupational Situs, Subjective Class Identification, and Political Affiliation," *ASR,* XXVI (1961), 383–392; Robert Hodge, "The Status Consistency of Occupational Groups," *ASR,* XXVII (1962), 336–343; Werner Landecker, "Class Crystallization and Class Consciousness," *ASR,* XXVIII (1963), 219–229.

8. *Cf.* Marcel Mauss, *Sociologie et anthropologie* (Paris: Presses Universitaires de France, 1960), *passim.* A classic statement of the social-psychological aspects of class is, of course, Maurice Halbwachs, *The Psychology of Social Class* (Glencoe: Free Press,

1958), but it does not contain a comprehensive approach to the question of mobility. For an interesting discussion of personality as it relates to economic change, *cf.* Joseph Bensman and Arthur Vidich, "Business Cycles, Class and Personality," *Psychoanalysis and the Psychoanalytic Review,* XLIX (1962), 30–52.

9. On the latter, *cf.* Lenski, *loc. cit.* (1954), and Landecker, *loc. cit.* (1960 and 1963).

10. Again, *cf.* the cited works of W. Lloyd Warner for discrepancies in the assignment of individuals through class criteria. Also *cf.* Lipset and Bendix, *op. cit.,* p. 42 ff.

11. *Cf.* Charles Cooley, *Human Nature and the Social Order* (New York: Scribner's, 1902), esp. Chap. V–VI; George Herbert Mead, *Mind, Self and Society* (Chicago: University of Chicago Press, 1934), esp. Chap. III.

12. On status and personality *cf.* Hans Gerth and C. Wright Mills, *Character and Social Structure* (New York: Harcourt, Brace & Co., 1953), p. 315 f; on transformation of identity *cf.* Anselm Strauss, *Mirrors and Masks* (Glencoe: Free Press, 1959), p. 89 ff; on problems of identity in mass society generally *cf.* Maurice Stein, Arthur Vidich and David White (eds.), *Identity and Anxiety* (Glencoe: Free Press, 1960), esp. p. 17 ff.

13. *Cf.* again references cited in Footnote 7; also, H.S. Sullivan, "Modern Conceptions of Psychiatry," *Psychiatry,* III (1940).

14. David Riesman, *The Lonely Crowd* (New Haven: Yale University Press, 1950); Arnold Gehlen, *Die Seele im technischen Zeitalter* (Hamburg: Rowohlt, 1957).

15. In this connection *cf.* the discussion of the functional changes in the American family in Talcott Parsons and Robert Bales, *Family, Socialization and Interaction Process* (Glencoe: Free Press, 1955).

16. On some aspects of "privatization" *cf.* Thomas Luckmann, *Das Problem der Religion in der modernen Gesellschaft* (Freiburg: Rombach, 1963).

17. On the inter-generational problem *cf.* S.N. Eisenstadt, *From Generation to Generation* (Glencoe: Free Press, 1956).

18. On the reinterpretation of past identity *cf.* Peter Berger, *Invitation to Sociology —A Humanistic Perspective* (Garden City, N.Y.: Doubleday-Anchor, 1963).

19. For a general discussion of anticipatory socialization *cf.* Robert Merton, *op. cit.,* Chap. VIII and IX, and especially pp. 265–269 and 293.

20. One likely consequence is "overconformity" to middle-class patterns (or, rather, their stereotypes). On the reasons of "overconformity" more generally, *cf.* Robert Merton, *op. cit.,* p. 182.

21. *Cf.* Alfred Schutz, *Collected Papers I—The Problem of Social Reality* (The Hague: Nijhoff, 1962).

22. *Cf.* C. Wright Mills, *White Collar,* (New York: Oxford University Press, 1951), p. 259 ff; William Whyte, *The Organization Man* (Garden City, N.Y.: Doubleday-Anchor, 1957), p. 3 ff. *Cf.* also Irwin Wyllie, *The Self-Made Man in America* (New Brunswick: Rutgers University Press, 1954); Robert Merton, *op. cit.* pp. 166–170; R. Richard Wohl, "The 'Rags to Riches' Story: An Episode of Secular Idealism," in Bendix and Lipset (eds.), *op. cit.* pp. 388–395.

23. As already Tocqueville observed, "The American Ministers of the Gospel do not attempt to draw or fix all the thoughts of man upon the life to come [. . .]. It is often difficult to ascertain from their discourses whether the principal object of religion is to procure eternal felicity in the other world or prosperity in this." For more recent evidence, *cf,* Louis Schneider and Sanford Dornbusch, *Popular Religion: Inspirational Books in America* (Chicago: The University of Chicago Press, 1958).

24. These remarks might be taken as a specific application of Marx's classic concept of the "fetishism of commodities." However, we would emphasize that what we have called the "sacramentalism of consumption" is not just a phenomenon of capitalism, but of industrialism and industrial societies in general.

SOCIAL MOBILITY AND EQUAL OPPORTUNITY

1. *The Big Business Executive/1964: A Study of His Social and Educational Background* (A study sponsored by the *Scientific American,* conducted by Market Statistics, Inc., of New York City, in collaboration with Dr. Mabel Newcomer). The study was designed to update Mabel Newcomer, *The Big Business Executive—The Factors that*

Made Him: 1900–1950 (New York: Columbia University Press, 1950). All comparisons in it are with materials in Dr. Newcomer's published work.

2. Stephan Thernstrom, "Migration and Social Mobility, 1880–1970: the Boston Case and the American Pattern," (Paper prepared for the Conference on Social Mobility in Past Societies, Princeton, N.J.: Institute for Advanced Study, June 15–17, 1972).

3. Peter Blau and O. D. Duncan, *The American Occupational Structure* (New York: John Wiley, 1967).

4. Edward Pessen, "The Egalitarian Myth and the American Social Reality: Wealth, Mobility, Morality in the 'Era of the Common Man,'" *The American Historical Review,* 76 (October 1971). Pessen cites many relevant recent historical works bearing on the intense forms of inequality in this period.

5. E. P. Hutchinson, *Immigrants and Their Children* (New York: John Wiley and Sons, 1956).

6. S. M. Lipset and Reinhard Bendix, *Social Mobility in Industrial Society* (Berkeley: University of California Press, 1959).

7. Albert Wohlstetter and Sinclair Coleman, *Racial Differences in Income* (Santa Monica: The Rand Corporation, October 1970).

8. See Richard Freeman, *Black Elite: Discrimination Against Qualified Black Workers* (New York: McGraw Hill, forthcoming, 1973).

CHAPTER 9 WHAT IS SOCIAL CONTROL? THE CASE OF EDUCATION

THE CENTRALITY OF SCHOOLING

1. Martin Trow, "The Democratization of Higher Education in America," *European Journal of Sociology,* III (1962), 231–262.

2. David Matza, "Position and Behavior Patterns of Youth," in R. E. L. Faris, ed., *Handbook of Modern Sociology* (Chicago: Rand McNally, 1964), p. 197.

3. Florian Znaniecki, "Educational Guidance," *Social Actions* (New York: Farrar and Rinehart, 1936), pp. 189–231.

YOUTH AND THE SOCIAL ORDER

1. See H. J. Eysenck, *The Scientific Study of Personality* (1952). [Musgrove's notes renumbered for consecutive sequence.—Ed.]

2. For evidence on introversion among the more successful Cambridge students see D. E. Broadbent, *Perception and Communication* (1958), ch. 7.

3. See W. D. Furneaux, 'The Psychologist and the University', *Universities Quarterly* (1962), 17. Thus 25.7 per cent of a borderline group of applicants for an engineering department were 'neurotic extraverts', but neurotic extraverts were only 6.2 per cent of those from this group who were admitted. The proportions of 'neurotic introverts' were 21.4 per cent and 37.5 per cent respectively.

4. See R. Lynn, 'Two Personality Characteristics Related to Academic Achievement', *British Journal of Educational Psychology* (1959), 29.

5. For a full examination of the vulnerability of this personality in some of the circumstances of modern professional life see the author's book, *The Migratory Élite* (1963), particularly ch. 7.

6. Aldous Huxley, *Island* (1962).

7. See Monica Wilson, *Good Company* (1951).

8. Cf. D. E. Broadbent, op. cit.: 'there is little doubt that anxiety, in some sense of the word, contributes to rapid conditioning and to good academic achievement. But the exact sense in which this is true is somewhat uncertain. The individual differences may

be due to drive or to reactive inhibition: to motivation or to fatigability, if one wishes to keep clear of Hullian terms' (p. 153).

9. R. Lynn, loc. cit.

10. Lewis M. Terman and Melita H. Oden, *The Gifted Group in Middle Life. Genetic Studies of Genius* (1959), vol. 5, p. 49, 8.3 per cent of (male) college graduates were 'seriously maladjusted', only 4.6 per cent of those who had not attended college (see Table 10). We urgently need more studies of university students matched with able individuals who did not attend universities. Some American research shows psychological deterioration among students, but no comparison with other populations is made. At Vassar, for example, 'A consistent trend is for seniors to be higher than freshmen on the following scales: Hypochondriasis, Depression, Hysteria, Psychopathic Deviate, Schizophrenia, and Mania.' (See Harold Webster *et al.*, 'Personality Changes in College Students' in Nevitt Sanford (ed.), *The American College* (1962).)

There are indications that American university students are more extraverted and less introverted than English university students, but that they suffer from higher levels of anxiety. Their degree of introversion seems to be similar to that found among English C.A.T. students. See R. B. Cattell and F. W. Warburton, 'A Cross-Cultural Comparison of Patterns of Extraversion and Anxiety', *British Journal of Psychology* (1961), 52.

11. Since this was written Ferdynand Zweig's depressing book, *The Student in the Age of Anxiety* (1963), has been published. This survey of students at Oxford and Manchester provides further support for the views advanced in this book. Dr Zweig finds the atmosphere of universities heavy and joyless, students anxious, harassed, guilt-ridden (even about their grants), unadventurous and conformist. 'They are not angry young men' (p. xiii), 'there is little doubt that the young are becoming old before their time' (p. xv). 'The students I interviewed did not strike me as young and carefree, on the contrary, they struck me as old, laden with responsibility, care and worry, with nightmares and horror dreams' (p. xiv).

12. See H. Jenkins and D. C. Jones, 'Social Class of Cambridge University Alumni of the 18th and 19th Centuries', *British Journal of Sociology* (1950), 1. Some 50 per cent of the sons of landowners failed to take a degree but assumed positions of leadership in later life.

13. See R. Heussler, *Yesterday's Rulers* (1963).

14. American experience should warn us against over-optimistic expectations that more worthwhile personalities will necessarily result from university education. Extensive American research on personality development at the university, while it often presents conflicting results, in general does not offer a re-assuring picture. In particular see the collation of research findings in P. E. Jacob, *Changing Values in College* (New York 1957). For discussions of later research see Nevitt Sanford (ed.), *The American College* (New York 1962). It would be singularly insular and arrogant to assume that we necessarily do better than, say, Vassar and Yale. We do not even bother to do the necessary research into the real consequence of 'educational' experiences. Numerous American inquiries show undergraduates developing somewhat more liberal values (e.g. Plant, 'Changes in Ethnocentricism during College', *Journal of Educational Psychology*, 1958) but declining in general mental health (e.g. Loomis and Green, 'Patterns of Mental Conflict in a Typical State University', *Journal of Abnormal and Social Psychology*, 1947). Deterioration e.g. in 'neurotic trends' has been frankly accepted as a prerequisite of academic education (e.g. Webster, 'Some Quantitative Results', *Journal of Social Issues*, 1956). The rigours of the 'real' world after college may promote a measure of recovery (e.g. Tate and Musick, 'Adjustment Problems of College Students', *Social Forces*, 1954).

THE MILITARY ACADEMY AS AN ASSIMILATING INSTITUTION

1. Robert E. Park and Ernest W. Burgess, *Introduction to the Science of Sociology* (Chicago: University of Chicago Press, 1921), pp. 735, 737.

2. Cf. Arnold Van Gennep, *Les Rites de Passage* (Paris: Emile Nourry, 1909).

Translated by Everett C. Hughes in *Anthropology-Sociology 240, Special Readings* (Chicago: University of Chicago Bookstore, 1948), Pt. II, p. 9.

3. Ralph H. Turner, "The Navy Disbursing Officer As a Bureaucrat," *American Sociological Review,* XII (June 1946), 344 and 348; Arnold Rose, "The Social Structure of the Army," *American Journal of Sociology,* LI (March 1946), 361.

4. Compare this viewpoint with that expressed in Hugh Mullan, "The Regular Service Myth," *American Journal of Sociology,* LIII (January 1948), 280, where hazing is viewed as the expression of "pent-up sadism." Such individualistic interpretations do not take into account the existence of an institutional structure, or else they give psychological interpretations to social processes.

5. "At each step of the ceremonies he feels that he is brought a little closer, until at last he can feel himself a man among men." A. R. Radcliffe-Brown, *The Andaman Islanders* (Glencoe, Illinois: The Free Press, 1948), p. 279.

6. Miriam Wagenschein, Reality Shock. Unpublished M.A. thesis, Department of Sociology, University of Chicago, 1950.

CHAPTER 10 BUREAUCRACY

BUREAUCRATIZATION IN INDUSTRY

1. Bureaucracy in government has been analyzed in great detail, frequently with the implication that waste and indolence were widespread. Many of these studies received their impetus from the scientific-management movement. Cf. Dwight Waldo, *The Administrative State,* New York, Ronald, 1948, pp. 47–61. The conventional term for industrial bureaucracy is "business administration." The implication of this terminology is obviously that waste and indolence are absent from business, that productivity is synonymous with efficiency, and that business administration involves technical know-how which the businessman has to keep secret in order to retain his competitive advantage. The semantics of this controversy have not been analyzed to my knowledge.

2. G. D. H. Cole, *The Life of Robert Owen,* London, Macmillan & Co., 1930, p. 70.

3. P. Sargant Florence, *The Logic of Industrial Organization,* London, Kegan Paul, Trench, Trubner & Co., 1933, pp. 159–160.

4. Erich Roll, *An Early Experiment in Industrial Organization, Being a History of the Firm of Boulton and Watt, 1775–1805,* New York, Longmans, 1930, pp. 250–251.

5. See William Miller, "The Business Elite in Business Bureaucracies," in William Miller (ed.), *Men in Business,* Cambridge, Mass., Harvard University Press, 1952, pp. 286–305.

6. See Lewis Corey, "The Middle Class," *Antioch Review,* Spring, 1945, pp. 73, 77.

7. See Chap. 13 [in *Industrial Conflict*], by Melville Dalton, for a detailed discussion of the foreman's modern role in industry.

8. See the detailed analysis of the rights and duties of subcontractors and foremen in England during the 19th century in William T. Delaney, "The Spirit and Structure of Labor Management in England, 1840–1940," Master's thesis, University of California (Berkeley), Chap. 3.

9. Ernest Dale, *The Development of Foremen in Management,* American Management Association Research Report 7, 1945, p. 9.

10. This comparative study cannot be reported here for reasons of space. But it should be mentioned that the economic meaning of this ratio depends in large part on the rate of capital investment. Industries which have a rate of increase in administrative overhead similar to that of the industries of the United States without, however, a similar increase in capital investment are obviously suffering from industrial bureaucratization far more than this country.

11. Alfred P. Sloan, *Adventures of a White-collar Man,* New York, Doubleday, 1941, p. 145.

12. Tom Girdler, *Boot Straps,* New York, Scribner, 1943, p. 177.

13. The belief in progress and individual success through effort has not changed basically for the last century and a half, though it is possible to note a gradual secularization. There is little evidence of God in the praise of "work and still more work [by which] we capitalize our unlimited opportunities . . ." (Alfred P. Sloan). This secularization of the Puritan ideology of individual success has been traced by A. W. Griswold, "The American Gospel of Success," Doctor's dissertation, Yale University, 1933.

14. Sloan, *op. cit.,* p. 153.

15. We speak here of intellectual trends and do not imply that such trends occur by design. On the other hand, the problems of large-scale industry are real problems, and any intellectualized concern with them will have meaning in the context of these problems, even if they are deliberately ignored. If, for example, industrial leaders are praised today in the same manner as they were in the 1870's, it would indicate an ideological distortion of the present situation, when the opportunities for social mobility consist, for the most part, in climbing the bureaucratic ladder in industry.

16. And it is possible that the ideology of the industrial bureaucrats, which was well summed up in Dale Carnegie's title *How to Win Friends and Influence People,* has affected the tycoon to the extent of changing his manners. See the suggestive comments on this point in David Riesman, *The Lonely Crowd,* New Haven, Conn., Yale University Press, 1950, esp. pp. 166–174, 236–239. Riesman posits a major change in character from the "inner-directed" personality of the "captains of industry" to the "other-directed" personality of the industrial bureaucrat, but I question whether one is warranted in inferring changes in "social character" from ideological and institutional trends. I question in particular whether white-collar workers of 1870 were inner-directed or whether industrial leaders of today are other-directed compared with their respective opposite numbers.

17. Samuel Smiles, *Self-help . . . ,* Chicago, Belford, Clarke and Co., 1881, p. vii (from the preface to the second edition, in which Smiles restated his major purpose, as he saw it, in response to the criticism which the book had received).

18. A new version of this book was published in 1936 under the title *How to Win Friends and Influence People.* It has sold a total of 4 million copies.

19. Dale Carnegie, *Public Speaking and Influencing Men in Business,* New York, Association Press, 1938, p. 509. Carnegie writes as follows: "I once asked a group of American businessmen in Paris to talk on *How to Succeed.* Most of them praised the homely virtues, preached at, lectured to, and bored their hearers. . . . So I halted this class and said something like this: 'We don't want to be lectured to. No one enjoys that. Remember you must be entertaining or we will pay no attention whatever to what you are saying. Also remember that one of the most interesting things in the world is sublimated, glorified gossip. So tell us the stories of two men you have known. Tell why one succeeded and why the other failed. We will gladly listen to that, remember it and possibly profit by it.' " *Ibid.,* p. 429. The shift from the praise of virtue to the description of "how to succeed" is well illustrated here, and if this passage is taken together with the one quoted in the text, it becomes clear that the techniques of "human relations" have superseded the idea of an emulation of virtues.

20. Lowell Thomas, "Introduction," in Carnegie, *op. cit.,* p. x.

21. In this context it would appear as if the "New Thought" movement, which Griswold has analyzed, is *not* the lineal descendant of the "gospel of success," which may be identified with the Puritan tradition and the Darwinian creed of the Gilded Age. The outstanding difference between New Thought and "self-improvement" as well as the personality cult is that the first remained a cultist belief, while the second called for action and promised success if the lessons had been learned well.

22. This ideological ambiguity is not confined to the managers of industry. In so far as salaried employees continue to believe in the free-enterprise system, they will tend to ignore their own dependent status and bureaucratic careers, which are the reverse of both freedom and enterprise (in the conventional economic meaning of these terms). Ideas are not abandoned readily under the pressure of circumstances. Indeed, there is some evidence to show that modifications of the ideal of individual success survive today among groups whose chances of individual success are minimal. Cf. Reinhard Bendix, "A Study of Managerial Ideologies, as It Bears on Drucker's Thesis," paper read before the American Sociological Society, Atlantic City, N.J., Sept. 3, 1952.

ASYLUMS

1. D. Cressey, "Achievement of an Unstated Organizational Goal: An Observation on Prisons," *Pacific Sociological Review,* I (1958), p. 43.

2. J. Bateman and H. Dunham, "The State Mental Hospital as a Specialized Community Experience," *American Journal of Psychiatry,* CV (1948–49), p. 446.

3. Belknap, *op. cit.,* p. 170.

4. Orwell, *op. cit.,* pp. 506–9.

5. *Ibid.,* p. 521.

6. See, for example, R. Lifton, " 'Thought Reform' of Western Civilians in Chinese Communist Prisons," *Psychiatry,* XIX (1956), especially pp. 182–84.

7. I derive this from Everett C. Hughes' review of Leopold von Wiese's *Spätlese,* in the *American Journal of Sociology,* LXI (1955), p. 182. A similar area is covered under the current anthropological term "ethnopsychology," except that the unit to which it applies is a culture, not an institution. It should be added that inmates, too, acquire a theory of human nature, partly taking over the one employed by staff and partly developing a countering one of their own. In this connection see in McCleery, *op. cit.,* pp. 14–15, the very interesting description of the concept of "rat" as evolved by prisoners.

8. Simon Raven, "Perish by the Sword," *Encounter,* XII (May 1959), pp. 38–39.

9. The engulfing character of an institution's theory of human nature is currently nicely expressed in progressive psychiatric establishments. The theories originally developed to deal with inmates are there being applied more and more to the staff as well, so that low-level staff must do its penance in group psychotherapy and high-level staff in individual psychoanalysis. There is even some movement to bring in consulting sociological therapists for the institution as a whole.

10. See the useful paper by Albert Biderman, "Social-Psychological Needs and 'Involuntary' Behavior as Illustrated by Compliance in Interrogation," *Sociometry,* XXIII (1960), pp. 120–47.

11. It would be quite wrong to view these "therapies" too cynically. Work such as that in a laundry or shoe-repair shop has its own rhythm and is managed often by individuals more closely connected with their trade than with the hospital; hence, very often, time spent at these tasks is much more pleasant than time spent on a dark, silent ward. Further, the notion of putting patients to "useful" work seems so captivating a possibility in our society, that operations such as shoe-repair shops or mattress-making shops may come to be maintained, at least for a time, at an actual cost to the institution.

12. Sister Mary Francis, P.C., *A Right to Be Merry* (New York: Sheed and Ward, 1956), p. 108.

13. *Ibid.,* p. 99. The application of an alternate meaning to poverty is of course a basic strategy in the religious life. Ideals of Spartan simplicity have also been used by radical political and military groups; currently, beatniks impute a special meaning to a show of poverty.

14. A good representation of this interpretative spread and thickness is given in Bernard Malamud's novel about management problems in a small grocery store: *The Assistant* (New York: New American Library, 1958).

15. *The Holy Rule of Saint Benedict,* Ch. 66.

16. For example, Harvey Powelson and Reinhard B. Bendix, "Psychiatry in Prison," *Psychiatry,* XIV (1951), pp. 73–86, and Waldo W. Burchard, "Role Conflicts of Military Chaplains," *American Sociological Review,* XIX (1954), pp. 528–35.

CHAPTER 11 YOUTH

POPULATION CHANGES AND THE STATUS OF THE YOUNG

1. See H. J. Habakkuk, 'English Population in the Eighteenth Century', *The Economic History Review* (1953), 4, 2nd series.

2. Cf. J. Hole, *Homes of the Working Classes* (1866), p. 17 for details of social-class differences in mortality in 1864.

3. C. Ansell, *Statistics of Families* (1874). Ansell's inquiry was among 54,635 upper and professional class children. The figure for the general population is from the Carlisle Tables.

4. Ibid.

5. D. V. Glass, 'Fertility and Economic Status in London', *Eugenics Review* (1938), 30. But cf. D. Heron, *On the Relations of Fertility in Man to Social Status* (1906): 'the intensity of the fertility-status relationship doubled between the middle of the nineteenth century and the beginning of the twentieth.' T. H. C. Stevenson thought that if analysis could be pushed far enough back, 'a period of substantial equality between all classes might have been met with.' 'The Fertility of Various Social Classes', *Journal of the Royal Statistical Society* (1920).

6. T. H. Marshall, 'The Population Problem during the Industrial Revolution', *Economic History: Economic Journal Supplement* (1929).

7. Jean Jousselin, *Jeunesse Fait Social Méconnu* (Toulouse 1959), p. 8.

8. T. S. Ashton, *The Industrial Revolution* (1948), p. 116.

9. Commercial clerks increased by 61 per cent 1861–71 and by 88 per cent 1871–81 while all occupied males increased by 13 per cent and 32 per cent.

10. *First Report of the Civil Service Inquiry Commission* (The Playfair) (1875) recommended an Upper Division recruited from the universities distinct from a Lower Division of routine clerks.

11. Until the Census of 1891 accountants were not distinguished from book-keepers. In 1880 the Institute of Chartered Accountants, in 1885 the Society of Accountants and Auditors were founded.

12. See N. J. Smelser, *Social Change in the Industrial Revolution* (1959), p. 185.

13. W. H. B. Court, *The Rise of the Midland Industries* (1938), pp. 230–32.

14. Andrew Ure, *Philosophy of Manufactures* (1861 ed.), p. 21.

15. G. Unwin, *Samuel Oldknow and the Arkwrights* (1924), ch. XI.

16. Charles Bray, 'The Industrial Employment of Women', *Transactions of the National Association for the Promotion of Social Science* (1857).

17. J. S. Wright, 'Employment of Women in Factories in Birmingham', *Transactions of the N.A.P.S.Sc.* (1857).

18. Vol. 9, p. 54.

19. Andrew Ure, op. cit., p. 290.

20. *First Report of the Commissioners* (*Mines*) (1842), pp. 40 ff.

21. N. J. Smelser, op. cit., p. 265. For the administrative complexities which the educational clauses entailed see A. A. Fry 'Report of the Inspectors of Factories on the Effects of the Educational Provisions of the Factories' Act', *Journal of the Statistical Society* (1839).

22. 34·4 births per 1,000 living in 1780, 35·4 1785–95; 34·2 1796–1806 cf. 31·1 in 1700, 27·5 in 1710 and 30·5 in 1720. See G. Talbot Griffith, op. cit., Table 5, p. 28.

23. Ibid., p. 103.

24. Ibid., p. 105. Talbot Griffith saw economic expansion as having a direct effect on the age of marriage and hence indirectly on the birth rate (p. 106).

25. T. H. Marshall, loc. cit., p. 454.

26. D. V. Glass, 'Changes in Fertility in England and Wales 1851 to 1931' in L. Hogben (ed.), *Political Arithmetic* (1938). 'Correlations between fertility and child labour yielded coefficients of $+0{\cdot}489 \pm 0{\cdot}116$ for 1851, $+0{\cdot}291 \pm 0{\cdot}140$ for 1871, and $+0{\cdot}043 \pm 0{\cdot}152$ for 1911. Of these coefficients only that for 1851 is significant.'

27. Arthur Young, *Political Arithmetic* (1774). 'People scarce—labour dear. Would you give a premium for population, could you express it in better terms? The commodity wanted is scarce, and the price raised; what is that but saying that the value of *man* is raised? Away! my boys—get children, they are worth more than ever they were.'

28. H. J. Habakkuk, loc. cit.

29. G. Talbot Griffith, op. cit., p. 66.

30. K. H. Connell, 'Land and Population in Ireland 1780–1840', *The Economic History Review* (1949), 2nd series, 2.

31. See Peter Quennell, *Mayhew's London* (1949), pp. 54 and 76.

32. Charles Booth, *Life and Labour of the People in London* (Final Volume 1903), p. 43.

33. T. H. C. Stevenson, 'The Fertility of Various Social Classes in England and Wales from the Middle of the Nineteenth Century to 1911', *Journal of the Royal Statistical Society* (1920), 83.

34. See P. H. J. Gosden, *The Friendly Societies of England 1815–1875* (1961). See also Charles Booth, *Life and Labour of the People in London* (1889), *1*, pp. 106–11. The Hearts of Oak charged a comparatively high subscription of 10 shillings a quarter; they provided £20 on a member's death, £10 on the death of a member's wife, sickness allowances beginning at 18 shillings a week, lying-in benefit of 30 shillings, and superannuation of 4 shillings a week.

35. See Sidney Webb, *The Decline of the Birth Rate* (1907), pp. 6–7.

36. G. Udny Yule, 'On the Changes in Marriage- and Birth-Rates in England and Wales during the Past Half Century', *Journal of the Royal Statistical Society* (1906), 69.

37. While the population increased by 11.7 per cent 1881–91 the working population aged 20–55 increased by 14 per cent; the increases 1891–1901 were 12·2 per cent and 19 per cent respectively.

38. 34·1 1851–60; 35·4 1871–80; cf. 29·9 1891–1900. See G. Talbot Griffith, op. cit., Table 5, p. 28.

39. S. Peller, 'Mortality, Past and Future', *Population Studies* (1948), 1.

40. 23 and 24 Vict. c. 151. See A. H. Robson, *The Education of Children in Industry in England 1833–1876* (1931), p. 159.

41. A. H. Robson, op. cit., pp. 204–5.

42. The Act of 1844 reduced the minimum age of employment from 9 to 8 and limited the daily hours of children to 6½ which could be worked either before or after the dinner hour.

43. A. H. Robson, op. cit., p. 133.

44. M. W. Thomas, *Young People in Industry* (1945), p. 85.

45. Charles Booth, 'Occupations of the People of the United Kingdom 1801–81', *Journal of the Statistical Society* (June 1886).

THE EMERGENCE OF AN ADOLESCENT SUBCULTURE IN INDUSTRIAL SOCIETY

1. An early statement of the general problem is that of Talcott Parsons, "Age and Sex in the Social Structure of the United States," reprinted in his *Essays in Sociological Theory, Pure and Applied* (Glencoe, Ill.: The Free Press, 1949).

2. These questions are in the fall questionnaire, which is included in the Appendix. In referring to questions, "I" or "II" preceding the question number will indicate the fall or spring questionnaire respectively. When both fall and spring responses are tabulated, the fall (I) question number will be used.

3. The results in Table 13 are based on responses from the students in the nine public schools. The nine-school totals will be used throughout the book rather than totals for all ten schools. This facilitates comparison between boys and girls, which would be confounded if the all-boys parochial school, St. Johns, were included.

4. Although the schools are not shown separately, each of the small-town schools is higher than any of the others in this percentage, with one exception: Elmtown, the most industrialized of the five small towns, scores lower than the upper-middle-class suburb, Executive Heights. It is interesting to note in passing that in the two working-class schools (Newlawn and Millburg) the son's choice shifts *away* from his father's occupation over the four years from freshman to senior; in Executive Heights, the son's choice shifts *toward* the father's occupation.

5. A teen-ager comments: As an adolescent, looking at our society from a distance, it seems to me to be merely an immature adult society. This immaturity is responsible for the "world of difference" between the culture of the teen-ager and the adult. Immaturity and lack of responsibility lower the goals and standards of an adolescent society. The adolescent borrows for his society the "glamorous and sophisticated" part of adult society. The high goals and worthwhile activities of the adult world are scorned because they involve responsibilities, which the adolescent is not ready to accept.

CHAPTER 12 WORK AND LEISURE

THE SHIP

1. From *A Sailor's Life* included in *The Call of the Sea* by Jan de Hartog. Copyright © 1955, 1956 by A. G. Littra. Copyright © 1955 by The Curtis Publishing Company. Reprinted by permission of Atheneum Publishers.
2. *Seafarer's Log,* March 4, 1966.
3. *A Sailor's Life,* by Jan de Hartog, p. 11.
4. *Mirror of the Sea,* by Joseph Conrad, p. 22. (First published 1906; permission granted by the Trustees of the Joseph Conrad Estate.) J. M. Dent & Sons Ltd., London.
5. *Ibid,* p. 56.
6. *A Sailor's Life,* by Jan de Hartog, p. 7–8.
7. *Mirror of the Sea,* by Joseph Conrad, p. 14.
8. *Ibid,* p. 15.
9. *A Sailor's Life,* by Jan de Hartog, p. 45.
10. *Collision Course,* by Alvin Moscow, p. 33–34.
11. *Mirror of the Sea,* by Joseph Conrad, p. 63.
12. *A Sailor's Life,* by Jan de Hartog, p. 141.
13. *The Monkeys Have No Tails in Zamboango,* by Wolfe Reese, Henry Regnery Co., Chicago, Ill. 1959, p. 48–49.
14. "Exciting Future," by G. Basil Hollas, *The Lookout,* Seamen's Church Institute of New York, Vol. 57, No. I, Jan. 1966, p. 5.
15. *The Monkeys Have No Tails in Zamboango,* by Wolfe Reese, p. 111–112.
16. *The Captain,* by Jan de Hartog, Atheneum Publishers, N.Y. 1966, p. 273.

FUN

1. See Martha Wolfenstein, "The Emergence of Fun Morality," *Journal of Social Issues,* Vol. 7, 1951, pp. 3–16; Nelson N. Foote, "Sex as Play," in Larrabee and Meyersohn, pp. 335–340.
2. Abraham Maslow, *Toward a Psychology of Being* (New York: Van Nostrand, 1962), pp. 97–108.
3. Irving Babbitt, *Rousseau and Romanticism* (Boston: Houghton Mifflin, 1919), pp. 31–34, 52–55, 63–64, 182–183, 216; Jacques Barzun, *Romanticism and the Modern Ego* (Boston: Little, Brown, 1968), pp. 111–113. The new right to be fulfilled seems to have bloomed in the feminist movement; Christopher Lasch notes the "cult of self-fulfillment" in the feminism of the early twentieth century. *The New Radicalism in America* (New York: Knopf, 1965), pp. 46–47. A recent version is found in promotional copy of *Cosmopolitan* magazine: "A girl can do almost anything she really wants to, don't you agree? . . . There's one magazine that seems to understand me—the girl who wants everything. I guess you could say I'm That Cosmopolitan Girl." *Time,* Feb. 9, 1968, p. 44.
4. The nature camps of the Club Méditerranée, Kathleen Halton, "Primitives of Cefalu," *The Times Magazine* (London), July 26, 1964, pp. 22 ff. William R. Burch explores rituals of "reindentification" in the American camping experience. "The Play World of Camping," *American Journal of Sociology,* Vol. LXX, March 1965, pp. 604–612.
5. Students of religion are beginning to explore the cultic aspects of sports—for example, Samuel Z. Klausner, "Religion and Emotional Valence: a Report on Sport Parachuting," delivered at meeting of Society for the Scientific Study of Religion, Berkeley, Calif., Dec. 27, 1965.
6. Sir John Hunt, interviewed by Kenneth Harris, *The Observer,* June 27, 1965.
7. Rod Pack interviewed by Donald Freeman, San Diego *Union,* March 5, 1965.
8. John Severson, Editor of *Surfer Magazine,* quoted in *Life,* Sept. 9, 1966, p. 37.
9. Interview with a San Diego, Calif., surfer, 1966. Other quotations from surfers illustrate the mystique: "I felt locked in life's bag."—"Just being out there with the

waves is enough. Your parents, your school and your girl could all be bugging you, but when you're surfing, the waves demand all your attention. After three or four hours out there you're so excited you can't talk. Your mind is blown out."—"You become part of the ocean. . . . I would lose track of time, find an inner space, and find myself in myself."—"It offers complete disassociation from society." *Newsweek,* Aug. 28, 1967, p. 41.

CHAPTER 13 POWER

"POWER ELITE" OR "VETO GROUPS"?

1. C. Wright Mills, *The Power Elite* (New York: Oxford University Press, 1956), p. 244.
2. David Riesman, *The Lonely Crowd* (New York: Doubleday Anchor Edition, 1953), pp. 257–258.
3. *Ibid.*, p. 239.
4. *Power Elite,* p. 270.
5. *Ibid.*, p. 270.
6. *Lonely Crowd,* p. 240.
7. *Ibid.*, p. 240.
8. *Power Elite,* p. 271.
9. *Ibid.*, p. 273.
10. C. Wright Mills, "The Power Elite," in Arthur Kornhauser (ed.), *Problems of Power in American Society* (Detroit: Wayne University Press, 1958), p. 161.
11. *Lonely Crowd,* pp. 246–247.
12. *Ibid.*, p. 256.
13. *Ibid.*, p. 247.
14. *Power Elite,* pp. 315–316.
15. *Lonely Crowd,* pp. 229–231.
16. *Power Elite,* p. 19.
17. *Lonely Crowd,* p. 247.
18. *Power Elite,* p. 19.
19. *Ibid.*, pp. 19–20.
20. C. Wright Mills, *White Collar* (New York: Oxford University Press, 1951).
21. C. Wright Mills, *The New Men of Power* (New York: Harcourt, Brace, 1948).
22. *Power Elite,* 302ff.
23. *Lonely Crowd,* p. 253.
24. *Ibid.*, p. 253.
25. *Ibid.*, p. 248.
26. *Lonely Crowd,* p. 249.
27. *White Collar,* p. 327.
28. David Riesman and Nathan Glazer, "Criteria for Political Apathy," in Alvin W. Gouldner (ed.), *Studies in Leadership* (New York: Harper, 1950).
29. *Power Elite,* 276ff.
30. *Lonely Crowd,* p. 257.
31. *White Collar,* pp. 342–350.
32. *Lonely Crowd,* p. 251.
33. "Criteria for Political Apathy," p. 520.
34. *Power Elite,* pp. 316–317.
35. *Lonely Crowd,* pp. 257–258.
36. *Power Elite,* p. 271.
37. *Lonely Crowd,* p. 257.
38. *Ibid.*, pp. 257, 248.
39. *Ibid.*, p. 255.
40. *Ibid.*, p. 260.

41. *Ibid.*, p. 275.

42. *Ibid.*, p. 247.

43. Riesman notes that "the concept of the veto groups is analogous to that of countervailing power developed in Galbraith's *American Capitalism,* although the latter is more sanguine in suggesting that excessive power tended to call forth its own limitation by opposing power . . ." (Riesman and Glazer, "The Lonely Crowd: A Reconsideration in 1960," in S. M. Lipset and Leo Lowenthal, editors, *Culture and Social Character,* New York: The Free Press of Glencoe, 1961, p. 449).

44. *Lonely Crowd,* p. 257.

45. Mills' narrow conception of power has been discussed by Talcott Parsons in his review of *The Power Elite* [Talcott Parsons, "The Distribution of Power in American Society," *World Politics,* X (October, 1957), 123–143]. Parsons notes that Mills uses a "zero-sum" notion, in that power is interpreted exclusively as a fixed quantity which is more or less unequally distributed among the various units in society. Power, however, also has a more general reference, to the political community as a whole, and to government as the agency of the total community. Viewed from this standpoint, power is a function of the integration of the community and serves general interests. Parsons argues, I think correctly, that Mills' sole concern for how power is used by some against others is associated with his tendency to exaggerate both the weight and the illegitimacy of power in the determination of social events.

CLASSES AND PARTIES IN AMERICAN POLITICS

1. See Herbert Hyman and Paul B. Sheatsley, "The Political Appeal of President Eisenhower," *Public Opinion Quarterly,* 17 (1953), pp. 443–60. They demonstrate this on the basis of poll results from 1947–48, which already indicated that Eisenhower could win the presidency under the banner of either party.

2. American Institute of Public Opinion news release, November 2, 1958. See Angus Campbell, Gerald Gurin, and Warren Miller, *The Voter Decides* (Evanston: Row, Peterson and Co., 1954), p. 211, for the 1952 electorate's perception of the support of each party.

3. Quote is cited by V. O. Key, Jr., in his *Politics, Parties, and Pressure Groups,* 4th ed. (New York: Crowell, 1958), p. 235.

4. Arthur Holcombe, *Our More Perfect Union* (Cambridge: Harvard University Press, 1950), p. 135. See also Samuel Lubell, *The Future of American Politics* (New York: Doubleday, Anchor Books, 1956), pp. 51–55, and Duncan MacRae, Jr., "Occupations and the Congressional Vote, 1940–1950," *American Sociological Review,* 20 (1955), pp. 332–40.

5. Dewey Anderson and Percy E. Davidson, *Ballots and the Democratic Class Struggle* (Stanford: Stanford University Press, 1943), pp. 118–47.

6. Leon D. Epstein, *Politics in Wisconsin* (Madison: University of Wisconsin Press, 1938), p. 186.

7. For a general study of the politics of the Jews, see Lawrence H. Fuchs, *The Political Behavior of American Jews* (Glencoe: The Free Press, 1956). See also Werner Cohn, "The Politics of the Jews," in Marshall Sklare, ed., *The Jews: Social Patterns of an American Group* (Glencoe: The Free Press, 1958), pp. 614–26. See Wesley and Beverly Allinsmith, "Religious Affiliation and Political-Economic Attitudes," *Public Opinion Quarterly,* 12 (1948), pp. 377–89; Paul F. Lazarsfeld, Bernard Berelson, and Hazel Gaudet, *The People's Choice* (New York: Columbia University Press, 1948), p. 22; W. F. Ogburn and N. S. Talbot, "A Measurement of the Factors in the Presidential Election of 1928," *Social Forces,* 8 (1929), pp. 175–83; H. F. Gosnell, *Grass Roots Politics* (Washington: American Council on Public Affairs, 1942), pp. 17, 33–34, 55, 102; S. J. Korchin, *Psychological Factors in the Behavior of Voters* (unpublished Ph.D. thesis, Department of Social Relations, Harvard University, 1946), Chap. V; Louis Harris, *Is There a Republican Majority?* (New York: Harper & Bros., 1954), p. 87; A. Campbell, G. Gurin, and W. Miller, *op. cit.,* pp. 71, 79; Bernard Berelson, Paul Lazarsfeld, and William McPhee, *Voting* (Chicago: University of Chicago Press, 1954), pp. 64–71; and

Oscar Glantz, "Protestant and Catholic Voting Behavior," *Public Opinion Quarterly*, 23 (1959), pp. 73–82.

8. Ellwood P. Cubberley, *Public Education in the United States* (Boston: Houghton Mifflin Co., 1954), p. 164.

9. *Loc. cit.*

10. *Ibid.*, p. 166.

11. *Ibid.*, pp. 164–65.

12. Manning Dauer, *The Adams Federalists* (Baltimore: The Johns Hopkins Press, 1953), pp. 24–27, 263.

13. Research is in process by Lee Benson of the Center for Advanced Study in the Behavioral Sciences.

14. Mabel Newcomer, *The Big Business Executive* (New York: Columbia University Press, 1955), p. 49. In 1928 a survey of those listed in *Who's Who* found that 87 per cent favored Herbert Hoover for President. See Jean-Louis Sevrin, *La structure interne des partis politiques Americains* (Paris: Librairie Armand Colin, 1953), p. 58.

15. See Wilfred E. Binkley, *American Political Parties, Their Natural History* (New York: Alfred A. Knopf, 1947), p. 163.

16. See Dixon Ryan Fox, "The Negro Vote in Old New York," *Political Science Quarterly*, 32 (1917), pp. 252–75; and Marvin Meyers, *The Jacksonian Persuasion: Politics and Beliefs* (Stanford: Stanford University Press, 1957), pp. 189–90.

17. See W. F. Ogburn and N. S. Talbot, *op. cit.*, p. 179.

18. See Rowland T. Berthoff, *British Immigrants in Industrial America, 1790–1950* (Cambridge: Harvard University Press, 1953), pp. 198–205.

19. In Wisconsin, the voters of German and Irish background voted Democratic from 1860 to World War I. See Leon Epstein, *op. cit.*, p. 36; similarly in Missouri, one survey reports that the Democrats dominated in counties with large German populations from the Civil War to the election of 1920, when these counties shifted overwhelmingly to the Republicans. See John H. Fenton, *Politics in the Border States* (New Orleans: The Hauser Press, 1957), pp. 162–63. See also Samuel Lubell, *op. cit.*, Chap. VII, for an attempt to analyze the sources of isolationism in terms of ethnic background and allegiances.

20. The Spanish Republicans were perhaps the foremost victims of the fact that American foreign policy reflects the social base of the political parties. Although Roosevelt and many of his closest advisers were personally strong supporters of the Loyalists, they felt that it would be politically impossible to antagonize the Catholic Democratic vote by aiding the Loyalists, who were regarded as Communists by the Church.

21. For an attempt to dissect the sources of support of the "radical right" in America, recently centering around McCarthyism, in terms of the politics of status-aspirations in times of prosperity, see S. M. Lipset, "The Sources of the 'Radical Right,'" in Daniel Bell, ed., *The New American Right* (New York: Criterion Books, 1955), pp. 166–235. The social base of this political tendency by no means looks solely to the Republican party as its standard-bearer, and, as the article points out, the attacks by the radical right on "modern Republicanism," combined with the Republican victory in 1952, were important elements leading to the downfall of its McCarthyite expression.

22. For data on regional and ethnic variations in 1952, from a national cross-sectional poll, see A. Campbell, G. Gurin, and W. Miller, *op. cit.*, pp. 69–83. In a 1952 Roper survey it was found that 85 percent of the upper-income Irish vote went to Eisenhower, and there was a strong trend at all income levels of the German vote toward Eisenhower. See Louis Harris, *Is There a Republican Majority?* (New York: Harper & Bros., 1954), pp. 87–94.

23. R. Hofstadter, *op. cit.*, p. 163.

24. E. D. Baltzell, *op. cit.*, p. 39.

25. An excellent discussion of sectionalism in American politics can be found in V. O. Key, Jr., *op. cit.*, pp. 250–79.

26. See Samuel Lubell, *op. cit.*, Chap. VI, for a discussion of the conflicts and changes in the South. Key points out that southern unity is based solely on the Negro issue, and that normal class politics, based on the old Populist supporters (the poor white farmers) against the plantation regions, emerges wherever the Negro issue does not supersede other issues. V. O. Key, Jr., *Southern Politics, op. cit.*, p. 302.

27. See L. Harris, *op. cit.*, pp. 68–73 and 134–36.

28. *Ibid.*, pp. 148–49. Sixty-two percent of the Catholic union members voted Democratic, while 62 percent of those Catholic families with no union member in the family voted Republican.

29. O. Glantz, "Unitary Political Behavior and Differential Political Motivation," *Western Political Quarterly*, 10 (1957), pp. 833–46.

30. Henry R. Mueller, *The Whig Party in Pennsylvania* (New York: privately printed, 1922), p. 143.

CHAPTER 14 DEVIANCE

PROBLEMS IN THE SOCIOLOGY OF DEVIANCE: SOCIAL DEFINITIONS AND BEHAVIOR

1. George Vold, *Theoretical Criminology* (New York: Oxford University Press, 1958), p. vi.

2. Clarence R. Jeffrey, "The Structure of American Criminological Thinking," *Journal of Criminal Law, Criminology, and Police Science*, 46 (January–February 1956), pp. 670–672; "Crime, Law and Social Structure," *Journal of Criminal Law, Criminology, and Police Science*, 47 (November–December 1956), pp. 423–425; and "An Integrated Theory of Crime and Criminal Behavior," *Journal of Criminal Law, Criminology, and Police Science*, 49 (March–April 1959), pp. 441–552. See also Donald J. Newman, "Legal Norms and Criminological Definitions," in Joseph S. Roucek (ed.), *Sociology of Crime* (New York: Philosophical Library, 1961), pp. 56–60; and "Sociology, Criminology, and Criminal Law," *Social Problems*, 7 (Summer 1959), pp. 43–45; and Richard Quinney, "Crime in Political Perspective," *American Behavioral Scientist*, 8 (December 1964), p. 19; and "Is Criminal Behaviour Deviant Behaviour?" *British Journal of Criminology* (April 1965), pp. 137–139.

3. Austin Turk, "Prospects for Theories of Criminal Behavior," *Journal of Criminal Law, Criminology, and Police Science*, 55 (December 1964), pp. 454–455; and "Conflict and Criminality," *American Sociological Review*, 31 (June 1966), pp. 338–352.

4. Vold, *op. cit.*, pp. 203–241; Thorsten Sellin, *Culture Conflict and Crime* (New York: Social Science Research Council, 1938); and Richard C. Fuller, "Morals and the Criminal Law," *Journal of Criminal Law and Criminology*, 32 (1942), pp. 624–630.

5. Vilhelm Aubert, "White Collar Crime and Social Structure," *American Journal of Sociology*, 58 (1952), p. 264.

6. Donald J. Newman, "White Collar Crime," *Law and Contemporary Problems*, 23 (1958), p. 746. See also Frank Hartung, "White Collar Crime: Its Significance for Theory and Practice," *Federal Probation*, 17 (1953), p. 31; Donald Cressy, "Foreword" to Edwin H. Sutherland, *White Collar Crime* (New York: Holt, Rinehart & Winston, 1961), p. xiii; and Earl R. Quinney, "The Study of White Collar Crime: Toward a Reorientation in Theory and Research," *Journal of Criminal Law, Criminology, and Police Science*, 55 (June 1964), p. 214.

7. Jack P. Gibbs, "Conceptions of Deviant Behavior: The Old and the New," *Pacific Sociological Review*, 9 (Spring 1966), p. 14.

8. Donald R. Cressey, "Epidemiology and Individual Conduct: A Case From Criminology," *Pacific Sociological Review*, 3 (Fall, 1960), pp. 47–58.

9. Albert K. Cohen, "The Study of Social Disorganization and Deviant Behavior," in Robert K. Merton, *et al.* (eds.), *Sociology Today* (New York: Basic Books, 1959), p. 461; *Deviance and Control* (Englewood Cliffs, New Jersey: Prentice-Hall, 1966), pp. 41–47; and "Review of Hermann Mannheim, Comparative Criminology," *Social Forces*, 45 (December 1966), pp. 298–299.

10. Homans seems to think that the argument is meaningful although he offers a different answer from the "levels" solution. See George C. Homans, "Bringing Men Back In," *American Sociological Review*, 29 (December 1964), pp. 808–818. He resolves the

issue by arguing that there are no general "sociological" propositions which cannot be derived from "psychological" principles. Thus, he feels that the two are not "levels" of independent analytic status. This is somewhat unenlightening, however, because by "psychological" Homans means simply propositions about men's observable behavior, and this is not what is usually meant by "psychological." The fact that Homans sees these propositions coming from the work of psychologists such as Skinner does not *ipso facto* make them psychological. Whether a theoretical framework be labelled sociological or psychological, then, is a definitional problem. I would agree with this concept: If the explanatory variables and processes in propositions about behavior are those contained in or arising out of social interaction (as contrasted, for instance, with intrapsychic, individual constitutional, personality, or unconscious variables) then one is justified in naming them sociological. If the independent variables are social-environmental, then we have a sociological theory, whether the dependent variables be individual or collective behavior. It is in this sense that both structural and processual theories can be sociological.

11. Donald Cressey, *Other People's Money* (Glencoe, Illinois: The Free Press, 1953); Alfred R. Lindesmith, *Opiate Addiction* (Bloomington, Indiana: Principia Press, 1947); Howard S. Becker, *Outsiders* (Glencoe, Illinois: The Free Press, 1963), Chaps. 3 and 4; and Edwin M. Lemert, "An Isolation and Closure Theory of Naive Check Forgery," *Journal of Criminal Law, Criminology, and Police Science,* 44 (September–October 1953), pp. 296–307. Quinney has presented a miniature "structural" theory of a form of white-collar crime. Earl R. Quinney, "Occupational Structure and Criminal Behavior: Prescription Violations by Retail Pharmacists," *Social Problems,* 11 (Fall 1963), pp. 179–185.

12. Edwin H. Sutherland and Donald R. Cressey, *Principles of Criminology* (6th ed.; Philadelphia: J. B. Lippincott, Co., 1960), pp. 77–79.

13. *Ibid.,* p. 80.

14. Donald R. Cressey, "The Theory of Differential Association: An Introduction," *Social Problems,* 8 (Summer 1960), p. 5.

15. Most other sociological perspectives are primarily structural in emphasis. The ones that have exerted the most influence are disorganization-anomie and conflict theory. Although there are now a number of variations on the disorganization-anomie theme, they all derive ultimately from Durkheim. The following are among the more careful and systematic statements on variants of this approach: Robert K. Merton, *Social Theory and Social Structure* (Glencoe, Illinois: The Free Press, 1957), pp. 131–194; Albert K. Cohen, *Delinquent Boys* (Glencoe, Illinois: The Free Press, 1955); Richard Cloward and Lloyd Ohlin, "Illegitimate Means, Anomie, and Deviant Behavior," *American Sociological Review,* 24 (April 1959), pp. 164–177, and *Delinquency and Opportunity* (Glencoe, Illinois: The Free Press, 1961); Arnold Rose, *Theory and Method in the Social Sciences* (Minneapolis: University of Minnesota Press, 1954), chap. 1; Cohen, "Study of Social Disorganization and Deviant Behavior"; Albert Cohen, "The Sociology of the Deviant Act: Anomie Theory and Beyond," *American Sociological Review,* 30 (February 1965), pp. 9–14; Marshall B. Clinard (ed.), *Anomie and Deviant Behavior* (New York: The Free Press of Glencoe, 1964); Clifford Shaw and Henry D. McKay, *Juvenile Delinquency and Urban Areas* (Chicago: University of Chicago Press, 1942); and Bernard Lander, *Towards an Understanding of Juvenile Delinquency* (New York: Columbia University Press, 1954). For the most thorough analysis, critique, and modification of that variant of anomie theory which emphasizes delinquent subcultures, see David Downes, *The Delinquent Solution* (New York: The Free Press, 1966). Conflict theories are found in Vold, *op. cit.;* Sellin, *op. cit.;* and Solomon Kobrin, "The Conflict of Values in Delinquency Areas," *American Sociological Review,* 16 (October 1951), pp. 653–661.

16. Other theories which attempt to answer the essentially processual questions of why particular individuals commit deviancy, although the "process" is not explicit in each case, are self-concept and role-commitment theories. The self-concept theory of Reckless and his associates and students at Ohio State is essentially a social control theory containing both structural and individual components, but the burden of this perspective is borne by the conceptualization of one's self-image as an individual selec-

tive mechanism which accounts for differential response to environment. Walter Reckless, "A New Theory of Delinquency and Crime," *Federal Probation,* 25 (December 1961), pp. 42–46; "The Self Component in Potential Delinquency and Potential Non-Delinquency," *American Sociological Review,* 22 (October 1957), pp. 566–570; "Self Concept as an Insulator Against Delinquency," *American Sociological Review,* 21 (December 1956), pp. 744–746; and "The Good Boy in a High Delinquency Area," *Journal of Criminal Law, Criminology, and Police Science,* 48 (August 1960), pp. 18–26; Frank R. Scarpitti, *et al.,* "Good Boy in a High Delinquency Area: Four Years Later," *American Sociological Review,* 25 (August 1960), pp. 555–558. Role commitment theories can be found in Howard S. Becker, "Notes on the Concept of Commitment," *American Journal of Sociology,* 66 (July 1960), pp. 32–40; David Matza, *Delinquency and Drift* (New York: John Wiley & Sons, 1964); Richard Korn and Lloyd W. McKorkle, *Criminology and Penology* (New York: Henry Holt Co., 1959), pp. 327–353; and Scott Briar and Irving Piliavin, "Delinquency, Situational Inducements, and Commitment to Conformity," *Social Problems,* 13 (Summer 1965), pp. 35–45. Of course, the impact of labelling conceptions also can be seen to be perspectives on role commitment. One should also view Short and Strodtbeck's conceptions of gang delinquency as explanations of individual actions. James F. Short, Jr., and Fred L. Strodtbeck, *Group Process and Gang Delinquency* (Chicago: University of Chicago Press, 1965). All of these are consistent with a general learning perspective. I omit multicausation, psychiatric, personal pathology, and other personality theories from discussion.

17. Robert L. Burgess and Ronald L. Akers, "A Differential Association-Reinforcement Theory of Criminal Behavior," *Social Problems,* 14 (Fall 1966), pp. 128–147.

18. Of a 32-item bibliography of studies using unofficial measures of delinquency I have collected, the following are cited: Ronald L. Akers, "Socioeconomic Status and Delinquent Behavior: A Re-Test," *Journal of Research in Crime and Delinquency,* 1 (January 1964), pp. 38–46; John P. Clark and Eugene P. Wenninger, "Goal Orientations and Illegal Behavior Among Juveniles," *Social Forces,* 42 (October 1963), pp. 49–60; Robert A. Dentler and Lawrence J. Monroe, "Social Correlates of Early Adolescent Theft," *American Sociological Review,* 26 (October 1961), pp. 733–743; Maynard L. Erickson and LaMar T. Empey, "Court Records, Undetected Delinquency, and Decision-Making," *Journal of Criminal Law, Criminology, and Police Science,* 54 (December 1963), pp. 456–469; Martin Gold, "Undetected Delinquent Behavior," *Journal of Research in Crime and Delinquency,* 3 (January 1966), pp. 27–46; David E. Hunt and Robert H. Hardt, "Developmental Stage, Delinquency, and Differential Treatment," *Journal of Research in Crime and Delinquency,* 2 (January 1965), pp. 20–31; Jay Lowe, "Prediction of Delinquency with an Attitudinal Configuration Model," *Social Forces,* 45 (September 1966), pp. 106–113; F. Ivan Nye and James Short, "Scaling Delinquent Behavior," *American Sociological Review,* 22 (June 1957), pp. 326–331; Austin Porterfield, "Delinquency and Its Outcome in Court and College," *American Journal of Sociology,* 49 (September 1943), pp. 199–204; John Finley Scott, "Two Dimensions of Delinquent Behavior," *American Sociological Review,* 24 (April 1959), pp. 240–243; James F. Short, Jr., "Differential Association with Delinquent Friends and Delinquent Behavior," *Pacific Sociological Review,* 1 (Spring 1958), pp. 20–25; James F. Short, Jr., and F. Ivan Nye, "Extent of Unrecorded Delinquency: Tentative Conclusions," *Journal of Criminal Law, Criminology, and Police Science,* 49 (November–December 1958), pp. 296–302; James F. Short, Jr., Ray A. Tennyson, and Kenneth I. Howard, "Behavior Dimensions of Gang Delinquency," *American Sociological Review,* 28 (June 1963), pp. 411–428; and Harwin L. Voss, "Socioeconomic Status and Reported Delinquent Behavior," *Social Problems,* 13 (Winter 1966), pp. 314–324.

19. Sutherland, *op. cit.;* Marshall B. Clinard, *The Black Market* (New York: Rinehart Co., 1952); and Frank Hartung, "White Collar Offenses in the Wholesale Meat Market Industry in Detroit," *American Journal of Sociology,* 56 (November 1950), pp. 325–342.

20. Mary O. Cameron, *The Booster and the Snitch: Department Store* (Glencoe, Illinois: The Free Press, 1964), makes use of department store detective files, James E. Price utilizes insurance statistics in "A Test of the Accuracy of Crime Statistics," *Social Problems,* 14 (Fall 1966), pp. 214–221. An imaginative combination of interview data

and data from the records of drug law enforcement agencies is found in Earl Quinney, "Occupational Structure and Criminal Behavior." In addition, see the rich, but unsystematically reported, cases taken from the files of his management consultant firm by Norman Jaspan, *The Thief in the White Collar* (Philadelphia: J. B. Lippincott Co., 1960).

21. Robert L. Burgess and Ronald. L. Akers, "Prospects for an Experimental Analysis in Criminology," paper read at the annual meeting of the American Sociological Association, August 1966.

22. Richard Quinney, "Crime in Political Perspective," pp. 19–21; Vold; *op. cit.*, pp. 207–209; Sellin, *op. cit.;* Clinard, *The Black Market,* p. 153; and F. James Davis, *et al., Society and the Law* (New York: The Free Press of Glencoe, 1962), pp. 69–71.

23. Of the vast literature that could be cited on this point, these are among the best: David Truman, *The Governmental Process* (8th ed.; New York: Alfred A. Knopf, 1962); Earl Latham, *The Group Basis of Politics* (New York: Octagon Books, 1965); Harmon Zeigler, *Interest Groups in American Society* (Englewood Cliffs, New Jersey: Prentice-Hall, 1964); V. O. Key, *Politics, Parties, and Pressure Groups* (New York: Thomas Y. Crowell, 1958), and *Public Opinion and American Democracy* (New York: Alfred A. Knopf, 1961), pp. 500–531; Henry W. Ehrmann (ed.), *Interest Groups on Four Continents* (Pittsburgh: University of Pittsburgh Press, 1958); Donald C. Blaisdell (ed.), *Unofficial Government: Pressure Groups and Lobbies,* Special Issue of *Annals of the American Academy of Political and Social Science,* 319 (September 1958); and Lester W. Milbrath, *The Washington Lobbyists* (Chicago: Rand McNally & Co., 1963).

24. Turk, "Prospects for . . . Criminal Behavior"; Richard Quinney, "Crime in Political Perspective," p. 19; Jeffery, "Structure of American Criminological Thinking," pp. 663–667; and Jeffery, "An Integral Theory of Crime and Criminal Behavior," p. 534.

25. Jerome Hall, *Theft Law and Society* (Indianapolis: The Bobbs-Merrill Co., 1952).

26. William Chambliss, "A Sociological Analysis of the Law of Vagrancy," *Social Problems,* 12 (Summer 1964), pp. 67–77.

27. Becker, *Outsiders,* pp. 135–146; Alfred R. Lindesmith, "Federal Law and Drug Addiction," *Social Problems,* 7 (Summer 1959), pp. 48–57.

28. Peter Odegard, *Pressure Politics: The Story of the Anti-Saloon League* (New York: Columbia University Press, 1928).

29. Ronald L. Akers, "Professional Organization, Political Power, and Occupational Laws," unpublished Ph.D. dissertation, University of Kentucky, 1966.

30. Jerome Skolnick, *Justice Without Trial* (New York: John Wiley & Sons, 1966.

31. Truman, *op. cit.,* p. 13; Marian D. Irish and James W. Prothro, *The Politics of American Democracy* (Englewood Cliffs, New Jersey: Prentice-Hall, 1959), p. 336.

32. Akers, "Professional Organization, Political Power and Occupational Laws."

33. William M. Evan, "Law as an Instrument of Social Change," *Estudio de Sociologia,* 2 (1962), pp. 167–175.

34. John I. Kitsuse, "Societal Reaction to Deviant Behavior: Problems of Theory and Method," *Social Problems,* 9 (Winter 1962), pp. 247–256; Elizabeth A. Rooney and Don C. Gibbons, "Social Reactions to 'Crimes without Victims,'" *Social Problems,* 13 (Spring 1966), pp. 400–410; J. L. Simmons, "Public Stereotypes of Deviants," *Social Problems,* 13 (Fall 1965), pp. 223–232; Donald J. Newman, "Public Attitudes Toward a Form of White Collar Crime," *Social Problems,* 4 (1957), pp. 228–232; Edwin M. Schur, "Attitudes Towards Addicts: Some General Observations and Comparative Findings," *American Journal of Orthopsychiatry,* 34 (January 1964), pp. 80–90; and Arnold Rose and Arthur E. Prell, "Does the Punishment Fit the Crime? A Study of Social Valuation," *American Journal of Sociology,* 24 (September 1955), pp. 247–259. For a historical treatment see Harris Isbell, "Historical Development of Attitudes Toward Opiate Addiction in the U.S.," in Seymour Farber and Roger Wilson (eds.), *Conflict and Creativity* (New York: McGraw-Hill Book Co., 1963), pp. 154–170.

35. Harry V. Ball, "Social Structure and Rent Control Violations," *American Journal of Sociology,* 65 (May 1960), pp. 598–604; Richard Lempert, "Strategies of Research Design in the Legal Impact Study," *Law and Society Review,* 1 (November

1966), pp. 111–132; and N. Walker and M. Argyle, "Does the Law Affect Moral Judgment?" *British Journal of Criminology*, 4 (October 1964), pp. 570–581.

36. Earl Quinney, "The Study of White Collar Crime," pp. 210–212.

37. See Paul Secord and Carl Backman, *Social Psychology* (New York: McGraw-Hill Book Co., 1964), pp. 323–351.

38. Becker, *Outsiders*, pp. 1–18; Howard S. Becker (ed.), *The Other Side* (New York: The Free Press of Glencoe, 1964), p. 3; Kai T. Erikson, "Notes on the Sociology of Deviance," in Becker, *The Other Side*, p. 12; Kitsuse, *op. cit.*, p. 247.

39. Becker, *Outsiders*, pp. 15–18. One example of what more is said about social definitions is Eliot Friedson, "Disability as Social Deviance," in Marvin R. Sussman (ed.), *Sociology and Rehabilitation* (Washington, D.C.: American Sociological Association, 1966), pp. 71–99. Friedson addresses the difficult problems of illness, disability, and handicaps as bases for labelling as a deviant. He offers a framework that could be applied not only to mental illness which has long been of interest to students of deviance but also to physical illness which has not received much attention. No special mention of the problem of illness as deviance has been made here, but it should be evident that this is another area awaiting further investigation. One can be socially defined as sick but not all sick roles are stigmatized. Friedson suggests that the curability of the illness and the extent to which the ill person is believed responsible for his disability or illness need to be considered in connection with stigmatized and nonstigmatized sick roles. Thus, he opens the way for research into the relationships among these variables and the question of why some illnesses are stigmatized and others are not. Another area of research which should prove profitable is deviation from the expectations of the sick role itself. Certainly, being sick or handicapped may prevent one from fulfilling his other role expectations, but it is also an acceptable excuse for doing so and may forestall sanctions for deviations from other roles. The question is what kinds of deviations from the norms of sickness, or under what conditions such deviations, are stigmatized or negatively sanctioned whether or not the person is labelled as a deviant.

40. Turk, "Conflict and Criminality."

41. Becker, *Outsiders*, pp. 31–39; Edwin M. Lemert, *Social Pathology* (New York: McGraw-Hill Book Co., 1951); Edwin M. Schur, *Narcotic Addiction in Britain and America: The Impact of Public Policy* (Bloomington: Indiana University Press, 1962); Harold Garfinkel, "Successful Degradation Ceremonies," *American Journal of Sociology*, 61 (January 1956), pp. 420–424; Richard D. Schwartz and Jerome H. Skolnick, "Two Studies of Legal Stigma," in Becker, *The Other Side*, pp. 103–117; Marsh Ray, "The Cycle of Abstinence and Relapse," in Becker, *ibid.*, pp. 163–179; Erikson *op. cit.*, pp. 16–17; Alfred R. Lindesmith and John Gagnon, "Anomie and Drug Addiction," in Clinard, *Anomie and Deviant Behavior*, pp. 158–188; and Edwin M. Schur, *Crimes Without Victims* (Englewood Cliffs, New Jersey: Prentice-Hall, 1965), pp. 1–7. For discussion of the impact of stigmatized physical characteristics on behavior of the stigmatized see Fred Davis, "Deviance Disavowal: The Management of Strained Interaction by the Visibly Handicapped," in Becker, *The Other Side*, pp. 119–138; and Erving Goffman, *Stigma* (Englewood Cliffs, New Jersey: Prentice-Hall, 1963).

42. Schur, *Crimes Without Victims*.

43. Lemert, of course, does recognize this point and consistently maintains the distinction between primary and secondary deviance. Edwin Lemert, *Human Deviance, Social Problems and Social Control* (Englewood Cliffs, New Jersey: Prentice-Hall, 1967), pp. 40–64.

44. Jeffery, "An Integrated Theory of Crime and Criminal Behavior."

LOWER CLASS CULTURE AS A GENERATING MILIEU OF GANG DELINQUENCY

1. The complex issues involved in deriving a definition of "delinquency" cannot be discussed here. The term "delinquent" is used in this paper to characterize behavior or acts committed by individuals within specified age limits which if known to official authorities could result in legal action. The concept of a "delinquent" individual has little or no utility in the approach used here; rather, specified types of *acts* which may be

committed rarely or frequently by few or many individuals are characterized as "delinquent."

2. A three-year research project is being financed under National Institutes of Health Grant M-1414 and administered through the Boston University School of Social Work. The primary research effort has subjected all collected material to a uniform data-coding process. All information bearing on some seventy areas of behavior (behavior in reference to school, police, theft, assault, sex, collective athletics, etc.) is extracted from the records, recorded on coded data cards, and filed under relevant categories. Analysis of these data aims to ascertain the actual nature of customary behavior in these areas and the extent to which the social work effort was able to effect behavioral changes.

3. Between 40 and 60 per cent of all Americans are directly influenced by lower class culture, with about 15 per cent, or twenty-five million, comprising the "hard core" lower class group—defined primarily by its use of the "female-based" household as the basic form of child-rearing unit and of the "serial monogamy" mating pattern as the primary form of marriage. The term "lower class culture" as used here refers most specifically to the way of life of the "hard core" group; systematic research in this area would probably reveal at least four to six major subtypes of lower class culture, for some of which the "concerns" presented here would be differently weighted, especially for those subtypes in which "law-abiding" behavior has a high overt valuation. It is impossible within the compass of this short paper to make the finer intracultural distinctions which a more accurate presentation would require.

4. The "brains-brawn" set of capacities are often paired in lower class folk lore or accounts of lower class life, e.g., "Brer Fox" and "Brer Bear" in the Uncle Remus stories, or George and Lennie in "Of Mice and Men."

5. Further data on the female-based household unit (estimated as comprising about 15 per cent of all American "families") and the role of one-sex groupings in lower class culture are contained in Walter B. Miller, "Implications of Urban Lower Class Culture for Social Work." *Social Service Review,* 1959, 33, No. 3.

CHAPTER 15 CHANGE

MIDDLETOWN FACES BOTH WAYS

1. See *Middletown,* p. 497.

2. Middletown receives ample encouragement in this congenial resolution of its problems. It read, for instance, on page one, column one, of its morning paper the following eloquent sermon by Arthur Brisbane on the prize fighter, Gene Tunney: "'Seest thou a man diligent in his business, he shall stand before kings.' Tunney was diligent in HIS BUSINESS, learned to know it thoroughly and now stands before kings, at least money kings. It is very important to know one thing thoroughly."

3. As pointed out in Chs. II and XII, Middletown's working class appears today to be less sure of many of the old values than is the business class; but in Middletown they have developed no ideology of their own, and they lack security on any basis of their own, such as labor organization. Hence, doubtful and uncertain, they tend to straggle after the wealthier, pace-setting fellow citizens in their affirmations of established values in the midst of confusion.

4. See the discussion of the handling of the relief problem in Ch. IV.

In keeping with this tendency to state its problems defensively and negatively, Middletown tends to avoid facing the implications of differences between its practices and those of other communities by recourse to the easy extenuations that "Our situation is different," or that a given problem "is just one of the peculiar problems that *our* community has always had to cope with." Such reasoning allows local practice to continue its course along the smooth grooves of past custom.

THE CAUSES FOR THE PROGRESS OF THE DIVISION OF LABOR

1. *Principles of Sociology*, II, p. 31.
2. *"Calunt diversi ac discreti,"* said Tacitus of the Germans; *"suum quisque domum spatio . . ." Germania xvi.*
3. See in Accarias, *Précis,* I, p. 640, the list of urban servitudes. Cf. Fustel de Coulanges, *La cité antique,* p. 65.
4. In reasoning thus, we do not mean to say that the development of density results from economic changes. The two facts mutually condition each other, and the presence of one proves the other's.
5. See Levasseur, *La Population française,* passim.
6. Tacitus, *Germania, xvi.*—Sohm, *Ueber die Entstehung der Städte.*
7. *Römische Alterthümer,* IV, 3.
8. See Dumont, *Dépopulation et Civilisation,* Paris, 1890, ch. viii, on this point, and Oettingen, *Moralstatistik,* pp. 273 ff.
9. Levasseur, *op. cit.*, p. 200.
10. We believe this is the opinion of Tarde in his *Lois de l'imitation.*
11. There are particular, exceptional cases, however, where material and moral density are perhaps not entirely in accord. See final note of this chapter.
12. *Principles of Sociology,* II, 23.
13. The village, which is originally only a fixed clan.
14. On this point, we can still rely on Comte as authority. "I must," he said, "now indicate the progressive condensation of our species as a last general concurrent element in regulating the effective speed of the social movement. We can first easily recognize that this influence contributed a great deal, especially in origin, in determining a more special division of human labor . . .

CHAPTER 16 OLD AGE, ILLNESS, AND DEATH

HUMAN OBSOLESCENCE

1. My discussion of Rosemont is based on 35 observation periods on each of two wards. Observations were made 'round the clock for one hour each day between the hours of 9 A.M. and 7 P.M; and thereafter for half an hour.

THE TREATMENT OF TUBERCULOSIS AS A BARGAINING PROCESS

1. The data on tuberculosis treatment on which this paper is based are drawn from systematic field notes kept while I was a patient (about one year in two hospitals), an attendant (three months in one hospital), and a sociological observer (about one and one-half years in two hospitals) in tuberculosis hospitals. In the last role I spent much of my time "hanging around" the hospital wards and offices, attending therapy conferences, making rounds of physicians, and listening to staff doctors discuss their "problems" with one another, and with their residents and nurses, conversations with staff persons at meals, in the halls, in their offices, and so on. I made a particular effort to follow up complete "incidents" over a period of time.

Some of my ideas about professional-client relationships derive from study with Everett C. Hughes at the University of Chicago. . . . The work of Erving Goffman . . . stimulated my thinking about the subtleties of social interaction as a bargaining process.

2. One of the advantages of the bargaining concept is the fact that it provides a bridge between the Cooley and Mead theory of human interaction and some of the ways in which present-day game theory is being applied to describe and analyze human behavior.

3. Specific therapeutic medicines, such as streptomycin or penicillin, are given in specific amounts at regular intervals as ordered by a physician. The kind of medicines I am discussing here are those intended to relieve distress—pain, cough, constipation, sleeplessness, difficult breathing, itching—and which are ordered to be given "when necessary."

4. Much the same is true in the case of the attendant in mental hospitals. . . .

5. Excision of lung tissue.

6. Removal of ribs to partially collapse pleural cavity.

7. In a five-class activity classification ranging from complete bed rest in Class 1 to maximum activity in Class 5.

8. Permission to go out on the hospital grounds during certain hours.

DEATH AND THE SOCIAL STRUCTURE

1. Many early anthropologists, including Malinowski, attributed human funerary customs to an alleged instinctive aversion to the corpse. Although there is no evidence for such an instinct, aversion to the corpse remains a widespread, if not universal, human reaction. See the extended discussion of the early theories in Goody, *Death, Property, Ancestors,* pp. 20–30; and for some exceptions to the general rule, Robert W. Habenstein, "The Social Organization of Death," *International Encyclopedia of the Social Sciences* [New York: Macmillan and Free Press, 1968), vol. 4, pp. 26–28].

2. Robert J. Lifton, "Psychological Effects of the Atomic Bomb in Hiroshima: The Theme of Death," *Daedalus,* XCII (1963), 462–97. Among other things, the dead body is too stark a reminder of man's mortal condition. Although man is the one species that knows he will eventually die, most people in most societies cannot live too successfully when constantly reminded of this truth. On the other hand, the exposure to the corpse has positive consequences for psychic functioning, as it contributes to the acceptance of the reality of a death on the part of the survivors. A study of deaths in military action during World War II found that the bereaved kin had particularly great difficulty in believing in and accepting the reality of their loss because they did not see the body and witness its disposal. T. D. Eliot, "Of the Shadow of Death." *Annals of the American Academy of Political and Social Science,* CCXXIX (1943), 87–99.

3. Statistics on the settings of death are not readily available. Robert Fulton reports that 53 per cent of all deaths in the United States take place in hospitals, but he does not give any source for this figure. See Fulton, *Death and Identity* (New York: John Wiley & Sons, Inc., 1965), pp. 81–82. Two recent English studies are also suggestive. In the case of the deaths of 72 working-class husbands, primarily in the middle years, 46 died in the hospital; 22 at home; and 4 at work or in the street. See Peter Marris, *Widows and Their Families* (London: Routledge & Kegan Paul, Ltd., 1958), p. 146. Of 359 Britishers who had experienced a recent bereavement, 50 per cent report that the death took place in a hospital; 44 per cent at home; and 6 per cent elsewhere. See Geoffrey Gorer, *Death, Grief, and Mourning* (London: The Cresset Press, Ltd., 1965), p. 149.

4. Max Weber, *Essays in Sociology,* trans. and ed., H. H. Gerth and C. Wright Mills (New York: Oxford University Press, 1953), pp. 196–98. See also, ———, *General Economic History,* trans. Frank H. Knight (New York: The Free Press, 1950).

5. Leroy Bowman reports that aversion to the corpse made this preparation an unpleasant task. Although sometimes farmed out to experienced relatives or neighbors, the task was still considered the family's responsibility. See Bowman, *The American Funeral: A Study in Guilt, Extravagance and Sublimity* (Washington, D.C.: Public Affairs Press, 1959), p. 71.

6. William K. Kephart, "Status After Death," *The American Sociological Review,* XV (1950), 635–43.

7. David N. Sudnow, "Passing On: The Social Organization of Dying in the County Hospital" (Doctoral thesis, University of California, Berkeley, 1965). Sudnow also includes comparative materials from a more well-to-do Jewish-sponsored hospital where he did additional field work; but most of his statements are based on the county institution. Barney G. Glaser and Anselm L. Strauss, *Awareness of Dying* (Chicago: Aldine Publishing Company, 1965).

8. See Sudnow, *op. cit.,* pp. 107, 109. This is even fewer than would be expected by the age-composition of mortality, because children's and teaching hospitals in the city were likely to care for many terminally ill children and younger adults.

9. *Ibid.,* pp. 49, 50.

10. J. K. Owen, *Modern Concepts of Hospital Administration* (Philadelphia: W. B. Saunders Co., 1962), p. 304; cited in Sudnow, *op. cit.,* p. 80. Such practice attests to the accuracy of Edgar Morin's rather melodramatic statement: "Man hides his death as he hides his sex, as he hides his excrements." See E. Morin, *L'Homme et La Mort dans L'Histoire* (Paris: Correa, 1951), p. 331.

11. See Sudnow, *op. cit.,* pp. 20–40, 49–50.

12. See Sudnow, "Passing On," p. 114.

13. See Glaser and Strauss, *Awareness of Dying,* pp. 142–43, 151–52.

14. On the doctor's attitudes toward death and the dying, see August M. Kasper, "The Doctor and Death," in *The Meaning of Death,* ed. Herman Feifel (New York: McGraw-Hill Book Company, Inc., 1959), pp. 259–70. Many writers have commented on the tendency of relatives to avoid the subject of death with the terminally ill; see, for example, Herman Feifel's "Attitudes toward Death in Some Normal and Mentally Ill Populations," *Meaning of Death,* pp. 114–32.

15. The most favorable situation for reducing isolation and meaninglessness would seem to be "where personnel and patient both are aware that he is dying, and where they act on this awareness relatively openly." This atmosphere, which Glaser and Strauss term an "open awareness context," did not typically predominate in the hospitals they studied. More common were one of three other awareness contexts they distinguished: "The situation where the patient does not recognize his impending death even though everyone else does" (closed awareness); "The situation where the patient suspects what the others know and therefore attempts to confirm or invalidate his suspicion" (suspected awareness); and "The situation where each party defines the patient as dying, but each pretends that the other has not done so" (mutual pretense awareness). See Glaser and Strauss, *Awareness of Dying,* p. 11.

16. See *Ibid.,* p. 129. Some patients, however, put up a struggle to control the pace and style of their dying; and some prefer to leave the hospital and end their days at home for this reason (see Glaser and Strauss, *op. cit.,* pp. 95, 181–83). For a classic and moving account of a cancer victim who struggled to achieve control over the conditions of his death, see Lael T. Wertenbaker, *Death of a Man* (New York: Random House, Inc, 1957).

For discussions of isolation, meaninglessness, and powerlessness as dimensions of alienation, see Melvin Seeman, "On the Meaning of Alienation," *The American Sociological Review,* XXIV (1959), 783–91; and Robert Blauner, *Alienation and Freedom: The Factory Worker and His Industry* (Chicago: University of Chicago Press, 1964).

17. Franz Borkenau, "The Concept of Death," *The Twentieth Century,* CLVII (1955), 313–29, reprinted in Fulton, *Death and Identity,* pp. 42–56.

18. The need to redefine the status of the departed is intensified because of tendencies to act toward him as if he were alive. There is a status discongruity inherent in the often abrupt change from a more or less responsive person to an inactive, nonresponding one. This confusion makes it difficult for the living to shift their mode of interaction toward the neomort. Glaser and Strauss report that relatives in the hospital often speak to the newly deceased and caress him as if he were alive; they act as if he knows what they are saying and doing. Nurses who had become emotionally involved with the patient sometimes back away from postmortem care because of a "mystic illusion" that the deceased is still sentient. See Glaser and Strauss, *Awareness of Dying,* pp. 113–14. We are all familiar with the expression of "doing the right thing" *for the deceased,* probably the most common conscious motivation underlying the bereaved's funeral preparations. This whole situation is sensitively depicted in Jules Romains's novel, *The Death of a Nobody* (New York: Alfred A. Knopf, Inc., 1944).

19. Arnold Van Gennep, *The Rites of Passage* (London: Routledge & Kegan Paul, Ltd., 1960 [first published in 1909]). See also, W. L. Warner, *The Living and the Dead* (New Haven: Yale University Press, 1959), especially Chapter 9; and Habenstein, "Social Organization," for a discussion of funerals as "dramas of disposal."

20. See Goody, *Death, Property, Ancestors,* for the specific material on the LoDagaa. For the general theoretical treatment, see Hertz, *Death and the Right Hand,* and also Émile Durkheim, *The Elementary Forms of the Religious Life* (New York: The Free Press, 1947), especially p. 447.

21. Hertz, *op. cit.,* took the multiple funerals of primitive societies as the strategic starting point for his analysis of mortality and social structure. See Goody *op. cit.,* for a discussion of Hertz (pp. 26–27), and the entire book for an investigation of multiple funerals among the LoDagaa.

22. I have been unable to locate precise statistics on the comparative frequency of funerals. The following data are suggestive. In a year and a half, Goody attended 30 among the LoDagaa, a people numbering some 4000 (see *op. cit.*). Of the Barra people, a Roman Catholic peasant folk culture in the Scottish Outer Hebrides, it is reported that "Most men and women participate in some ten to fifteen funerals in their neighborhood every year." See D. Mandelbaum, "Social Uses of Funeral Rites," in *The Meaning of Death,* p. 206.

Considering the life expectancy in our society today, it is probable that only a minority of people would attend one funeral or more per year. Probably most people during the first 40 (or even 50) years of life attend only one or two funerals a decade. In old age, the deaths of the spouse, collateral relations, and friends become more common; thus funeral attendance in modern societies tends to become more age-specific. For a discussion of the loss of intimates in later years, see J. Moreno, "The Social Atom and Death," in *The Sociometry Reader,* ed. J. Moreno (New York: The Free Press, 1960), pp. 62–66.

23. For a discussion of funerals among the Romans and early Christians, see Alfred C. Rush, *Death and Burial in Christian Antiquity* (Washington, D.C.: Catholic University of America Press, 1941), especially Part III, pp. 187–273. On funerals in the medieval and preindustrial West, see Bertram S. Puckle, *Funeral Customs* (London: T. Werner Laurie, Ltd., 1926).

24. See Eissler, *Psychiatrist,* p. 144: "The religious dogma is, with relatively rare exceptions, not an essential help to the psychiatrist since the belief in the immortality of the soul, although deeply rooted in man's unconscious, is only rarely encountered nowadays as a well-integrated idea from which the ego could draw strength." On the basis of a sociological survey, Gorer confirms the psychiatrist's judgment: ". . . how small a role dogmatic Christian beliefs play . . ." (see Gorer, *Death, Grief, Mourning,* p. 39). Forty-nine per cent of his sample affirmed a belief in an afterlife; twenty-five per cent disbelieved; twenty-six per cent were uncertain or would not answer (*Ibid.,* p. 166).

25. The problem of sacred institutions in an essentially secular society has been well analyzed by Robert Fulton. See Fulton and Gilbert Geis, "Death and Social Values," pp. 67–75, and Fulton, "The Sacred and the Secular," pp. 89–105, in *Death and Identity.*

26. LeRoy Bowman interprets the decline of the American funeral primarily in terms of urbanization. When communities were made up of closely knit, geographically isolated groups of families, the death of an individual was a deprivation of the customary social give and take, a distinctly felt diminution of the total community. It made sense for the community as a whole to participate in a funeral. But in cities, individual families are in a much more limited relationship to other families; and the population loses its unity of social and religious ideals. For ethical and religious reasons, Bowman is unwilling to accept "a bitter deduction from this line of thought . . . that the death of one person is not so important as once it would have been, at least to the community in which he has lived." But that is the logical implication of his perceptive sociological analysis. See Bowman, *American Funeral,* pp. 9, 113–15, 126–28.

27. Edmund Volkart, "Bereavement and Mental Health," in *Explorations in Social Psychiatry,* Alexander H. Leighton, John A. Clausen, and Robert N. Wilson, eds. (New York: Basic Books, Inc., Publishers, 1957), pp. 281–307. Volkart suggests that bereavement is a greater crisis in modern American society than in similar cultures because our family system develops selves in which people relate to others as persons rather than in terms of roles (see pp. 293–95).

28. In a study of bereavement reactions in England, Geoffrey Gorer found that 30

of a group of 80 persons who had lost a close relative were mourning in a style he characterized as *unlimited.* He attributes the inability to get over one's grief "to the absence of any ritual, either individual or socially."

CHAPTER 17 VALUES AND ULTIMATE MEANINGS

BELIEF, UNBELIEF, AND RELIGION

1. Cf. J. Milton Yinger, *Religion, Society and the Individual: An Introduction to the Sociology of Religion* (New York: Macmillan Co., 1957).
2. Cf. Guy E. Swanson, *The Birth of the Gods: The Origin of Primitive Beliefs* (Ann Arbor, Mich.: University of Michigan Press, 1960).
3. Emile Durkheim, *De la division du travail social* (Paris: Presses Universitaires, 1967).
4. Robert Redfield, *The Primitive World and its Transformations* (New York: Cornell University Press, 1957).

Index

Abolitionist Movement, 158
Abortion: as victimless crime, 387; working-class attempts at, 106
Abstraction in research, 39–40
Acculturation: African slaves and, 158, 162; inner city and, 138, 139, 141
Achieved status, definition of, 201
Achievement ethos, class system and, 201
Administration: bureaucracy compared with, 260; capitalism and development of, 33–34
Adolescence (*See* Youth)
Adolescent Society (Coleman), 291
Adult-centered families, 106
Adult-directed families, 106–107
Advertising executives, 315, 330–333
Affectional expression, 109
African slaves, 158, 159, 162
Afterlife, belief in, 463
Age, stratification through, 192
Aged: class differences in treatment of, 101; generation gap and, 463; inner city residency by, 140; segregation of, from family, 447–449
Agnew, Spiro, 496
Agricultural Children Act of 1873 (Great Britain), 302
Agricultural societies, extended families in, 100
Agriculture, child labor in, 302
Akers, Ronald L., 385, 392–400
Alienation: dying patient and, 461; working-class, 177, 181
American Dream, organization man and, 272–273
American Indians: settlement patterns of, 432; tribal wars of, 412
American Institute of Banking, 268
American Medical Association, 189
American Motors, 369
American Sociological Association, 496
Amish people, 306
Andrews, P. B. S., 95
Annapolis Naval Academy, 249
Anslinger, Harry J., 398
Anticipatory socialization, 206–207
Anti-intellectualism, working-class, 179–180
Anti-poverty programs, 113
Anti-trust actions, 368
Anti-white sentiments, 163
Apathy, political, 356
Apprenticeships, 292, 293, 297, 298
Aquinas, Thomas, 442
Ariès, Philippe, 291, 292–295, 440
Armed Love (Katz), 291
Armistice Day celebrations, 469
Ascetism, capitalism based on ideas of, 34–36
Ascribed status, definition of, 201
Asquith, Margot, 95
Assimilation, definition of, 249
Astral bodies, 483
Astrology: problems in hip relationships explained through, 119; student interest in, 483, 484, 485, 490
Astronomy, development of, 31
Asylums, social control through, 284–288
A. T. & T., 370
Athens, ancient, 153, 432
Atlantic, The (magazine), 389
Attitudes, assumption of, 66–67
Audemars, Mlle., 92
Aunts, alternatives for kin terms for, 83
Auras, 483
Authoritarianism, working-class, 175, 178
Authority: definition of, 349; high school teaching about, 241–245; industrial management centralization and, 261; institutions and, 77; white-collar stratification through, 192
Autobiography of an Ex-Coloured Man, The (Johnson), 3
Autoerotic incidents, changing ideas on children's, 70, 71
Automation, 442
Automobiles, change in culture through, 306
Autonomy: lower-class concern with, 402, 407–408; professional group, 450

Babies (*see* Children)
Babylonia, 31, 32
Back-to-Africa Movement, 160
Bacon, Francis, 79
Baldwin, James Mark, 92

Baltimore (Maryland), 145
Baltzell, E. Digby, 379
Banking: rise of capitalism and, 32; white-collar employees in, 187
Baptists, political preferences of, 375
Bardo Thodol, 483
Bargaining relationships between doctor and patient, 449–458
Bavelas, Alex, 320–321
Baxter, Richard, 36
Beard, Charles, 374
Behaviorism, 69
Belgium, dispersion of wages in, 224, 225
Belknap, Ivan, 285
Bendix, Reinhard, 229–230, 259, 260–270
Bennett, J. W., 157
Berger, Bennett M., 99, 114–119, 143
Berger, Brigitte, 415, 439–445
Berger, Peter L., 3, 4–5, 20, 41–46, 99–100, 119–123, 201, 202–207, 415, 439–445, 495–503
Berguer, Mlle., 93
Berkeley (California), hippies in, 388, 389–390
Besant, Walter, 95
Bethlehem Steel, 368
Betjeman, John, 94
Bettelheim, Bruno, 441
Bewusstein als Verhaengnis (Seidel), 499
Bewusstmachung, 498
Bible, 34, 35, 51
Bidwell, Charles E., 233, 234–236
Biography, relation of history to, 12–13
Biological phenomena, 25
Birmingham, England, 298
Birth control: appearance of, 294; Catholic Church opposition to, 106; superabundance of young and, 296
Birth rates, child labor and, 299–301
Black Muslims, 163
Blacks: bases for sociological examination of, 156–158; Boston school curricula and, 237, 239–240; cause of inadequate education of, 111; Chicago, 131, 135; childhood experiences of, 3, 6–8; civil rights movement and (*see* Civil rights movement); color-caste system and, 172; education and mobility among, 230–231; employment discrimination against, 371; ethnogenesis of, 158–163; limitations of compensatory education for, 112–114; lower-class style of life of, 181; mutual insult game among, 405; occupation mobility among, 229–231; passing as white by, 6–8, 52–54; physical characteristics of, 172; police harassment of, 388; power structure and, 389; relations with whites of, 157, 160–161, 162–163; revolution among, 439; service occupations and, 176; as social entity, 157; socializing influences of family among, 99, 111–114; Southern childhood of, 49, 50–55; white-collar class and, 191
Black women, income level of, 231
Blasé attitude, metropolis and, 150–151
Blau, Peter, 228, 230
Blauner, Robert, 447, 458–464
Blind children, language learning by, 84–90
Bohemians, in Chicago, 132, 135–136, 139
Booth, Charles, 300
Borch, Fred, 370
Boston: career patterns in, 227; education in, 236, 238, 239; stratification in, 166; upward mobility in, 227; working-class Italian families in, 99, 100–108
Bott, Elizabeth, 103
Boulton and Watt factory, 262
Bourgeois class: childhood structure transformed by, 49, 440–441; class struggle and, 421–430; Puritan beliefs and economic ethic of, 35–36; rise of capitalism and, 33
Brandeis, Louis, 379
Brazil, 161
Breaking points, 286, 287
Bricker, John, 382
British West Indies, 161
Brothers: alternatives for kin terms for, 83; Boston Italian working-class communication among, 104, 105; friends as, in communes, 115
Brown, Michael E., 385, 386–391
Brown rice diets, 483
Bryce, Murray D., 37
Bureaucracy, 258–288; asylums and, 284–288; of death, 458–462; definition of, 260–261; differences of interest between population and, 259; industrial, 260–270; military academy and, 252–253; organization man in, 259, 270–278; Parkinson's law on, 278–283; rationalization of, 435–439; special examination system and, 435–436; upward mobility and, 225
Burgess, Robert L., 394
Burlington (Vermont), 380
Burnham, James, 358
Businessmen: political preferences of, 374, 376, 381; power among, 349, 353, 355, 361

Calculability, metropolis and, 149–150
California: government employees of, 364; hippies in, 388, 390

Calling, Protestant Ethic emphasis on, 34–35, 36
Cambridge (Massachusetts), hippies in, 388, 389
Camus, Albert, 344, 502
Canada: income distribution in, 224, 225; work time for meal in, 222
Canterbury (England), 296
Capitalism: background on rise of, 20, 30–36; bureaucratization of, 436; Marxian ideas on, 415; Protestant Ethic and, 20, 30–36; Puritan beliefs and rise of, 34–36; as pursuit of profit, 32
Carnegie, Andrew, 266
Carnegie, Dale, 268
Cartography, 21
Caste: blacks as, 156; race and, 165, 172; stratification through, 202
Castro, Fidel, 444
Catholic Church, 293; birth control opposition by, 106; freedom of children and, 294; medieval disputes between state and, 481; political preferences of members of, 375–378, 380
Caucasians (*see* Whites)
Celibacy, women in communes and, 116
Celtic clan feuds, 412
Census Bureau, 217, 229, 230
Center for International Studies, 373
Ceylon, income distribution in, 224
Change, 414–445; American cultural, 439–445; causes for progress of division of labor and, 430–434; Middletown community study of, 416–422; rationalization of education and training and, 415, 435–439
Chankras, 483
Charity schools, 294
Chemistry, development of, 31
Chicago: communities within, 129, 130–136; outer city of, 145; sociological study of, 139; urban sociology school of, 129, 137, 139, 315
Childless residents of inner city, 139, 140, 141
Child-rearing: analysis of literature on, 49, 69–74; Boston working-class Italian, 100, 101, 104, 106–108; communal, 115–116, 117; education of parents and approaches to, 107; experts on, as identity-marketing agency, 205; upper-middle class (Crestwood Heights), 109
Child-Rearing Practices in the Communal Family (Berger, Hackett, and Millar), 99
Children: accepted forms of social behavior and, 110; analysis of data from everyday activities of, 57–60; bourgeoisie transformation of, 440–441; changes in conception of basic impulses of, 70–73; communal families and, 99, 114–119; culture of, as subject for study, 79; deaf-mute (Helen Keller), and language learning, 84–90; families centered around, 106; family types and roles of, 106–107; game as social situation for, 65–66; history through, 500; imposition of modes of behavior through education of, 26–27; location finding in society by, 4–5; lore and language of, 77, 78–81; mortality rates of, 296, 300–301; parents' relationship with (*see* Child-rearing); racial attitudes of, 3, 6–8; respect for older people among, 101; socialization of (*see* Socialization); society experienced through language by, 77; as stranger in world of adults, 49; transformation in patterns of, 49; transition from ego-centrism to social awareness through language by, 77, 90–93; upward social mobility preparation of, 110; in white American South, 49, 50–55; working-class, 178, 181; (*see also* Youth)
Children's Bureau, 69, 70
China: ancient, 31, 32; population density in, 433; special examination system in, 436, 437, 438; "thought reform" camps in, 286
Christianity: business ideas and, 266, 267; systematic theology developed by, 31
Chrysler Corporation, 367–368
Church, Joseph, 81
Cicourel, Aaron V., 56
Cities: bourgeoisie and, 424; definition of, 137; distinction between personal troubles and public issues in study of, 15; inner, 138–142; intensification of nervous stimulation in, 147–148; outer, 142–143; as prototype of mass society, 138; race riots in, 160; sociological conception of, 137; suburb compared with, 143–146; urban renewal activities in, 146; (*see also* Communities)
Civil rights movement, 209; black strength through, 113; solidarity and, 157–158
Clan feuds, 412
Class: abstract, 64; achievement ethos in, 201; appearance of concept of, 295; class struggle and, 421–430, 444; concrete, 63–64; family pattern differences between, 99; generalities of American, 172–175; kinds of, 63–64; identity and, 202–207; language usage and distinctions between, 78, 93–97; looseness of, 203–204; political preferences and, 372–

Class (*continued*)
376, 380; stratification within (*see* Stratification); technological society and changes in, 443–444; (*see also specific classes*)
Clay, Henry, 383
Cleaver, Eldridge, 391
Clerks, bureaucratic increase in work load of, 264–265
Cleveland, Grover, 377
Clinard, Marshall B., 395
Coast Guard Academy, 233, 248–256
Coch, Lester, 320, 321
Cohen, Albert, 412
Coleman, James, 291, 302–307
Coleman, Sinclair, 230–231
Collective behavior, mass emotions in crowds and, 25–26, 27, 28
College education: Boston Italian working-class attitudes toward, 107; bureaucratic mobility and, 227; cosmopolite residency and, 146; psychological impact of, 245–248; resurgence of interest in religion during, 482–492; social mobility through, 110, 201, 211–217; upper-middle class children and, 168; work ethic and, 339
Collision Course (Moscow), 325
Commercialization, 32
Committee on Cultural Freedom, 94
Communes: division of labor in, 117–118; family structure in, 99, 114–119; religious, 119; rural, 115, 117–118, 119; splitting phenomenon in, 115–116; urban, 115, 116, 118; women's liberation movement and, 118–119
Communication: density of society and, 432; Loop, Chicago, strengthened by development of, 130; power elite and, 354–355; white-collar employees in, 187
Communist hunt of McCarthy, 378
Communist Manifesto, The (Marx and Engels), 415
Communities: anti-poverty programs and, 113; within Chicago, 129, 130–136; differentiation between rural and urban, 129; effects on individuals of, 129, 147–156; individuals and life within, 128–163; Middletown study of change in, 415, 416–422; outer city, 142–143; personality development and, 66–68; settlement types in, 137–138; as social entity, 157; stratification in (*see* Stratification); suburban (*see* Suburbs); types of inner city residents in, 140–142; unity of self through, 62; (*see also* Cities)
Community Apprentice Experiment, 113
Compensatory education, limitations of, 113–114
Compulsory education: college education as, 339; introduction of, 301–302
Comte, Auguste, 12, 19, 415, 496
Concientización, 498
Congregationalists, political preferences of, 375, 376
Congress, 351, 361, 362, 365, 381
Connecticut: political preferences in, 379; tax income of, 364
Conrad, Joseph, 324, 325
Consciousness, definition of, 68
Consciousness as Doom (Seidel), 499
Construction industry, increase of industrial bureaucrats in, 262–263
Consumption: distribution of goods and, 229; free time related to, 336; lack of imagination in, 365; sacramentalism of, 207
Contraceptives, 106
Control (*see* Social control)
Conversations: analysis of, between adult and child, 57–60; Boston Italian working-class, 102, 104; children's, 92; dominance of marital, 122, 123
Cooley, Charles, 204
Cornell University, 366
Corn Is Green, The (Williams), 210
Corporations: criticism of, 368–369; elite in control of, 369–370; power in, 349–350, 363–372; size of, 364; social effects of, 370–372; (*see also* Industry)
Cosmopolites, inner city residency of, 139–140, 141, 146
Cost, bureaucratic increase in, 263–264
Counterculture, 291, 387, 441
Couples (Updike), 100
Cousins, alternatives for kin terms for, 83
Cox, O. C., 157, 162, 163
Credit: rise of capitalism and, 32; white-collar employees in, 187
Cressey, Donald R., 393
Crestwood Heights family, 99, 108–110
Crime: dimensions of, 392; labelling of, 397–399; political nature of, 395–396; repression through law of, 387; victimless, 387
Crowd, mass emotions in, 25–26, 28
Cuba, 161
Cubberley, Ellwood, P., 375
Cultural mobility, 215–216
Culture: child-rearing and transmission of, 109; children's, as subject for study, 79; definition of, 21; role of sociologist as stranger in study of, 19, 20–24; stratification in (*see* Stratification); thinking as usual in, 23–24
Culture against Man (Henry), 447

Culture shock, sociologist's experiencing of, 45–46
Czechoslovakia, dispersion of wages in, 224, 225

Darwin, Charles, 39, 486
Darwinian morality, 269
Daughters: alternatives for kin terms for, 83; in Boston Italian working-class families, 101
David, Allison, 157, 171, 175
Death, 447, 458–464; bureaucratization of, 458–462; disappearance of public event nature of, 463–464; disruptive impact of, 458–459; rationalization of, 460, 461
Death at an Early Age (Kozol), 233, 236–240
Death of a Salesman (Miller), 487
De Grazia, Sebastian, 315, 334–342
Delinquency: definition of, 400; focal concerns of, 408–410; lower-class culture and, 385, 400–413; working-class, 175, 182
Delinquent Boys (Cohen), 412
Democracy, 435, 437
Democratic party: class and, 372–383; ethnic group support for, 377–379; one-party states and, 379–380
Denmark: income distribution in, 223, 224; work time for meal in, 222–223
Density: city versus suburban, 143, 144–145; division of labor and population, 430–434; urban life shaped by, 137, 138, 139, 141–142
Depersonalization of death, 460–461
Desegregation in education, benefits from, 113–114
Destruction of the European Jew, The (Hilberg), 387
Deviance, 384–413; behavioral questions about, 393–395; definition of, 385; labelling of, 397–399; relativity of definitions of, 385; social definitions of, 395–397, 399
Dewey, John, 23
Dewey, Thomas E., 382
Dewhurst, J. F., 186
Differential association theory of deviance, 393, 394
Differentiation, 415
Discrimination: blasé attitude and blunting of, 151; racial (*see* Racial discrimination)
Disraeli, Benjamin, 298
Distribution systems, 186–187
Division of labor: causes for progress of, 430–434; in cities, 155; communes and, 117–119; complexity of, 315; ethnic group development and, 159; group development and, 153; husband-wife relationships and, 103–104; Puritan beliefs as background of, 34–35; specialization of abstract skills within, 185
Division of Labor in Society (Durkheim), 415
Divorce: distinction between personal troubles and public issues in study of, 15; economic basis for, 99; working-class, 178
Doctors: bargaining relationship between patients and, 449–458; terminally ill and, 461
Donner, Frederic, 370
Dornbusch, Sanford M., 233, 248–256
Double standard of behavior, 102
Douglas Aircraft, 367
Downward social mobility: inner city residence and, 139, 140, 141; limitations on, 212; rates of, 227
Dress regulations, high school, 243
Dropouts, educational, 213, 214
Drug usage: addiction and, 398–399; cultural revolution and, 440; hippie emphasis on, 486; psychedelic, 483; victimless crime and, 387
Druids, Californian, 483, 491
DuBois, W. E. B., 163
Duncan, O. D., 228, 230
Durkheim, Emile, 3, 8–9, 12, 19, 25–30, 77, 112, 203, 205, 272, 326, 393, 397, 415, 430–434, 467, 496

East Village, New York City, 388, 389, 390
East Village Other (newspaper), 388
Economics: class status and, 173–174; class struggle and change in, 421–430; division of labor within cities and, 155; personal troubles and public issues in study of, 16; segmentation of, 205; separation of family from processes of, 99; stratification and, 165
Economist (London), 228
Economist, sociologist compared with, 43
Edsel (automobile), 365, 368
Education: benefits from integrated, 113–114; black challenges and changes in, 113; black children and, 112–114; black occupational mobility and, 230–231; Boston Italian working-class attitudes toward, 107; cause for inadequate black, 111; centrality of, for social control, 234–236; compensatory, limitations of, 112–114; compulsory, 301–302; contra-

Education (*continued*)
dictory tendencies in, 235–236; as control access to upper-middle class, 213–215; corporate impact on, 365–366; correlation between occupational status and, 212–213; historical review of, 292–295; imposition of modes of behavior in, 26–27; leisure temperament and, 342; medieval, 292; mobility through, 174; power patterns and, 354; psychological impact of, 245–248; rationalization of, 435–439; social control through, 232–256; socialization as object of, 27; specialized examination system in, 435–436; universal, 234; white-collar, 191; working-class attitudes toward, 177, 178, 179, 180; (*see also* Schools)
Eells, Kenneth, 165, 166–175
Ego-centrism, children's language reflecting, 90–93
Egypt, ancient, 31
Eisenhower, Dwight D., 372, 379, 380, 381, 382
Eliade, Mircea, 488
Eli Lilly and Company, 370
El Salvador, income distribution in, 224
Emancipation of slaves (in the United States), 159–161
Emotions, mass, in crowds, 25–26, 28
Empirical knowledge, 31, 486
Employment: child mortality rates and, 296; increased black, 111, 113
Encounter (magazine), 94
Encounter groups, 488
Endogamy, 162, 172
Engels, Friedrich, 415, 421–430
England: class structure changes in, 444; income distribution in, 223, 224, 225; language and class distinctions in, 78, 93–97; London's place in history of, 149; lore and language of schoolchildren in, 77, 78–81; middle-class husband-wife relationships in, 103; mother-daughter relationships among working class in, 101; political parties in, 372; population changes and child labor in, 296–302; population density in, 431; proletariat reforms in, 428, 429; ratio of administrative and production personnel in, 264; special examination system in, 435, 436, 437; work time for meal in, 222–223
Entertainment facilities, 144
Episcopalians: political preferences of, 375, 376; power patterns and, 354
Equality (Tawney), 221
Equality of opportunity, social mobility and, 225–231
Erogenous zones, changing ideas on child's handling of, 70, 71
"Escalation of Violence Through Legitimation, The" (Westley), 387
Estates, stratification through, 202
Ethnic groups: Boston Italian working-class, 99, 100–108; inner-city residence of, 139, 140–141; neighborhood settlement by, 145; political preferences of, 377–379; process of development of, 159; qualities of, 162; as social entity, 157; (*see also* Blacks)
Ethnogenesis of blacks, 158–163; definition of, 159; phases of process of, 159–161
Ethnologist, sociologist compared with, 43
Etiquette, the Blue Book of Social Usage (Post), 96–97
"Everyday World of the Child, The" (Speier), 49, 77
Exactness, metropolis and, 149–150
Excitement: lower-class concern with, 402, 405–406; working-class, 180
Excretory functions, openness about, in communes, 118
Executives: advertising, 315, 330–333; junior, 271; political preferences of, 374, 376; social mobility of, 226–227
Executive Suite (motion picture), 204
Expanded families, definition of, 100
Experience: marriage and new modes of, 121; structures in society and, 3
Extended families: black, 113; Boston Italian working-class, 105; commune as, 117; generational limitations on, 101; nuclear family differentiated from, 100; psychotherapy for, 112; working-class, 177, 178
Extraversion, 245–246, 248
Eysenck, H. J., 245

Factory Acts Extension Act of 1867 (Great Britain), 302
Factory Commissioners, 298
Families, 98–126; adult-centered, 106; adult-directed, 106–107; aged separated from, 447–449; appearance of concept of, 295; Boston Italian working-class, 99, 100–108; bourgeoisie separation from, 423; child-centered, 106; communal, 99, 114–119; conjugal, 119; construction of reality through, 119–123; Crestwood Heights, 99, 108–110; culture transmission through, 109; distinction between personal troubles and public issues in study of, 15, 17; economic production separated from, 99; employers and employees as, 270; expanded, 100; extended

(*see* Extended families); female-based, lower-class, 408; first experience of power within, 349; funerals and, 463; house as symbol of solidarity in, 109; identity generation and, 206; income level and, 217–221; kin term usage in, 81–84; mobility patterns of, 226–228; nuclear (*see* Nuclear families); orientation toward future in, 110; separation of illness and death from, 459, 460; size of, among Italian working-class, 106; social control within, 233; socialization in (*see* Socialization); transformation of, 99; types of, 206–207; upper-middle class (Crestwood Heights), 99, 108–110; upward social mobility and, 110; working-class, 177, 178
Far West, class stratification in, 169–171
Farmer, James, 163
Fate, lower-class concern with, 402, 406–407
Fathers: alternatives for kin terms for, 83; changing role of, 74
Featherbedding, white-collar, 367–368
Federalist party, 376, 377
Feeding, changing ideas on children's, 72
Female-male relationships (*see* Male-female relationships; Marriage)
Feudal age, 153–154, 422
Feuer, Lewis, 441
FHA, 143
Folk society, 138
Ford Motor Company, 367, 368
Foremen, bureaucratic changes in, functions of, 263, 265
Foresters (friendly society), 301
Fox, Renée, 451
France, 495; development of sociology in, 19; dispersion of wages in, 224, 225; proletariat struggles in, 429; ratio of administrative and production personnel in, 264; special examination system in, 435; third estate in, 422; voting percentages in, 372
Francis, E. K., 162
Frankfurt School of sociology, 497
Frazier, E. F., 160–161, 163
Freedmen, 158, 159, 162, 421
Freedom Riders, 161
Freedom within metropolis, 153–155
Free time, projected increase in, 334–335, 337, 339
French, John R. P., Jr., 320, 321
French-Canadians: class membership of, 168; occupational mobility among, 229–230
Freud, Sigmund, 39, 499
Freudian psychology, 441
Friedenberg, Edgar Z., 233, 241–245
Friendly societies, 301
Funerals, decline of, 462–464
Fun morality, child-rearing and, 49, 69–74
Future: family orientation toward, 110; hip relationships and uncertainty of, 114, 115

Galbraith, John Kenneth, 242, 369
Gallup Poll, 373, 374
Game: difference between play and, 62; learning of society through, 61; social situation exemplified by, 65–66
Gang Act of 1867 (Great Britain), 302
Gang delinquency: definition of, 400; focal concerns of, 408–410; lower-class culture and, 385, 400–413; working-class, 175, 182
Gans, Herbert J., 99, 100–108, 129, 137–146, 441
Gardner, Burleigh B., 171
Gardner, Mary R., 171
Garofalo, Raffaele, 397
Garvey Movement, 160
Gay, John, 79
Geertz, Hildred, 82
Gehlen, Arnold, 205, 206, 501
General Electric, 364, 370
General Motors, 364, 368, 370
Generation gap: aged and, 463; economic bases for, 99; among sociologists, 496
Genitals, changing ideas on children's self-play with, 70, 71, 72
Geographic mobility, 202, 207
Geometry, development of, 31
German immigrants: Chicago, 131; class membership of, 171; political preferences of, 377, 378, 379
Germany, 387, 495; development of sociology in, 19; early settlement patterns in, 432; feudal nobility in, 422; New Left in, 498; proletariat struggles in, 429; ratio of administrative and production personnel in, 264; schools of research in, 37; special examination system in, 436, 437; voting percentages in, 372
Gesell, Arnold, 69
Gesellschaft, 138, 139
Ghosts, 463
Gibbs, Jack P., 392
Girdler, Tom, 267
Glantz, Oscar, 381
Glaser, Barney G., 460, 461
Glass, D. V., 296, 299
Goffman, Erving, 213, 259, 284–288, 498
Gold Coast, Chicago, 132, 133–134, 136, 139

Gold Coast and the Slum, The (Zorbaugh), 129
Goldfinger (motion picture), 344
Government officials: free time for, 335; functioning of society dependent on, 31; higher education as basis for selection of, 247–248; increased numbers of, 188
Grandparents: alternatives for kin terms for, 83; Boston Italian working-class, 101
Great Britain (*see* England)
Greece, ancient, 31, 32, 292, 437, 438
Greek immigrants to Chicago, 131, 135
Greeley, Andrew M., 467, 482–492
Greene, Graham, 94–95
Green revolution, 439, 441, 444
Greer, Scott, 139
Griffith, Talbot, 299, 300
Group association, marriage and pressures from, 122
Group-dynamics, 488
Group marriage, 116–117
Guaranteed annual wage, 342
Guerrilla theater, 390
Gulf Oil, 364
Gurdjieff, Georges, 483

Habakkuk, H. J., 299, 301
Habits, unconscious self and, 67–68
Hacker, Andrew, 349–350, 363–372
Hackett, Bruce M., 99, 114–119
Haight-Ashbury (San Francisco), 389
Haiti, 161
Handicapped children, language learning by, 84–90
Harding, Warren G., 378
Harlem, Haryou Report on pathology of, 112
Harriman, W. Averell, 382
Harrington, Michael, 209, 228
Harris, Mark, 389
Hartog, Jan de, 324, 325
Hartung, Frank, 395
Harwood Manufacturing Company, 315, 319–322
Haryou Report, 112
Havinghurst, Robert J., 175, 176
Hawthorne plant research program, Western Electric, 315, 316–319
Hazing system, military academies, 251, 252–253, 255
Hearts of Oak (friendly society), 301
Hedonism, 344–345, 440
Heilbroner, Robert L., 209
Heilswissen, 496
Hellenism, 31, 292, 437, 438
Henry VIII, King of England, 79
Henry, Jules, 447–449
Heterogeneity, urban life shaped by, 137, 138, 139, 141–142
Higher education (*see* College education)
High schools, control mechanisms in, 241–245
Hilberg, Raul, 387
Hippies, 386–391; culture of, 486, 487; persecution of, 388–391
Hip relationships, 114–119
Hiroshima, 459
Historians, sociologists compared with, 44
Historicity, institutions and, 77
History: through children, 500; individual lives and course of, 10–11; relation of biography to, 12–13
Hitchhiking, 389
Hoffman, Abbie, 391
Hofstadter, Richard, 379
Holcombe, Arthur, 374
Holiday (magazine), 343
Hollas, G. Basil, 328
Homes: suburban ownership of, 143; as symbol of prosperity and solidarity, 109
Homogeneity, city versus suburban, 143, 145–146
Homo Ludens (Huizinga), 485
Homosexuality, 387, 398
Honor system in military academies, 251
Horoscopes, 484, 488
Hospitals, bargaining relationships in, 449–458
Hostility, sexual, 103
Housing, city versus suburban, 143, 144–145, 146
Howard University, 113
"How to Get on in Society" (Betjeman), 94
Hughes, Everett C., 212, 255
Huizinga, J., 485
Humanists, 293
Hungary, dispersion of wages in, 224, 225
Hunting societies, extended families in, 100
Husband-wife relationships (*see* Marriage)
Husserl, Edmund, 23
Hutchinson, E. P., 229
Huxley, Aldous, 246

I.B.M., 366
I Ching, 483, 484, 485, 487
Ideal types, 41
Identification: family provision of models for, 109; social categories and selection of, 77, 81–84
Identity: mobility and, 206–207; play and, 343–346; social processes and, 204–205
Illegitimacy: black, 111; lower-lower class, 169

Illinois, government employees in, 364
Imagination, sociological, 3–4, 10–17
Imitation, children's, 92
Immigrants: to Chicago, 131, 135, 139; corporate locations and, 371; occupational distribution among, 229; political preferences of, 375, 376; relationships between adults and their parents among, 101; as social category, 157; as stranger in new society, 20–24
Incest, 169
Income: black female, 231; class stratification by, 189–191, 193; comparison of United States with other countries for, 223–225; distribution of, 217–225, 228; earnings from education gap in, 230–231; equal opportunity and social mobility and, 225–231; families, by level, 217–221; separation of work from, 342; unequal distribution of, 221–223
India, ancient, 31, 32
Indiana, voting records in, 374
Indianapolis (Indiana), 227
Indians (American): settlement patterns of, 432; tribal wars of, 412
Indifference, threat to personal values and, 16–17
Individuality: evolution of, 153; organization man and, 272, 278
Individuals: course of history and, 10–11; effect of metropolis on, 129, 147–156; phase of biography of, 291; questions in social study of, 12–13; social constraints on, 3, 8–9, 25, 29, 31; social phenomena and, 25–30; social structure and, 14–15, 29–30; stratification of interest of, 21–23; threats to values of, 16–17
Industrialization: politics and, 380; social mobility and, 225, 226, 228
Industrial revolution: family transformations caused by, 99; physical setting for family before, 119
Industrial society: pre-industrial society compared with, 137–138; religion in, 481; youth in, 291
Industrial sociology, 315
Industrial therapy, 287
Industry: bureaucratization of, 260–270; decentralization of, 137, 146; proletariat in, 426–427; suburban development and, 137; working conditions in, 316–322; (*see also* Corporations)
Infant Care (Children's Bureau), analysis of various editions of, 70–74
Infants (*see* Children)
"In Flanders Fields" (McCrae), 471
Inner city, 138–142; types of residents in 139–140
Institutions: bureaucratic control in, 284; characteristics of, 77; contradictions in, 14, 15; goals of, 284; perception of, 233; redefinition of self-direction in, 286; segmentation of, 205; values in, 467; (*see also specific institutions*)
Insurance, beginnings of, 300–301
Integrated education, benefits from, 113–114
Intellectuality, metropolis and, 148–149, 150, 154
Intensity, working-class, 179
Interaction: analysis of, 57–60; in military academies, 253–254; socialization as acquisition of components for, 56; understanding, 3
Intercourse, Boston Italian working-class attitudes toward, 103
Intermarriage, 162, 172
Internalization, 49, 478
International Labor Office, 222
Introversion, 245–246
Invitation to Sociology (Berger), 3
Ireland, birth rate in, 299–300
Irish immigrants: to Chicago, 131, 135; class membership of, 168, 171; political preferences of, 377, 378, 379
Iroquois Indians, 432
Islam, 31, 436
Isohypses, 21
Israel, income distribution in, 224
Italian immigrants: Boston working-class, 100–108; Chicago, 131, 135, 136; class membership of, 168; political preferences of, 378
Italy: attitudes toward extremes of wealth and poverty in, 295; early settlement patterns in, 432; feudal nobility in, 422; income distribution in, 224; voting percentages in, 372; work time for meal in, 222–223

Jackson, Andrew, 352, 376
Jamaica, 161
James, Henry, 94–95
James, William, 22
Janet, Pierre, 92
Janowitz, Morris, 482
Japan, income distribution in, 224
Java, kin term usage in, 82
Jefferson, Thomas, 376
Jeffersonian democracy, 376
Jencks, Christopher, 201, 211–217
Jesuits, 293
Jews, 391; Chicago, 139; cultural revolution and, 439; military academies and, 250; persecution of, 387; political preferences of, 375, 377; social mobility

Jews (*continued*)
among, 201, 208–211; tribal formation among, 433
Johnson, James Weldon, 3, 6–8
Jones, Ernest, 17
Jousselin, Jean, 297
Junior executives, 271
Juvenile delinquency: definition of, 400; focal concerns of, 408–410; lower-class culture and, 385, 400–413; working-class, 175, 182

Kahl, Joseph A., 176
Kama Sutra, 246
Kappel, Frederick, 370
Karate, 483
Katz, Elia, 291, 307–312
Kefauver, Estes, 382
Keller, Helen, 77, 84–90
Kellner, Hansfried, 99–100, 119–123
Kennedy, John F., 238
Kierkegaard, Sören, 499
Killers of the Dream (Smith), 49
Kindergarten, organization of play in, 61
King, Martin Luther, Jr., 163
Kinship: ethnic groups in inner city and, 140; friends and, 115; religious institutions and, 478–479; use of terms for, 82
Klapp, Orrin E., 315, 343–346
Knowledge: empirical, 31; kinds of, 22
Knowledge industry, 443
Kolko, Gabriel, 228
Korean War, 287, 381
Kornhauser, William, 349, 350–363
Kozol, Jonathan, 233, 236–240
Kravis, Irving, 223–224
Kube, Ella, 139
Kundalini, 483

Labeling theory: deviance and, 385; hospitals and, 447
Labor force: class struggle and, 421–430; corporate control of, 365; division of (*see* Division of labor); rise of capitalism and organization of, 32; percentage of, in white-collar jobs, 184–185, 186; population changes and youth in, 297–302
La Dolce Vita (motion picture), 345
Lafendel, Mlle., 92
Lamb, Charles, 79
Landecker, Werner, 203
Language, 76–97; characteristics of, 77; child's first experience of society through, 77; class determination through, 64; class distinctions embodied in, 78, 93–97; deaf-mute child (Helen Keller) and learning of, 84–90; double standard of behavior revealed in, 102; hip black, 113; identification and social categories imposed through, 77, 81–84; learning of society through, 61; perfecting of self through, 66, 67; schoolchildren's, 77, 78–81; transition from ego-centrism to social awareness through, 77, 90–93
Language and Thought of the Child, The (Piaget), 77
Law: capitalism and development of, 33–34; social definitions of deviance and, 395–397; special examination system and, 435
Lawrence, D. H., 182
Leary, Tim, 391
Lecky, W. E. H., 12
Lefebvre, Henri, 497
Legitimacy, definition of, 349
Leibniz, Gottfried von, 23
Leisure: definition of, 334; disaffection and, 339–340; distinctions between personal troubles and public issues in study of, 17; identity and, 343–346; objectives in creating, 341–342; preparation for, 337–339; problems and promises of, 334–342; projected increase in free time and, 334–335, 337, 339; rationalization of, 315; rhetoric of, 340–342
Lenski, Gerhard, 203
Levittown (New Jersey), 143
Lewin, Kurt, 319
Lewis, Ian, 315, 330–333
Liberalism, upper-class, 379
Life expectancy rates, 447
Limits of American Capitalism, The (Heilbroner), 209
Lincoln, Abraham, 221, 473
Lincoln Center, New York City, 390–391
Lippitt, Ronald, 112
Lipset, Seymour Martin, 202, 225–231, 350, 372–383
Literature: as identity-marketing agency, 204; printing and, 31; self-image of blacks and, 161
Liturgy, 485
Loans, rise of capitalism and, 32
Lobbies, congressional, 353
LoDagaa people of West Africa, 462
London, 149; child labor in, 298; fertility rates in, 296; insurance in, 300
Lonely Crowd, The (Riesman), 350–363
Lookout (magazine), 328
Loop, Chicago, 130–131, 136
Looseness of class structure, 203–204
Loosley, E. W., 99, 108–110
Lord's Prayer, 78, 79
Lore, schoolchildren's, 77, 78–81

Lore and Language of Schoolchildren, The (Opie and Opie), 77
Los Angeles, 171
Louis XIV, King of France, 95
Lower classes: extended kin groups among, 117; focal concerns of, 401–408; gang delinquency among, 400–413; middle-class ideas through education and, 240; use of pronouns by, 5; (*see also* Working classes)
Lower East Side, New York City, 388, 389, 390, 391
Lower-lower class: Middle and Far West, 171; New England, 169
Lower-middle class: child-centered families of, 106, 107; cosmopolite settlement in suburbs of, 139; green revolution and, 439; home ownership by, 143; Middle and Far West, 171; New England, 168; youth culture changes in, 442
LSD-25, 399
Luckmann, Thomas, 201, 202–207, 467, 476–482
Lutherans, political preferences of, 375
Lynchings, 160
Lynd, Helen M., 415, 416–422, 498
Lynd, Robert S., 23, 415, 416–422, 498
Lynes, Russell, 77, 93–97

McCarthy, Joseph, 378
McCrae, John, 471
Machlup, Fritz, 443
McKinley, William, 352, 353
Macrobiotic diets, 483
Macroscopic style of research: definition of, 36–37; differences between molecular research and, 37, 39–40
Madigan, Sheriff, 390
Magic, 483, 484
Maher Baba, 483
Mailer, Norman, 344
Maine, political preferences in, 379
Making It (Podhoretz), 201
Male-female relationships: as backbone of communes, 115; Boston Italian working-class, 101–103; male sexual fears and, 102–103; military cadets and, 254; splitting phenomenon of men in hip relationships and, 115–116; (*see also* Marriage)
Malthusianism, 294
Managers: changes in functions of, 261–265; free time for, 335; ideology of, 265–270; organization man as, 259, 270–278; percentage of labor force as, 184–185, 186
Manchester, England, 297
Manchester (New Hampshire), 380
Manchester Unity (friendly society), 301
Man in the Grey Flannel Suit, The (motion picture), 204
Manipulation, power and, 354
Mannheim, Karl, 12, 37
Manufacturing, increase in industrial bureaucrats in, 262–263
Mao Tse-tung, 500
Marjorie Morningstar (motion picture), 204
Marquardt, Joachim, 432
Marriage: between blacks and whites, 162; Boston Italian working-class, 103–106; changing social context for, 119–120; class factors in, 174; construction of reality and, 119–123; distinction between personal troubles and public issues in study of, 15; experts on, as identity-marketing agencies, 205; fictional account of common meanings in, 100, 123–126; group, 116–117; ideal-type process of, 120–121; importance of relationship in, 109; lower-class, 409; mobility through, 174; patriarchal authority within, 102; special examination system and eligibility for, 436; support for nuclear family through, 114; (*see also* Male-female relationships)
Marshall, T. H., 299
Marshall, Thurgood, 163
Marxist theories: American politics and, 382; class struggle and, 415; contradictions in institutional arrangements and, 14; economic conditions of 1930s seen through, 16; stratification and, 165; values and ideas in, 467
Marx, Karl, 12, 16, 37, 225, 228, 358, 371, 415, 421–430, 444, 463
Masculinity, 403
Mass, 485
Massachusetts: political preferences in, 379; tax income of, 364
Massachusetts Institute of Technology, 373
Mass media: anticipatory socialization through, 206–207; as identity-marketing agency, 204, 205; as instrument of manipulation, 354; small-town life affected by, 306
Mass-production techniques, 186
Mass-transit, suburban life and shrinkage of, 143, 144
Masturbation, changing ideas on, 70, 71, 72, 73
Materialism, 207
Matza, David, 235
Mauss, Marcel, 203
Maxwell Street Ghetto, Chicago, 139
Maya Indians, 478

Mayhew, Henry, 300
Mayo, Elton, 316, 317
M'Connel and Kennedy (firm), 297
Mead, George Herbert, 19, 49, 61–68, 204
Mechanics, development of, 31
Medicine: development of, 31; life expectancy rise through, 447; special examination system and, 435
Medieval civilization, 292
Meditation, 483
Meecker, Marchia, 165, 166–175
Melman, Seymour, 264
Melting pot effect, 138, 141
Memorial Day celebration, 468–476
Men: career as validation of role of, 109; college education of, by class, 216; communal child-caring responsibilities of, 115–116; increased educational participation of, among blacks, 112; ladies', 103; marriage and changes in relationships among, 121–122; sexual fears of, 102–103; as social category, 157; splitting phenomenon of, in communes, 115–116
Mental hospitals, social control through, 284–288
Merchant seamen, 315, 322–329
Merton, Robert K., 19, 203, 385
Methodists, political preferences of, 375
Methodology, frame of reference and, 41–42
Metropolis (*see* Cities)
Mexicans: class status of, 172; occupational mobility among, 229–230
Michels, Robert, 37, 358
Middle Ages, 153, 292, 463, 481
Middle classes: birth rate among, 301; commerce expansion and need for youth of, 297; educational promulgation of ideas of, 240; husband-wife relationships among, 103; impulse-inhibition of, 175; language usage among, 93–97; lower-class conflict with, 401; lower-class standards as reversal of, 413; male-female relationships among, 101; mobility rates of, 227; political preferences of, 381; prohibition and, 377; proletariat and, 426–427, 428; promiscuity and monogamous couplings among, 116; secession from society of, 295; suburban settlement by, 145–146; treatment of older generation by, 101; working-class move into, 182; (*see also* White-collar class)
Middletown (Lynd and Lynd), 498
Middletown, community study of, 416–422
Middletown-spirit, 23
Middle West, class stratification in, 169–171
Military academies, social control through, 233, 248–256
Military-industrial complex, 349
Military life, breaking points in, 186, 187
Millar, R. Mervyn, 99, 114–119
Millay, Edna St. Vincent, 95
Miller, Arthur, 487
Miller, Herman P., 201, 217–225
Miller, S. M., 165, 175–183
Miller, Walter B., 385, 400–413
Mills, C. Wright, 3–4, 10–17, 20, 36–40, 165, 183–193, 349, 497; power elite concept of, 350–363
Milwaukee (Wisconsin), 374
Mind, Self and Society (Mead), 49
Mines Act of 1860 (Great Britain), 302
Mining industry, increase of industrial bureaucrats in, 262–263
Minority groups: mobility and, 229–231; (*see also specific groups*)
Mirror of the Sea (Conrad), 324, 325
Mitford, Nancy, 77, 93–97
Mobility (*see* Downward social mobility; Social mobility; Upward social mobility)
Mobil Oil, 368
Modernization, 415
Molecular style of research: definition of, 37; differences between macroscopic style and, 37, 39–40; institutionalization of, 38–39
Money: as common denominator in values, 151; metropolis development and, 148–149, 151, 155; mobility through, 173–174; social functions of anonymity of, 207
Monod, Auguste, 95
Monogamy, hip relationships and, 115, 116
Moral density, 431, 433
Moralization of society, 293
Mortality rates: of children, 296, 300–301; in primitive cultures, 463
Mortuary industry, 458, 459
Mosca, Gaetano, 358
Moscow, Alvin, 325
Mothers: alternatives for kin terms for, 83; Boston Italian working-class, 101, 104
Motor development theories, 69
Mountain climbing mystique, 345
Mourning, 462–463, 464
Moynihan, Daniel, 99, 111–114
Muggeridge, Malcolm, 95
Muncie, Indiana (Middletown), studies on, 415, 416–421
Musgrove, F., 233, 245–248, 291, 296–302
Mute children, language learning by, 84–90
Myrdal, Gunnar, 157, 160, 162, 163, 172
Mysticism, 484

Naegele, Kaspar, 486
Narcotics Bureau, 398
National Association for the Advancement of Colored People, 160
National Deposit Friendly Society, 301
National Guard, 390
National Industrial Conference Board, 222
National Institute of Credit, 268
National Urban League, 160
Natural sciences, development of, 31
Near North Side, Chicago, 131–132
Needs, changing ideas on children's, 71–72
Negroes (*see* Blacks)
Neighborhoods (*see* Communities)
Netherlands, income distribution in, 224, 225
Never on Sunday (motion picture), 344
Newcomer, Mabel, 376
New Deal, 352, 353
New England: economic activity in, 366; Republican party in, 379; stratification in, 166–169
New Hampshire, political preferences in, 379, 380
New Left, 387, 391
Newman, Donald J., 392
New Streets and Roads (reader), 239
Newsweek (magazine), 388, 389
Newton, Douglas, 79
Newton, Huey, 391
New York City: economic activity in, 366; hippies in, 388, 389, 390–391; income distribution in, 228; mobility and class stratification in, 201, 208–211
New York State: government employees in, 364; voting behavior in, 376, 377
New York Telephone Company, 268
New York Times, 389
New York Times Magazine, The, 388
Niagara Movement, 160
Nietzsche, Friedrich, 147, 150
Noblesse Oblige (Mitford), 77–78, 93
Normal, definition of, 385
North Side, Chicago, 131–133
Norwegian Lutherans, 171
Nuclear families: extended families differentiated from, 100; objectivation within, 120; (*see also* Families)
Nudity, openness about, in communes, 118
Number, urban life shaped by, 137, 138, 139, 141–142
Nursery rhymes, children's language and, 78–81
Nurses, shortage of, 371
Nyakyusa people of Tanganyika, 246

Oberlin College, 367
Objectivation, social relationships and, 120
Objectivity: institutions and, 77; metropolis, 156; research, 39
Occupations: changes in, 183–185; correlation between education and status in, 212–213; definition of, 185; educational control over, 234, 235; increase of industrial bureaucracy and, 262–263; major shifts in, 185–188; percentages of, as white-collar jobs, 184–185, 186; political preferences and, 373–374; sociology of, 315; stratification and transformation of, 202
Oddfellows (society), 301
"Of-course" assumptions, 23
Office workers, percentage of labor force as, 184–185
Officials: free time for, 335; functioning of society dependent on, 31; higher education as basis for selection of, 247–248; increased numbers of, 188
Ohio, government employees of, 364
Old-family class: Deep South, 171; Middle and Far West, 169–170, 171; New England, 166, 167, 168; power patterns in, 354
Oldknow, Samuel, 298
Old people (*see* Aged)
Old Testament, 35
One-party states, 379–380
Opie, Iona, 77, 78–81
Opie, Peter, 77, 78–81
Opportunity, equality of, and social mobility, 225–231
Oral rhymes, children's language and, 78–81
Oratorians (religious order), 293
Organization man, 259, 270–278; influence of corporations on, 366–367
Organization Man, The (Whyte), 259
Orientals, class status of, 172
Origin of the Species, The (Darwin), 39
Ortega y Gasset, José, 344
Orwell, George, 285–286
Ostrogorski, Moisei, 37
Other America, The (Harrington), 209
Other-directedness, 206, 355
Our World Today (textbook), 236–237
Outer city, 142–143
Overhead, bureaucratic increase in, 263–264
Owen, Robert, 261

Pacifism, 484
Packard, Vance, 204
Pageants, play processes in, 61
Paideia, 292
Paranoia, hippies and, 389
Parapsychology, 483

Parents: alternatives for kin terms for, 83; changing ideas on, 74; children and (*see* Child-rearing)
Pareto, Vilfredo, 349, 358, 444, 496
Paris, 155
Park, R. E., 158
Parkinson, C. Northcote, 259, 278–283
Parkinson's Law, 259, 278–283
Parsimony, working-class, 180–181
Parsons, Talcott, 19, 181, 496
Péguy, Charles, 221
Pennsylvania, government employees of, 364
Pennsylvania State University, 367
Persia, Chicago immigrants from, 131, 135
Person-centeredness, working-class, 179, 181
Pessen, Edward, 228
Phenomenological sociology, 19
Phenomenon of Man, The (Teilhard de Chardin), 491
Philadelphia, 145, 459
Philadelphia Gentleman (Baltzell), 379
Philanthropy, mobility through, 174
Physicians, bargaining relationships between patients and, 449–458
Physics, development of, 31
Piaget, Jean, 77, 90–93
Pitkin, Donald, 100
Play: Boston Italian working-class attitudes toward, 108; breakdown in barriers between work and, 69; changing ideas on children's, 72–73; cultic tendencies in, 345–346; differences between game and, 62; identity and, 343–346; language in, 92; learning of society through, 61; new romanticism in, 343–345; role-taking by children in, 66
Playboy ideal, 343
Podhoretz, Norman, 201, 208–211
Poland, dispersion of wages in, 224, 225
Police brutality, 388
Polis, small-town character of, 153
Polish immigrants: Chicago, 135; class membership of, 171
Political parties: development of, 152, 153; self-identification with, 63
Political scientists, sociologists compared with, 43
Politics: absence of effective action in, 362; apathy and, 356; centralization and bureaucratization of, 361; classes and parties in, 372–383; community participation in, 144; decline in meaning of, 356; nationalization of, 380–383; one-party states and, 379–380; power structures within, 349; segmentation of, 205
Poor: Democratic party and, 374; discovery of, 209; inner city residence by, 140; political power of, 381
Poor Clares (religious order), 287–288
Poor Law (Great Britain), 298, 299, 300
Population, division of labor and density of, 430–434
Post, Emily, 96–97
Poughkeepsie (New York), 227
Poverty: medieval attitude toward, 295; Puritan beliefs about, 34
Power, 348–383; authority and, 349; bases of, 354–356; changes in structure and, 352–353; comparison of power elite and veto group concepts of, 350–363; consequences of, 356–357; constraints on, 358–360; corporate, 349–350, 363–372; exercise of, by small group, 349; legitimacy of, 349; operation of, 353–354; stratification and, 165; structure of, 351–352
Power Elite, The (Mills), 350–363
Pragmatism, working-class, 179–180
Pre-industrial society, industrial society compared with, 137–138
Presbyterians: political preferences of, 375, 376; power patterns and, 354
President's Commission on Civil Disorders, 389
Prestige, class stratification by, 191–192, 193
Priests, dying patients and, 461
Primary-group relationships, 137
Primitive cultures: funerals in, 462–463; play processes among, 61
Primitive fusion concept, 478
Princeton University, 369
Printing, development of, 31
Privacy: communes and, 117–118; teaching about, 243
Productivity, factory research programs in, 316–322
Professions: associations for, 189; autonomy of, 450; congressional power of, 361; special examination system and, 435
Profit: capitalism as pursuit of, 32; Puritan beliefs about, 34–35
Progressive movement, 379
Promiscuity, communal life and, 116
Prohibition, 377
Proletariat, class struggle and, 421–430
Pronouns, lower-class usage of, 5
Property: class stratification and ownership of, 183, 185, 189, 190, 191, 192–193; economic and social status through ownership of, 109; increase in wealth through use of, 266; responses to attitudes toward, 66–67

Protestant Ethic: functioning of society and, 443; mobility ethos and, 207; organization man and, 272, 273; spirit of capitalism and, 20, 34–36
Protestant Ethic and the Spirit of Capitalism, The (Weber), 19–20
Providence (Rhode Island), 166
Prudential Life, 369
Psychedelic drugs, 483
Psychoanalysis: child-rearing ideas influenced by, 69; New Left vocabulary derived from, 498; study of contemporary problems through, 17
Psychokinesis, 483
Psychological phenomena, 25
Psychological theories, child-rearing ideas influenced by, 69–70
Psychologists, sociologists compared with, 43
Psychotherapy, 204; low-income populations and, 112
Public officials: free time for, 335; functioning of society dependent on, 31; higher education as basis for selection of, 247–248; increased numbers of, 188
Public Speaking and Influencing Men in Business (Carnegie), 268
Puerto Ricans: inner city residence of, 140; lower-class style of life of, 181; occupational mobility among, 229–230; service occupations and, 176
Puerto Rico, income distribution in, 224
Punctuality, metropolis and, 149–150
Purdue University, 367
Puritan Ethic: barriers between play and work in, 69; bureaucratization and, 267, 268; bourgeois culture and, 441; development of capitalism and, 34–36

Quasi-primary relationships, 138, 142, 143, 146
Quatorzième (occupation), 155

Race, appearance of concept of, 295
Race riots, 160
Racial discrimination: inner city, 141; Moynihan Report on consequences of, 111; peak of, 160; stratification and, 165; during white American Southern childhood, 50–55
Radicalism, 390, 440
Rationalization, 415; of death, 460, 461; definition of, 259; of education and training, 435–439; leisure and, 315
Readers, bias in, 112
Reading curriculum, relevancy of, 236–240
Reagan, Ronald, 390
Reality, marriage and construction of, 119–123
Reality shock in military academies, 255–256
Rebel (Camus), 502
Rechabites (friendly society), 301
Reconstruction, 159–160, 380
Recreational space, increased free time and, 335
Redfield, Robert, 478
Redlich, Frederick C., 176
Reese, Wolfe, 327, 329
Refinement, class stratification and, 165, 193–198
Reformation, 485
Regionalism in politics, 379–380
Relationships: astrological explanations of problems in, 117; bargaining, between doctor and patient, 449–458; hip, 114–119; metropolis and reserve in, 151–152; primary group, 137; quasi-primary, 138, 142, 143, 146; (*see also* Family: Male-female relationships; Marriage)
Relevance, individual strata of, 21–23
Religion: class stratification and, 167, 171; decline in authority of, 460, 467; funerals and, 463; institutional specialization of, 477–481; political preferences and, 374–375; resurgence of interest in, 467, 482–492
Religious communes, 119
Religious orders: founding of, 293; neighborhoods and, 145; work perspectives in, 287–288
Renaissance, 31, 481
Repression, social control and, 386–387
Republican party: class and, 372–383; ethnic group support for, 377–379; one-party states and, 379–380
Rerum Novarum (encyclical), 341
Research: macroscopic, 36–37, 39–40; molecular, 37, 38–40; two styles of, 20, 36–40
Reserve, metropolis and, 151–152, 154
Respectability, rentals and land values measurement of, 132
Responses, as structure for self, 66–67
Rhode Island, political preferences in, 379
Rhymes: children's language and, 78–81; classes of, 79
Rhythm method of birth control, 106
Richard II, King of England, 95
Rich Man, Poor Man (Miller), 201–202
Riesman, David, 201, 205, 206, 211–217; veto groups concept of, 350–363
Riessman, Frank, 99, 111–114, 165, 175–183
Rockefeller, Nelson, 379

Rockefeller Foundation, 369
Roethlisberger, F. J. R., 316
Roles: children and taking of, in games, 65–66; community membership through assumption of, 67
Roll, Erich, 262
Roman Catholic Church (*see* Catholic Church)
Romanticism, play and, 343–345
Romantic love: Boston Italian working-class attitudes toward, 104; hip relationships and, 115
Rome, ancient, 32, 421, 432, 433
Roosevelt, Franklin D., 353, 377, 378, 379, 380
Ross, Alan S. C., 94
Ross, E. A., 12
Rossi, Peter H., 490
Roszak, Theodore, 387
Roth, Julius A., 447, 449–458
Rousseau, Jean Jacques, 334, 440
Rules of Sociological Method, The (Durkheim), 3, 19
Rural communes, 115, 117–118, 119
Rural communities, urbanization of, 129
Ruskin, John, 150
Russia: foreign relations with, 377, 378; population density of, 433; power structure in, 357

Sailor's Life, A (Hartog), 324
Saint-Simon, Louis de Rouvroy, duc de, 95
Salem (Massachusetts), 166
Salespeople, white-collar class and, 184, 191
Salford Unity (friendly society), 301
Samuelson, Paul, 217–218
San Francisco, 171; hippies in, 388–389; party registration in, 374
Santa Clara County, California, 374
Saroyan, William, 165, 193–198
Schegloff, Emanuel A., 59
Scheler, Max, 23, 496
Schneider, D., 82–83
Schools: benefits from desegregation of, 113–114; Boston Italian working-class attitudes toward, 108; control mechanisms in, 241–245; culture transmission through, 109; development of, 293–294; racial attitudes of children in, 6–8; relevancy of reading curriculum in, 236–240; specialized examination system in, 435–436; (*see also* Education)
Schumpeter, Joseph, 12
Schutz, Alfred, 19, 20–24, 46, 498
Science: death of, 485–486, 491; Western foundation for, 31
Scientific American, 226–227
Scott Foresman (publishers), 239
Seamen, 315, 322–329
Seamen's Church Institute of New York, 328
Séances, 484
Searle and Company, 370
Secularization, 467
Security, working-class, 177–178
Seeley, J. R., 99, 108–110
Seeley, John, 441
Segregation: of aged, from families, 447–449; melting pot effect and, 138; Moynihan Report on consequences of, 111; during white American Southern childhood, 50–55
Seidel, Alfred, 499
Self: general stages in, 64–65; learning of society by, 49, 61–68
Self-consciousness, definition of, 68
Self-determinism, 25
Self-help with Illustrations of Character, Conduct and Perseverance (Smiles), 268
Service industries, 364
Settlement types, analysis of, 137–138
Sex, stratification through, 192
Sexual hostility, 103
Sexual intercourse, Boston Italian working-class attitude toward, 103
Sexuality, deviance in, 385
Sexual mores: Boston Italian working-class, 102–103; communes and openness about, 118; family regulation through, 109; lower-lower class, 169, 402, 405–406; youth culture and, 442
Shakespeare, William, 79, 95
Sherar, Mariam G., 315, 322–329
Sherman Law, 368
Shipping Out (Sherar), 315
Siblings (*see* Brothers; Sisters)
Sim, R. A., 99, 108–110
Simmel, Georg, 37, 120, 129, 147–156, 207
Singer, Lester, 129, 156–163
Sisters: alternatives for kin terms for, 83; Boston Italian working-class communication among, 104; friends as, in communes, 115
Sit-ins, 161
Six-Day School, 484
Sky-diving mystique, 345
Skyscrapers, effect on centralization of Loop, Chicago, 130–131
Slavery, 158, 159, 162, 377, 421
Slav immigrants to Chicago, 135
Sloan, Alfred P., 267
Slums, Chicago, 131, 134–136
Small Business Administration, 368

Smartness, lower-class concern with, 402, 404–405
Smiles, Samuel, 268
Smith, Al, 380
Smith, Huston, 483
Smith, Lillian, 49, 50–55
Snobbery, language and, 96
Social category, definition of, 157
Social control: bargaining relationships in hospitals and, 449–458; bureaucratic means of, 284–285; centrality of schooling for, 234–236; of death, 458–464; definition of, 233; education as, 232–256; high school mechanisms for, 241–245; inmate identification as, 285; military academy and, 233, 248–256; psychological impact of, 245–248; repression through, 386–387; teaching of reading as example of, 236–240
Social entity, definition of, 157
Social Ethic, organization man and, 273–276
Social facts, 77; definition of, 3, 8–9
Socialism, 147
Socialist countries, dispersion of wages in, 224–225
Socialization, 48–74; analysis of data from everyday childhood activities and, 57–60; anticipatory, 206–207; definition of, 56; black family and, 99, 111–114; as developmental process, 56; identity generation and, 206; marriage and, 121, 122–123; as object of education, 27; referential procedures for kin identification in, 82–83, 84; religious, 481–482; transformation in patterns of, 49; transition through language from ego-centrism to, 90–93; in white American South, 49, 50–55
Social map, location finding on, 3, 4–5
Social mobility: cultural mobility and, 215–216; degree of, 225; educational system and, 201, 211–217; equal opportunity and, 225–231; green revolution and, 439; home ownership and, 145; inner-city neighborhoods dispersed through, 139; lack of, 386; male splitting phenomenon in communes and, 115–116; military academies and, 254–255; minority groups and, 229–231; in New York City, 201, 208–211; personal identity and, 202–207; psychological consequences of, 206–207; racial attitudes and, 172; strata of society and, 201
Social phenomenon: definition of, 25–26; universality of, 27, 29; ways of being as, 29–30
Social psychology, 19
Social structures: ethnic group development and, 159; personal troubles and, 14–15; urban, 137
Society: child's experience of, through language, 77; city as prototype of mass, 138; classes within (*see* Class); common sense view of, 5; condensation of, 431–433; constraints on individuals in, 3, 8–9, 25, 29, 31; deviance in (*see* Deviance); distinctions between personal troubles and public issues in, 14–15; experience of, 3–17; imposition through language of categories of, 77, 81–84; industrial compared with pre-industrial, 137–138; location in, 4–5; location of religion in, 481–482; moralization of, 293; organization man's need for meaning within, 273–278; psychiatric description of problems of, 17; rationalization of, 259; relation of biography and history within, 12–13; role of sociologist as stranger in study of, 19, 20–24; schoolchildren's rhymes reflecting experience in, 77, 78–81; as social entity, 157; stratification in (*see* Stratification)
Sociological Imagination, The (Mills), 3
Sociologists: culture shock and, 45–46; definition of, 41; other professional observers compared with, 43–44; perspectives of, 41–46; as strangers in own society, 19, 20–24; voyeuristic tendencies of, 43
Sociology: Chicago School of, 129, 137, 139; conservatism of, 499–502; development of, in different countries, 19; discipline of, 18–46; imagination in study of, 3–4, 10–17; as liberating discipline, 497; of occupations, 315; origin of word, 19; rules of method in, 19, 25–30; schools of, 19; scientific rules in, 41; significance of, 495–503; systemization of insight through, 3; two styles of research in, 20, 36–40; urban, 315
Solemn High Mass, 485
Solidarity, military academy and development of, 251–252
Sons, alternatives for kin terms for, 83
Sons and Lovers (Lawrence), 182
Sorcery, 483
South: benefits from school desegregation in, 114; class stratification in, 171–172; new economic activity in, 366; political regionalism in, 379, 380; white childhood in, 49, 50–55
Southey, Robert, 79
Soviet Union: foreign relations with, 377, 378; population density in, 433; power system in, 357

Spain, New Left in, 498
Spanish Americans, class status of, 172
Special examination system, 435–436
Specialization: bureaucracy and, 266; institutional, of religion, 477–481; in responsibility for dying and care of body, 459; services within cities and, 155
Speenhamland System, 298
Speier, Matthew, 49, 55–60, 77, 81–84
Spencer, Herbert, 12, 26, 431, 433
Spender, Stephen, 94
Sperry Rand, 366
Sphinx (bookstore), 484
Spiegel, John, 112
Spiritualism, 484
Splitting phenomenon of men in communes, 115–116
Sports, cultic tendencies in, 345–346
Stability, working-class, 177–178
State Department, 378
Statistics: molecular style of research and, 39; use of, 42
Status: adult status among children, 411–412; anxiety and, 204; definition of, 201; politics and, 373; property ownership and, 109; stratification and, 165; suppression of, in military academy, 249–250; uncertainty of, through looseness of class, 203–204
Status Seekers, The (Packard), 204
Stevenson, Adlai, 372, 382
Stevenson, T. H. C., 300, 301
Stockholders, 367, 369
Story of My Life (Keller), 77
Stranger: child as, in world of adults, 49; definition of, 20; role of sociologist as, in own society, 19, 20–24
Stranger, The (Schutz), 19
Stratification, 164–196, 200–231; approach to, 165; Deep South, 171–172; features of, 202; Middle and Far West, 169–171; New England, 166–169; New York City, 201, 208–211; personal income distribution and, 217–225; refinements and, 165, 193–198; religious representation and, 480; within white-collar class, 165, 183–193
Strauss, Anselm L., 460, 461
Streets: playground use of, among blacks, 113; police sweeps of, 388, 390
Structural-functionalist school, 19, 385, 415
Strutt, Anthony, 298
Students for a Democratic Society, 482, 492
Subjectivity, research, 39
Suburbs: Boston Italian working-class move to, 106, 107; bridging of rural and urban features by, 129; city compared with, 143–146; class in, 174; decentralization of industry and, 137
Success, as reward for virtue, 267, 269
Sudnow, David N., 460, 461
Suicide, 204; among black population, 112; Chicago rate of, 132, 135; mobility and, 203
Sullivan, Anne Mansfield, 77, 84–90
Sunset (magazine), 343
Superstition, 490
Surfing mystique, 345–346
Surgery, bargaining relationships and, 453
Survey Research Center, 219
Sutherland, Edwin H., 393, 394, 395
Swarthmore College, 367
Sweden, income distribution in, 223, 224, 225
Swedish immigrants in Chicago, 131
Swift, Jonathan, 79
Sybil (Disraeli), 298
Systematic theology, 31

Tactical Police Force, New York City, 388
Taft, William Howard, 382, 383
Tanganyika, 246
Tantra, 483
Tarot cards, 483, 484
Tawney, R. H., 221
Taxes, family income levels and, 219–221
Taylor, Zachary, 383
Teachers: occupational destinies of students and, 235; social control through, 236–240; status of, 242; white-collar class and, 184, 185
Technology: administration of industry and, 260–261; capitalism and development of, 33; free time and, 336; occupational change through, 186; population changes and youth affected by, 297–298, 302; youth revolution and, 442–443
Teilhard de Chardin, Pierre, 491
Terman, Lewis M., 247
Texaco, 367
Texas, government employees in, 364
Textbooks, curricula relevancy and, 236, 239–240
Textile industry, child labor in, 298–299, 300, 302
Theology, systematic, 31
Thernstrom, Stephan, 227
Third estate, 422
Thomas, Dylan, 80
Thomas, Lowell, 268
Thomas, W. I., 24
Thumb-sucking, changing ideas on, 70, 71, 74
Tibet, 483
Time, metropolis development and sense of, 149–150

Tocqueville, Alexis de, 228, 272, 357, 383
Toennies, Ferdinand, 138
Toilet-training, 49
Tom Jones (motion picture), 344
Tompkins Square Park, New York City, 388, 390
Toughness, lower-class concern with, 402, 403–404
Toynbee, Philip, 93–95
Trade: rise of capitalism and, 32; white-collar employees in, 187
Trade-unions, 266
Traditionalism: military academy, 250; working-class, 178–179, 181
Training, rationalization of, 435–439
Transience, inner city residency and, 141
Transportation: city outskirts expanded by, 130; density of society and, 432; small-town life affected by, 306; suburban life and, 143, 144, 146; white-collar employees in, 187
Trapped population of inner city, 139, 140, 141
Tribal wars, American Indian, 412
Triviality, 501
Trouble, lower-class concern with, 402–403
Truman, Harry S., 382
Tuberculosis hospitals, bargaining relationships in, 449–458
Tumin, M. M., 157

U.F.O.'s, 483
Uncles, alternatives for kin terms for, 83
"Uncle Toms," 163
Unconscious self, habits and, 67–68
Underground Railroad, 158
Uneasiness, threat to personal values and, 16–17
Unemployment, distinction between personal troubles and public issues in study of, 14, 15
United Charities (Chicago), 134
United States: class stratification by geographic regions of, 165, 166–175; cultural change in, 439–445; development of sociology in, 19; dispersion of wages in, 224, 225; language and class distinctions in, 78, 95, 96–97; ratio of administrative and production personnel in, 264; research schools in, 37; work time for meal in, 222
United States Coast Guard Academy, 233, 248–256
United States Steel, 364, 371
Universal education, 234
University education (*see* College education)
University of Chicago, 482, 492
Unknown Soldier, 472, 473
Unmarried residents of inner city, 139, 140, 141
Updike, John, 100, 123–126
Upper classes: educational advantages of children of, 212; home ownership by, 143; language usage among, 93–97; liberalism of, 379; New England, 166–167; power structure and, 354
Upper-lower class: Middle and Far West, 171; New England, 168–169
Upper-middle class: adult-directed families of, 106–107; class stratification and, 170–171; Crestwood Heights family, 99, 108–110; cultural revolution and youth of, 440; educational advantages of children of, 212; education as control access to, 213–215; home ownership by, 143; New England, 167, 168; suburban life of, 144
Upper-working class, home ownership by, 143
Upper-upper class (*see* Old-family class)
Upward social mobility: equality of opportunity and, 225–226; green revolution and, 439; limitations on, 212; upper-middle class family and, 110; youth revolution and, 443
Urban communes, 115, 116, 118, 390
Urban communities (*see* Cities)
"Urbanism as a Way of Life" (Wirth), 137
Urbanization: as force for change, 129; politics and, 380; upward mobility and, 225
Urban renewal activities, 146
Urban Villagers (Gans), 99
Ure, Andrew, 298

Values: contemporary threats to, 16–17; money as common denominator in, 151; ultimate meanings and, 467–492
Veblen, Thorstein, 12, 358
Vegetarian diets, 484
Vermont, political preferences in, 379, 380
Veterans' Day celebrations, 469
Veto groups concept of power, 350–363
Vietnam, 388, 487
Vold, George, 392

Wallace, Henry, 382
Wall Street Journal, 113
Wants, changing ideas on children's, 71–72
War: distinctions between personal troubles and public issues in study of, 14–15; personal income during, 190
Warlocks, 483

Warner, W. Lloyd, 157, 165, 166–175, 467, 468–476
Washington, Booker T., 160, 163
Washington, George, 237–238, 473
Wealth: class stratification through, 166–167, 170; corporate, 367; medieval attitudes toward, 295; power and, 361; Puritan beliefs about, 34; pursuit of, 491; Republican party and, 374; through use of property, 266
Weaning, 49, 292
Weber, Max, 12, 19–20, 30–36, 37, 39, 42, 165, 205, 259, 272, 349, 358, 415, 435–439, 440, 441, 459, 467, 491, 496, 499
Webster's Collegiate Dictionary, 260
Weimar culture, 154
Welfarism, utilization of black strengths to reduce, 114
Well-being, threat to personal values and, 16
Wesley, John, 44
West End, Boston, Italian working-class families, 99, 100–108
Western Electric Research Program, 315, 316–318
Western Germany: class structure changes in, 444; dispersion of wages in, 224, 225; work time for meal in, 222–223
Westley, W. A., 387
West Point Academy, 249
Whig party, 376, 377, 383
White Brotherhood, 484
White-collar class, 165, 183–193; corporate influence on, 366–367; crime among, 392, 393, 395; expansion of commerce and need for, 297; featherbedding among, 367–368; occupational change and, 183–185; percentage of labor force in, 184–185, 186; political and economic weakness of, 355; pyramids of ranks of, 189–192; shifts in occupations of, 185–189; stratification and, 202; (*see also* Middle classes)
Whitehead, Alfred North, 273
White House Conference on Civil Rights, 111
Whites: blacks' defensiveness with, 157; blacks in power structure and, 389; exodus of, from city to suburbs, 137; experience of blacks passing as, 6–8, 52–54; mobility of, 230; physical features of, 172; relations with blacks, 157, 160–161, 162–163; Southern childhood of, 49, 50–55
White supremacy, 160
Whitty, Dame May, 96
Whyte, William Foote, 315, 316–322
Whyte, William H., Jr., 259, 270–278
Wife-husband relationships (*see* Marriage)
Wilensky, Harold L., 139
Willkie, Wendell, 378, 379, 382
Wilson, Woodrow, 377, 378
Wilson and Company, 367
Winooski, Vermont, 380
Wirth, Louis, 137–142, 146
Wisconsin, tax income in, 364
Wise, Frank, 388
WITCH (Women's International Terrorists Corps from Hell), 482, 483, 488, 491
Witchcraft, 483, 485
Wohlstetter, Albert, 230–231
Wolfenstein, Martha, 49, 69–74
Women: black, income level of, 231; college education of, by class, 216; family as validation of role of, 109; primary communal child-care responsibility of, 115–116, 117; romantic images in hip relationships of, 115; as social category, 157; status of, in communes, 117, 118–119
Women's liberation movement, communes and, 118–119
Work: breakdown in barriers between play and, 69; distinction between personal troubles and public issues in study of, 17; incentives for, 221; institutional perspectives on, 287–288; length of time for meal, by country, 222–223; plant conditions for, 316–322; separation of income from, 342; as therapy, 287
Working class, 165, 175–183; adult-centered families of, 106; Boston Italian, 99, 100–108; bureaucratic success and higher education of children of, 227; definition of, 176; Democratic party preferences of, 372–373, 380, 382; English, 101; extended kin groups among, 117; green revolution and, 439; Middletown study of change in, 418; occupational mobility of, 230; stable style as goal of, 182–183; themes of, 177–181; treatment of older generation by, 101; variations in, 181–182; youth culture changes in, 442; (*see also* Lower classes)
World history, individual lives and course of, 10–11
World War I, 378
World War II, 129, 353, 378

Yippies, 388, 390, 391
YMCA schools, 268
Yoga, 483
Young, Arthur, 299
Youth, 290–312; autonomy from adult world by, 291; cultural revolution and,

439–445; definition of, 291; emergence of subculture of, 302–307; historical review of status of, 292–295; population changes and status of, 296–302; as social category, 235; socialization in, 122–123
Youth and the Social Order (Musgrove), 291

Yugoslavia, dispersion of wages in, 224, 225
Yule, G. Udny, 301

Zen, 483
Znaniecki, Florian, 236
Zorba the Greek (motion picture), 344
Zorbaugh, Harvey Warren, 129, 130–136